Ulrich Mohr (Editor)

International Classification of Rodent Tumors: The Mouse

Springer-Verlag Berlin Heidelberg GmbH

WHO · International Agency
for Research on Cancer

Ulrich Mohr (Editor)

International Classification of Rodent Tumors

The Mouse

With 402 Figures in Color

Springer

Professor Dr. Ulrich Mohr

Institut für Experimentelle Pathologie
Medizinische Hochschule Hannover
Carl-Neuberg-Strasse 1
30625 Hannover Germany

ISBN 978-3-642-08422-5 ISBN 978-3-662-07973-7 (eBook)
DOI 10.1007/978-3-662-07973-7

Library of Congress Cataloging-in-Publication Data
International classification of rodent tumors. The mouse/[edited by] Ulrich Mohr. p. cm.
Includes bibliogaphical references and index.

1. Tumors-Classification. 2. Tumors-Animal models. 3. Mice-Diseases. 4. Tumors in animals-Classification. I. Mohr, U. (Ulrich)
RC258.I4462 2001 616.99'2'0012-dc21 00-064092

http://www.springer.de

Originally published by Springer-Verlag Berlin Heidelberg New York in 2001

Cover-Design: Erich Kirchner, Springer-Verlag Heidelberg
Typesetting: Data conversion by Springer-Verlag Heidelberg

Printed on acid-free paper. SPIN 10678562 24/3130 ih - 5 4 3 2 1 0

Preface

The standardized nomenclature of murine neoplasms presented in this monograph continues the series that was initiated by WHO/IARC with the publication of the *International Classification of Rodent Tumors, Part I, The Rat.* As with the previous volumes, the purpose is to provide diagnostic criteria and a standardized, widely accepted nomenclature for neoplastic and pre-neoplastic lesions. The classification reflects the consensus of an international panel of expert pathologists and of science societies from many countries that worked together in the Organ System Committees.
The major objective of this classification is to harmonize and improve tumor diagnosis in murine carcinogenesis studies, since the rodent bioassay is still a major investigative tool for prediction of carcinogenic risk to humans of a large variety of exposures, including drugs, pollutants, and food additives. It provides an internationally accepted uniform standard for establishing contemporary databases that are necessary for an accurate interpretation of lesions induced in experimental carcinogenesis studies.
A standardized nomenclature and diagnostic criteria will also facilitate the identification of tumors and related proliferative lesions in genetically engineered transgenic and knock-out mice that are increasingly used in biomedical research.
IARC is pleased to present Part II of the International Classification of Rodent Tumors. We are confident that it will contribute to significantly reduce the traditional inter-pathologist variability in the histopathological diagnosis of murine tumors that has confused much work in the past.

Dr. Ulrich Mohr
Director, Institute of Experimental Pathology, Hannover Medial School

Dr. Paul Kleihues
Director, International Agency for Research on Cancer World Health Organization

Introduction

History

A long-term goal of scientists working in the field of toxicologic pathology has been the standardization and harmonization of nomenclature and diagnostic criteria for lesions observed in laboratory rats and mice. Initiatives were started in the late 1980s in the United States by the STP (Society of Toxicologic Pathologists) and in Europe by the RITA data base group (Registry of Industrial Toxicology Animal-data). These initiatives resulted in the internationally-recognized publications (*SSNDC: Guides for Toxicologic Pathology*, and *WHO/IARC: International Classification of Rodent Tumors, Part I, The Rat*).
In 1994 the Joint STPs/ILSI Committee on International Harmonization of Nomenclature and Diagnostic Criteria in Toxicologic Pathology was established. The objective of the committee was to coordinate the international activities for the standardization of diagnostic criteria of tumors and other proliferative lesions in mice and to reconcile the already published nomenclature of proliferative lesions in rats. The committee consisted of representatives from the European RITA database group (Registry of Industrial Toxicology Animal-data), the STP (Society of Toxicologic Pathologists, USA), the BSTP (British Society of Toxicologic Pathologists), JSTP (Japanese Society of Toxicologic Pathologists), and the GTP (Gesellschaft für Toxikologische Pathologie – German-speaking Society of Toxicologic Pathology). The steering committee established organ system committees for the development of nomenclature and diagnostic criteria for the mouse, largely through the STP (USA) and the RITA database group (Europe). To obtain world-wide comments and acceptance, successive drafts of manuscripts were reviewed according to procedures developed by the STP in the USA (poster and platform presentations), and were reviewed by the RITA database group and representatives of the Societies of Toxicologic Pathology in Japan and the UK. This monograph represents the product of all that effort.

Description

Included in this classification are tumors and potentially pre-neoplastic lesions of the mouse organized according to organ systems. Most of the lesions are illustrated, but in a few instances satisfactory images were not available.
Each chapter is divided into "data sheets", each of which represents the essential information on a particular lesion. Furthermore, each data sheet

is prepared in a standard layout, always starting on a new page with a header section. On the left side of each header section, the lesion name (the preferred term) appears followed in parenthesis by an indication of the biological behavior of the lesion, according to the following codes:

(*H*) hyperplastic and pre-neoplastic lesion
(*B*) benign tumor
(*M*) malignant tumor
(*S*) systemic tumor

If modifiers are defined for a particular lesion, they are printed in italics below the lesion name. On the right-hand side of the header, the organ in which the lesion occurs is mentioned. If the same criteria are used for a lesion at several topographical sites, all these organ sites are included here. All lesion names used in the headers and in the synonym(s) and "Differential Diagnosis" sections are presented with the pathological disposition (such as hyperplasia, tumor, adenoma, etc.) placed first, followed by terms describing a sub-topography, a growth pattern, a cell type, etc., separated by commas. This ordering of words constitutes the structure of preferred terms. Use of this terminology facilitates the identification and grouping of lesions with the same biological behavior, especially when using a computerized pathology data system. In the text and the figure legends, the ordering of words is usually changed to the more common "speaking form"; thus if the preferred term is called "adenoma, bronchiolo-alveolar" this lesion is referred to in the text as "bronchiolo-alveolar adenoma".
The descriptions of the diagnostic features comprise only the main histopathological features of the specified lesion, in the form of a concise list. In several cases, modifiers are included for a more precise subclassification to define a specific growth pattern (e.g., papillary, solid, cystic, etc.) or to subdivide findings by cell type (e.g., pheochromocytoma type, small cell type, etc.). The criteria which are specific for modifiers are listed under subheadings. Because illustrations are essential in pathology, most lesions are documented with at least one micrograph. These illustrations have been selected from either spontaneous or induced lesions.
The section "Differential Diagnosis" is also kept as short as possible and includes only the main diagnostic criteria used in distinguishing lesions.
In the reference section of each data sheet only numbers are printed. These refer to the numbered list of references at the end of the chapter. Only the most recently published and important papers are included in the literature references.
An 'electronic' version of this classification with a more extensive range of coloured illustrations will be issued once the monograph is published. This will enhance the speed and facility of access to the information required for accurate diagnosis.

Acknowledgements

The Editorial Board thanks the following colleagues for providing additional illustrative material and helping in the preparation of the manuscripts and figure legends:

Dr. L. Anderson, NCI, USA; Dr. G. Boorman, NTP, USA; Dr. A. Bube, Aventis Pharma GmbH, Germany; Dr. C. Dawe, Harvard Medical School, USA; Dr. L. Elcock, Bayer Corporation, USA; Dr. M. Enomoto, Biosafety Research Center, Japan; Dr. H. Ernst, Fraunhofer Institute of Toxicology and Aerosol Research, Germany; Dr. S.R. Frame, DuPont, USA; Dr. K. Heider, Novartis AG, Switzerland; Dr. H. Iwata, Biosafety Research Center, Japan; Dr. M.P. Jokinen, NTP archive (Pathology Associates International), USA; Dr. A. Luz, GSF-Forschungszentrum für Umwelt und Gesundheit GmbH, Germany; Dr. D. Morton, Monsanto, USA; Dr. Y. Nomura, Biosafety Research Center, Japan; Dr. A. Nyska, NIEHS, USA; Dr. S. Rittinghausen, Fraunhofer Institute of Toxicology and Aerosol Research, Germany; Dr. B. Stuart, Bayer Corporation, USA; Dr. S. Yamamoto, Biosafety Research Center, Japan

Particular thanks are due to Dipl.-Ing. Gerd Morawietz (Fraunhofer Institute of Toxicology and Aerosol Research, Germany) for establishing the structure of the nomenclature, suitable for electronic data processing, for preparing the proofs, and for carrying out the technical editing.

Contents

Contributors

Ackerman, Larry J., Dr.
Pharmakon Research International Inc., PO Box 2365,
Andrews, NC 28901, USA

Anver, Miriam R., Dr.
Pathology/Histopathology Laboratories, SAIC/NCI-FCRDC,
Frederick, MD 21702, USA

Bader, Rainer, Dr.
Boehringer Ingelheim Pharma KG, Abteilung für Experimentelle
Pathologie und Toxikologie, Birkendorferstrasse 65,
88397 Biberach, Germany

Betton, Graham R.S., Dr.
AstraZeneca, Mereside, Alderley Park, Macclesfield,
Cheshire SK10 4TG, UK

Boothe, Arliss, Dr.
International Research and Development Corporation,
500 North Main Street, Mattawan, MI 49071, USA

Brown, Roger, Dr.
Glaxo Wellcome Inc., Five Moore Drive,
Research Triangle Park, NC 27709, USA

Bruner, Richard, Dr.
Pathology Associates Intl., 6717 Centre Park Drive,
West Chester, OH 45069, USA

Bube, Axel, Dr.
Aventis Pharma Deutschland GmbH, Department of Toxicology/Pathology,
Mainzer Landstrasse 500, 65795 Hattersheim, Germany

Capen, Charles C., Dr.
Department of Veterinary Biosciences, Ohio State University,
1925 Coffey Road, Columbus, OH 43210, USA

Carlton, William W., Dr.
Department of Veterinary Pathobiology, School of Veterinary Medicine,
Purdue University, West Lafayette, IN 47907, USA

Cary, Maurice, Dr.
Novartis Pharma AG, Toxicology, 4002 Basel, Switzerland

Cattley, Russel C., Dr.
Amgen Inc., One Amgen Center Drive, Thousand Oaks, CA 91320, USA

Courtney, Cynthia, Dr.
Parke-Davis Pharmaceutical Research, Division of Warner-Lambert,
2800 Plymouth Road, Ann Arbor, MI 48105-2403, USA

Creasey, Diane, Dr.
Huntingdon Life Sciences Inc., PO Box 2360, Mettlers Road,
East Millstone, NJ 08875, USA

Davis, Barbara, Dr.
NIEHS MD-B3-06, PO Box 12233, Research Triangle Park, NC 27709, USA

Deschl, Ulrich, Dr.
Bayer AG, Pharma Forschungszentrum, Institut für Toxikologische
Pathologie, Aprather Weg, 42096 Wuppertal, Germany

Dungworth, Donald L., Dr.
6260 Cape George Road, Port Townsend, WA 98368, USA

Durchfeld-Meyer, Beate, Dr.
Aventis Pharma Deutschland GmbH, Department of Toxicology/Pathology,
Mainzer Landstrasse 500, 65795 Hattersheim, Germany

Elwell, Michael, Dr.
Pfizer Inc., Central Research, Eastern Point Road, Groton, CT 06340, USA

Enomoto, Makoto, Dr.
Biosafety Research Center, Foods, Drugs and Pesticides,
582-2 Arahama Shioshinden, Fukude, Iwats, Shizuoka 437-12, Japan

Ernst, Heinrich, Dr.
Fraunhofer Institute of Toxicology and Aerosol Research,
Nikolai-Fuchs-Strasse 1, 30625 Hannover, Germany

Ettlin, Robert A., Dr.
Novartis Pharma AG, WKL-125.1513, 4002 Basel, Switzerland

Eustis, Scot L., Dr.
SmithKline Beecham Pharmaceuticals, 709 Swedeland Road, PO Box 1539,
King of Prussia, PA 19406, USA

Everitt, Jeffrey, Dr.
Chemical Industries Institute of Toxicology, 6 Davis Drive,
Research Triangle Park, NC 27709, USA

Fix, Andrew S., Dr.
Procter and Gamble Co., Miami Valley Laboratories,
Cincinnati, OH 45253, USA

Foley, George L., Dr.
Pfizer Inc., Central Research, Eastern Point Road, Groton, CT 06340, USA

Frith, Charles H., Dr.
Toxicology Pathology Associates, 14201 Clarborne Court,
Little Rock, AR 72211, USA

Garman, Robert H., Dr.
Consultants in Veterinary Pathology, PO Box 68, Murrysville, PA 15668, USA

George, Catherine, Dr.
Pfizer Centre de Recherche, Pathology Department, B.P. 159,
37401 Amboise Cedex, France

Germann, Paul-Georg, Dr.
Novartis Pharma AG, Pathology, 4002 Basel, Switzerland

Glaister, John, Dr.
Hazleton Laboratories, Otley Road,
Harrogate, North Yorkshire, HG3 1PY, UK

Gopinath, Chirukandath, Dr.
Huntingdon Research Centre, PO Box 2, Huntingdon, Cambs PE17 5DP, UK

Greaves, Peter, Dr.
MRC Toxicology Unit, Hodgkin Building, University of Leicester,
Lancaster Road, Leicester LE1 9HN, UK

Hailey, James R., Dr.
103 Norham Drive, Cary, NC 27513, USA

Halm, Sabine, Dr.
Knoll AG, Institut für Toxikologie, Knollstrasse 2,
67061 Ludwigshafen/Rhein, Germany

Harada, Takanori, Dr.
Mitsukaido Laboratories, Institute of Environmental Toxicology,
4321 Uchimoriya-cho, Mitsukaido-shi, Ibaraki 303-0043, Japan

Hard, Gordon C., Dr.
American Health Foundation, 1 Dana Road, Valhalla, NY 10595, USA

Hardisty, Jerry F., Dr.
Experimental Pathology Laboratories Inc., PO Box 12766,
Research Triangle Park, NC 27709, USA

Harkema, Jack R., Dr.
212 Food Safety and Toxicology Building, Michigan State University,
East Lansing, MI 48824, USA

Harleman, J. Hans, Dr.
Novartis Pharma AG, Toxicology, 4002 Basel, Switzerland

Hartig, Franz, Dr.
Kastellweg 13a, 68526 Ladenburg, Germany

Hayashi, Yuzo, Dr.
National Institute of Health Sciences, 1-18-1 Kamiyoga, Setagaya-Ku,
Tokyo 158, Japan

Heinrichs, Martin, Dr.
Aventis Pharma Deutschland GmbH, Department of Toxicology/Pathology,
Mainzer Landstrasse 500, 65795 Hattersheim, Germany

Holland, J. Michael, Dr.
34111 Callita Drive, Dana Point, CA 92629, USA

Inoue, Tohru, Dr.
Pathology/School of Medicine, Yokohama City University, 3-9 Fukuura, Kanazawa-ku, Yokohama 236, Japan

Isaacs, Kevin R., Dr.
Toxicological Path Consultant, 14 Rossett Park Road, Harrogate, North Yorks, HG2 9NP, UK

Jacobs, Maren, Dr.
Merck KGaA, Toxikologie, Postfach 4119, 64271 Darmstadt, Germany

Jortner, Bernard S., Dr.
Laboratories for Neurotoxicity Studies, Virginia Tech, College of Veterinary Medicine, Blacksburg, VA 24061, USA

Kanno, Jun, Dr.
Department of Pathology, Tokyo Medical and Dental University, Yushima 1-5-45 Bunkyo-ku, Tokyo 113, Japan

Karbe, Eberhard, Dr.
Langendorfer Strasse 17, 42489 Wülfrath, Germany

Kaufmann, Wolfgang, Dr.
BASF Aktiengesellschaft, Department of Product Safety, Regulations, Toxicology and Ecology, Z 470, 67056 Ludwigshafen/Rhein, Germany

Kittel, Birgit, Dr.
BASF Aktiengesellschaft, Department of Product Safety, Regulations, Toxicology and Ecology, Z 470, 67056 Ludwigshafen/Rhein, Germany

Konishi, Yoichi, Dr.
Department of Oncological Pathology, Cancer Center, Nara Medical College, 840 Shijo-cho, Kashihara, Nara 634, Japan

Krieg, Kurt, Dr.
Geisenheimer Strasse 90, 60529 Frankfurt, Germany
(formerly Hoechst AG, Frankfurt)

Krinke, Georg, Dr.
Novartis Crop Protection AG, WST-453.E.24, 4332 Stein, Switzerland

Kulwich, Barbara, Dr.
RW Johnson Pharmaceutical Research, Drug Safety Evaluation, Welsh and McKean Roads, Spring House, PA 19477, USA

Küttler, Karin, Dr.
BASF Aktiengesellschaft, Department of Product Safety, Regulations, Toxicology and Ecology, Z 470, 67056 Ludwigshafen/Rhein, Germany

Landes, Christian, Dr.
Novartis Crop Protection AG, WST-453.E.24, 4332 Stein, Switzerland

Leblanc, Bernard, Dr.
Pfizer Inc., Central Research, Eastern Point Road, Groton, CT 06340, USA

Leininger, Joel R., Dr.
3M Pharmaceuticals, 3M Center, Building 270-3S-05,
St. Paul, MN 55144-1000, USA

LeNet, Jean-Loic, Dr.
Pfizer Centre de Recherche, Pathology Department, BP 159,
37401 Amboise Cedex, France

Lenz, Barbara, Dr.
Hoffmann-LaRoche AG, Toxikologie, Bau 73, 401, Grenzacherstrasse 124,
4002 Basel, Switzerland

Lewis, David, Dr.
Huntingdon Research Centre, Woolley Road,
Alconbury, Huntingdon, Cambs PE18 6ES, UK

Long, Philip H., Dr.
Pathology Associates Int'l, 6217 Centre Park Drive,
West Chester, OH 45069-3866, USA

Longeart, Loic, Dr.
Pfizer Centre de Recherche, Pathology Department, BP 159,
37401 Amboise Cedex, France

Maekawa, Akihiko, Dr.
Department of Pathology, Sasaki Institute, 2-2 Kanda-Surugadai,
Chiyoda-ku, Tokyo 101-0062, Japan

Marsman, Daniel S., Dr.
Procter and Gamble Co., C1N19 SWTC, 11530 Reed Hartman Highway,
Cincinnati, OH 45241, USA

McConnell, Robert F., Dr.
Consultant Pathology Services, 19255 Wildwood West Drive,
Penn Valley, CA 95946, USA

Mesfin, G. M., Dr.
Pharmacia and Upjohn Inc., 301 Henrietta Street,
Kalamazoo, MI 49007, USA

Miller, Rodney A., Dr.
Battelle Pacific Nothwest Laboratories, Box 999, Battelle Blvd,
Richland, WA 99352, USA

Mitsumori, K., Dr.
National Institute of Health Sciences, Kamiyoga 1-18-1, Setagaya-ku,
Tokyo 158, Japan

Mohr, Ulrich, Dr.
Hannover Medical School, Institute of Experimental Pathology,
Carl Neuberg-Strasse 1, 30625 Hannover, Germany

Morgan, Kevin T., Dr.
Glaxo Wellcome, Main T, 1128, Research Triangle Park, NC 27709, USA

Mori, Hideki, Dr.
Department of Pathology, Gifu University School of Medicine,
Tsukasa-cho 40, Gifu 500, Japan

Nolte, Thomas, Dr.
ASTA Medica AG, Institut für Toxikologie, Kantstrasse 2,
33790 Halle, Germany

Parish, William E., Dr.
Unilever Research, Colworth House, Sharnbrook, Bedford, MK44 1LQ, UK

Paulson, Ivar, Dr.
AstraZeneca, Safety Assessment, Pathology Department,
15185 Soedertaelje, Sweden

Püschner, Hubert, Dr.
Boehringer Ingelheim Pharma KG, Abteilung für Experimentelle
Pathologie und Toxikologie, Birkendorferstrasse 65, 88397 Biberach, Germany

Rehm, Sabine, Dr.
SmithKline Beecham Pharmaceuticals, 709 Swedeland Road, PO Box 1539,
King of Prussia, PA 19406, USA

Reiland, Sven, Dr.
AstraZeneca, Safety Assessment, Pathology Department,
15185 Soedertaelje, Sweden

Render, James, Dr.
Pfizer Inc., Central Research, Eastern Point Road, Groton, CT 06340, USA

Reznik (†), Gerd, Dr.
Formerly: Byk Gulden Pharmazeutika, Institut für Pathologie
und Toxikologie, Friedrich-Ebert-Damm 101, 22047 Hamburg, Germany

Rinke, Matthias, Dr.
Bayer AG, Pharma Forschungszentrum, Institut für Toxikologische
Pathologie, Aprather Weg, 42096 Wuppertal, Germany

Rittinghausen, Susanne, Dr.
Fraunhofer Institute of Toxicology and Aerosol Research,
Nikolai-Fuchs-Strasse 1, 30625 Hannover, Germany

Robinson, Mervyn, Dr.
Central Toxicology Laboratory, Alderely Park,
Macclesfield, Cheshire, SK10 4TJ, UK

Sander, Erich, Dr.
Bayer AG, Pharma Forschungszentrum, Institut für Toxikologische
Pathologie, Aprather Weg, 42096 Wuppertal, Germany

Sandusky Jr., George, Dr.
Lilly Research Laboratories, PO Box 708, Greenfield, IN 46140, USA

Schwartz, Lester, Dr.
SmithKline Beecham Pharmaceuticals, 709 Swedeland Road, PO Box 1539,
King of Prussia, PA 19406, USA

Seely, John C., Dr.
Experimental Pathology Laboratories Inc., 615 Davis Drive, Suite 500, Durham, NC 27713, USA

Shirai, Tomoyuki, Dr.
1st Department of Pathology, Nagoya City University Medical School, 1 Kawasumi Mizuho-cho Mizuho-ku, Nagoya 467, Japan

Short, Brian G., Dr.
Allergan Inc., 2525 Dupont Drive, Irvine, CA 92623, USA

Slayter, Michael V., Dr.
Dupont Merck Pharmaceutical Co., PO Box 30 Elkton Road, Stine Haskell Research Center, Building 320, Newark, DE 19714, USA

Stromberg, Paul C., Dr.
Department of Veterinary Biosciences, Ohio State University, 1925 Coffey Road, Columbus, OH 43210, USA

Tsuda, H., Dr.
Chemotherapy Division, National Cancer Center Research Institute, 1-1 Tsukiji, S-chome Chuo-ku, Tokyo 104, Japan

Tuch, Klaus, Dr.
Byk Gulden Pharmazeutika, Institut für Pathologie und Toxikologie, Friedrich-Ebert-Damm 101, 22047 Hamburg, Germany

Tucker, Mary, Dr.
AstraZeneca, Safety of Medicines Group, Mereside, Alderley Park, Macclesfield, Cheshire, SK10 4TG, UK

Turusov, V. S., Dr.
Laboratory of Carcinogenic Substances, Cancer Research Center, Kashirskoye Sh. 24, 115478 Moscow, Russia

Wadsworth, Peter F., Dr.
AstraZeneca, Mereside, Alderley Park, Macclesfield, Cheshire SK10 4TG, UK

Ward, Jerrold M., Dr.
NCI Veterinary/Tumor Pathology Section, National Cancer Institute, NCI-FCRDC Fairview 201, Frederick, MD 21701-1201, USA

Weisse (†), Ingo, Dr.
Formerly: Boehringer Ingelheim KG, Experimentelle Pathologie und Toxikologie, Postfach 200, 55216 Ingelheim/Rhein, Germany

Whiteley, Laurence O., Dr.
Procter and Gamble Co., Miami Valley Labs, PO Box 398707, Cincinnati, OH 45253, USA

Wright, Jayne A., Dr.
Underhill Housen Limited, Woolhope Cockshoot, Putley, HR8 2QR Ledbury, Hertfordshire, UK

Integumentary System

R. Bruner[1] · K. Küttler[2,3] · R. Bader[3] · W. Kaufmann[3] · A. Boothe
M. Enomoto · J.M. Holland · W.E. Parish

[1]Chair; [2]Co-chair; [3]Initial draft.

Hyperplasia, Squamous Cell *(H)*

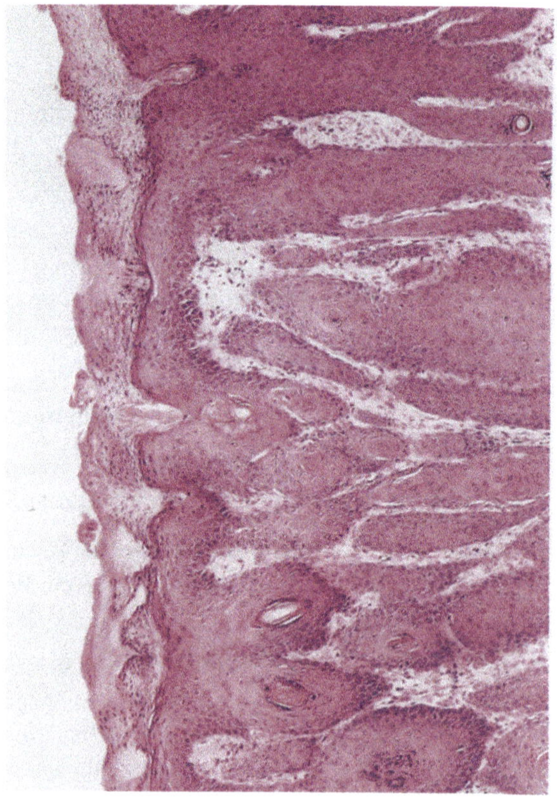

Synonym(s). Hyperplasia, epidermal

Histogenesis

Epidermal cells.

Diagnostic Features

- Growth pattern is variable, it may be regular, irregular to papillary
- Width of non-keratinized layers of the epidermis is increased
- Hyperkeratosis is present
- Number of epidermal cells is increased, especially in prickle cell layer
- Rete ridge formation is often present

Differential Diagnosis

Papilloma, Squamous Cell
Growth pattern is exophytic.

Carcinoma, Squamous Cell
Invasive growth is present or
Mitotic figures are numerous and atypical and
Nuclear atypia is present and
No evidence of intercellular bridges

Comment

Squamous cell hyperplasia is found primarily in cases of chronic and inflammatory irritation of the skin.

References

(See 1, 20, 21)

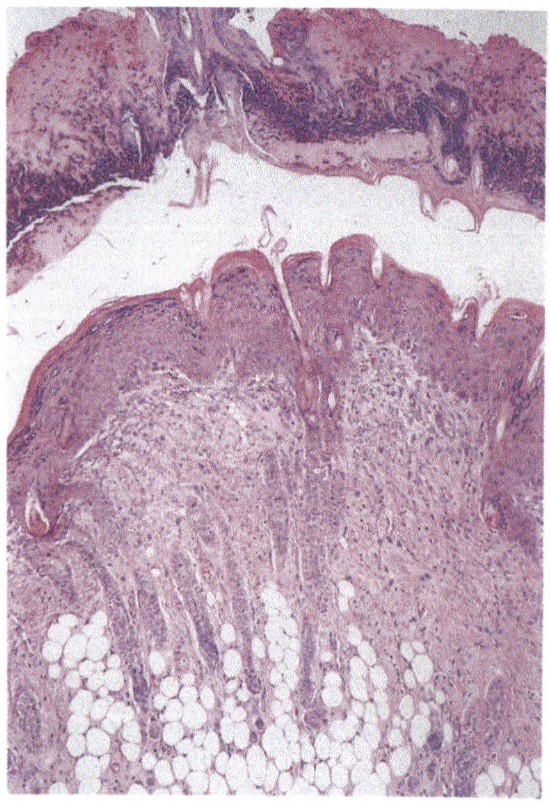

Fig. 1 *(above).* Squamous cell hyperplasia of the skin, B6C3F1 mouse. H&E

Fig. 2 *(below).* Squamous cell hyperplasia of the skin, p53 wildtype mouse, male, age 107 weeks. H&E

Papilloma, Squamous Cell *(B)*

Pedunculated • Flat

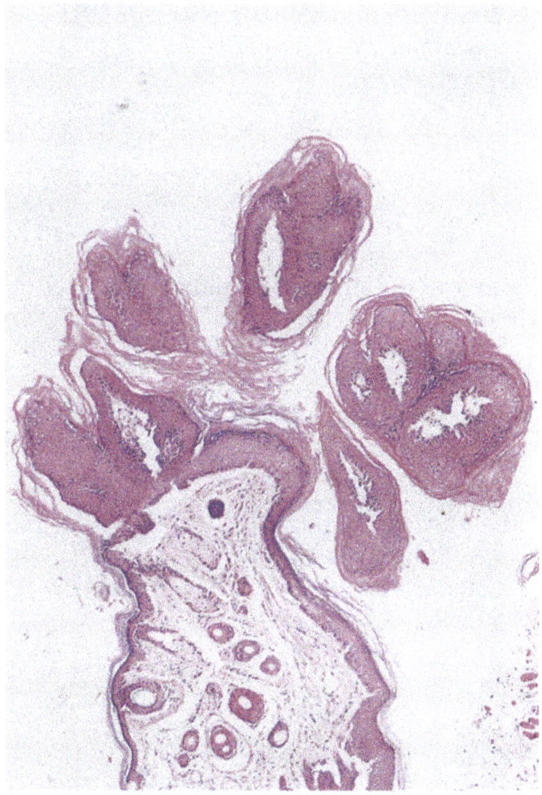

Fig. 3. Squamous cell papilloma of the skin, p53 homozygous mouse, female, age 37 weeks. H&E

Histogenesis

Epidermal cells.

Diagnostic Features

- Basal cell borderline is distinct
- Growth pattern is exophytic or papillary
- Well vascularized stroma is lined by acanthotic epithelium with small foci of keratinization
- Exuberant hyperkeratosis and some parakeratosis may be present
- Dyskeratosis may be present with premature keratinization in individual cells
- Cells may be dark-staining
- Cells may be fusiform or columnar representing basal portions of the papillary epithelium
- Mitotic figures are rare

Pedunculated

- A narrow stalk is present at the base.

Flat ("Sessile Papillomas")

- No evidence of a stalk
- Covers a fairly large area of the skin
- Stroma is indistinctly demarcated from the underlying dermis
- A continuous even transition is present between neoplastic epithelium and adjacent hyperplastic epidermis
- Hyperplastic sebaceous glands may be enclosed

Differential Diagnosis

Keratoacanthoma

Single or multiple cup-shaped cavities are present and
Cavities are filled with concentrically arranged keratin material

Carcinoma, Squamous Cell

Invasive growth is present or
Mitotic figures are numerous and atypical and
Nuclear atypia is present and
No evidence of intercellular bridges

Fibroma (fibropapilloma)

Exophytic growth of connective tissue is covered with regular epithelial lining

References

(See 1, 4, 8, 10, 14, 18, 21)

Carcinoma, Squamous Cell *(M)*

Keratinizing • Pseudo-glandular
Non-keratinizing

Synonym(s). Carcinoma, epidermoid

Histogenesis

Epidermal squamous cells.

Diagnostic Features

- Ulceration may be present
- Cellular squamous differentiation is variable
- Nuclear atypia is present
- Size and staining of nuclei are variable
- Loss of intercellular bridges is present
- Mitotic figures are numerous
- Atypical, bizarre mitotic figures are present
- Invasion is present of the dermis and striated muscle by nests or cords of squamous cells
- Basal lamina is penetrated by invasive growth

Keratinizing
- Cells are arranged in cords or whorls with central keratinization, often resembling lamellated horn pearls
- Dyskeratosis is present
- Parakeratosis is present
- Keratinization may be decreased
- Buds and nests of cells may resemble basal cells without any signs of keratinization
- Nuclear atypia varies from slight to severe

Pseudo-glandular
- Acantholysis may result in a tubular or alveolar pseudo-glandular pattern
- Lumen of pseudo-glandular structures are filled with desquamated acantholytic cells
- Many cells are partially or fully keratinized
- One or several layers of epithelial cells line the lumen

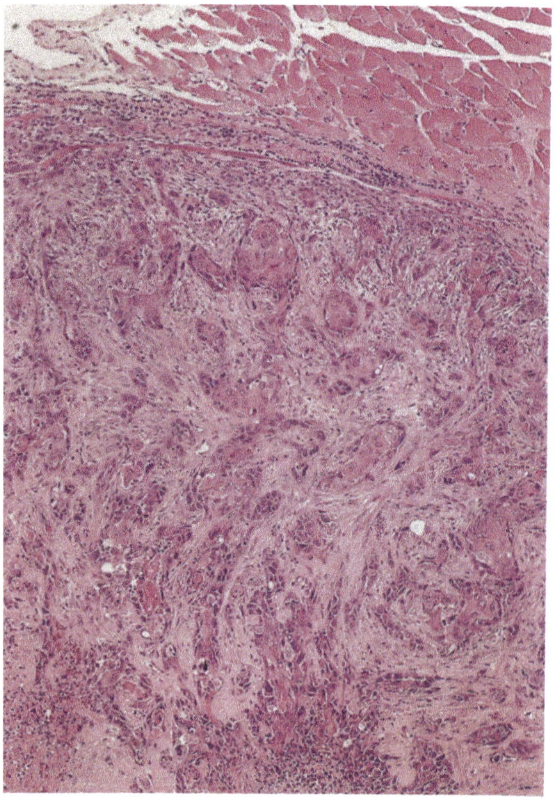

Fig. 4. Squamous cell carcinoma of the skin, B6C3F1 mouse, female, age 105 weeks. H&E (courtesy of Dr. Shinji Yamamoto)

Non-keratinizing
- Hardly any keratinization is present
- Cellular atypia is present
- Cells may be spindle-shaped
- Nuclei exhibit severe atypia
- Giant, fragmented, or multiple nuclei are present
- Mitotic figures are numerous
- Atypical, bizarre mitotic figures are present

Differential Diagnosis

Papilloma, Squamous Cell
Basal cell border line is distinct and
No evidence of invasive growth

Keratoacanthoma
Well demarcated and
No evidence of irregular downward growth

Adenoacanthoma, Malignant (Mammary Gland)
Both, glandular and squamous differentiation is present

Carcinoma, Basal Cell
Nests and cords of cells resembling basal cells of the epidermis are well demarcated

References

(See 1, 4, 8, 10, 13, 15, 18, 21, 24, 27)

Tumor, Basal Cell, Benign *(B)*

Histogenesis

Basal cells of the epidermis and epidermal appendages.

Diagnostic Features

- Well circumscribed
- Distinctly demarcated
- Uniform sheets and strands of cells are present
- Cells are closely packed
- Transition into the basal layer of the hyperplastic epidermis is often continuous
- Keratinization, sebaceous cells or hair follicle formation may be present
- Areas of sebaceous cell differentiation or hair follicle formation are rare
- Cells resemble basal cells of the epidermis and its appendages
- Cells are intensely stained and are slightly basophilic
- Cytoplasm is scant
- Nuclei are round or oval
- Mitotic figures are rare

Differential Diagnosis

Carcinoma, Basal Cell
Invasion is present or
No evidence of distinct demarcation and
Cytological and growth pattern is heterogeneous and
Mitotic figures are numerous

References

(See 3, 7, 12, 16, 21)

Carcinoma, Basal Cell *(M)*

Histogenesis

Basal cells of the epidermis and epidermal appendices.

Diagnostic Features

- More or less demarcated
- Heterogeneous sheets and strands of cells are present
- Palisading cells are present at the periphery
- Cells are closely packed
- Transition into squamous, hair follicle, or sebaceous components may be present
- Areas of sebaceous cell differentiation or hair follicle formation are rare
- Cells resemble basal cells of the epidermis or its appendages
- Cells are intensely stained and relatively small and are slightly basophilic
- Cytoplasm is scant
- Nuclei are dark blue
- Mitotic figures are numerous
- Extensive local invasion is present
- No evidence of metastasis

Differential Diagnosis

Tumor, Basal Cell, Benign
No evidence of invasion and
Well demarcated and
Cytological pattern is uniform and
Mitotic figures are rare

Tumor, Hair Follicle, Benign
Predominant growth pattern are hair follicles and
Cytoplasm is eosinophilic

Carcinoma, Sebaceous Cell
Predominant growth pattern is glandular

References

(See 3, 7, 12, 21)

Hyperplasia, Sebaceous Cell *(H)*

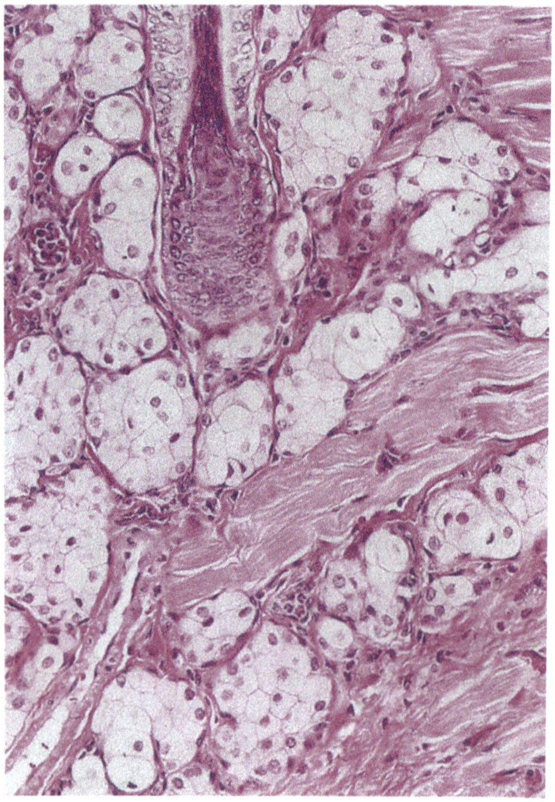

Fig. 5. Sebaceous hyperplasia of the skin, CD-1 mouse. H&E

Histogenesis

Sebaceous cells.

Diagnostic Features

- Regular sebaceous gland architecture is maintained

Differential Diagnosis

Adenoma, Sebaceous Cell

Regular sebaceous gland architecture is not maintained and
Large numbers of immature germinative, basaloid cells are present and
Growths pattern may be exophytic

Comment

There is no special description for a "hyperplasia, sebaceous cell" for mice found in the literature. Therefore the criteria mentioned in the rat manuscript are also used for mice.

References

(See 1, 21)

Adenoma, Sebaceous Cell *(B)*

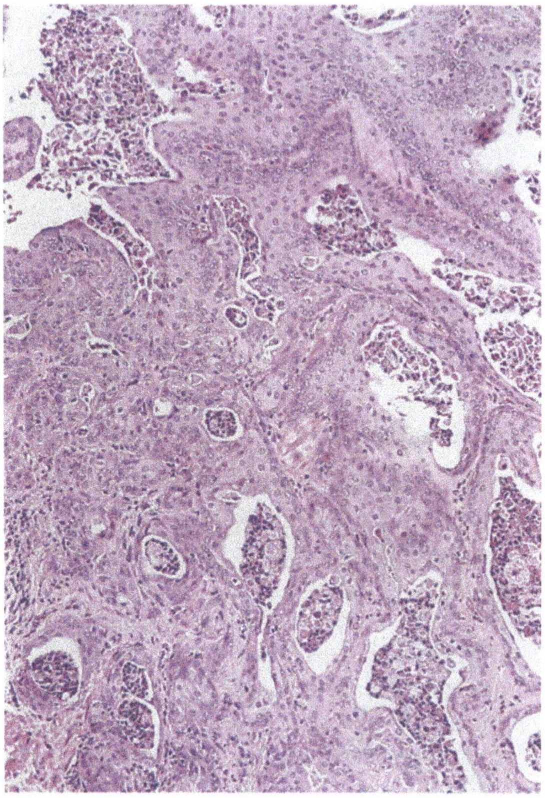

Fig. 6. Sebaceous adenoma of the skin, v-Ha-ras mouse, female, age 69 weeks. H&E

Histogenesis

Sebaceous gland cells.

Diagnostic Features

- Regular sebaceous gland architecture is not maintained
- Glandular acino-lobular pattern is maintained
- Endophytic growth exhibits small flat subepidermal nodules
- Exophytic growth is protruding as papillomatous nodules
- In the center of the acini, cells in all stages of maturation are present
- In the periphery, large numbers of immature germinative, basaloid cells are present
- Cells are small
- Cytoplasm is abundant, foamy and clear
- Nuclei are partly pyknotic
- Immature germinative, basaloid cells may be predominant in some acini or lobules exhibiting mitoses
- Mitotic figures may be present

Differential Diagnosis

Hyperplasia, Sebaceous Cell
Regular sebaceous gland architecture is maintained and
Cells are mature glandular cells and only few immature germinative cells are present

Carcinoma, Sebaceous Cell
Invasive growth or metastasis is present or
Cells are poorly differentiated and
Cellular and nuclear atypia are present

Tumor, Basal Cell, Benign
Growth pattern is non-glandular and
Areas of sebaceous cell differentiation are rare

References

(See 3, 7, 16, 21)

Carcinoma, Sebaceous Cell *(M)*

Histogenesis

Sebaceous cells.

Diagnostic Features

- Regular sebaceous gland architecture is distorted
- Glandular acino-lobular pattern is maintained
- Cells are poorly differentiated, irregular, varying in shape and size, but may be focally well differentiated
- Nuclear atypia is often severe
- Squamous cell differentiation may be present
- Variability in the size of the intracytoplasmic lipid vacuoles is high
- Invasive growth is present
- Metastasis may be present

Differential Diagnosis

Adenoma, Sebaceous Cell
No evidence of invasive growth or metastasis and
Cells are well differentiated and
No evidence of nuclear atypia.

Carcinoma, Basal Cell
Areas of sebaceous cell differentiation are rare

References

(See 3, 7, 16, 21)

Keratoacanthoma *(B)*

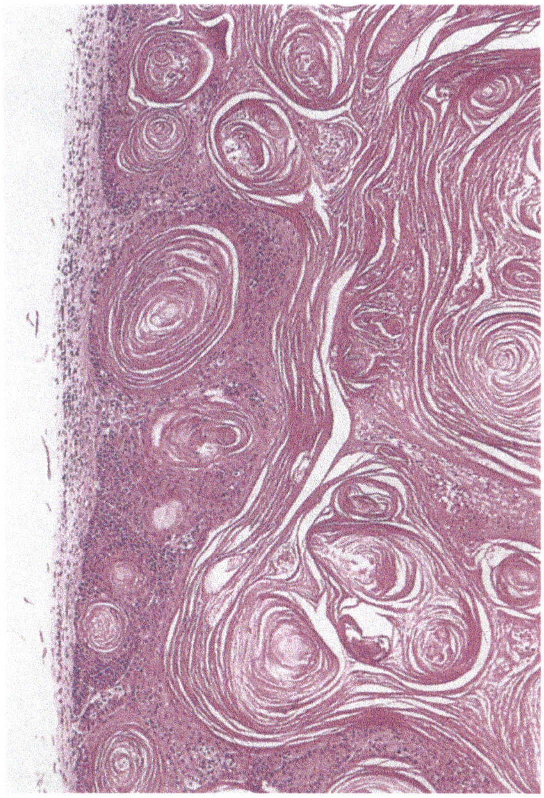

Fig. 7. Keratoacanthoma of the skin. H&E

Histogenesis

Is thought to arise from follicular epithelium of hair follicles.

Diagnostic Features

- Well-demarcated
- Single or multiple cup-shaped cavities are present
- Cavities are filled with lamellated, concentrically arranged or homogenous keratin masses
- Cavities are lined by well differentiated or acanthotic squamous epithelium of several layers with distinct features of abortive hair follicle formation
- Stratum granulosum is often lacking
- Whorls with central keratinization may be present, often resembling lamellated horn pearls
- Central areas may fuse to a form a keratin-loaded central cavity
- Nuclear to cytoplasmic ratio is low
- Mitotic figures are present

Differential Diagnosis

Cysts
Any kind of cyst of the skin, epidermal, dermoid, follicular, cyst with epithelial proliferation, are lined by a regular squamous epithelium

Tumor, Hair Follicle, Benign
In the trichofolliculoma type mature hair follicle formation and trichohyaline granules are present or
In the pilomatricoma type cysts filled with material representing different stages of hair follicle formation are present

Papilloma, Squamous Cell
Growth is exophytic

Carcinoma, Squamous Cell
Downward invasion is irregular with penetration through the basal lamina and Nuclear atypia is present

Comment

There are subtle differences between keratoacanthoma with keratohyaline granules and trichofolliculoma. In contrast to rats, induced keratoacanthomas tend to regress.

References

(See 1, 4, 8, 11, 18, 21, 23)

Tumor, Hair Follicle, Benign *(B)*

Trichofolliculoma type
Pilomatricoma type • Trichoepithelioma type
Tricholemmoma type • NOS

Histogenesis

Pilosebaceous unit.

Diagnostic Features

- Well-demarcated
- Different stages of hair formation are present
- Single or multiple cysts may be present
- Various differentiation stages of the typical cellular structures of the hair follicle are expressed

Trichofolliculoma Type
- All components of the regular anagen hair cycle are present
- Cysts and many mature hair shafts are present

Pilomatricoma Type
- All components of the regular anagen hair cycle are present
- Cysts and many ghost cells are present

Trichoepithelioma Type
- No evidence of an infundibulum, or mature hair shafts, or ghost cells
- Keratinization is abrupt

Tricholemmoma Type
- Only cells present are basal cells, hair matrix cells, and cells of the outer sheath. The latter cells are vacuolated and have glycogen

Differential Diagnosis

Cysts
Any kind of cyst of the skin, epidermal, dermoid, follicular, cyst with epithelial proliferation, are lined by a regular squamous epithelium

Tumor, Basal Cell, Benign
Focal hair follicle differentiation is present only sporadically

Keratoacanthoma
Cavities invaginating from the skin surface are predominantly filled with keratin masses and
Only remnants of hair follicles may be present

Carcinoma, Basal Cell
No evidence of mature hair follicle differentiation

References

(See 3, 7, 16, 21)

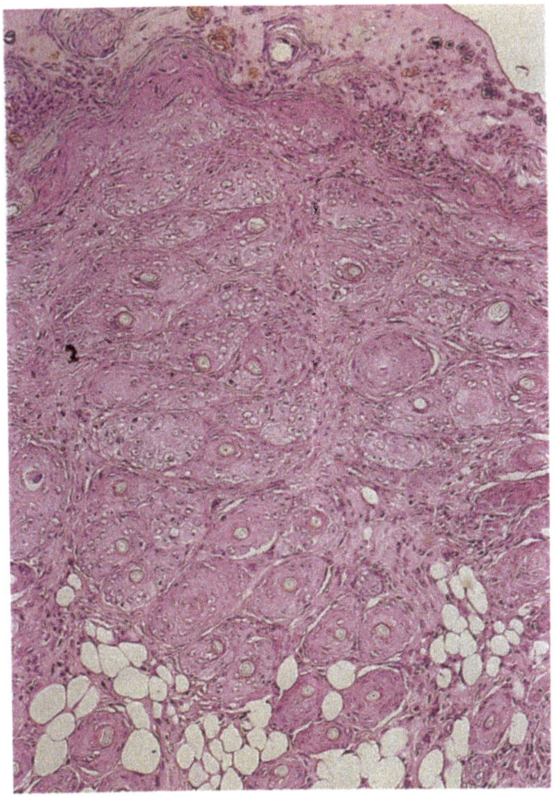

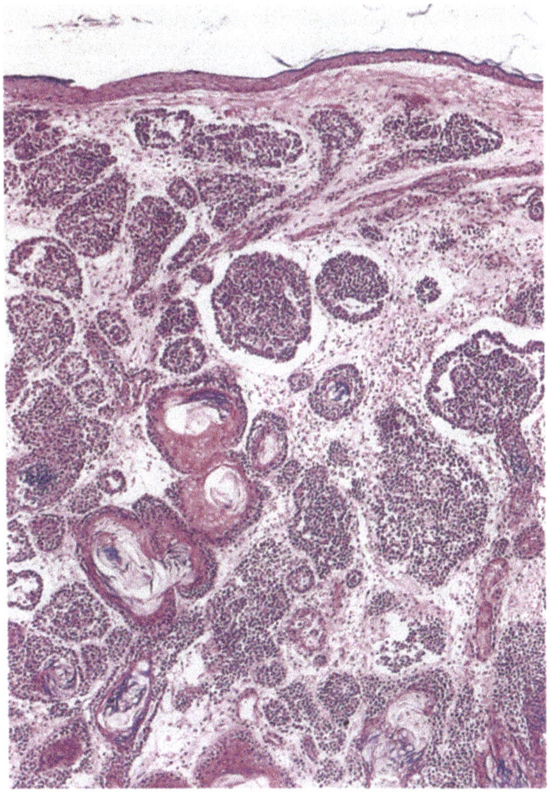

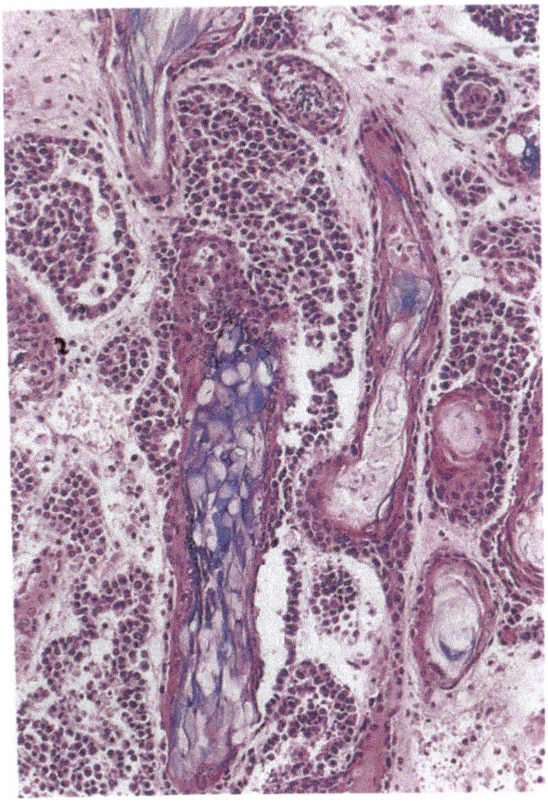

Fig. 8 *(upper left)*. Benign hair follicle tumor, trichoepithelioma type, of the skin. H&E

Fig. 9 *(upper right)*. Benign hair follicle tumor, trichoepithelioma/trichofolliculoma type, of the skin, CD-1 mouse. H&E

Fig. 10 *(below)*. Benign hair follicle tumor, trichoepithelioma/trichofolliculoma type, of the skin, CD-1 mouse. H&E

Melanoma, Malignant *(M)*

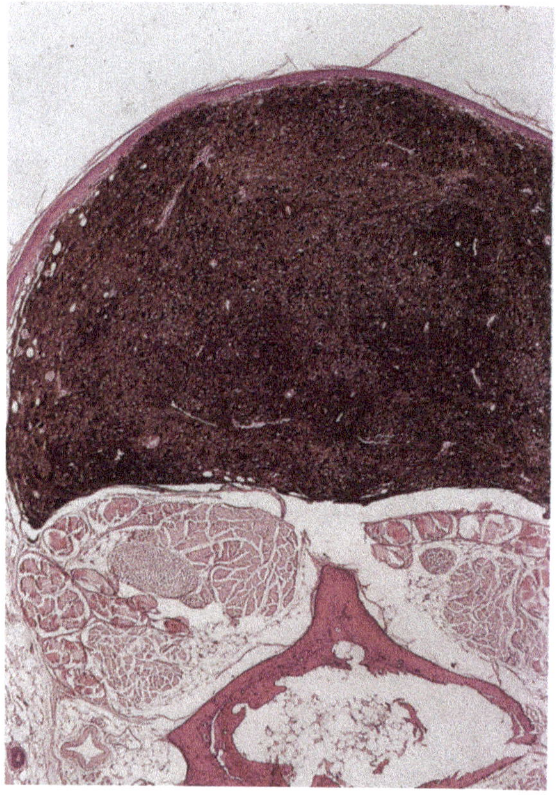

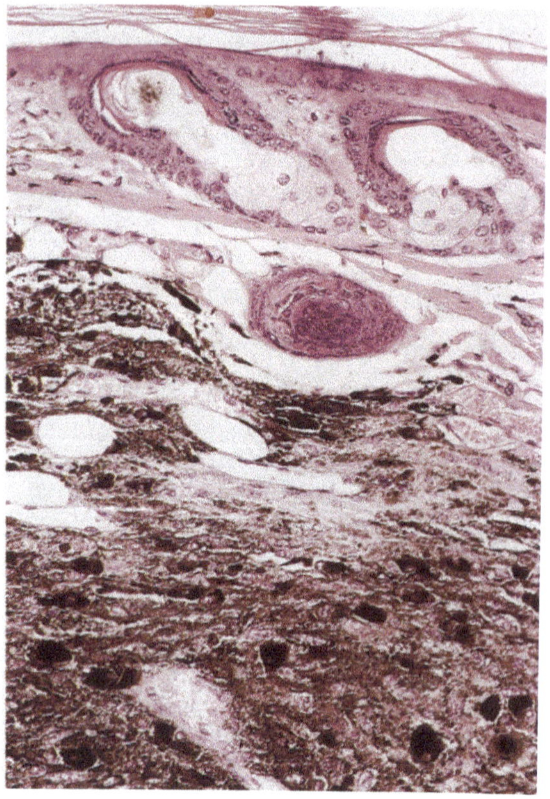

Synonym(s). Melanosarcoma, melanocytic tumor

Histogenesis

Neuroectodermal origin.

Diagnostic Features

- Dense aggregates of dermal melanocytes are present
- Intracytoplasmic dark brown pigment granules are present
- Cellular pleomorphism is present
- Cell types are spindle cells, epithelioid cells, or anaplastic cells
- Spindle cell lesions are the most common ones
- Invasive growth and penetration of skin muscle may be present

Differential Diagnosis

Carcinoma, Basal Cell

Palisading cells are present at the periphery and

No evidence of intracytoplasmic pigment

Comment

Melanomas are exceedingly rare incidental tumors in laboratory mice, but there are some reports about induced malignant melanomas and pigmented lesions.

References

(See 1, 4, 17, 21, 22)

Fig. 11 *(above).* Malignant melanoma of the skin, CBA/S mouse. H&E

Fig. 12 *(below).* Malignant melanoma of the skin, CBA/S mouse. H&E

Hyperplasia *(H)*

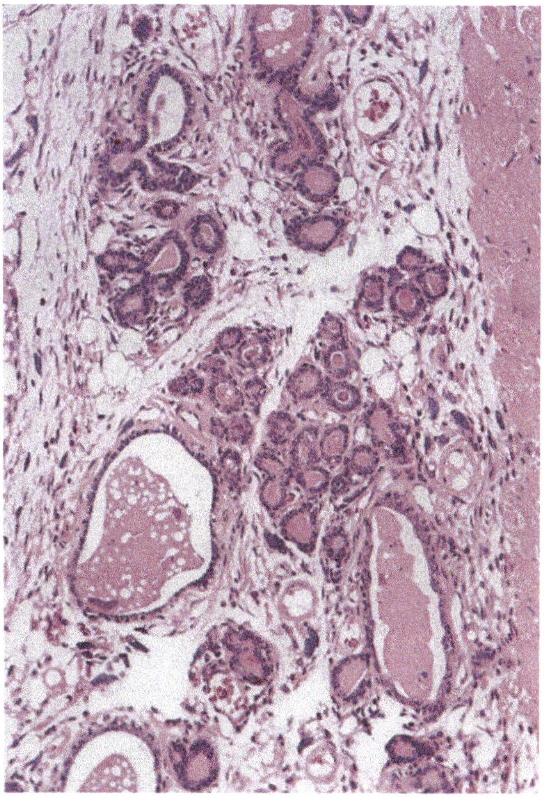

Fig. 13. Moderate hyperplasia of the mammary gland. H&E

Histogenesis

Mammary gland epithelium.

Diagnostic Features

- Regular lobular architecture is maintained
- Alveoli or duct epithelium is present
- Lining of alveoli is single layered
- Cystic dilation may be present
- No evidence of pleomorphism
- Cells lining the alveoli are cuboidal
- Cells lining the ducts vary from columnar to flattened

Diffuse
- Mammary fat pads are completely filled with glands
- Lumen are filled with eosinophilic material as evidence of secretory activity

Focal
- Single or multiple
- Islets of glandular tissue are embedded in fatty stroma
- Alveoli and ducts may be distended with eosinophilic material as evidence of secretory activity

Differential Diagnosis

Adenoma

Well demarcated from the adjacent tissue and

Regular glandular architecture is not maintained

References

(See 2, 5, 6, 7, 9, 10, 16, 19, 25, 26, 28)

Adenoma *(B)*

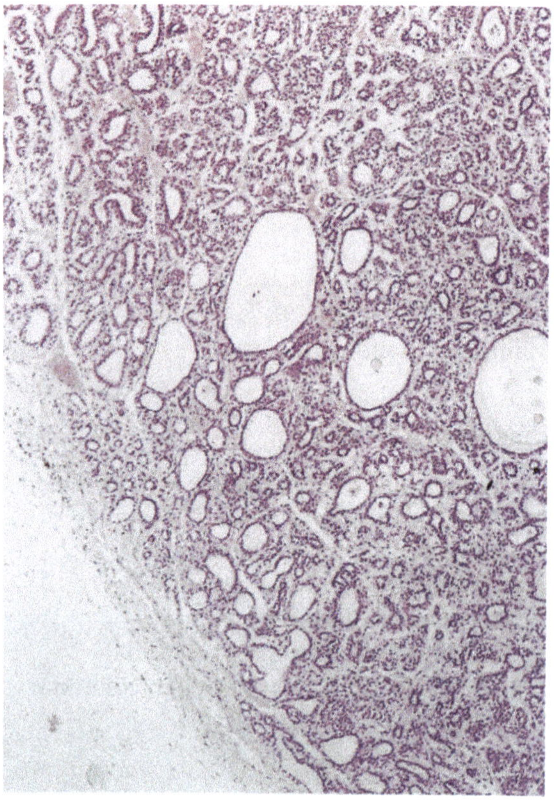

Fig. 14. Adenoma of the mammary gland, B6C3F1 mouse. H&E

Histogenesis

Mammary gland epithelium.

Diagnostic Features

- Well demarcated
- Regular lobular architecture is not maintained
- Small solid acinar structures are uniform in size and closely packed
- Lining of acinar structures is single layered
- Scanty connective tissue is present
- Cells are well differentiated
- Cells lining the alveolar structures are cuboidal
- Cells may have secretory vacuoles
- Mitotic figures may be present

Differential Diagnosis

Hyperplasia
Lobular pattern is maintained

Fibroadenoma
Amount of connective tissue enclosing proliferating mammary tissue is prominent

Adenocarcinoma
Invasion of adjacent tissues or metastases may be present or
Cellular pleomorphism is present and
Mitotic figures are numerous and
Different growth patterns may be present within a single lesion

References

(See 2, 5, 6, 7, 9, 10, 16, 19, 25, 26, 28)

Adenocarcinoma *(M)*

Histogenesis

Mammary gland epithelium.

Diagnostic Features

- Regular lobular architecture is not maintained
- Variability in growth pattern is high, tubular, papillary, cystic, solid, comedo, undifferentiated
- Different growths pattern may be present within a single lesion
- Intracystic papillary projections, solid cords, sheets, nests or tubes with no signs of glandular differentiation may be present
- Acinar areas are frequently associated with cysts filled with blood, or fluid as evidence of secretory activity
- Squamous areas cover not more than 25% of the lesion
- Amount of stroma is variable
- Glandular structures may be embedded in a loose myxomatous stroma resembling myoepithelium
- Necrosis and hemorrhages may be present
- Cellular pleomorphism is present
- Mitotic figures are numerous
- Invasion of adjacent tissues, muscle and skin or metastasis is present

Differential Diagnosis

Adenoacanthoma, Malignant
Squamous cell areas cover more than 25% of the lesion

Adenoma
Well demarcated and
Well differentiated and
Mitotic figures are not numerous and
Acini are uniform in size and pattern

References

(See 2, 5, 6, 7, 9, 10, 16, 19, 25, 26, 28)

Figs. 15-18 see p. 18.

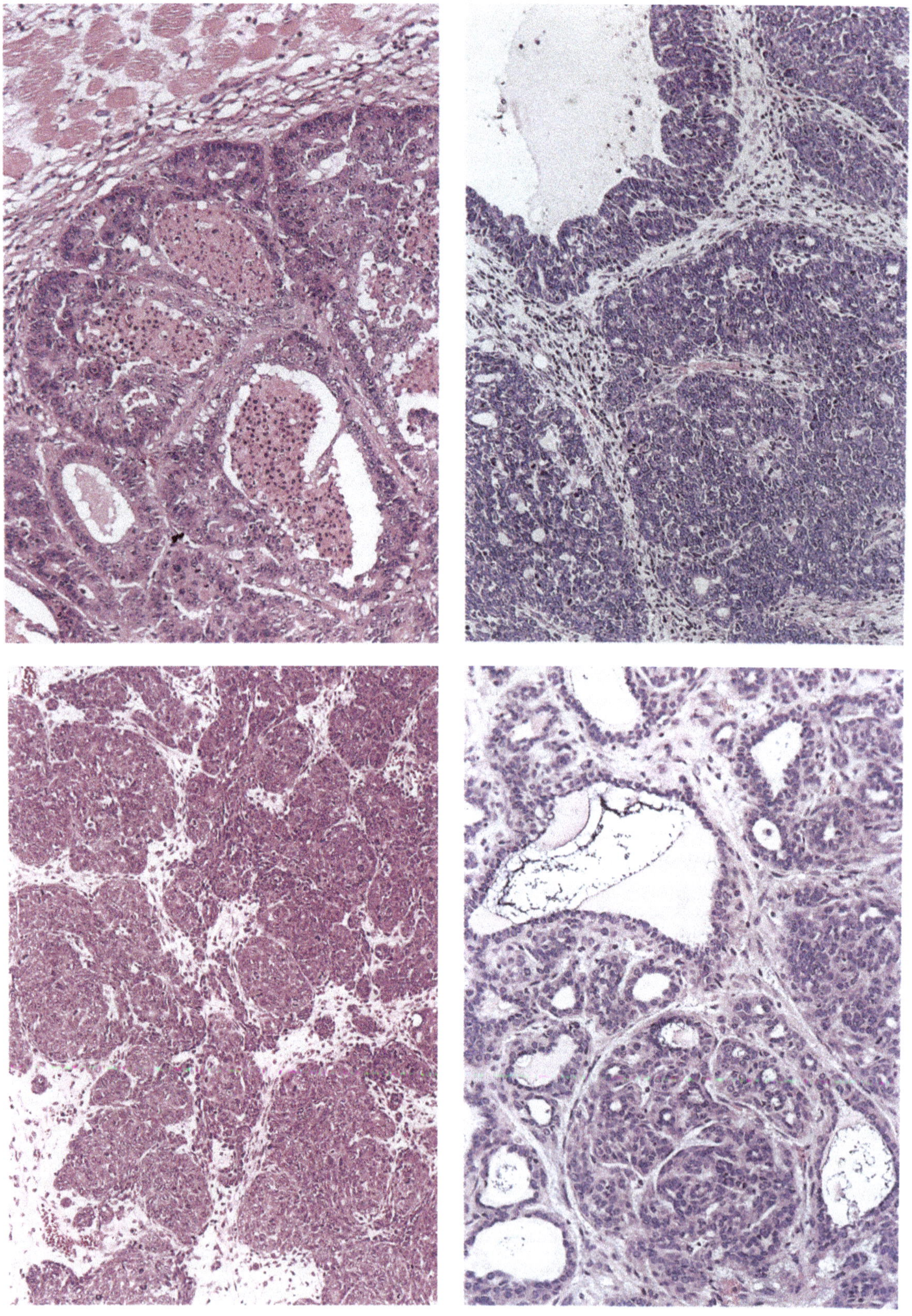

Fibroadenoma *(B)*

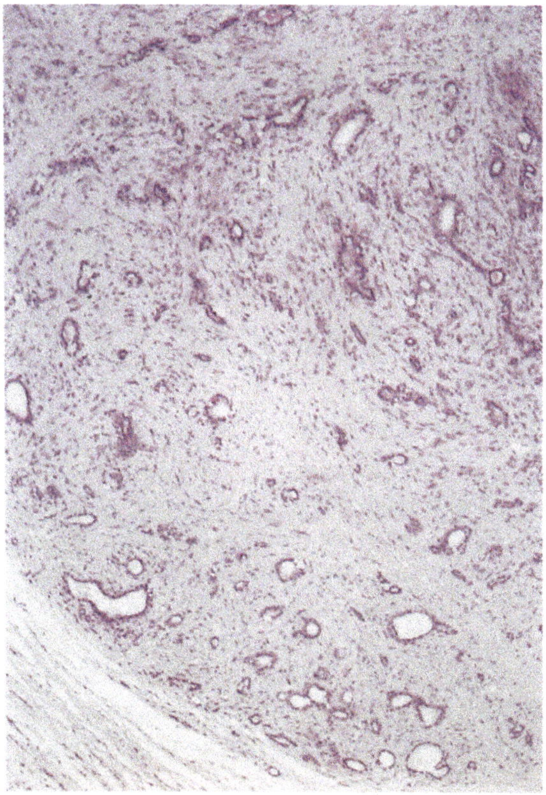

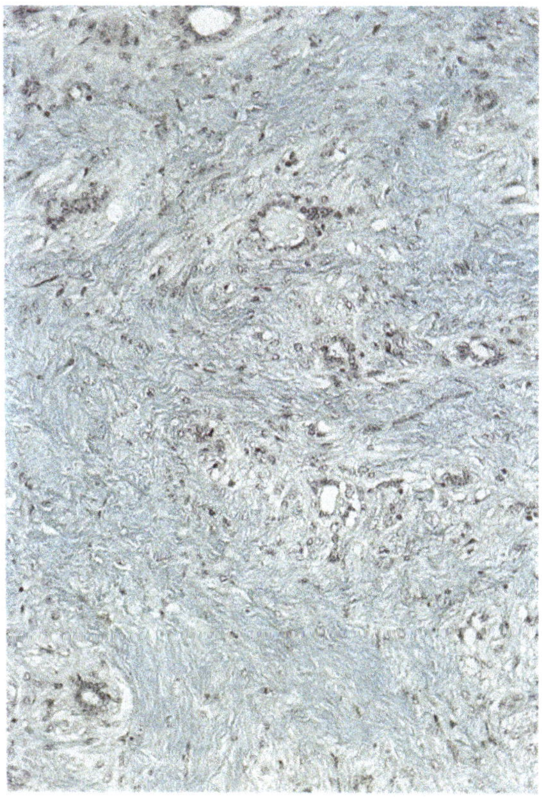

Histogenesis

Mammary gland epithelium and connective tissue.

Diagnostic Features

- Clearly demarcated
- A fibrous capsule may be present
- Discrete alveoli or ducts are present within a prominent, fairly dense fibrous stroma
- Amount of connective tissue enclosing glandular structures is variable and may be unobtrusive
- Lumen of glandular structures may be filled with eosinophilic material as evidence of secretory activity
- Mitotic figures are rare

Differential Diagnosis

Adenoma

No evidence of increased amount of stromal connective tissue

Fibroma

No evidence of epithelial elements

References

(See 2, 5, 6, 7, 9, 10, 16, 19, 25, 26, 28)

◀ **Fig. 15** *(upper left)*. Adenocarcinoma of the mammary gland. H&E

Fig. 16 *(upper right)*. Adenocarcinoma of the mammary gland, B6C3F1 mouse. H&E

Fig. 17 *(lower left)*. Adenocarcinoma of the mammary gland, B6C3F1 mouse. H&E

Fig. 18 *(lower right)*. Adenocarcinoma of the mammary gland, B6C3F1 mouse. H&E

Fig. 19 *(above)*. Fibroadenoma of the mammary gland, B6C3F1 mouse. H&E ▶

Fig. 20 *(below)*. Fibroadenoma of the mammary gland, B6C3F1 mouse. Elastica Masson Goldner

Adenoacanthoma, Malignant *(M)*

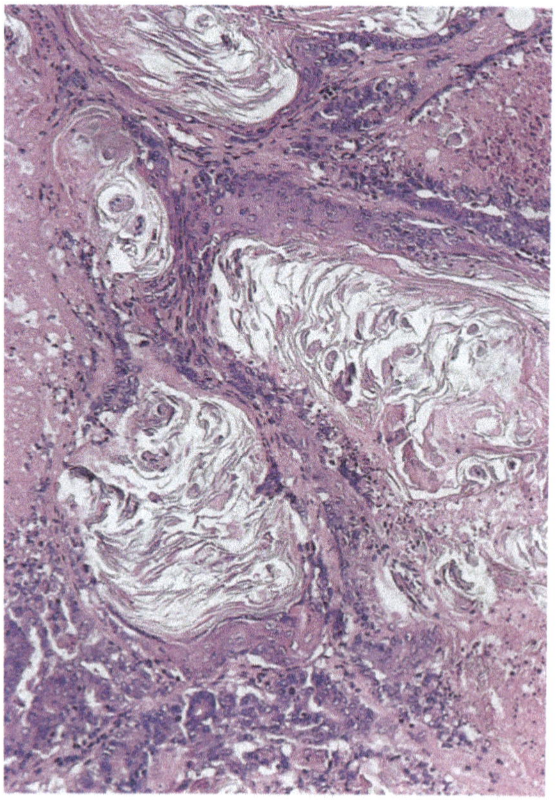

Fig. 21. Adenoacanthoma of the mammary gland. H&E

Histogenesis

Mammary gland epithelium.

Diagnostic Features

- May be well circumscribed
- Differentiation is glandular and squamous
- Glandular and squamous patterns are present in varying proportions
- Squamous areas are well-differentiated
- Squamous cell areas cover more than 25% of the lesion
- Areas of gradual transition may be present from round or polygonal epithelial cells to flattened squamous cells
- Squamous cells exhibit intracytoplasmic keratin and pearl formation
- Metastasis exhibits squamous or glandular tissue

Differential Diagnosis

Adenocarcinoma
Squamous areas may be present, but cover not more than 25% of the lesion

Carcinoma, Squamous Cell (Skin)
Only squamous differentiation is present
and
No evidence of glandular pattern

References

(See 2, 5, 6, 7, 9, 10, 16, 19, 25, 26, 28)

References

1. Bader R, Gembardt C, Kaufmann W, Kuettler K, Mann PC, van Zwieten MJ, Zurcher C (1993) 5. Integumentary system. In: Mohr U, Capen CC, Dungworth DL, Griesemer RA, Ito N, Turusov VS (eds) International classification of rodent tumours, Part I: The rat. IARC Scientific Publications No. 122, Lyon, pp 1–21
2. Benirschke K, Garner FM, Jones TC (1978) Pathology of laboratory animals, vol II. Springer, Berlin
3. Bogovski P (1979) Tumours of the skin. In: Turusov VS (ed) Pathology of tumours in laboratory animals. vol II. Tumours of the mouse. IARC Scientific Publications No. 23, Lyon, pp 1–41
4. Bogovski P (1994) Tumours of the skin. In: Turusov VS, Mohr U (eds) Pathology of tumours in laboratory animals, vol 2. Tumours of the mouse, 2nd edn. IARC Scientific Publications No. 111, Lyon, pp 1–45
5. Dunn TB (1959) In: Homburger F (ed) Physiopathology of cancer. 2nd edn. Hoeber, New York, pp 38–84
6. Ebbenhorst Tengbergen WJ (1970) Morphological classification of mammary tumours in the mouse. Pathol Eur 5: 260–272
7. Faccini JM, Abbott DP, Paulus GJJ (1990) Mouse histopathology. A glossary for use in toxicity and carcinogenicity studies. Elsevier, Amsterdam
8. Faccini JM, Abbott DP, Paulus GJJ (1990) Mouse histopathology. A glossary for use in toxicity and carcinogenicity studies. I. Integumentary System. Elsevier, Amsterdam, pp 1–17
9. Faulkin LJ, Mitchell DJ, Young LJ, Morris DW, Malone RW, Cardiff RD, Gardner MB (1984) Hyperplastic and neoplastic changes in the mammary glands of feral mice free of endogenous mouse mammary tumor virus provirus. J Natl Cancer Inst 73: 971–982
10. Frith CH, Ward JM (1988) Color atlas of neoplastic and non-neoplastic lesions in aging mice. Elsevier, Amsterdam
11. Ghadially FN (1961) The role of hair follicles in the origin and evolution of some cutaneous neoplasms of man and experimental animals. Cancer 14: 1801–1816
12. Hasegawa R, Miyakawa Y, Sato H (1989) Basal cell tumor, skin, mouse. In: Jones TC, Mohr U, Hunt RD (eds) Monographs on pathology of laboratory animals. Integument and mammary glands. Springer, Berlin Heidelberg New York Tokyo, pp 52–55
13. Hasegawa R, Miyakawa Y, Sato H (1989) Squamous cell carcinoma, skin, mouse. In: Jones TC, Mohr U, Hunt RD (eds) Monographs on pathology of laboratory animals. Integument and mammary glands. Springer, Berlin Heidelberg New York Tokyo, pp 31–38
14. Hasegawa R, Sato H, Miyakawa Y (1989) Squamous cell papilloma, skin, mouse. In: Jones TC, Mohr U, Hunt RD (eds) Monographs on pathology of laboratory animals. Integument and mammary glands. Springer, Berlin Heidelberg New York Tokyo, pp 19–24
15. Hirose M (1989) Squamous cell carcinoma, skin, rat. In: Jones TC, Mohr U, Hunt RD (eds) Monographs on pathology of laboratory animals. Integument and mammary glands. Springer, Berlin Heidelberg New York Tokyo, pp 25–30
16. Jones TC, Mohr U, Hunt RD (1989) Monographs on pathology of laboratory animals. Integument and mammary glands. Springer, Berlin Heidelberg New York Tokyo
17. Kanno J (1989) Melanocytic tumors, skin, mouse. In: Jones TC, Mohr U, Hunt RD (eds) Monographs on pathology of laboratory animals. Integument and mammary glands. Springer, Berlin Heidelberg New York Tokyo, pp 63–70
18. Maita K, Hirano M, Harada T, Mitsumori K, Yoshida A, Takahashi K, Nakashima N, Kitazawa T, Enomoto A, Inui K, Shirasu Y (1988) Mortality, major cause of moribundity, and spontaneous tumors in CD-1 mice. Toxicol Pathol 16: 340–349
19. Medina D (1976) Mammary gland as a morphological end point in carcinogenesis studies. J Toxicol Environ Health 1: 551–560
20. Muller GH, Kirk RW, Scott DW (1983) Small animal dermatology. W.B. Saunders, Philadelphia London Toronto, pp 56–58
21. Peckham JC, Heider K (1999) Skin and subcutis. In: Maronpot RR, Boorman GA, Gaul BW (eds) Pathology of the mouse. Reference and atlas. Cache River Press, Vienna, pp 555–612
22. Ramon y Cajal S, Suster S, Halaban R, Filvaroff E, Dotto GP (1991) Induction of different morphologic features of malignant melanoma and pigmented lesions after transformation of murine melanocytes with bFGF-cDNA and H-ras, myc, neu, and E1a oncogenes. Am J Pathol 138: 349–358
23. Ramselaar CG, Ruitenberg EJ, Kruizinga W (1980) Regression of induced keratoacanthomas in anagen (hair growth phase) skin grafts in mice. Cancer Res 40: 1668–1673
24. Rehm S, Ward JM, Devor DE (1989) Squamous cell carcinoma arising in induced papilloma, skin, mouse. In: Jones TC, Mohr U, Hunt RD (eds) Monographs on pathology of laboratory animals. Integument and mammary glands. Springer, Berlin Heidelberg New York Tokyo, pp 38–42

25. Sass B, Dunn TB (1979) Classification of mouse mammary tumors in Dunn's miscellaneous group including recently reported types. J Natl Cancer Inst 62: 1287–1293
26. Seely JC, Boorman GA (1999) Mammary gland and specialized sebaceous glands (Zymbal, preputial, clitoral, anal). In: Maronpot RR, Boorman GA, Gaul BW (eds) Pathology of the mouse. Reference and atlas. Cache River Press, Vienna, pp 613–635
27. Squire RA, Goodman DG, Valerio MG, Fredrickson T, Strandberg JD, Levitt MH, Lingeman CH, Harshbarger JC, Dawe CJ (1978) Tumors. In: Benirschke K, Garner FM, Jones TC (eds) Pathology of laboratory animals, vol II. Springer, Berlin Heidelberg New York Tokyo, pp 1051–1283
28. Turusov VS (1979) Pathology of tumours in laboratory animals, vol II Tumours of the mouse. IARC Scientific Publications No. 23, Lyon

Gastrointestinal Tract

G.R. Betton[1,3] · L.O. Whiteley[2] · M.R. Anver · R. Brown
U. Deschl[3] · M. Elwell · P.-G. Germann[3] · F. Hartig · K. Küttler[3]
H. Mori · T. Nolte[3] · H. Püschner[3] · K. Tuch[3]

[1]Chair; [2]Co-chair; [3]Initial draft.

Hyperplasia, Squamous Cell *(H)*

Histogenesis

Squamous epithelial cells of mucosal surface.

Diagnostic Features

- Focal or multifocal increase in well-differentiated squamous cells with basal to squamous maturation
- Well-circumscribed
- Single or multiple finger-like projections are present, each with a simple non-branching fibrovascular connective tissue core
- Hyperkeratosis may be present

Differential Diagnosis

Papilloma, Squamous Cell
Projections have a single complex branching fibrovascular connective tissue core

Carcinoma, Squamous Cell
Evidence of invasion in underlying tissue, stroma or muscle
Cellular pleomorphism or dyskeratosis is present

References

(See 16, 17)

Papilloma, Squamous Cell *(B)*

Histogenesis

Squamous epithelial cells of the mucosal surface.

Diagnostic Features

- Lesion is solitary or multifocal
- Well circumscribed
- Single or multiple finger-like projections with a complex branching fibrovascular connective tissue core are present
- Invagination of epithelium appearing as cellular nests may be present but no true invasion
- Progression towards a less orderly cellular arrangement may be present if traumatized or inflamed
- Epithelial cells are well differentiated
- Cells are often hyperplastic or hyperkeratotic but with normal epithelial maturation
- No evidence of invasion of the stroma by epithelial cells

Differential Diagnosis

Hyperplasia, Squamous Cell
Projections have simple non-branching fibrovascular epithelial connective tissue cores

Carcinoma, Squamous Cell
Evidence for true invasion of underlying tissue, stroma or muscle
Cellular pleomorphism or dyskeratosis is present
Irregular cords, clusters or nests of epithelial cells are present

References

(See 16, 17, 19)

Carcinoma, Squamous Cell *(M)*

Histogenesis

Squamous epithelial cells of the mucosal surface.

Diagnostic Features

- Irregular cords, clusters and nests of epithelial cells
- Cells are pleomorphic
- Dyskeratosis and keratin pearls may be present
- Invasion of the underlying connective tissue or stroma is present

Differential Diagnosis

Papilloma, Squamous Cell
No true invasion of underlying tissue
No evidence of dyskeratosis of keratin pearls

Hyperplasia, Squamous Cell
No invasion of underlying tissue
Projections have a simple non-branching fibrovascular connective tissue core

Odontoma, Odontogenic Cysts
Dental epithelium shows basal palisading
Dentine or enamel stroma produced
No keratin produced

References

(See 16, 17, 31)

Adenoma *(B)*

Histogenesis

Ductal and acinar epithelium of main salivary glands and minor glands of tongue and pharynx.

Diagnostic Features

- Well-demarcated
- Compression of the adjacent parenchyma or connective tissue may be present
- Surrounded partially or completely by a thin capsule but without marked desmoplasia
- Growth pattern may be acinar or papillary
- Cells are small and contain secretory granules
- Nuclei are hyperchromatic and basally oriented when cytoplasmic mucus present
- Mitotic figures are rare

Differential Diagnosis

Regenerative Hyperplasia
Chronic inflammation or fibroplasia present
Cytoplasmic basophilia with small foci of acinar or ductular proliferation

Adenocarcinoma
Cellular pleomorphism and cytoplasmic amphophila present
Mitotic figures are numerous
Growth in whorls or spindle patterns

Tumor, Mixed, Benign
Consist of two proliferative tissues, one myoepithelium and the other glandular acinar epithelium

Adenomas of the Mammary Gland (Anterior Part)
Glandular structure is not as compact
Staining characteristics of epithelium different

Comment

The distinction between regenerative hyperplasia and adenomas may be very difficult.

References

(See 8, 10, 13)

Adenocarcinoma *(M)*

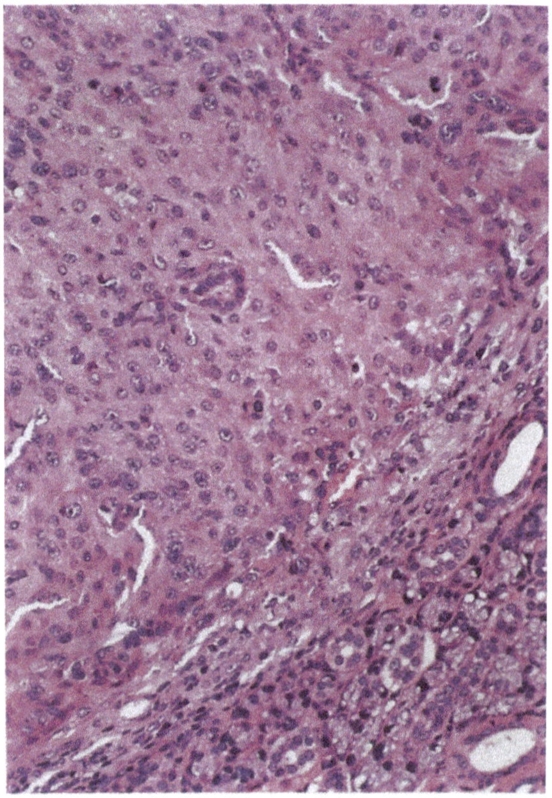

Fig. 1. Solid adenocarcinoma of the submandibular salivary gland. H&E

Histogenesis

Ductal and acinar epithelium of main salivary glands and minor salivary glands of tongue and pharynx.

Diagnostic Features

- Growth patterns are solid, papillary or mixed
- Grows in whorls or spindle pattern
- Necrosis may often be present
- Cells are large, pleomorphic, and polyhedral with amphophilic cytoplasm
- Nuclear: cytoplasm ratio high
- Nuclei are large vesicular and have multiple nucleoli
- Mitotic figures numerous
- Invasion of the surrounding tissue present
- Metastasis in the lung may be present

Differential Diagnosis

Regenerative Hyperplasia
No evidence of invasion
Chronic inflammation or fibroplasia present

Adenoma
Minimal cellular pleomorphism and
Acinar cells well differentiated with secretory granules and
Mitotic figures rare

Adenocarcinomas of the Mammary Gland
Cytoplasmic staining differences and lipidvacuolation may be present

References

(See 8, 10)

Myoepithelioma, Malignant *(M)*

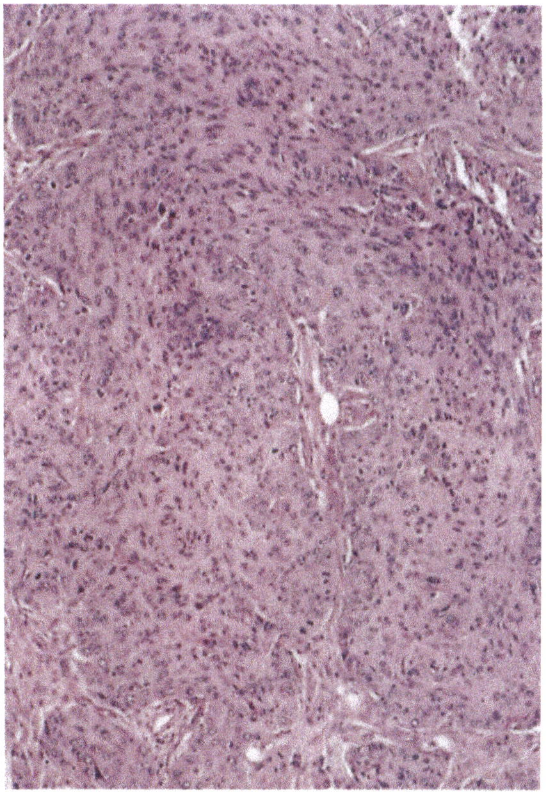

Fig. 2. Myoepithelioma of salivary gland. H&E (courtesy of Dr. M. Jokinen, NTP archive)

Histogenesis

Myoepithelial cells of the salivary glands.

Diagnostic Features

- Unencapsulated tumor, may show tongues of solid tissue extending into adjacent structures
- Spindle or polymorphic patterns with no acinar or ductular structures
- Adjacent to vessels tumor cells tend to align themselves in epithelial fashion
- Squamous forms without keratinization or pearl formation may occur
- Mucoid material accumulates in pseudocysts as a result of degeneration
- Cells pleomorphic and can resemble epithelial, mesenchymal or spindle-shaped cells
- Multiple necrotic areas are present
- Invasive into surrounding tissue present; large tumors may metastasize to the lung
- Infiltrates of lymphocytes and plasma cells are not present (unlike polyomavirus neoplasms)

Differential Diagnosis

Pleomorphic Tumors
Can be solely epithelial, pure mesenchymal or mixed epithelial/mesenchymal

Adenoma
Formation of acini present

Adenocarcinoma
Formation of acini present in more differentiated areas

Comment

Positive for actin and cytokeratin. Polyomavirus-induced salivary gland tumors may be epithelial, mesenchymal or mixed types.

References

(See 5, 8, 11)

Tumor, Mixed, Benign *(B)*

Synonym(s). Tumor, composite, benign

Histogenesis

Glandular epithelial and myoepithelial mesenchymal components of the salivary glands.

Diagnostic Features

- Features of both a fibroma and an adenoma present
- Cellular pleomorphism is not present

Differential Diagnosis

Tumor, Mixed, Malignant
  Features of both a sarcoma and carcinoma present
  Cellular pleomorphism present

Adenoma
  Lack of a neoplastic mesenchymal component

Fibroma
  Lack of a neoplastic epithelial component

Comment

Mouse polyoma virus causes salivary gland tumors including a mixed epithelial-mesenchymal type.

References

(See 8, 10)

Tumor, Mixed, Malignant *(M)*

Synonym(s). Tumor, composite, malignant

Histogenesis

Glandular epithelial and myoepithelial mesenchymal cells.

Diagnostic Features

- Cellular pleomorphism present
- Features of both a sarcoma and a carcinoma are present

Differential Diagnosis

Tumor, Mixed, Benign
Features of both a fibroma and an adenoma are present and
Cellular pleomorphism is not present
No local invasion

Sarcoma, Undifferentiated
Lack of a malignant epithelial component

Schwannoma, Malignant
Lack of a malignant epithelial component

References

(See 8, 10)

Hyperplasia, Squamous Cell *(H)*

Synonym(s). Hyperkeratosis

Histogenesis

Arises from stratified squamous epithelium of the esophagus or non-glandular stomach.

Diagnostic Features

- Focal or diffuse thickening of squamous epithelium with convoluted germinal layer and basal cell proliferation with increased severity
- Papillomatous projections overlie convoluted lamina propria but have no single fibrovascular connective tissue core
- Hyperplastic epithelium may show exophytic or endophytic growth. Endophytic growth may displace the muscularis mucosae and in a keratoacanthomatous growth pattern
- Epithelial cells are often parakeratotic or hyperkeratotic

Differential Diagnosis

Reactive Hyperplasia
Squamous epithelial hyperplasia in response to irritants or ulceration around foci of erosion, inflammation or edema
Reactive hyperplastic epithelium may show exophytic or endophytic growth

Papilloma, Squamous Cell
Exophytic or endophytic solitary or multiple lesions
Projections have a simple or complex branching fibrovascular connective tissue core

References

(See 17)

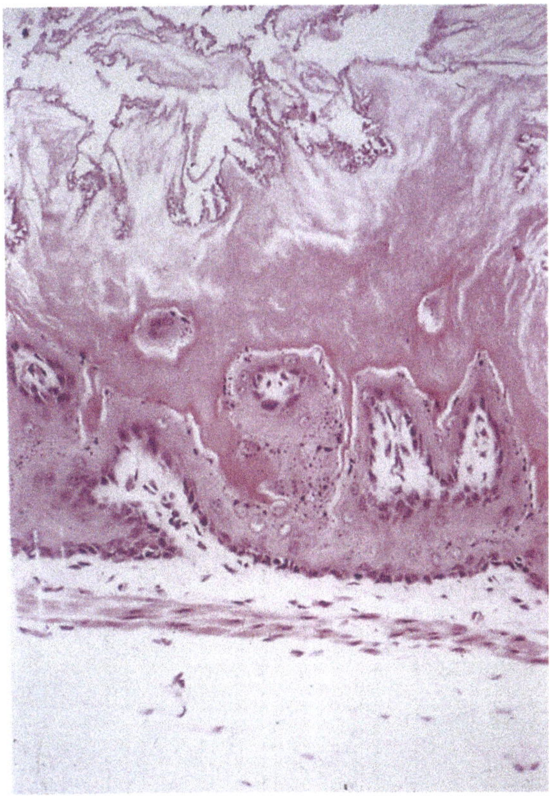

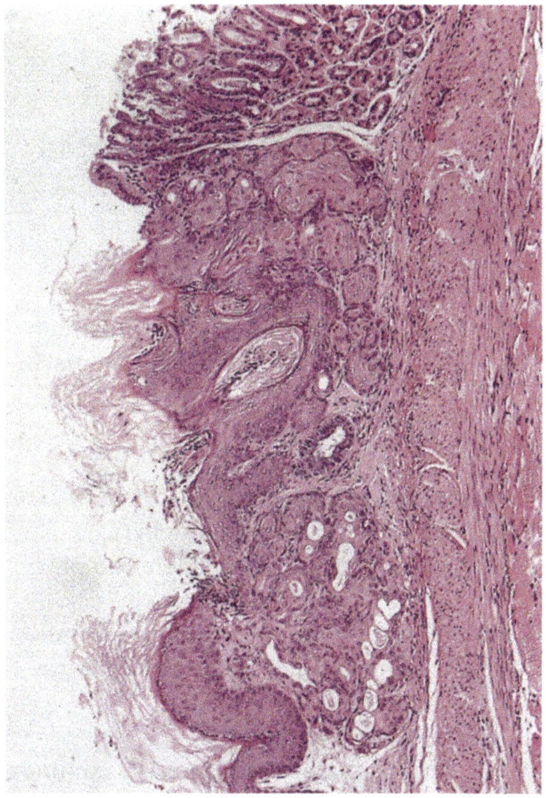

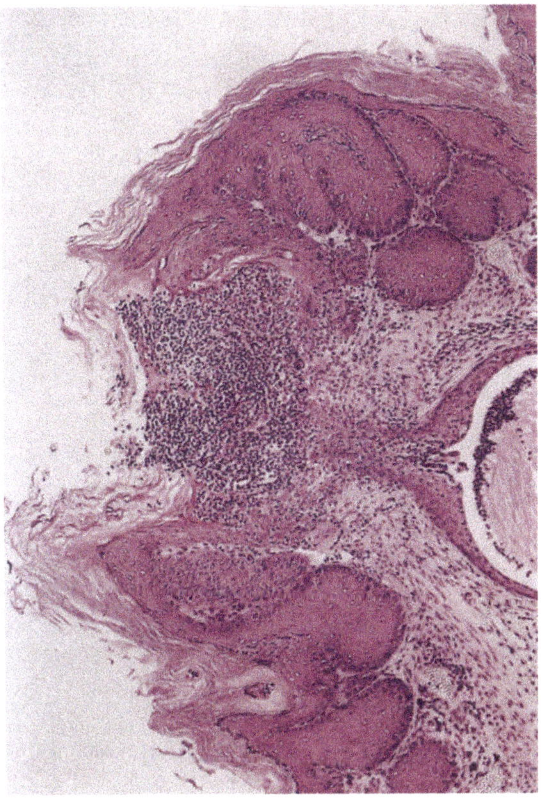

Fig. 3 *(upper left)*. Squamous cell hyperplasia of esophagus with convolution of germinal epithelium and hyperkeratosis. H&E

Fig. 4 *(upper right)*. Squamous epithelial hyperplasia of the forestomach. H&E

Fig. 5 *(below)*. Focal ulceration of the forestomach surrounded by reactive squamous epithelial hyperplasia. H&E

Papilloma, Squamous Cell *(B)*

Histogenesis

Arises from stratified squamous epithelium of the esophagus or non-glandular stomach.

Diagnostic Features

- May be solitary of multiple
- Growth is exophytic or endophytic
- Exophytic tumors have a single complex branching fibrovascular core
- Epithelial cells are often hyperkeratotic and acanthotic but show normal maturation
- Mitotic figures are rare
- Local inflammatory reactions is a common feature

Differential Diagnosis

Hyperplasia, Squamous Cell
No evidence of a solitary stalk or complex branching

Carcinoma, Squamous Cell
Invasion of muscularis mucosae present
Invasion of fibrovascular connective tissue stalk is present in carcinomas arising in a papilloma

References

(See 16, 17)

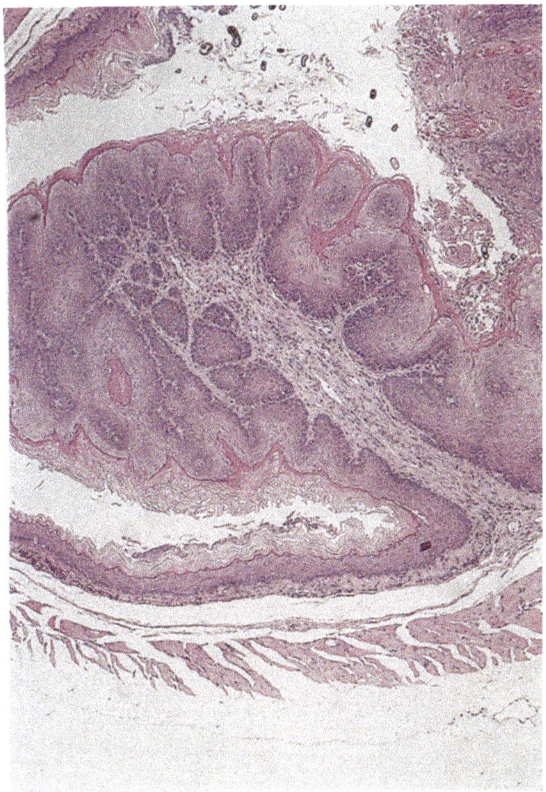

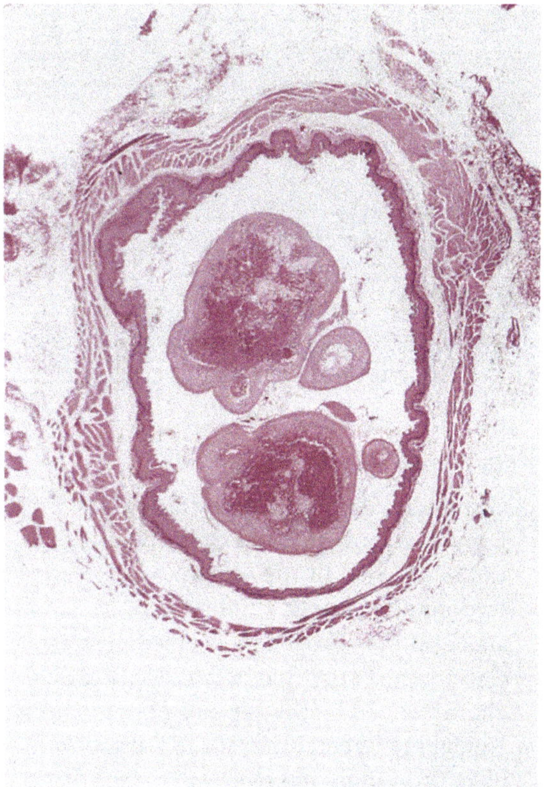

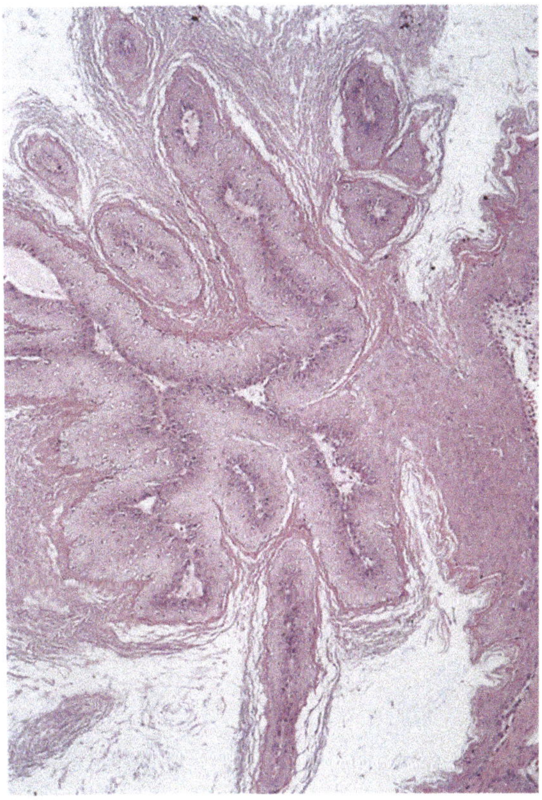

Fig. 6 *(upper left).* Esophageal squamous cell papilloma, B6C3F1 mouse, female, age 95 weeks (courtesy of Dr. Shinji Yamamoto). H&E

Fig. 7 *(upper right).* Esophageal squamous cell papilloma, CD-1 mouse (courtesy of Dr. Barry Stuart). H&E

Fig. 8 *(below).* Squamous cell papilloma of the forestomach, V-Ha-ras Oncomouse, female, age 69 weeks. H&E

Carcinoma, Squamous Cell *(M)*

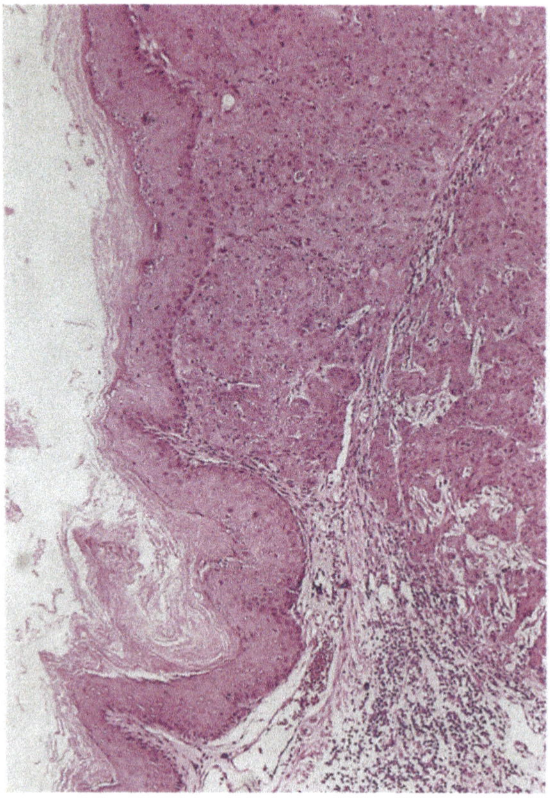

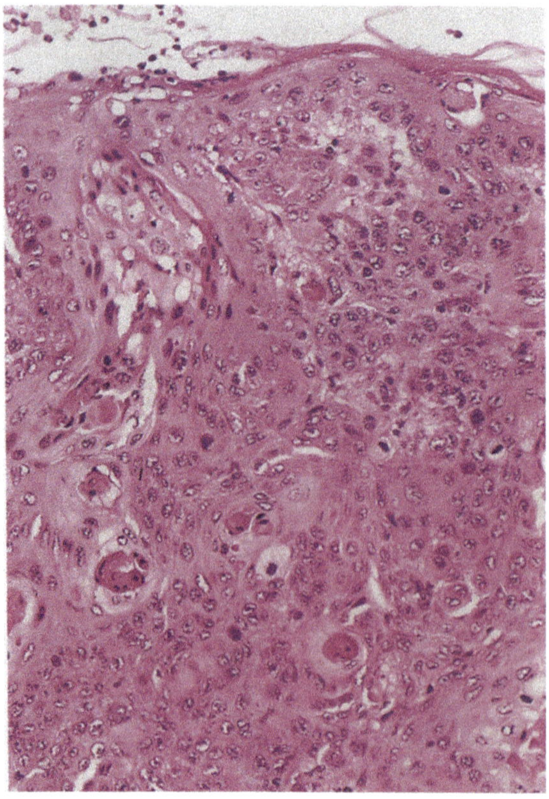

Histogenesis

Arises from stratified squamous epithelium of the esophagus or non-glandular stomach.

Diagnostic Features

- Growth is exophytic or endophytic
- Exophytic growth can exhibit partial or total occlusion of the lumen
- Endophytic growth exhibits invasion of the wall and may show superficial ulceration
- Carcinomas arising in a papilloma are characterized by invasion of fibrous stalks
- Nuclei are hyperchromatic, enlarged and have prominent nucleoli
- Various degrees of keratinization and keratin pearl formation present
- Mitotic figures are numerous
- Superficial ulceration and inflammation or fibroplasia of the lamina propria or submucosa are common features
- Invasive growth is present. Penetration of esophagus and direct spread into the lung may occur in carcinomas of the thoracic portion

Fig. 9 *(above)*. Squamous cell carcinoma of the forestomach. H&E

Fig. 10 *(below)*. Squamous cell carcinoma of the forestomach. H&E

Differential Diagnosis

Hyperplasia, Squamous Cell
No evidence of true invasive growth although muscularis mucosae may be displaced

Papilloma, Squamous Cell
No evidence of invasive growth
Endophytic tumors may displace muscularis mucosae
Inflammation and fibroplasia absent unless ulcerated

Epithelial Cyst
Submucosal cysts are lined by a thin stratified squamous epithelium
Probably developmental in origin, seen on limiting ridge or in adjacent glandular mucosa

Ulceration
Squamous epithelial hyperplasia reactive to traumatic ulceration by dosing cannula may be associated with subepithelial deposits of proliferating epithelium, inflammation or fibroplasia

References

(See 16, 17, 26, 31)

Hyperplasia, Fundic Mucosal *(H)*

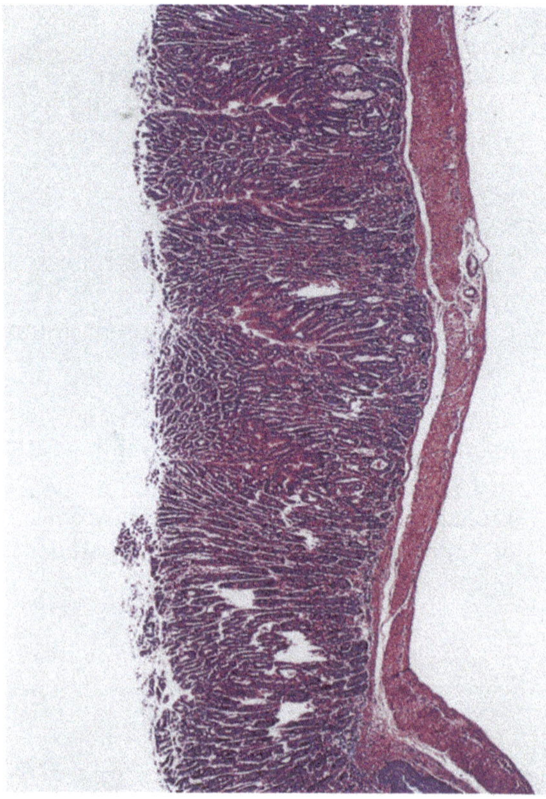

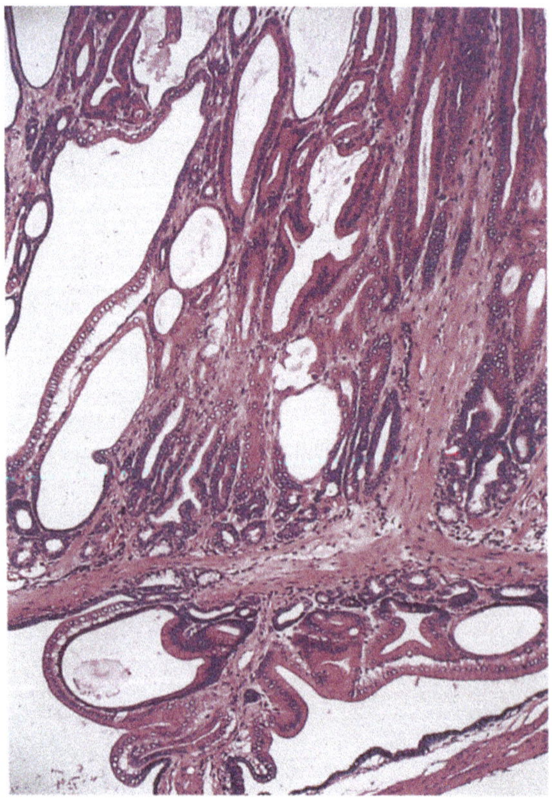

Synonym(s). Hyperplasia, adenomatous; hyperplasia, gastric

Histogenesis

Origin from the proliferative zone with increase in both foveolar and oxyntic glandular differentiation.

Diagnostic Features

- No significant compression is present
- Glandular architecture is maintained
- No single cell type predominant
- Pits become elongated and more basophilic
- Slight grade shows increased gastric mucosal height resulting from increase in pit and gland length
- Moderate grade also shows mucosal folding but differentiation patterns are retained
- Severe grade shows focal penetration of muscularis mucosae which may occur through crypt herniation, but basement membrane always not penetrated
- Submucosal cysts may form from in-folded epithelium showing foveolar differentiation
- Epithelial cells are hypertrophied
- Increased cytoplasmic basophilia present

Fig. 11 *(above)*. Fundic mucosal hyperplasia (moderate) of the glandular stomach with in-folding of oxyntic mucosa. H&E

Fig. 12 *(below)*. Fundic mucosal hyperplasia (severe) of the glandular stomach with submucosal cyst formation and infolded foveolar epithelium. H&E

- Increased number of nuclei
- Nuclei are enlarged and have prominent nucleoli
- Mucus content of foveolar cells is decreased

Differential Diagnosis

Adenoma
Glandular pattern atypical and distorted and
Glandular polarity is lost and
Compression of adjacent tissue is present

Adenocarcinoma
Unequivocal penetration of basement membrane and invasion through muscularis mucosae as glandular or solid cords and
Uniform polarity of cells is lost and
Cells show pleomorphism and
Marked increase of mitotic figures

Glandular Atypia ("Dysplasia")
Hyperplasia of solitary foci of crypts of antral glands
Cellular basophilia is increased
Multilayering of nuclei sometimes present
Often associated with ulceration and regenerative hyperplasia

Hyperplastic Gastritis
Generalized condition with less ordered differentiation
Widely distributed hyperplasia which may also penetrate muscularis mucosae and form submucosal cysts

Comment

Fundic mucosal hyperplasia is a frequent finding in mice treated with anti-secretory compounds which cause hypergastrinemia. Gastrin is a trophic hormone for the fundic mucosa. Neuroendocrine cell hyperplasia and neoplasia may also be associated. Also seen as a spontaneous lesion in strain I mice and other strains as an influence of diet or housing.

References

(See 3, 12, 24, 25, 29, 30)

Hyperplasia, Neuroendocrine Cell *(H)*

Histogenesis

Origin from the neuroendocrine cells of the gastric mucosa. The enterochromaffin-like (ECL) cell is the predominant neuroendocrine cell of the oxyntic mucosa.

Diagnostic Features

- Arises in the fundic mucosa with increased numbers of solitary pale-staining argyrophilic cells (diffuse hyperplasia)
- Within areas of diffuse neuroendocrine cell hyperplasia, clusters of neuroendocrine cells may develop, filling the gastric gland but remaining within basement membrane (focal hyperplasia)
- May show infiltration of muscularis mucosae by solitary neuroendocrine cells
- Special stains (argyrophil, chromogranin A, neuron-specific enolase) give strong staining reaction
- No more than three contiguous glands may be replaced by focal neuroendocrine cell hyperplasia

Differential Diagnosis

Tumor, Neuroendocrine Cell, Benign
More than three contiguous glands filled by neuroendocrine cells

Tumor, Neuroendocrine Cell, Malignant
Nodular growth in lamina propria is present or
Invasion of the muscularis mucosae present

Comment

Neuroendocrine cell hyperplasia and neoplasia may be seen in response to hypergastrinemia following anti-secretory drug treatment. Fundic mucosal hyperplasia usually also evident.

References

(See 3)

Tumor, Neuroendocrine Cell, Benign *(B)*

Histogenesis

Origin from the neuroendocrine cells of the gastric mucosa. The enterochromaffin-like (ECL) cell is the predominant neuroendocrine cell of the oxyntic mucosa.

Diagnostic Features

- Neuroendocrine cell tumors arise in the fundic mucosa within areas of diffuse or focal neuroendocrine cell hyperplasia
- Growth is intramucosal as nodules or groups of four or more glands filled with pale-staining neuroendocrine cells
- No evidence of nodular invasion through muscularis mucosae
- Special stains (argyrophil, chromogranin A, neuron-specific enolase) may give a weaker staining reaction than non-neoplastic neuroendocrine cells

Differential Diagnosis

Hyperplasia, Neuroendocrine Cell (Diffuse, Focal)
  No more than three contiguous glands are filled by neuroendocrine cells
  Invasion of the muscularis mucosae by solitary neuroendocrine cells may be present

Tumor, Neuroendocrine Cell, Malignant
  Nodular invasive growth of neuroendocrine cells through muscularis mucosae or
  Serosal deposits or distant metastases

Comment

Neuroendocrine cell hyperplasia and neoplasia may be seen in response to hypergastrinemia following anti-secretory drug treatment. Fundic mucosal hyperplasia is usually also evident.

References

(See 24, 30)

Tumor, Neuroendocrine Cell, Malignant *(M)*

Histogenesis

Origin from the neuroendocrine cells of the gastric mucosa. The enterochromaffin-like (ECL) cell is the predominant neuroendocrine cell of the oxyntic mucosa.

Diagnostic Features

- Neuroendocrine cell tumors arise in the fundic mucosa within areas of diffuse or focal neuroendocrine cell hyperplasia
- Growth is intramucosal as nodules of pale-staining or eosinophilic cells, sometimes with larger submucosal nodules
- Nodular invasion through muscularis mucosae may be present
- Metastasis to regional lymph nodes or liver may be present
- Special stains (argyrophil, chromogranin A, neuron-specific enolase) may give a weaker staining reaction than in non-neoplastic or benign neuroendocrine cells
- Mitotic figures may be present

Differential Diagnosis

Hyperplasia, Neuroendocrine Cell (Diffuse, Focal)
No more than three contiguous glands are filled by neuroendocrine cells and
Invasion of the muscularis mucosae by solitary neuroendocrine cells may be present

Tumor, Neuroendocrine Cell, Benign
No nodular invasion through muscularis mucosae is present
Mitotic figures not present

Comment

Neuroendocrine cell hyperplasia and neoplasia may be seen in response to hypergastrinemia following anti-secretory drug treatment. Fundic mucosal hyperplasia usually also evident.

References

(See 24, 30)

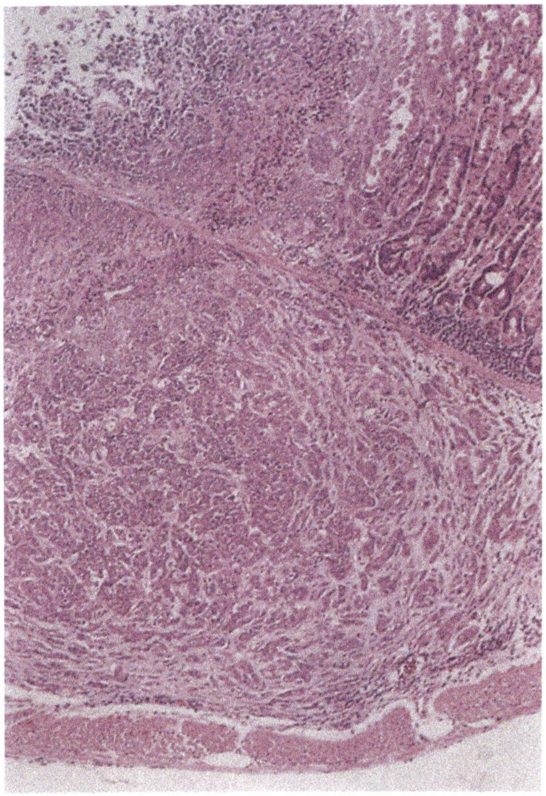

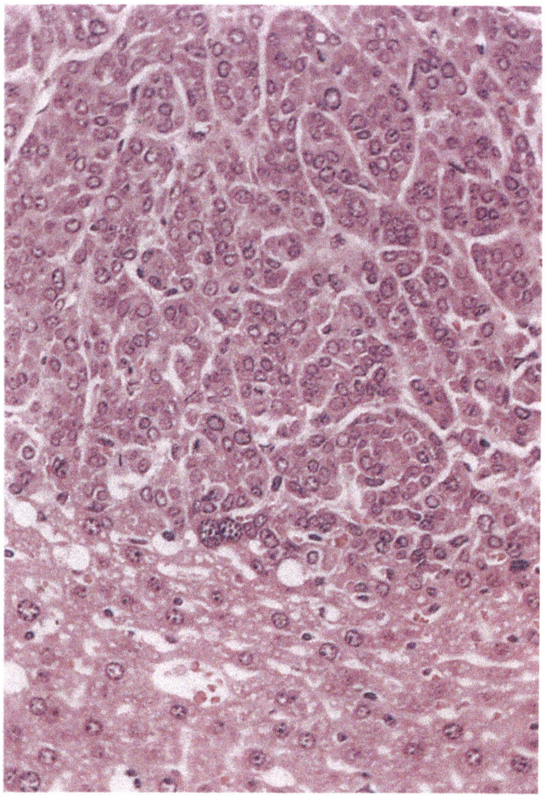

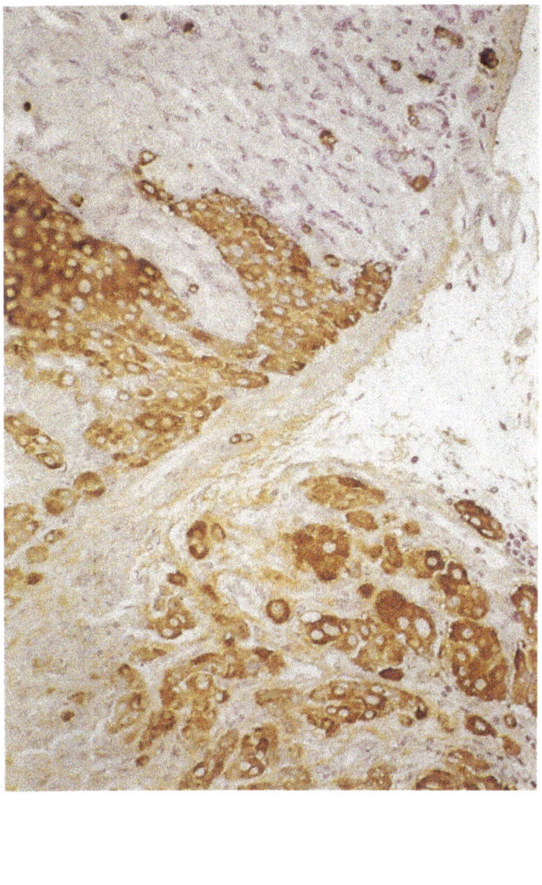

Fig. 13 *(upper left)*. Malignant neuroendocrine cell tumor of the glandular stomach showing solid nests of ECL cells invading lamina propria. H&E

Fig. 14 *(upper right)*. Liver metastasis of gastric malignant neuroendocrine cell tumor. H&E

Fig. 15 *(below)*. Malignant neuroendocrine cell tumor of the glandular stomach showing chromogranin A-positive cells replacing glands and invading submucosa. Chromogranin A

Adenoma *(B)*

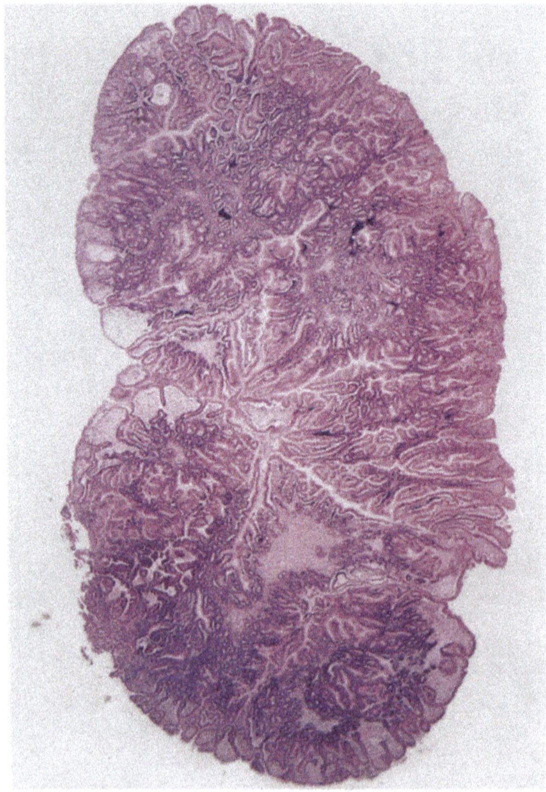

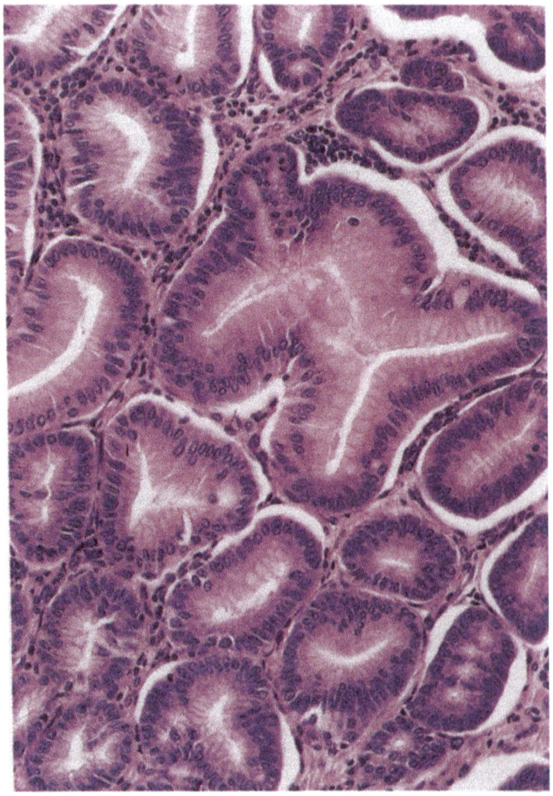

Synonym(s). Adenomatous polyp

Histogenesis

Origin from the exocrine cells of the gastric fundic or antral mucosa.

Diagnostic Features

- Well demarcated with compression present
- Gland architecture is atypical or distorted and glandular polarity is lost
- Branching glands or tubular crypt proliferation or both are present
- In polypoid adenomas of the antral mucosa, gland and crypt architecture may be retained
- One cell type is predominant
- Focal penetration of muscularis mucosae by crypt herniation may be present but basement membrane is always not penetrated
- Polypoid adenomas may have a long stromal stalk which may be branching. Depending on the development of the stroma they show many different patterns of growth
- A stromal core is not observed in sessile adenomas

Fig. 16 *(above).* Adenoma of the glandular stomach showing well-differentiated columnar epithelium. H&E

Fig. 17 *(below).* Adenoma of the glandular stomach showing well-differentiated columnar epithelium. H&E

- Epithelial nuclei are unilayered or organized with varying degrees of stratification
- Cells are columnar and eosinophilic or basophilic
- Nuclei are hyperchromatic
- Cellular atypia present to variable degrees but polarity retained
- Mitotic figures may be slightly increased

Differential Diagnosis

Hyperplasia, Fundic Mucosal
Glandular structure and polarity is maintained
No evidence of compression

Hyperplasia, Reactive
Increase in mitotic figures is present and cells exhibit basophilia
Slight multilayering of glandular epithelial cell nuclei may be present

Adenocarcinoma
Invasion of the fibrovascular connective tissue core of a polyp or
Unequivocal penetration of the basement membrane and
Invasion through the muscularis mucosae
Loss of polarity of cells or pleomorphism of cells or marked increase of mitoses

Comment

The term adenomatous polyp should be avoided, because of implied uncertainty about the neoplastic nature of this proliferative lesion. Polyps arising in the pylorus may prolapse into the lumen of the duodenum.

References

(See 33)

Adenocarcinoma (*M*)

Signet ring cell type

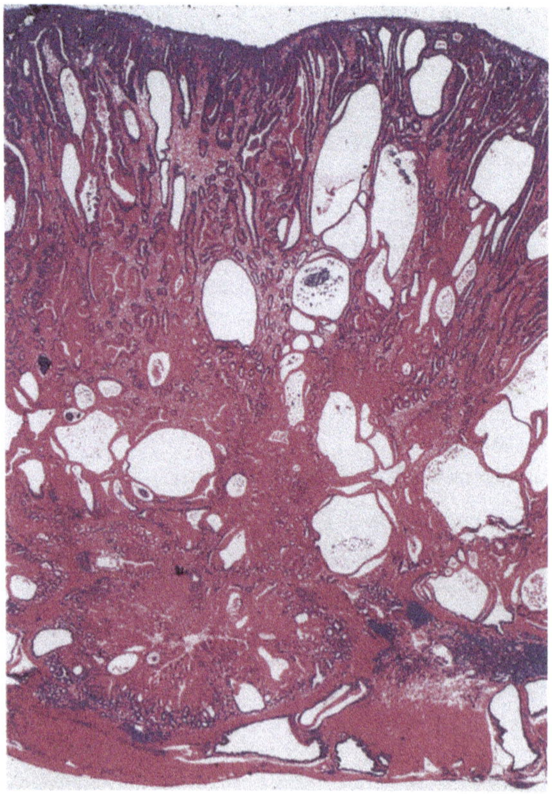

Fig. 18. Adenocarcinoma arising in the fundic mucosa of the glandular stomach within a region of fundic mucosal hyperplasia. Solid nodule invading submucosa. H&E

Histogenesis

Origin from the exocrine cells of the gastric fundic or antral mucosa.

Diagnostic Features

- Predominantly found in the antrum
- Penetration of basement membrane and invasion of muscularis mucosae obligatory
- Regular glandular structure is lost
- Growth may be exophytic or endophytic
- Either tubular endophytic or nodular growth of varying stage of differentiation may be present
- Cystic or solid areas may also be present
- Occasional neuroendocrine cells (NSE-positive) may be present in epithelium
- Marked pleomorphism of cells may be present
- Polarity of cells may be lost
- Mitotic figures markedly increased

Signet Ring Cell Type

- The mucus-producing tumor cells show a tendency to intracellular accumulation of mucus which leads to the characteristic appearance of the cells
- Single tumor cells and small groups of cells may be randomly distributed between pre-existing connective tissue strands
- Penetration of basement membrane and invasion of the submucosa by glandular or solid groups of tumor cells present
- Invasion of adjacent organs and lymphatic or hematogenous metastasis may be present

Differential Diagnosis

Adenoma

No penetration of basement membrane and muscularis mucosae and Cellular polarity is maintained

Hyperplasia, Neuroendocrine Cell or Tumor, Neuroendocrine Cell, Benign

All cells positive for neuroendocrine cell markers, even if pseudo-glandular structures present

Glandular Atypia ("Dysplasia")

Solitary crypts of antral glands may show hyperplasia, increased basophilia and some multilayering of nuclei. This is often associated with ulceration and represents reparative hyperplasia. There is no evidence for this being a pre-neoplastic lesion

Submucosal Glandular Cyst

Single or multiloculated glandular cysts and

Cells are well-differentiated with low cuboidal appearance

In-folding of foveolar epithelium may be present

Associated with regenerative hyperplasia of glandular atypia in most cases

Comment

Following moderate or severe mucosal hyperplasia or hyperplastic gastritis, proliferative zone cells may be trapped in the submucosa and can produce multiloculated cysts lined by low cuboidal well differentiated epithelium. These are not considered to be neoplasms. Adenocarcinomas may develop from adenomas or directly from atypical crypts or mucosal hyperplasia. Even in highly malignant carcinomas, the epithelium may be unilayered and well differentiated but not as much as in regenerative hyperplasia or submucosal cysts. Induced and spontaneous adenocarcinomas arise most frequently in the antrum. Some cytotoxic gastric carcinogens may cause such changes as a result of ulceration, sometimes admixed with adenocarcinoma. Intestinal metaplasia is not seen in the mouse stomach.

References

(See 16, 17)

Hyperplasia (*H*)

Atypical

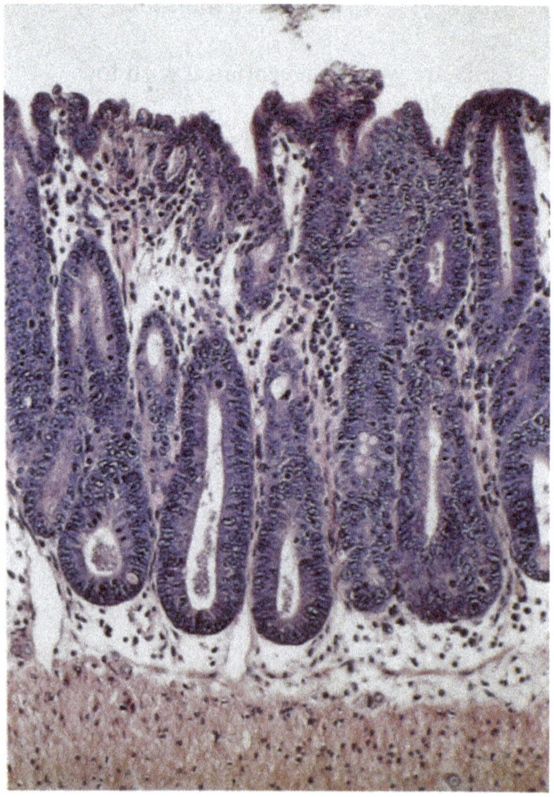

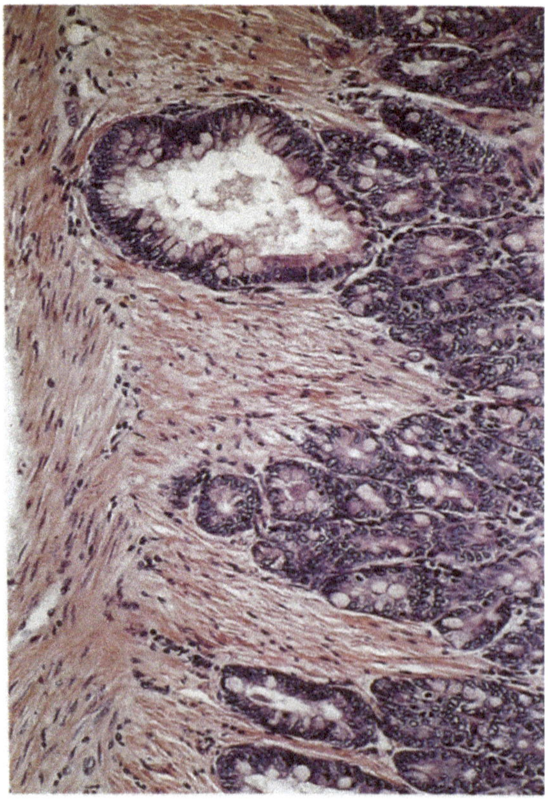

Histogenesis

Origin from the enterocytes of the intestinal mucosa.

Diagnostic Features

- No evidence of compression
- Focal penetration of muscularis mucosae may occur through crypt herniation, but basement membrane always not penetrated
- Villous and glandular architecture is maintained
- A single cell type is not predominant
- Goblet cells may be decreased in number
- Epithelial cells are hypertrophied
- Increased cytoplasmic basophilia is present
- Increased number of enlarged nuclei with prominent nucleoli are present

Atypical

- Villous and glandular structure is maintained, but nuclear stratification may occur
- Focal penetration of muscularis mucosae may be present, but basement membrane is not penetrated
- Cytoplasmic basophilia may be present
- High nucleus:cytoplasm ratio with prominent nucleoli may be present
- Cellular atypia and loss of polarity may be present

Fig. 19 *(above)*. Typical diffuse hyperplasia of duodenal mucosa with basophilia. H&E

Fig. 20 *(above)*. Atypical hyperplasia of the jejunum with loss of structure and cellular atypia. H&E

Differential Diagnosis

Adenoma
Glandular pattern is distorted and atypical and
Compression of adjacent tissue is present

Adenocarcinoma
Penetration of basement membrane and invasion through muscularis mucosae may be present
Cellular pleomorphism is present and
Polarity of cells may be absent and
Mitotic figures are markedly increased

Comment

Focal atypical hyperplasia is a frequent finding in mice treated with compounds which affect the intestinal mucosa. Synonyms used for focal atypical hyperplasia are: early neoplastic lesions, atypical or dysplastic crypts and dysplastic foci. These changes have not been described in untreated animals so far. Colon crypts exhibiting focal atypia ("aberrant crypt foci") can be demonstrated using methylene blue staining of colon whole mounts. Multifocal or diffuse reactive hyperplasia is seen in the caecum, colon and rectum of immuno-compromised mice following *Helicobacter* infection. Diffuse reactive hyperplasia is observed in the distal colon of mice infected with the gram negative bacterium *Citrobacter freundii* (transmissible murine colonic hyperplasia). Occasionally the transverse and ascending colon and the caecum are affected. In such cases of hyperplasia the surface epithelium is covered by myriads of cocobacillary bacteria. The lamina propria shows diffuse leukocytic infiltration.

References

(See 1, 2, 6, 14, 15)

Hyperplasia, Avillous (*H*)

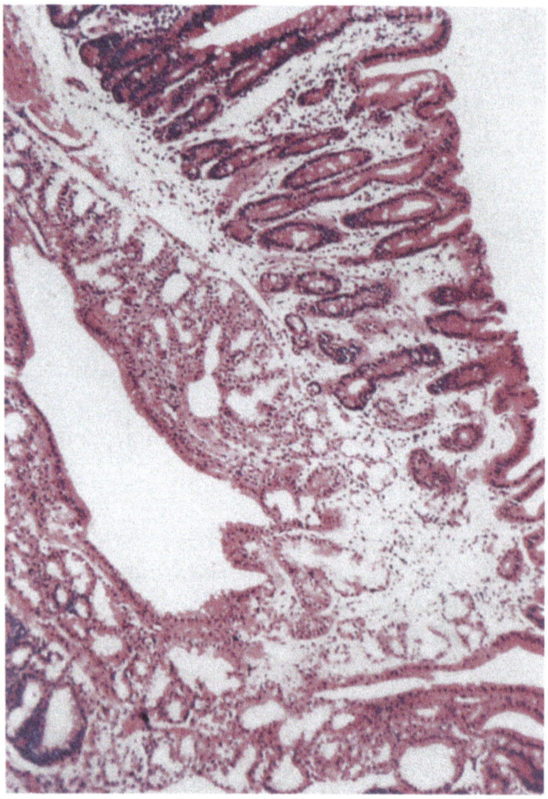

Fig. 21. Avillous hyperplasia of the duodenum over ampulla of Vater with hyperplasia of submucosal glands. H&E

Histogenesis

Simultaneous origin from all epithelial cells of the intestinal mucosa and in many cases also from the Brunner's glands.

Diagnostic Features

- Focal lesion
- Smooth luminal surface lacking intestinal villi
- Hyperplastic crypts may often be interspersed between hyperplastic Brunner's glands
- Mixture of cell types typically populate the mucosa
- Paneth cells in the crypt epithelium are reduced in number
- Number of goblet cells may be unchanged, decreased or increased
- Crypt dilatation may be present in larger lesions
- Associated with submucosal edema and inflammatory cell infiltrates

Differential Diagnosis

Hyperplasia

Villous and glandular structure is maintained and
Hyperplasia of Brunner's glands is not pronounced and
Inflammatory cell infiltrates are not pronounced

Adenoma

Proliferation of one predominant cell type and
Inflammatory cell infiltrates are not pronounced

Comment

The term "plaque" is used as a synonym for avillous hyperplasia. It refers to the gross appearance of this lesion: circular projections of the mucosa into the duodenal lumen, rounded, with well defined edges, flat on the surface and broad at the base. The term "plaque" is avoided here because of its primarily use for the description in macroscopy.
The incidence of avillous hyperplasia is about 86% in female and 30% in male DBA-mice aged 13–20 months. Incidences of about 20% are reported for both sexes of C57BL mice. The avillous hyperplasia is reactive in nature, as evidenced by inflammatory cell infiltrates and simultaneous hyperplasia of more than one cell type.

References

(See 13, 27, 28)

Atypical Cysts with Glandular Hyperplasia *(H)*

Synonym(s). Colitis cystica profunda

Histogenesis

Glandular epithelial cells of the rectal mucosa.

Diagnostic Features

- Focal lesions at ano-rectal junction
- Intramucosal basophilia of rectal glands
- Adjacent mucosa may be ulcerated
- Hyperplastic submucosal glandular cysts with goblet cells and extracellular mucous secretion
- Mucous cysts may contain macrophages and may be ruptured
- No cellular atypia present
- Cyst lined by columnar, cuboidal and goblet epithelial cells.
- Associated with submucosal edema and inflammatory cell infiltrates if ruptured

Differential Diagnosis

Hyperplasia
Glandular structure is maintained and lacks mucous component
Inflammatory cell infiltrates and edema are not pronounced

Adenoma
Proliferation of one predominant cell type and
Inflammatory cell infiltrates are not pronounced

Adenocarcinoma
Uniform cell type predominant
Mucous types usually signet ring not goblet cell morphology
Cellular pleomorphism or atypia present
Invasive growth pattern or metastasis

Comment

Typically seen at the ano-rectal junction accompanying rectal prolapse or inflammation.

Adenoma *(B)*

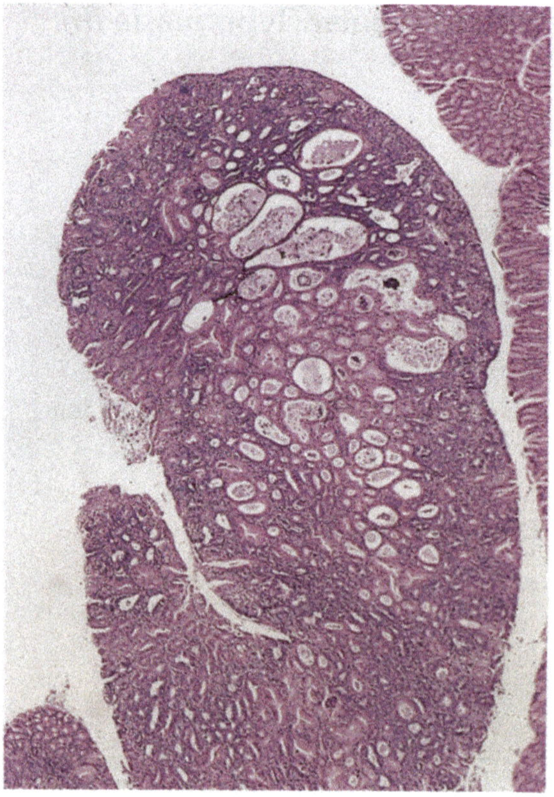

Histogenesis

Origin from the enterocytes of the intestinal mucosa.

Diagnostic Features

- Compression is present
- Glandular architecture distorted
- Branching villi or tubular crypt proliferation can be present
- Polypoid adenomas have a long stromal stalk which is partly branching
- Many different patterns of growth and stroma are found
- A stromal core is not observed in sessile adenomas
- Squamous metaplasia and focal mineralization may be present
- One cell type is predominant and goblet cells are reduced in number
- Nuclei of epithelium are unilayered or have varying degrees of stratification
- Cells are columnar and basophilic
- Cellular atypia may be present
- Nuclei are hyperchromatic and may be polymorphic with loss of polarity
- Mitotic figures slightly increased
- Focal penetration of muscularis mucosae and crypt herniation may be present but basement membrane intact

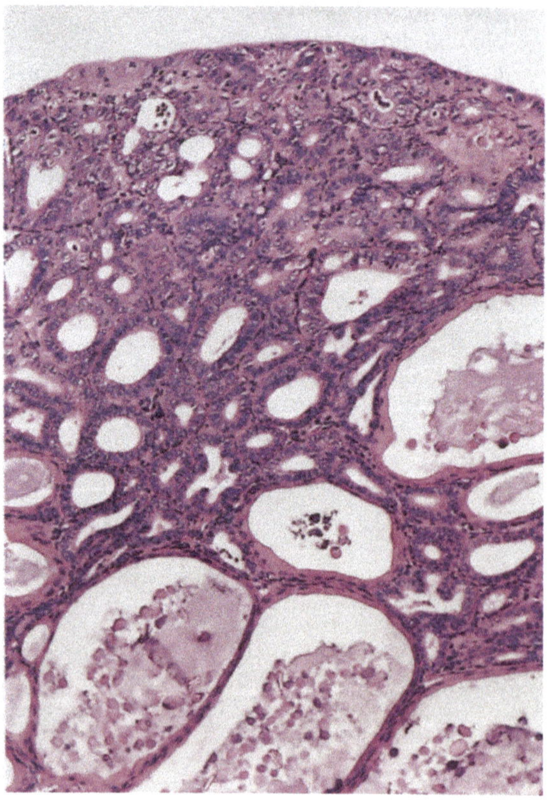

Fig. 22 *(above)*. Adenoma of the large intestine. H&E

Fig. 23 *(below)*. Adenoma of the large intestine. H&E

Differential Diagnosis

Hyperplasia
Glandular architecture is maintained
No evidence of compression

Adenocarcinoma
Penetration of the basement membrane and invasion of the muscularis mucosae is present and
Cellular pleomorphism is present and
Polarity of cells varies and
Mitotic figures are markedly increased

Comment

The term "polyp" was avoided, because of implied uncertainty about the neoplastic nature of this proliferative lesion. Benign neoplasms may develop dysplastic foci and evolve into malignant neoplasms.

The frequency of spontaneous intestinal adenomas varies with the intestinal segment. Incidences up to 51% are reported for the duodenum and anterior part of the jejunum. In the other intestinal segments the incidence of adenomas is below 5%. However, the detection frequency is highly dependent on the mode of macroscopic examination of the intestines, e.g., random blocks, naked eye inspection of mucosal surface, transillumination or Swiss roll technique.

References

(See 4, 6, 9, 20, 21, 28)

Adenocarcinoma (*M*)

Signet ring cell type • Umbilicate lesion

Histogenesis

Origin from the enterocytes of the intestinal mucosa.

Diagnostic Features

- Regular glandular structure is lost
- Growth is tubular endophytic or nodular with growth of varying stages of differentiation
- Cystic and solid areas may also be present
- Cells are frequently multilayered
- Fibrosis and foci of lymphocytic infiltrates may be present
- Squamous metaplasia and focal mineralization is not present
- Cells are columnar or polygonal and are basophilic
- Nuclei are polymorphic and may show loss of polarity
- Mucus-producing cells may be interspersed between absorptive cells
- Mucus may be present extracellularly or stored intracellularly
- Mitotic figures are markedly increased
- Penetration of the basement membrane and invasion of the lamina propria or muscularis mucosae is present
- Invasion of submucosa, stalk or adjacent organs may be present
- Lymphogenous metastasis may be present

Signet Ring Cell Type

- Mucus producing cells show a tendency to intracellular accumulation of mucus which leads to the characteristic appearance of the cells
- High local invasive growth is present; therefore single tumor cells and small groups of cells are randomly dispersed between pre-existing connective tissue strands

Umbilicate Lesion

- Central depression pronounced and margins elevated
- The depressed fibroblastic or fibrous center is infiltrated from the margins by malignant epithelial cells

Differential Diagnosis

Adenoma

No unequivocal penetration of basement membrane or muscularis mucosae
In polypoid lesions, no invasion of the stalk via basement membrane destruction present
Squamous metaplasia and focal mineralization may be present
Cellular atypia may be present

Hyperplasia

Glandular architecture is maintained and
No unequivocal penetration of basement membrane

Fig. 24 *(upper left).* Adenocarcinoma, umbilicate lesion, with submucosal invasion by tumor glands. H&E ▶

Fig. 25 *(upper right).* Highly malignant adenocarcinoma, signet ring cell type, invading muscularis externa with desmoplastic reaction. H&E

Fig. 26 *(lower left).* Adenocarcinoma of the duodenum. H&E

Fig. 27 *(lower right).* Adenocarcinoma of the duodenum. H&E

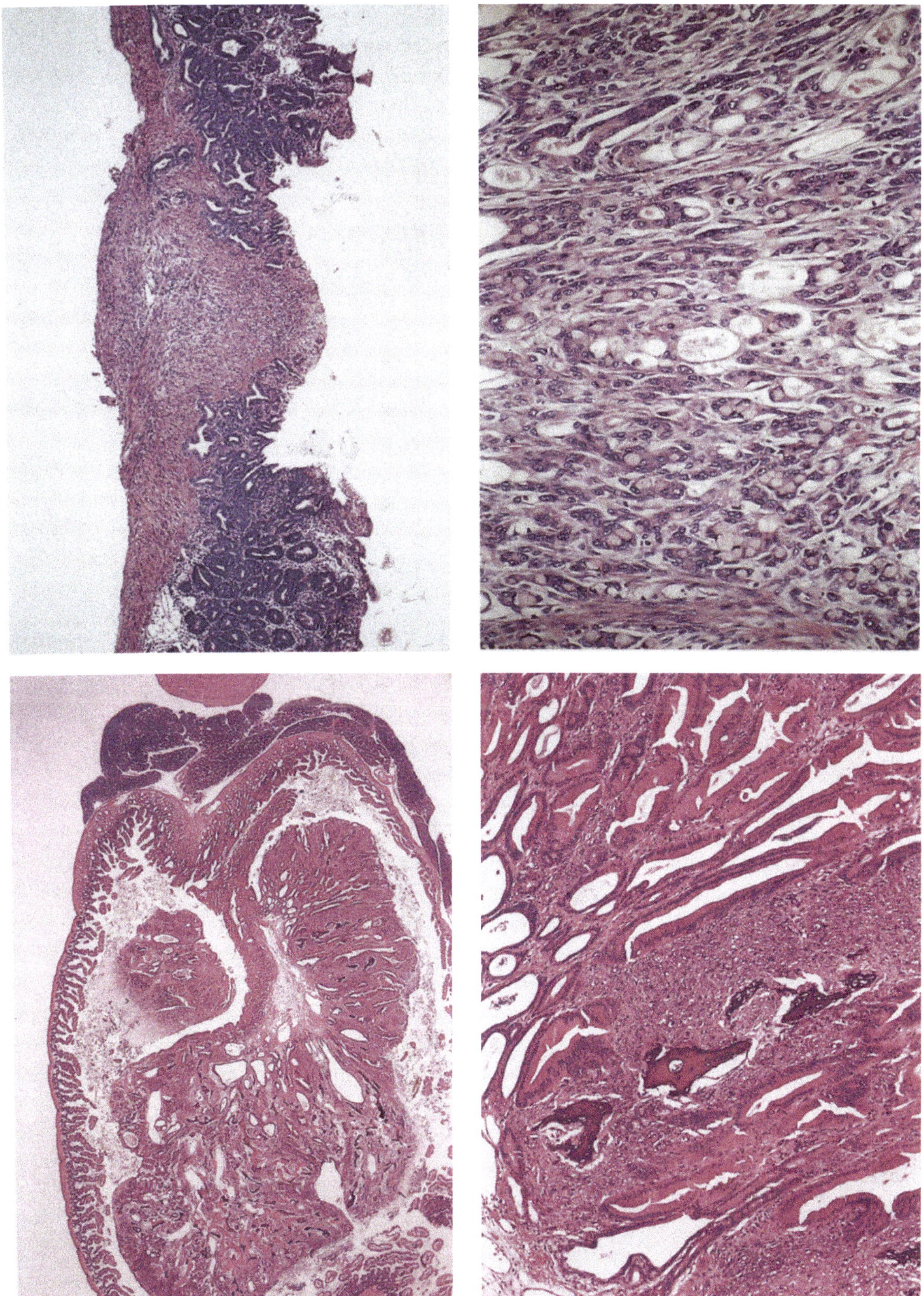

Comment

Even in highly malignant carcinomas, epithelium may be unilayered and well differentiated. Adenocarcinomas may develop from adenomas or directly from atypical crypts or hyperplasia. Penetration of the muscularis mucosae in the latter lesions is often multiple and lacks penetration of the basement membrane. In contrast to man, spontaneous adenocarcinomas show a predilection for the small intestine, especially the jejunum and ileum, rather than the large intestine. The descriptions available indicate that the morphology of spontaneous lesions does not obviously differ from the induced ones, except of their preferred localization in the distal colon and their high invasiveness combined with a low tendency to metastasize.

References

(See 7, 18, 20, 21, 22, 23, 27, 32, 34)

References

1. Barthold SW (1985) *Citrobacter freundii* infection, colon, mouse. In: Jones TC, Mohr U, Hunt RD (eds) Monographs on pathology of laboratory animals. Digestive system. Springer, Berlin Heidelberg New York Tokyo, pp 337–340
2. Barthold SW, Coleman GL, Bhatt PN, Osbaldiston GW, Jonas AM (1976) The etiology of transmissible murine colonic hyperplasia. Lab Anim Sci 26: 889–894
3. Betton GR, Dormer CS, Wells T, Pert P, Price CA, Buckley P (1988) Gastric ECL-cell hyperplasia and carcinoids in rodents following chronic administration of H2-antagonists SK&F 93479 and oxmetidine and omeprazole. Toxicol Pathol 16: 288–298
4. Bomhard E (1993) Frequency of spontaneous tumours in NMRI mice in 21-month studies. Exp Toxicol Pathol 45: 269–289
5. Burger GT, Frith CH, Townsend JW (1985) Myoepithelioma, salivary glands, mouse. In: Jones TC, Mohr U, Hunt RD (eds) Monographs on pathology of laboratory animals. Digestive system. Springer, Berlin Heidelberg New York Tokyo, pp 185–189
6. Chang WW (1984) Histogenesis of colon cancer in experimental animals. Scand J Gastroenterol 104: 27–43
7. Chang WW, Mak KM, MacDonald PD (1979) Cell population kinetics of 1,2-dimethylhydrazine-induced colonic neoplasms and their adjacent colonic mucosa in the mouse. Virchows Arch B 30: 349–361
8. Dawe CJ (1979) Tumours of the salivary and lacrymal glands, nasal fossa and maxillary sinuses. In: Turusov VS (ed) Pathology of tumours in laboratory animals, vol II. Tumours of the mouse. IARC Scientific Publications No. 23, Lyon, pp 91–134
9. Faccini JM, Abbott DP, Paulus GJJ (1990) Intestines. Mouse histopathology. A glossary for use in toxicity and carcinogenicity studies. Elsevier, Amsterdam, pp 105–115
10. Frith CH, Heath JE (1985) Adenoma, adenocarcinoma, salivary gland, mouse. In: Jones TC, Mohr U, Hunt RD (eds) Monographs on pathology of laboratory animals. Digestive system. Springer, Berlin Heidelberg New York Tokyo, pp 190–192
11. Frith CH, Ward JM (1988) Color atlas of neoplastic and non-neoplastic lesions in aging mice. Elsevier, Amsterdam, pp 7–8
12. Greaves P, Boiziau JL (1984) Altered patterns of mucin secretion in gastric hyperplasia in mice. Vet Pathol 21: 224–228
13. Hare WV, Stewart HL (1956) Chronic gastritis of the glandular stomach, adenomatous polyps of the duodenum, and calcareous pericarditis in strain DBA mice. J Natl Cancer Inst 16: 889–911
14. James JT, Shamsuddin AM, Trump BF (1983) Comparative study of the morphologic, histochemical, and proliferative changes induced in the large intestine of ICR/Ha and C57BL/Ha mice by 1,2-dimethylhydrazine. J Natl Cancer Inst 71: 955–964
15. Kubota Y, Funaki J, Tanaka M, Sasaki H (1989) Electron microscopic histochemistry of lectin binding sites in 1,2-dimethylhydrazine-induced colon cancer in mice. J Clin Electron Microsc 22: 415–434
16. Leininger JR, Jokinen MP (1994) Tumours of the oral cavity, pharynx, oesophagus and stomach. In: Turusov VS, Mohr U (eds) Pathology of tumours in laboratory animals, vol 2. Tumours of the mouse, 2nd edn. IARC Scientific Publications No. 111, Lyon, pp 167–193
17. Leininger JR, Jokinen MP, Dangler CA, Whiteley LO (1999) Oral cavity, esophagus, and stomach. In: Maronpot RR, Boorman GA, Gaul BW (eds) Pathology of the mouse. Reference and atlas. Cache River Press, Vienna, pp 29–48
18. Lorenz E, Stewart HL (1940) Intestinal carcinoma and other lesions in mice following oral administration of 1,2,5,6-dibenzanthracene and 20-methylcholanthrene. J Natl Cancer Inst 1: 17
19. Maita K, Hirano M, Harada T, Mitsumori K, Yoshida A, Takahashi K, Nakashima N, Kitazawa T, Enomoto A, Inui K, Shirasu Y (1988) Mortality, major cause of moribundity, and spontaneous tumors in CD-1 mice. Toxicol Pathol 16: 340–349
20. Mizutani T, Yamamoto T, Ozaki A, Oowada T, Mitsuoka T (1984) Spontaneous polyposis in the small intestine of germ-free and conventionalized BALB/c mice. Cancer Lett 25: 19–23
21. Nakamura S, Kino I (1985) Morphogenesis of experimental colonic neoplasms induced by dimethyl hydrazine. In: Pfeiffer CJ (ed) Animal models for intestinal disease. CRC Press, Boca Raton, pp 99–122
22. Nyska A, Waner T, Paster Z, Bracha P, Gordon EB, Klein B (1990) Induction of gastrointestinal tumors in mice fed the fungicide folpet: possible mechanisms. Jpn J Cancer Res 81: 545–549
23. Oomen LC, van der Valk MA, Emmelot P (1984) Stem cell carcinoma in the small intestine of mice treated transplacentally with N-ethyl-N-nitrosourea: some quantitative and histological aspects. Cancer Lett 25: 71–79
24. Poynter D, Selway SAM, Papworth SA, Riches SR (1986) Changes in the gastric mucosa of the

mouse associated with long lasting unsurmountable histamine H2 blockade. Gut 27: 1338–1346
25. Rehm S, Sommer R, Deerberg F (1987) Spontaneous nonneoplastic gastric lesions in female Han:NMRI mice, and influence of food restriction throughout life. Vet Pathol 24: 216–225
26. Reuber MD (1979) Neoplasms of the forestomach in mice ingesting dihydrosafrole. Digestion 19: 42–47
27. Rowlatt C, Franks LM, Sheriff MU, Chesterman FC (1969) Naturally occurring tumors and other lesions of the digestive tract in untreated C57BL mice. J Natl Cancer Inst 43: 1353–1364
28. Seronde J (1970) Focal avillous hyperplasia of the mouse duodenum. J Pathol 100: 245–248
29. Stewart HL, Andervont HB (1938) Pathologic observations on the adenomatous lesion of the stomach of mice of strain I. Arch Pathol 26: 1009–1022
30. Streett CS, Robertson JL, Crissman JW (1988) Morphologic stomach findings in rats and mice treated with the H2 receptor antagonists, ICI 125,211 and ICI 162,846. Toxicol Pathol 16: 299–304
31. Tamano S, Hagiwara A, Shibata MA, Kurata Y, Fukushima S, Ito N (1988) Spontaneous tumors in aging (C57BL/6 N x C3H/HeN)F1 (B6C3F1) mice. Toxicol Pathol 16: 321–326
32. Toth B, Malick L, Shimizu H (1976) Production of intestinal and other tumors by 1,2-dimethylhydrazine dihydrochloride in mice. I. A light and transmission electron microscopic study of colonic neoplasms. Am J Pathol 84: 69–86
33. Tucker MJ, Kalinowski AE, Orton TC (1996) Carcinogenicity of cyproterone acetate in the mouse. Carcinogenesis 17: 1473–1476
34. Wadsworth PF, Jackson DG (1991) Epithelial tumours of the gastro-intestinal tract in C57BL/10 J mice. Toxicol Pathol 19: 623

Liver, Gallbladder, and Exocrine Pancreas

U. Deschl[2,3] · R.C. Cattley[1] · T. Harada[2] · K. Küttler[3] · J.R. Hailey · F. Hartig[3]
B. Leblanc[3] · D.S. Marsman · T. Shirai

[1]Chair; [2]Co-chair; [3]Initial draft.

Focus of Cellular Alteration *(H)*

Basophilic • Eosinophilic
Basophilic/eosinophilic mixed • Clear cell

Synonym(s). Hyperplastic focus, enzyme altered focus

Histogenesis

Hepatocytes.

Diagnostic Features

- Distinguished by virtue of tinctorial variation, size of hepatocytes and textural appearances from surrounding parenchyma
- Size may range from less than one lobule to multiple lobules in diameter
- Liver architecture is not obviously disrupted
- Sinusoids within focus may be compressed, so typical parenchymal plates are difficult to be detected
- Liver plates within focus are partially or completely continuous with surrounding hepatic parenchyma
- Limited compression of the adjacent parenchyma may be present
- Size of cells and cytoplasmic tinctorial variation depend on type of focus
- The choice of the modifier depends on the predominant cell type

Basophilic
- Consist of cells larger or smaller than normal hepatocytes, in general they are smaller
- Cytoplasm exhibits distinct basophilia due to polyribosomes or rough endoplasmic reticulum
- Vascular pseudo-invasion may be present
- Eosinophilic cytoplasmic inclusions may be found occasionally within hepatocytes

Eosinophilic
- Cells are usually enlarged, but can also be of the same size or appear smaller than normal hepatocytes
- Cytoplasm is distinctly eosinophilic with granular or ground-glass appearance
- Eosinophilia may result from an increase in smooth endoplasmic reticulum or mitochondria

Basophilic/Eosinophilic Mixed
- Focus consists of heterogeneous cell types
- Exhibits a combination of eosinophilic/clear cell and basophilic staining hepatocytes

Clear Cell
- Composed of cells with clear cytoplasm
- Nuclei are often small and dense, prominent and centrally located, sometimes exhibiting increased volume
- Storing glycogen in excess

Differential Diagnosis

Hyperplasia, Hepatocellular, Regenerative
Hepatocytes appear tinctorially unaltered but may exhibit a slightly basophilic cytoplasm. Evidence of prior or ongoing liver damage is present

Adenoma, Hepatocellular
Compression of adjacent liver tissue. Liver plates often impinge perpendicularly or obliquely on the surrounding parenchyma. Lobular architecture is not maintained resulting in an irregular growth pattern

Carcinoma, Hepatocellular
Distinct trabecular or adenoid or poorly differentiated growth is present. Lobular and plate architecture is not maintained. Cellular atypia and invasive growth may be present

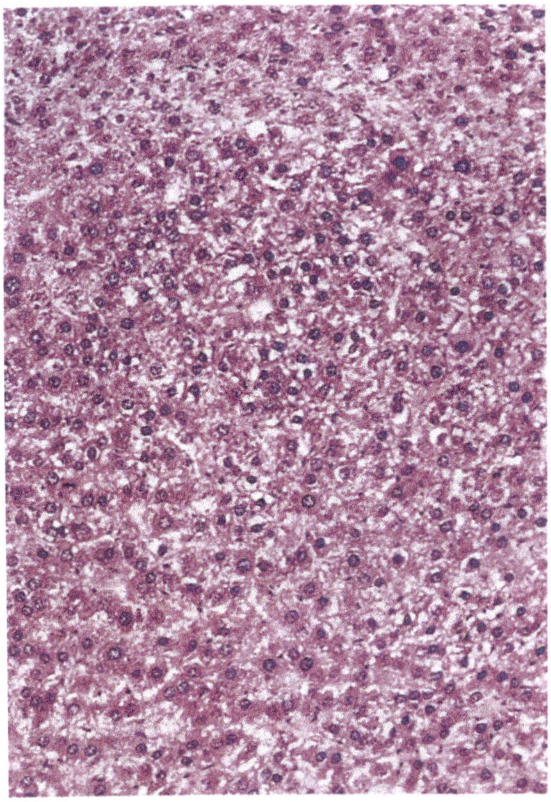

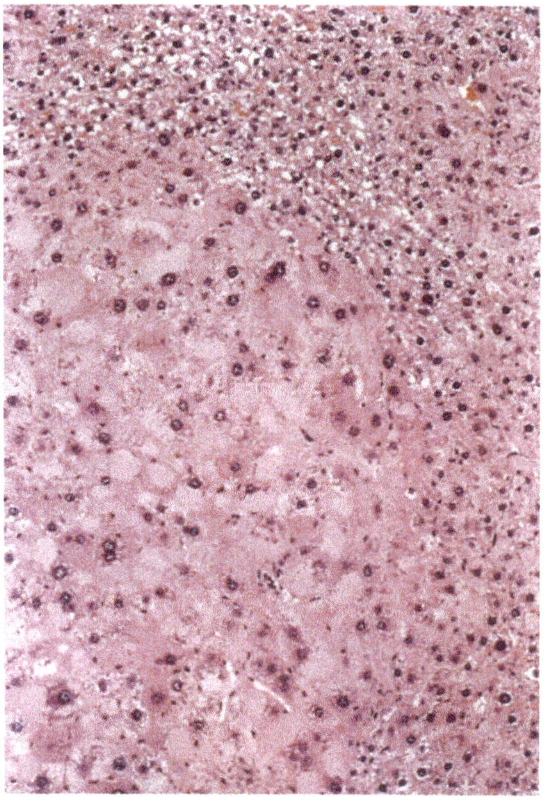

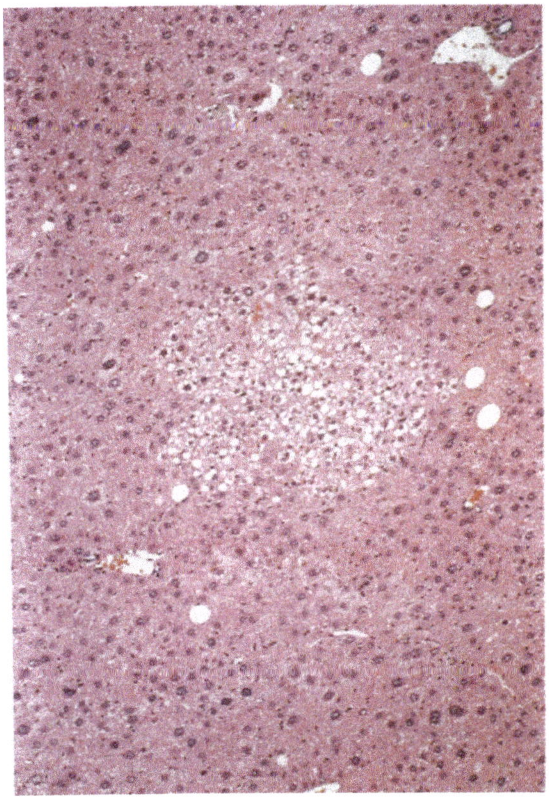

Fig. 1 *(upper left)*. Basophilic focus of the liver. H&E

Fig. 2 *(upper right)*. Eosinophilic focus of the liver. Enlarged eosinophilic hepatocytes with mild compression. B6C3F1 mouse, male, age 18 months. H&E

Fig. 3 *(below)*. Clear cell focus of the liver. Note clear cytoplasm and dense, small nuclei. B6C3F1 mouse, male, age 18 months. H&E

Comment

Foci of cellular alteration in mice are similar to those described in rats but are less common.
The impingement of foci of cellular alteration on hepatic veins should not be construed as evidence of malignancy.
An area/s of fatty change should not be considered as foci of cellular alteration.
The presence of some vacuolated or clear cells within foci that are otherwise basophilic or eosinophilic should not result in a classification as mixed focus.
Basophilic foci exhibit a high degree of tinctorial variability and may sometimes have the appearance of being amphophilic.

References

(See 1, 10, 11, 13, 14, 15, 16, 19, 21, 26, 29, 33, 37, 38, 39, 41, 40, 42)

Hyperplasia, Hepatocellular, Regenerative *(H)*

Synonym(s). Hyperplasia, hepatocellular; hyperplasia, regenerative.

Histogenesis

Hepatocytes.

Diagnostic Features

- Lesion may reach several millimeters in diameter
- Usually occurs multifocal
- Evidence of prior or ongoing hepatocellular damage is present, e.g., acute or chronic inflammation, necrosis, fibrosis, atrophy, collapse of parenchyma, congestion or tumor invasion
- Fibrous encapsulation may be present
- Bile duct and oval cell proliferation is generally present
- Hepatic plate structure is present, though lobular architecture may be distorted
- Portal triads and central veins are present
- Hepatocytes appear unaltered, but may have slightly basophilic cytoplasm and/or prominent nucleoli

Differential Diagnosis

Focus of Cellular Alteration
Tinctorial variation is present. Generally not associated with liver damage, though foci may occur in damaged liver. Liver plates usually merge imperceptibly with surrounding hepatic parenchyma. No evidence of or only minimal compression

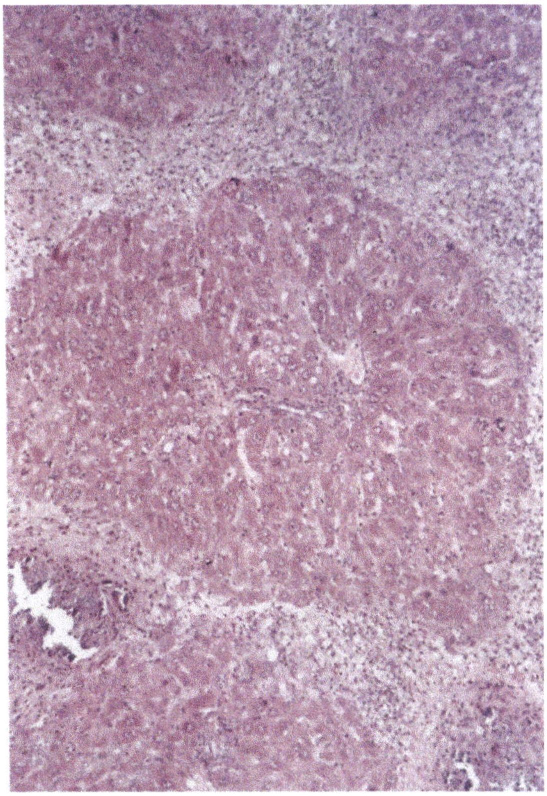

Fig. 4. Hepatocellular regenerative hyperplasia of the liver. Note distinct inflammatory reaction. Plate structure is present, though architecture is distorted. NMRI mouse, female, age 18 months. H&E

Adenoma, Hepatocellular
Generally no evidence of liver damage. Lobular architecture is not maintained resulting in an irregular growth pattern. Liver plates often impinge perpendicularly or obliquely on the surrounding parenchyma

Carcinoma, Hepatocellular
Distinct trabecular or adenoid or poorly differentiated growth pattern is present. Lobular and plate architecture is not maintained. Invasive growth and cellular atypia may be present

References

(*See* 12, 16, 26)

Adenoma, Hepatocellular *(B)*

Synonym(s). Hepatoma, benign; liver cell tumor, benign; hepatic adenoma, benign; tumor, liver cell, benign; nodule, type A

Histogenesis

Hepatocytes.

Diagnostic Features

- Well-demarcated from surrounding parenchyma
- Lesion may be grossly visible
- Distinct compression of adjacent parenchyma particularly in large tumors may be present
- Lesion may bulge from liver surface
- Lobular architecture is not maintained resulting in an irregular growth pattern
- Portal triads are principally absent, but entrapped triads may be present
- Cells are uniform, well-differentiated and closely packed with regular nuclei
- Cells may be larger or smaller or unchanged in size
- Cytoplasm appears eosinophilic, basophilic, clear or vacuolated, resembling cells in foci
- Hepatocyte plates are regular and consist of one to two cell layers
- Plates may often impinge at right angles at surrounding normal liver cell plates
- Pseudo-invasion of blood vessels may be present
- Necrotic lesions are usually absent
- Mitotic figures are rare
- Eosinophilic cytoplasmic inclusions may be found occasionally within hepatocytes

Differential Diagnosis

Carcinoma, Hepatocellular
Distinct trabecular, adenoid or poorly differentiated solid growth. Plates consist of multiple and irregular cell layers. Invasive growth and cellular atypia may be present

Hyperplasia, Hepatocellular, Regenerative
Evidence of prior or ongoing hepatocellular damage is present. Lobular architecture is maintained, albeit distorted

Focus of Cellular Alteration
No evidence of compression or only minimal compression. Plates usually merge imperceptibly with surrounding parenchyma. Lobular architecture is maintained

References

(See 7, 11, 12, 14, 16, 25, 26, 30, 38, 40, 41, 42)

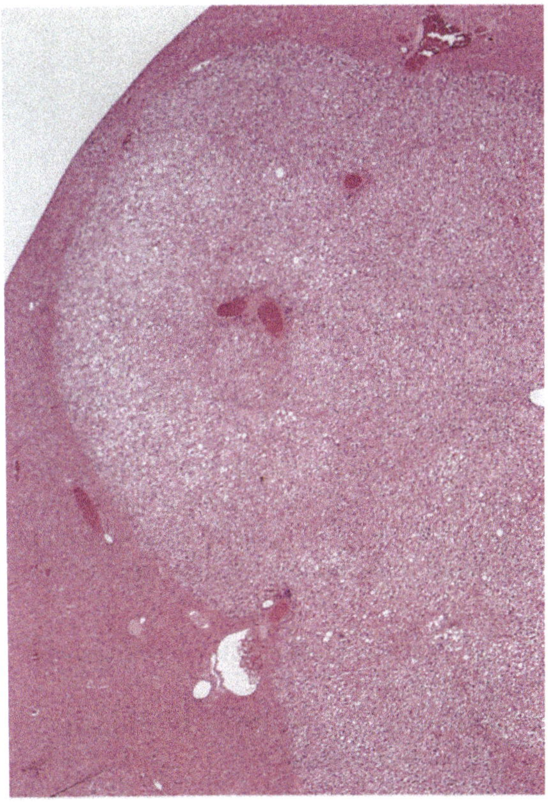

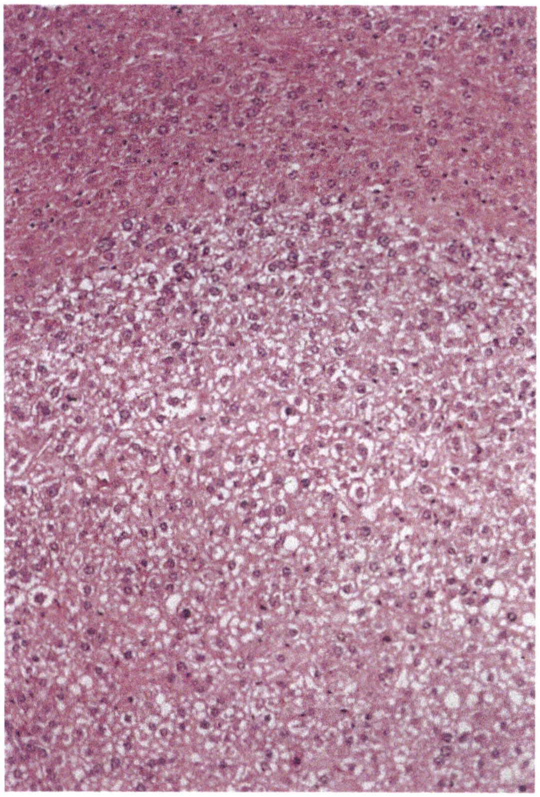

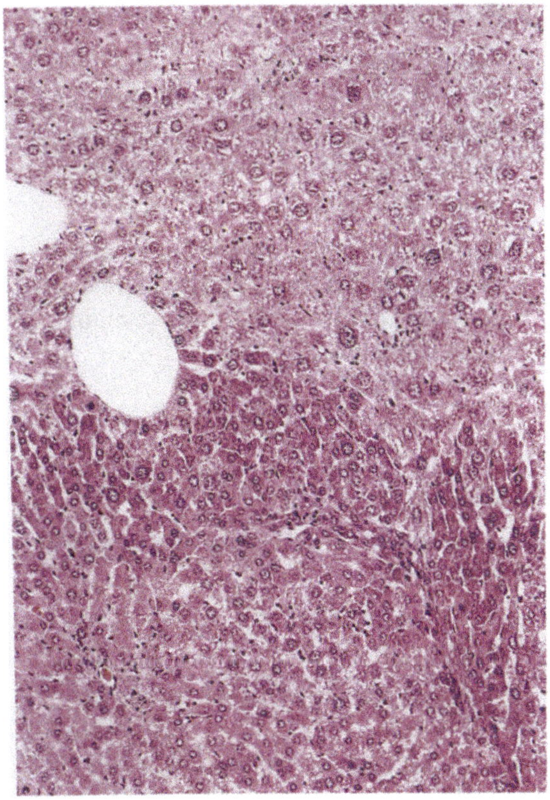

Fig. 5 *(upper left)*. Hepatocellular adenoma of the liver. H&E

Fig. 6 *(upper right)*. Hepatocellular adenoma of the liver. H&E

Fig. 7 *(below)*. Hepatocellular adenoma of the liver. H&E

Carcinoma, Hepatocellular *(M)*

Trabecular • Solid • Adenoid

Synonym(s). Hepatoma, malignant; carcinoma, trabecular; nodule, type B

Histogenesis

Hepatocytes.

Diagnostic Features

- May be grossly visible
- Growth pattern is trabecular, adenoid or solid
- Distinctly demarcated, even in poorly differentiated carcinomas
- Border may be disrupted by cells penetrating into the surrounding tissue
- Lobular architecture is not maintained
- Cellular differentiation may range from well to poorly differentiated cells
- Cellular polymorphism may be pronounced including large cells with vacuolated cytoplasm
- Marked cellular atypia may be present in poorly differentiated tumors
- Mitotic figures may be numerous
- Cysts of various size, hemorrhage and necrosis may be present
- Eosinophilic cytoplasmic inclusions may be found within hepatocytes

Trabecular
- Growth pattern is distinctly trabecular
- Trabecula comprise multiple cell layers
- Thick trabecula consisting of several cell layers may coexist with small plates consisting of merely one cell layer
- Tumors with well-preserved trabecular structure may resemble hepatocellular adenoma. Sinusoids are dilated, progressive sinusoidal dilation may disrupt regular trabecular structure or result in pseudoglandular or papillary-like structure

Solid
- Distinct polymorphism of cells and nuclei is present
- Large or even giant cells with large hyperchromatic nuclei are common, sometimes coexisting with neoplastic cells, which are smaller than normal hepatocytes

Adenoid
- A layer of neoplastic cells surround luminal structures
- Lining cells usually consist of strongly basophilic cuboidal cells

Differential Diagnosis

Adenoma, Hepatocellular
No evidence of distinct trabecular, adenoid or poorly differentiated growth. Cells are uniform and well-differentiated. No evidence of invasion

Hyperplasia, Hepatocellular, Regenerative
Evidence of prior or ongoing hepatocyte damage. Lobular architecture is maintained, albeit distorted

Fig. 8 *(upper left).* Hepatocellular carcinoma of the liver. H&E ▶

Fig. 9 *(upper right).* Hepatocellular carcinoma of the liver. H&E

Fig. 10 *(lower left).* Hepatocellular carcinoma of the liver. H&E

Fig. 11 *(lower right).* Hepatocellular carcinoma of the liver. H&E

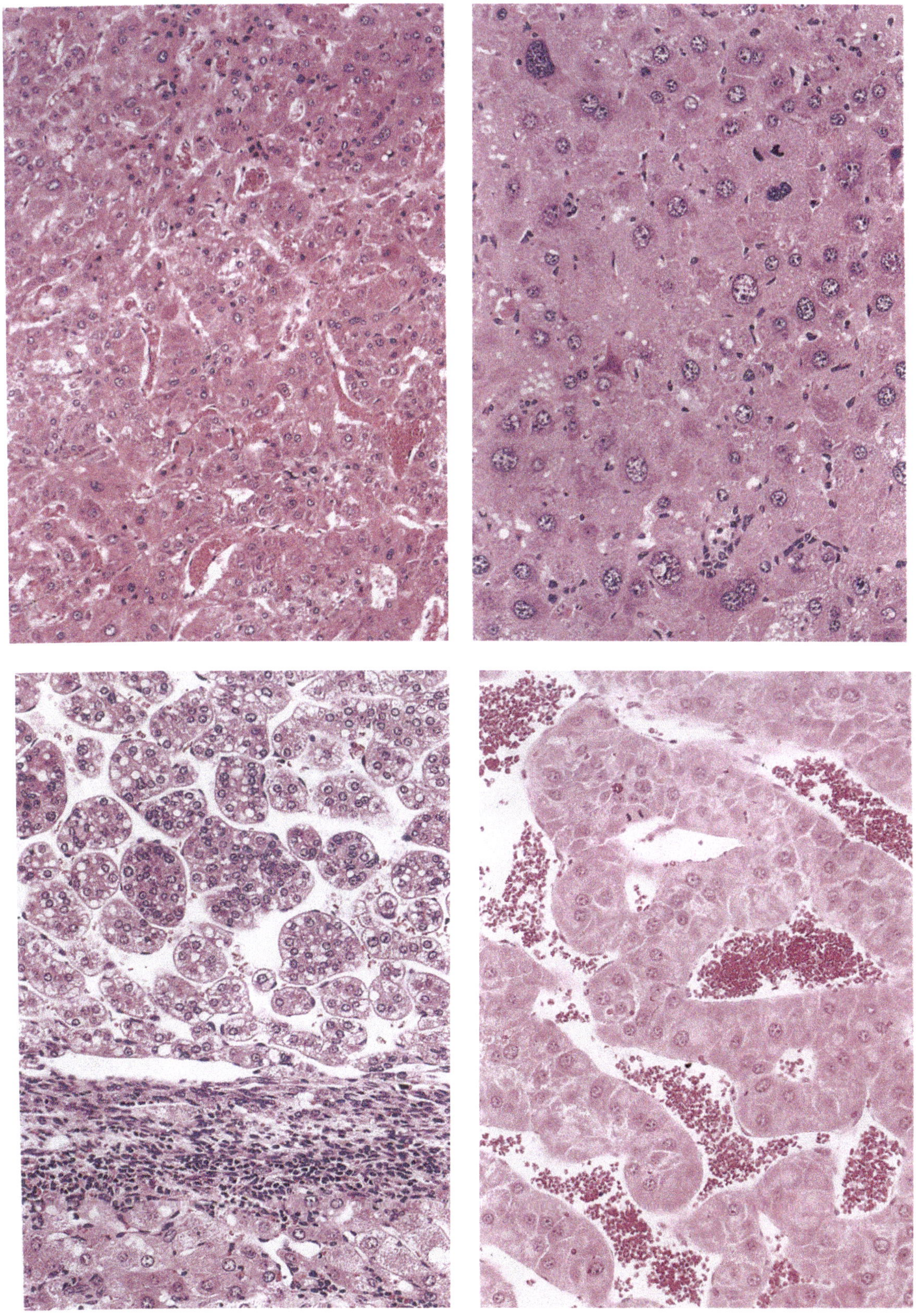

Focus of Cellular Alteration
No distinct trabecular, adenoid or poorly differentiated growth. Liver plates usually merge imperceptibly with surrounding parenchyma. Lobular architecture is maintained

Cholangiocarcinoma
Mucus production is present

Comment

The trabecular type of hepatocellular carcinoma is the most common one. Very often hepatocellular carcinomas appear to arise within pre-existing adenomas. While some authors refer to these lesions as a focus of carcinoma in adenoma, such lesions should be diagnosed as hepatocellular carcinoma. These carcinomas exhibit the same morphological appearance as the carcinomas described above and hint towards a progression of adenoma into carcinoma. Attention should be paid to hemorrhagic areas, where proliferation of endothelial cells may lead to the erroneous diagnosis of a vascular tumor.

References

(See 7, 11, 12, 16, 23, 25, 26, 31, 36, 38, 43)

Adenoma, Hepatocholangiocellular *(B)*

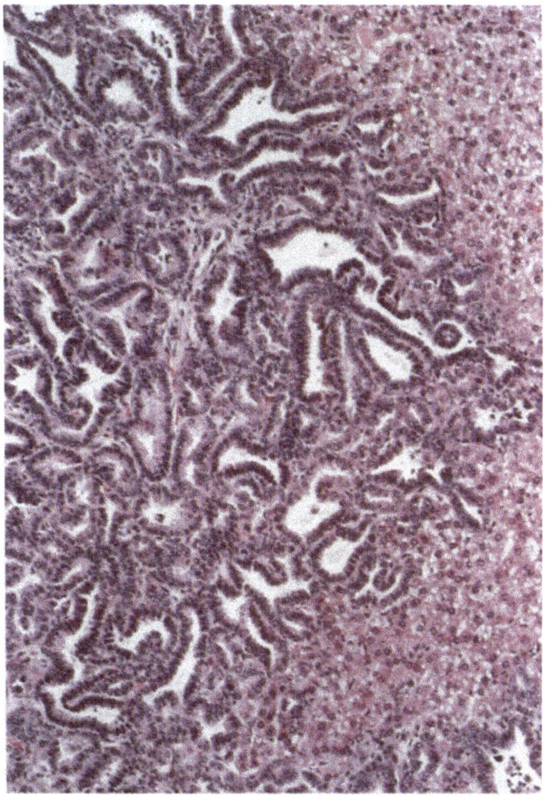

Fig. 12. Hepatocholangiocellular adenoma of the liver. Neoplastic hepatocytes and formation of ducts lined by well-differentiated neoplastic biliary epithelial cells. B6C3F1 mouse, male, age 18 months. H&E

Histogenesis

Unknown.

Diagnostic Features

- Features of hepatocellular adenoma and cholangioma are present
- Areas composed of cords of neoplastic hepatocellular cells merge with areas composed of ducts lined by neoplastic epithelium that resembles bile duct epithelium
- Mitotic figures are rare
- Stromal component may be absent
- Hepatocytes may have some alteration in staining (eosinophilic, basophilic, or clear cell) compared to surrounding parenchyma
- Stratification and atypia of the biliary epithelium is minimal or absent

Differential Diagnosis

Adenoma, Hepatocellular or Cholangioma
Composed of either neoplastic hepatocytes or duct epithelium, but not both

Carcinoma, Hepatocholangiocellular
Hepatocytes arranged in solid or trabecular patterns. Stratification and atypia of biliary epithelial component is present

Comment

Rare spontaneous lesion.

References

(See 12, 16)

Carcinoma, Hepatocholangiocellular *(M)*

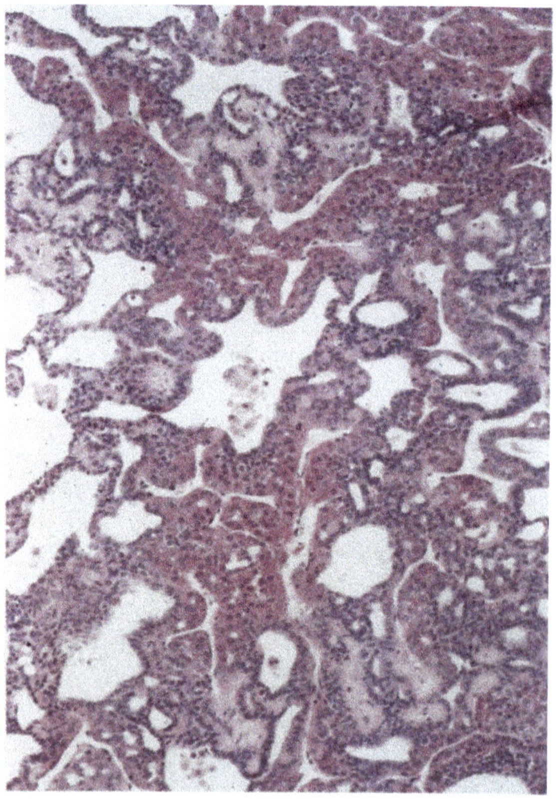

Fig. 13. Hepatocholangiocellular carcinoma of the liver. Note trabeculae of neoplastic hepatocytes and formation of irregular ducts lined by neoplastic biliary epithelial cells and hepatocytes. B6C3F1 mouse, male, age 18 months. H&E

Histogenesis

Unknown.

Diagnostic Features

- Features of hepatocellular carcinoma and cholangiocarcinoma are present
- The hepatocytic component may be arranged in solid patterns, columnar cords or nests and ducts
- The biliary component may exhibit stratification, cyst formation and cellular atypia
- Areas of hemorrhage and necrosis may be present
- Mitotic figures may be numerous
- Occasionally ducts lined by both cell types, hepatocytic and biliary epithelial, may be present

Differential Diagnosis

Carcinoma, Hepatocellular or Cholangiocarcinoma
Composed of either malignant hepatocytes or duct epithelium

Adenoma, Hepatocholangiocellular
Hepatocytic component has typical cord-like arrangement of cells. Evidence of stratification and atypia of biliary epithelial component is minimal or absent

Comment

Rare spontaneous lesion.

References

(See 11, 16)

Hepatoblastoma *(M)*

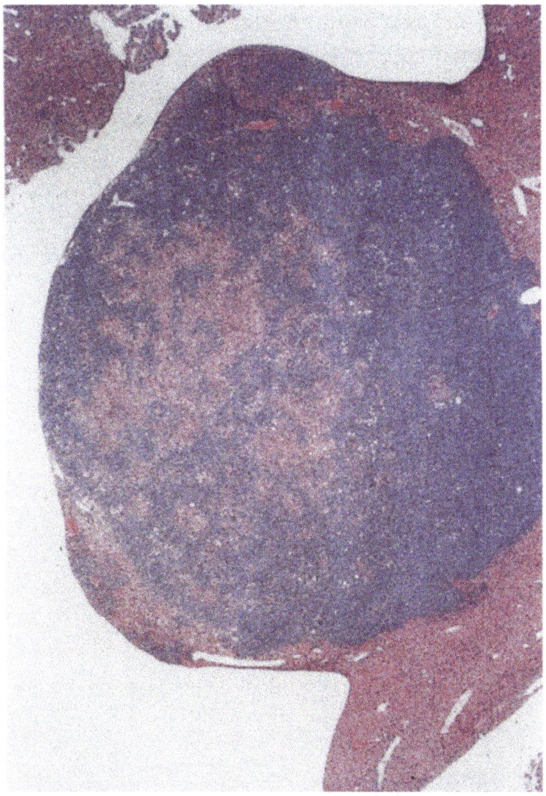

Synonym(s). Tumor, mixed, poorly differentiated

Histogenesis

Unknown; liver blastema cells, neoplastic hepatocytes, oval cells, and biliary epithelial cells have been proposed.

Diagnostic Features

- Well-circumscribed nodule
- Distinct encapsulation with variable structure may be present
- Organoid structures are lined by vascular cavities filled with blood
- The channels are surrounded by several layers of tumor cells
- The cells are arranged either radially or concentrically forming rosettes, trabecula or pseudo-glandular structures
- The center of the rosettes sometimes contains amphophilic material or small, endothelium-lined vessel
- Cells are small, strongly basophilic and elongated. Sometimes they may be more or less eosinophilic and have smaller, rounder and less hyperchromatic nuclei
- Numerous mitotic figures are present
- Areas of large hemorrhage, pigmentation, fibrosis or necrosis are present

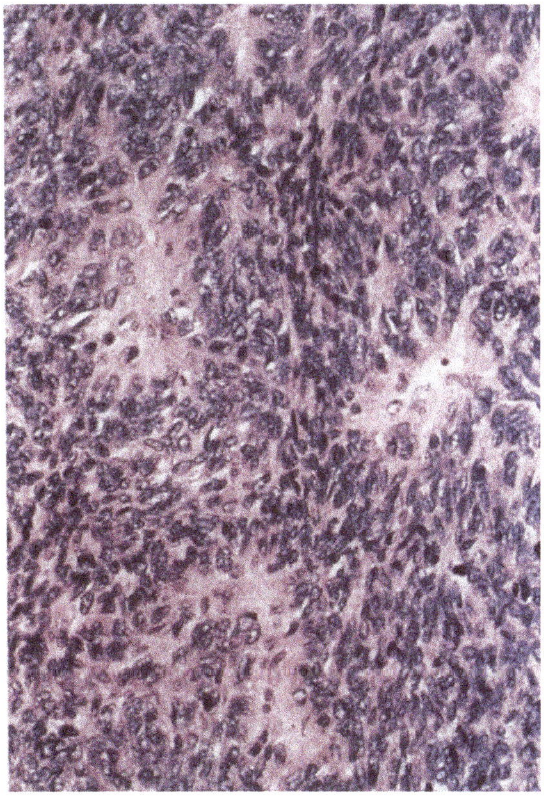

Fig. 14 *(above)*. Hepatoblastoma of the liver. Note marked basophilia and cellularity. B6C3F1 mouse, male, age 18 months. H&E

Fig. 15 *(below)*. Hepatoblastoma of the liver. Note small hyperchromatic cells arranged in trabeculae and rosettes. B6C3F1 mouse, male, age 18 months. H&E

- Osteoid, bone, squamous differentiation and extramedullary hematopoietic foci may be present

Differential Diagnosis

Cholangiocarcinoma (Poorly Differentiated)
Glandular structures with mucus production are present. Often evidence is present of extensive fibrosis, but not of osteoid, or bone formation

Sarcoma, NOS
Only mesenchymal structures are present

Carcinoma, Hepatocellular
Only sparse mesenchymal structures are present. Hepatic cell differentiation is obvious

Comment

Hepatoblastoma is generally seen within or adjacent to hepatocellular neoplasm. In these cases, some preference has been expressed for a single diagnosis of hepatoblastoma, rather than two diagnoses (the hepatoblastoma and the hepatocellular neoplasm). If this preferred convention of using the single diagnosis of hepatoblastoma is not used, than the alternative convention will need to be defined by the pathologist.

References

(See 3, 16, 27, 28, 34, 35)

Hyperplasia, Bile Duct *(H)*

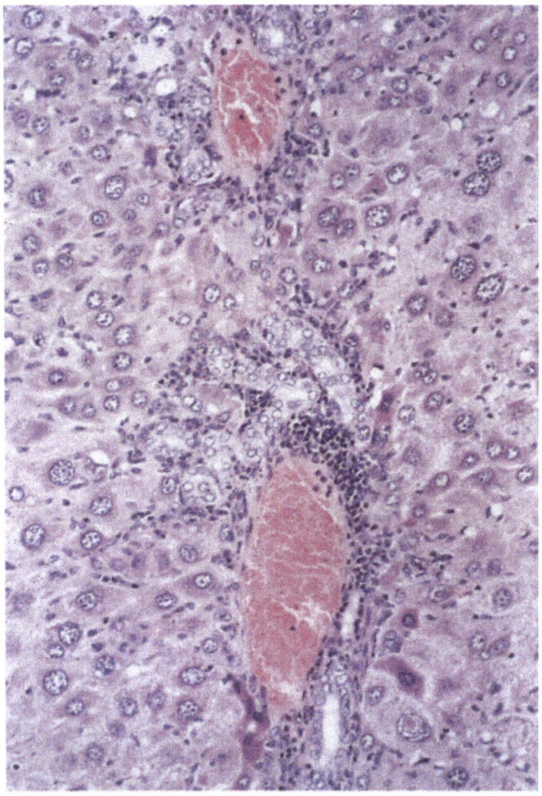

Fig. 16. Bile duct hyperplasia of the liver. Numerous bile ducts arising in portal region associated with periductular cell infiltration. NMRI mouse, female, age 18 months. H&E

Histogenesis

Bile duct epithelium.

Diagnostic Features

- Not visible grossly
- Several small bile ducts arising in the portal region
- Biliary epithelium is generally well-differentiated
- May be associated with periductular fibrosis, periductular cell infiltration, evidence of hepatic injury and repair

Differential Diagnosis

Cholangioma
Distinct demarcation is present; no evidence of an obvious portal affiliation. Lesion is usually solitary

Cholangiocarcinoma
Local invasion of adjacent liver parenchyma is present. Cells are pleomorphic

Hyperplasia, Oval Cell
Composed of small cells with scant cytoplasm. Any duct-like structures are rudimentary

Comment

Often associated with evidence of hepatic injury and repair and obstruction of bile flow.

References

(See 16, 22)

Cholangioma *(B)*

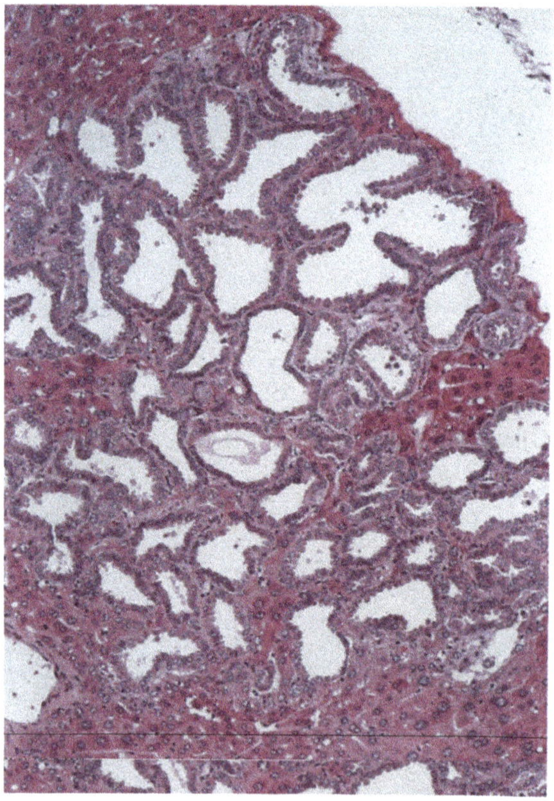

Fig. 17. Cholangioma of the liver. Note lining of ductular structures by single-layered well-differentiated cuboidal epithelial cells. B6C3F1 mouse, male, age 82 weeks. H&E

Histogenesis

Bile duct epithelium.

Diagnostic Features

- Well-demarcated lesion
- Compression of adjacent hepatic parenchyma may be present
- Ductal structures are lined by single-layered well-differentiated epithelial cells
- Multiple lesions are rare
- No evidence of an obvious portal affiliation
- Cells are uniform and usually well-differentiated

Differential Diagnosis

Hyperplasia, Bile Duct
Distribution is multifocal. No evidence of distinct demarcation or compression present

Cholangiocarcinoma
Local invasion, cellular pleomorphism, mucus production, fibrosis, and necrosis may be present

Comment

Rare lesion.

References

(See 11, 16, 22)

Cholangiocarcinoma *(M)*

Histogenesis

Bile duct epithelium.

Diagnostic Features

- May grossly be visible
- Local invasion of adjacent hepatic tissue is present
- Growth pattern is glandular, papillary or solid
- Cells are pleomorphic epithelial cells, with prominent nuclei and basophilic cytoplasm
- Mucus production usually occurs, may be intracellular or extracellular
- Areas of necrosis and fibrous connective tissue may be present
- Mitotic figures may be numerous

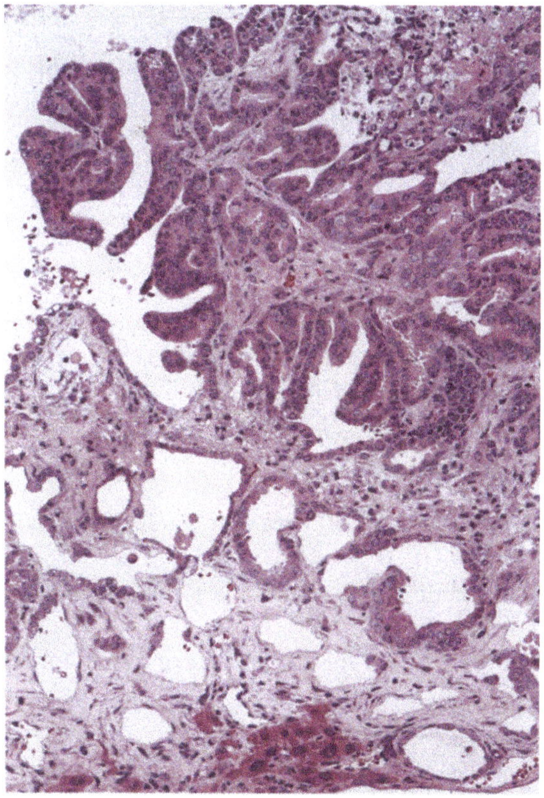

Fig. 18. Cholangiocarcinoma of the liver. Irregular duct structures lined by atypical epithelial cells. CD-1 mouse, female, age 72 weeks. H&E

Differential Diagnosis

Cholangioma

No evidence of local invasion, sclerosis or fibrosis

Cells are uniform and well-differentiated

Carcinoma, Hepatocellular

Composed of malignant hepatocytes. Usually, no evidence of mucus production

Cholangiofibrosis

Atypical ducts are strongly basophilic within distinct fibrous stroma

Comment

Cholangiofibrosis is an extremely rare and controversial lesion. It is less common in the mouse than in the rat. Only once published by Kimbrough and Linder following Aroclor 1254 application.

References

(See 11, 16, 20, 22)

Tumor, Ito Cell, Benign *(B)*

Histogenesis

Fat-storing perisinusoidal cells, the so-called Ito-cell (presumptive).

Diagnostic Features

- Unicentric or multicentric unencapsulated mass
- Focal or diffuse accumulation of tumor cells
- May grow in sheets, clusters or along cords of hepatocytes
- Cells vary in size and shape and are vacuolated
- Multiple cytoplasmic fat droplets of different size occur
- The nuclei are ovoid or round and may be indented by cytoplasmatic lipid droplets
- Frequently evidence of atrophy of adjacent liver tissue is present
- Modest amount of collagenous matrix may be present
- Unencapsulated

Differential Diagnosis

Liposarcoma
Various types of fat cells, foam cells, giant cells, myxoid cells or fibroblast-like cells may be present

Comment

Ito cell neoplasm are extremely rare. As a consequence the histogenesis and the biological behavior of these tumors is not well established.

References

(See 5, 6, 32)

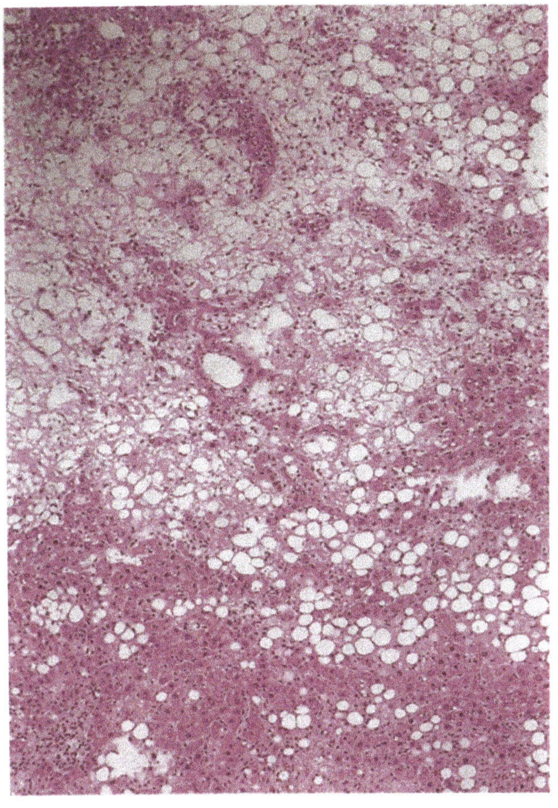

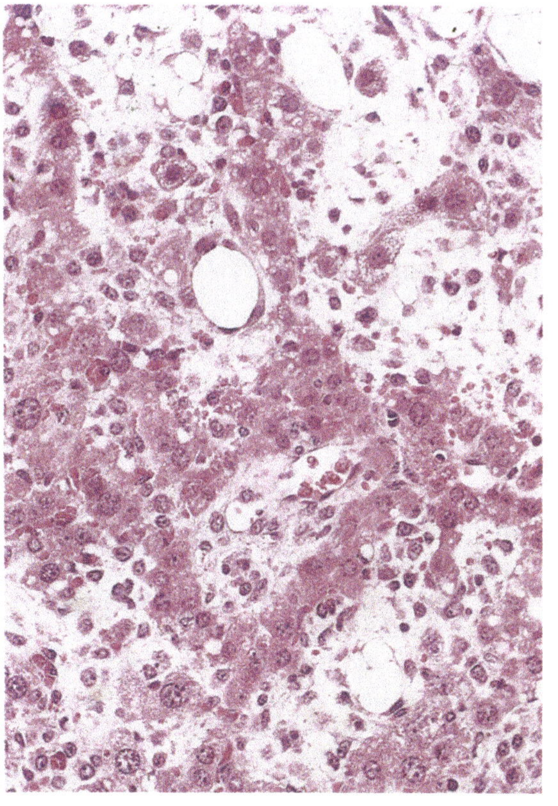

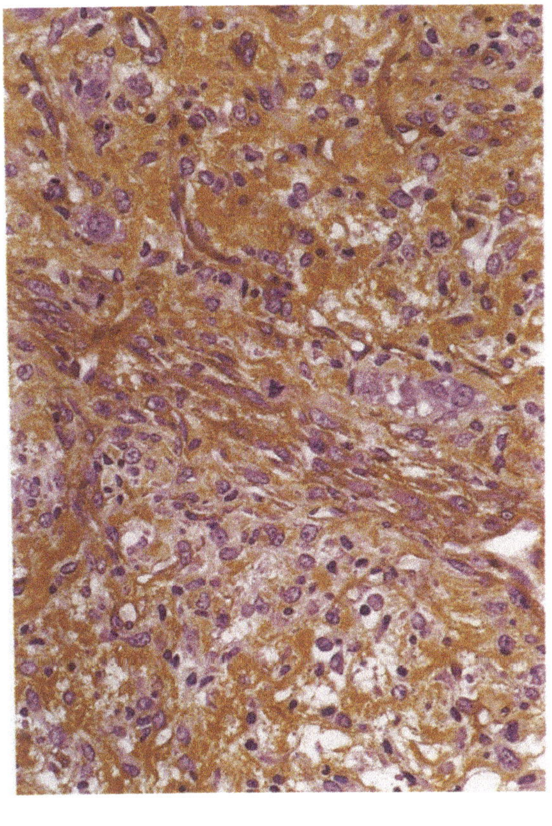

Fig. 19 *(upper left)*. Ito cell tumor of the liver. H&E

Fig. 20 *(upper right)*. Ito cell tumor of the liver. H&E

Fig. 21 *(below)*. Ito cell tumor of the liver. Laminin

Hyperplasia, Oval Cell *(H)*

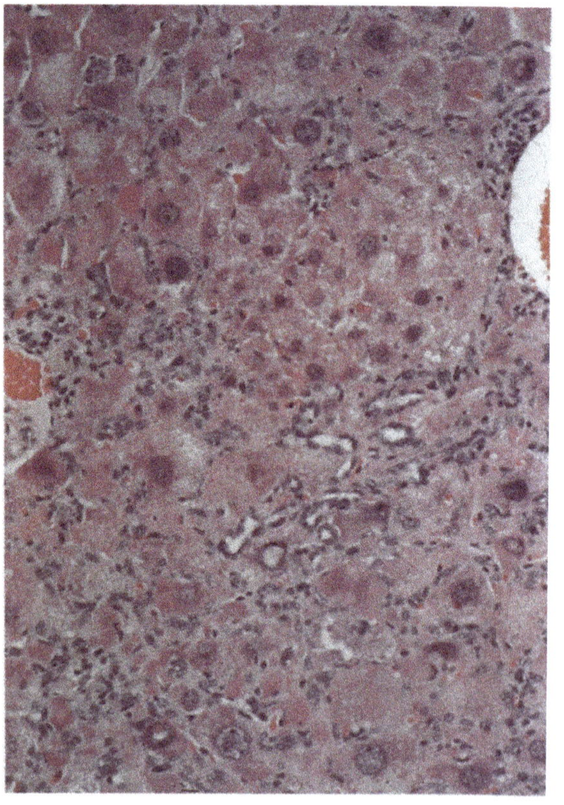

Fig. 22. Oval cell hyperplasia of the liver. Note oval cell accumulation between cords of hepatocytes with formation of incomplete duct-like structures. B6C3F1 mouse, male. H&E

Histogenesis

Uncertain, proposed to arise from terminal ductule epithelial cells.

Diagnostic Features

- Oval cells accumulate between cords of hepatocytes, often with a predominant periportal distribution
- Nuclei are oval-shaped and heterochromatic. Amount of cytoplasm is scant
- Formation of incomplete duct-like structures may be present
- Cells are small

Differential Diagnosis

Hyperplasia, Bile Duct
Periportal distribution with minimal extension along cords of hepatocytes. Cells have typical epithelial (cuboidal-columnar) appearance, forming well-defined ducts

Comment

Often associated with evidence of severe hepatocellular injury, chronic active hepatitis caused by *Helicobacter hepaticus* and the treatment with various hepatocarcinogens. Furthermore hyperplastic oval cells occur in association with a high incidence of hepatocellular neoplasm and may play an important role in hepatocarcinogenesis. Some authors support the concept that oval cells may participate in the lineage of hepatocellular and cholangiocellular carcinomas.

References

(See 8, 9, 11, 16, 18, 44, 45.)

Hyperplasia *(H)*

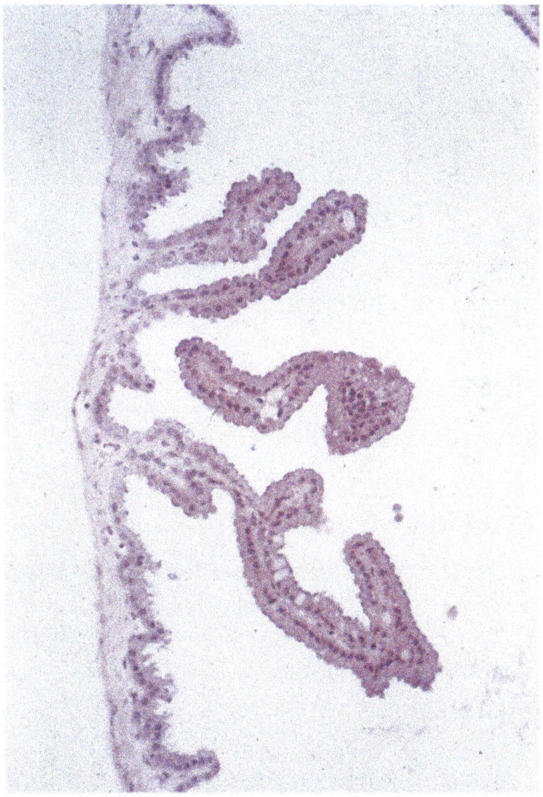

Fig. 23. Hyperplasia of the gallbladder. Note papillary projections with well-differentiated and single-layered epithelial cells. C57BL mouse, male, age 18 months. H&E

Histogenesis

Epithelium of gallbladder.

Diagnostic Features

- Lesion varies from few cells on papillary folds to small papillary projections
- Lesion is small
- Cells are well-differentiated
- Little atypia may be present
- Epithelium is usually single layered

Differential Diagnosis

Adenoma

Exhibits moderate evidence of atypia
Growth pattern is disordered

Adenocarcinoma

Invasion of the wall of the gallbladder or adjacent tissue

Adenomatoid Change

Distinct eosinophilic crystals are present in the cytoplasm or in the lumen of glands or the gallbladder. Formation of glands with thickening of the mucosa

Comment

Hyperplasia should be differentiated from adenomatoid change, which is noted in aging mice of different strains. Adenomatoid change consists of a thickened mucosa with gland formation. The hypertrophic cells contain eosinophilic crystalline inclusions, which may also be found in the lumen of the glands and in the gallbladder. The cells are well-differentiated with no or only little atypia.

References

(See 16, 17, 46, 47)

Adenoma *(B)*

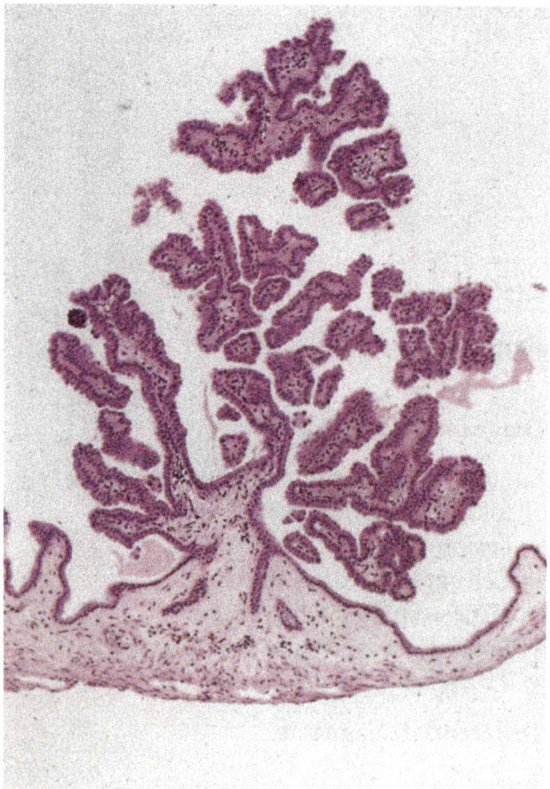

Fig. 24. Adenoma of the gallbladder. Note tumor mass in lumen of gallbladder with papillary arrangement of epithelial cells lining a fibrovascular stroma. C57BL mouse, male, age 18 months. H&E

Synonym(s). Adenoma, papillary

Histogenesis

Epithelium of gallbladder.

Diagnostic Features

- Growth is disordered, predominantly papillary- or cauliflower-like
- Amount of fibrovascular stroma is variable
- Epithelium is single-layered, but may also be multilayered
- Cells may exhibit moderate atypia, with enlarged nuclei or even giant nuclei with one or two nucleoli
- Mitotic figures are present
- Often inflammatory cellular infiltration and focal mineralization of the stroma may be present

Differential Diagnosis

Hyperplasia
Small lesion. Exhibits only little evidence of atypia

Adenocarcinoma
Increased cellular atypia and disordered growth pattern. Invasion of the wall of the gallbladder or adjacent tissue

Adenomatoid Change
Cells are well-differentiated with little or no cellular atypia. Distinct eosinophilic crystals are present in the cytoplasm, or in the lumen of glands or gallbladder

Comment

Adenoma should be differentiated from adenomatoid change, which is noted in aging mice of different strains. Adenomatoid change consists of a thickened mucosa with gland formation. The hypertrophic cells contain eosinophilic crystalline inclusions, which may also be found in the lumen of the glands and in the gallbladder. The cells are well-differentiated with no or only little atypia.

References

(See 16, 17, 22, 46, 47)

Adenocarcinoma *(M)*

Histogenesis

Epithelium of the gallbladder.

Diagnostic Features

- May be a sessile broad-based mass or a diffusely thickening of the mucosa
- Invasion of the wall of the gallbladder or adjacent tissue may occur
- Growth pattern is disordered
- Cellular atypia is present. Cytoplasm is scant and basophilic
- Nuclei are enlarged
- Mitotic figures are present

Differential Diagnosis

Hyperplasia
Lesion is small. Exhibits only little evidence of atypia. No evidence of invasive growth

Adenoma
No evidence of invasion of the wall of the gallbladder. Less cellular atypia and less disordered growth pattern

Adenomatoid Change
Cells are well-differentiated with little or no cellular atypia. No evidence of invasion of the wall of the gallbladder

Comment

Adenocarcinoma should be differentiated from adenomatoid change, which is noted in aging mice of different strains. Adenomatoid change consists of a thickened mucosa with gland formation. The hypertrophic cells contain eosinophilic crystalline inclusions, which may also be found in the lumen of the glands and in the gallbladder. The cells are well-differentiated with no or only little atypia.

References

(See 16, 17, 22, 46, 47)

Hyperplasia, Acinar Cell *(H)*

Histogenesis

Acinar cells.

Diagnostic Features

- Usually larger than unchanged islets
- May exhibit tinctorial variation from surrounding tissue
- Merge usually imperceptibly with adjacent tissue, nevertheless very mild compression or displacement of adjacent acinar tissue possible
- Usually non-encapsulated acinar architecture may be slightly altered
- Low to high mitotic index nuclear polymorphism and nuclear crowding possible

Differential Diagnosis

Adenoma, Acinar Cell
Usually distinctly demarcated with compression and sometimes encapsulation

References

(See 2, 4, 24)

Adenoma, Acinar Cell *(B)*

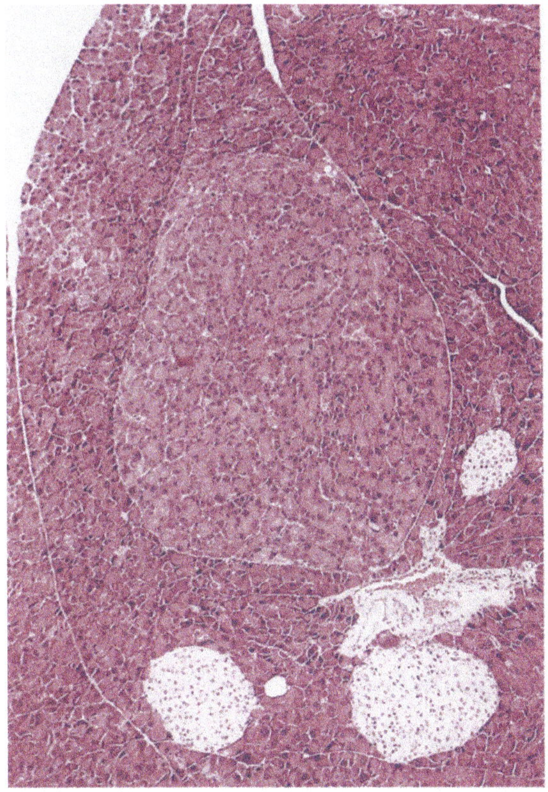

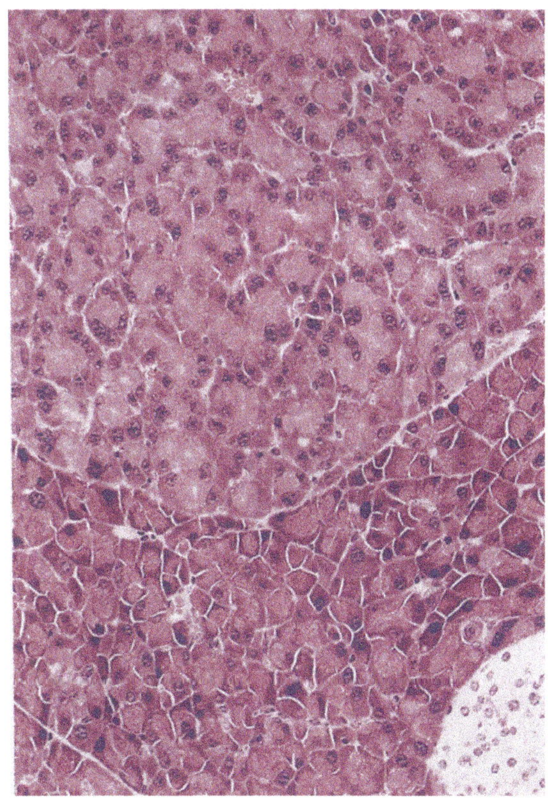

Histogenesis

Acinar cells.

Diagnostic Features

- Compression of adjacent tissue often present
- Encapsulation may occur
- Architecture is altered with increasing size
- Cells usually appear well-differentiated
- Exhibit variable degree of mitotic index and nuclear polymorphism

Differential Diagnosis

Hyperplasia, Acinar Cell
No encapsulation or sharp demarcation

Adenocarcinoma, Acinar Cell
Local invasion or metastases

References

(See 2, 4, 24)

Fig. 25 *(above)*. Acinar cell adenoma of the pancreas. H&E

Fig. 26 *(below)*. Acinar cell adenoma of the pancreas. H&E

Adenocarcinoma, Acinar Cell *(M)*

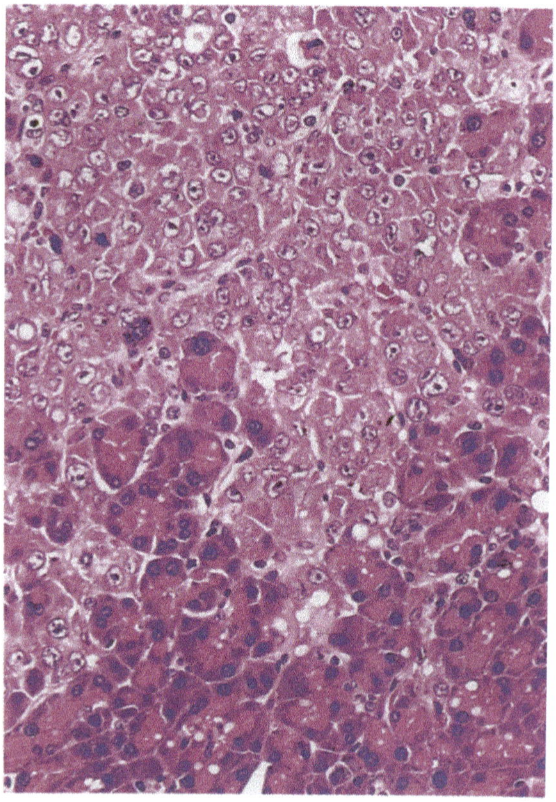

Fig. 27. Acinar cell adenocarcinoma of the pancreas. H&E

Histogenesis

Acinar cells.

Diagnostic Features

- Local invasion or distant metastases
- Glandular, trabecular or solid growth pattern
- Loss of acinar architecture
- Cellular pleomorphism and anaplasia

Differential Diagnosis

Adenoma, Acinar Cell
No local invasion or metastases

Comment

May be associated with scirrhous reactions. Squamous cell carcinomas in the acinar pancreas of mice are so far only described as experimentally induced lesions, sharing the histological features of squamous cell carcinomas described in other tissues.

References

(See 2, 4, 24)

References

1. Bannasch P, Mayer D, Venske G (1979) Pränatale Induktion von hepatocellulären Glykogenspeicherarealen und Tumoren bei Mäusen durch Äthylnitrosoharnstoff. Virchows Arch B 30: 143–160
2. Boorman GA, Sills RC (1999) Exocrine and endocrine pancreas. In: Maronpot RR, Boorman GA, Gaul BW (eds) Pathology of the mouse. Reference and atlas. Cache River Press, Vienna, pp 185–205
3. Diwan BA, Ward JM, Rice JM (1989) Promotion of malignant 'embryonal' liver tumors by phenobarbital: increased incidence and shortened latency of hepatoblastomas in (DBA/2 x C57BL/6)F1 mice initiated with N-nitrosodiethylamine. Carcinogenesis 10: 1345–1348
4. Dixon D, Maronpot RR (1994) Tumours of the pancreas. Pathology of tumours in laboratory animals, vol 2. Tumours of the mouse, 2nd edn. IARC Scientific Publications No. 111, Lyon, pp 281–303
5. Dixon D, Yoshitomi K, Boorman GA, Maronpot RR (1994) "Lipomatous" lesions of unknown cellular origin in the liver of B6C3F1 mice. Vet Pathol 31: 173–182
6. Enanz H (1985) Proliferation of Ito cells (fat-storing cells) in acute carbon tetrachloride liver injury. A light and electron microscopic autoradiographic study. Acta Pathol Jpn 35: 1301–1308
7. Evans JG, Collins MA, Lake BG, Butler WH (1992) The histology and development of hepatic nodules and carcinoma in C3H/He and C57BL/6 mice following chronic phenobarbitone administration. Toxicol Pathol 20: 585–594
8. Factor VM, Radaeva SA (1993) Oval cells – hepatocytes relationships in Dipin-induced hepatocarcinogenesis in mice. Exp Toxicol Pathol 45: 239–244
9. Factor VM, Radaeva SA, Thorgeirsson SS (1994) Origin and fate of oval cells in dipin-induced hepatocarcinogenesis in the mouse. Am J Pathol 145: 409–422
10. Frith CH, Dooley K (1976) Hepatic cytologic and neoplastic changes in mice given benzidine dihydrochloride. J Natl Cancer Inst 56: 679–682
11. Frith CH, Ward JM (1979) A morphologic classification of proliferative and neoplastic hepatic lesions in mice. J Environ Pathol Toxicol 3: 329–351
12. Frith CH, Wiley L (1982) Spontaneous hepatocellular neoplasms and hepatic hemangiosarcomas in several strains of mice. Lab Anim Sci 32: 157–162
13. Frith CH, Baetcke KP, Nelson CJ, Schieferstein G (1980) Sequential morphogenesis of liver tumors in mice given benzidine dihydrochloride. Eur J Cancer 16: 1205–1216
14. Frith CH, Ward JM, Turusov VS (1994) Tumours of the liver. Pathology of tumours in laboratory animals, vol 2. Tumours of the mouse, 2nd edn. IARC Scientific Publications No. 111, Lyon, pp 223–248
15. Hanigan MH, Winkler ML, Drinkwater NR (1993) Induction of three histochemically distinct populations of hepatic foci in C57BL/6J mice. Carcinogenesis 14: 1035–1040
16. Harada T, Maronpot RR, Enomoto A, Tamano S, Ward JM (1996) Changes in the liver and gallbladder. In Mohr U, Dungworth DL, Capen CC, Carlton WW, Sundberg JP, Ward JM (eds) Pathobiology of the aging mouse, vol. 2. ILSI Press, Washigton DC, pp 207–241
17. Harada T, Enomoto A, Boorman GA, Maronpot RR (1999) Liver and gallbladder. In: Maronpot RR, Boorman GA, Gaul BW (eds) Pathology of the mouse. Reference and atlas. Cache River Press, Vienna, pp 119–183
18. Hoover KL (1985) Oval cell hyperplasia, liver, mouse, rat. In Jones TC, Mohr U, Hunt RD (eds) Digestive system. ILSI Monograph, ILSI Press, Washington DC, pp 125–127
19. Kakizoe S, Goldfarb S, Pugh TD (1989) Focal impairment of growth in hepatocellular neoplasms of C57BL/6 mice: a possible explanation for the strain's resistance to hepatocarcinogenesis. Cancer Res 49: 3985–3989
20. Kimbrough RD, Linder RE (1974) Induction of adenofibrosis and hepatomas of the liver in BALB-cJ mice by polychlorinated biphenyls (Aroclor 1254). J Natl Cancer Inst 53: 547–552
21. Koen H, Pugh TD, Goldfarb S (1983) Hepatocarcinogenesis in the mouse. Combined morphologic-stereologic studies. Am J Pathol 112: 89–100
22. Lewis DJ (1984) Spontaneous lesions of the mouse biliary tract. J Comp Pathol 94: 263–271
23. Lipsky MM, Tanner DC, Hinton DE, Trump BF (1984) Reversibility, persistence, and progression of safrole-induced mouse liver lesions following cessation of exposure. In: Popp JA (ed) Mouse liver neoplasia. Current perspectives. Hemisphere Publishing, Washington DC, pp 161–177
24. Longnecker DS (1986) Experimental models of exocrine pancreatic tumours. In: Go VLW, Gardener JD, Brooks FP, Lebenthal E, Diamagno EP, Scheele GA (eds) The exocrine pancreas: biology, pathobiology and diseases. Raven Press, New York, pp 443–458

25. Malarkey DE, Devereux TR, Dinse GE, Mann PC, Maronpot RR (1995) Hepatocarcinogenicity of chlordane in B6C3F1 and B6D2F1 male mice: evidence for regression in B6C3F1 mice and carcinogenesis independent of ras proto-oncogene activation. Carcinogenesis 16: 2617–2625
26. Maronpot RR, Haseman JK, Boorman GA, Eustis SL, Rao GN, Huff JE (1987) Liver lesions in B6C3F1 mice: the National Toxicology Program, experience and position. Arch Toxicol Suppl 10: 10–26
27. Nonoyama T, Fullerton F, Reznik G, Bucci TJ, Ward JM (1988) Mouse hepatoblastomas: a histologic, ultrastructural, and immunohistochemical study. Vet Pathol 25: 286–296
28. Nonoyama T, Reznik G, Bucci TJ, Fullerton F (1986) Hepatoblastoma with squamous differentiation in a B6C3F1 mouse. Vet Pathol 23: 619–622
29. Standeven AM, Wolf DC, Goldsworthy TL (1994) Interactive effects of unleaded gasoline and estrogen on liver tumor promotion in female B6C3F1 mice. Cancer Res 54: 1198–1204
30. Stinson SF, Hoover KL, Ward JM (1981) Quantitation of differences between spontaneous and induced liver tumors in mice with an automated image analyzer. Cancer Lett 14: 143–150
31. Thorpe E, Walker AIT (1973) The toxicology of dieldrin (HEOD). II. Comparative long-term oral toxicity studies in mice with dieldrin, DDT, phenobarbitone, beta-BHC and gamma-HBC. Food Cosmet Toxicol 11: 433–442
32. Tillmann T, Kamino K, Dasenbrock C, Germann PG, Kohler M, Morawietz G, Campo E, Cardesa A, Tomatis L, Mohr U (1999) Ito cell tumor: immunohistochemical investigations of a rare lesion in the liver of mice. Toxicol Pathol 27: 364–369
33. Tsuji S, Ogawa K, Takasaka H, Sonoda T, Mori M (1988) Clonal origin of gamma-glutamyl transpeptidase-positive hepatic lesions induced by initiation-promotion in ornithine carbamoyltransferase mosaic mice. Jpn J Cancer Res 79: 148–151
34. Turusov VS, Day NE, Tomatis L, Gati E, Charles RT (1973a) Tumors in CF-1 mice exposed for six consecutive generations to DDT. J Natl Cancer Inst 51: 983–997
35. Turusov VS, Deringer MK, Dunn TB, Stewart HL (1973b) Malignant mouse-liver tumors resembling human hepatoblastomas. J Natl Cancer Inst 51: 1689–1695
36. Vesselinovitch SD, Mihailovich N, Rao KVN (1978) Morphology and metastatic nature of induced hepatic nodular lesions in C57BL x C3H F1 mice. Cancer Res 38: 2003–2010
37. Ward JM (1980) Morphology of hepatocellular neoplasms in B6C3F1 mice. Cancer Lett 9: 319–325
38. Ward JM (1984a) Morphology of potential preneoplastic hepatocyte lesions and liver tumors in mice and a comparison with other species. In: Popp JA (ed) Mouse liver neoplasia, current perspectives. Hemisphere, Washington, pp 1–26
39. Ward JM (1984b) Pathology of toxic, preneoplastic, and neoplastic lesions. In: Douglas JF (ed) Carcinogenesis and mutagenesis testing, Humana Press, Clifton NJ, pp 97–130
40. Ward JM, Rice JM, Creasia D, Lynch P, Riggs C (1983) Dissimilar patterns of promotion by di(2-ethylhexyl)phthalate and phenobarbital of hepatocellular neoplasia initiated by diethylnitrosamine in B6C3F1 mice. Carcinogenesis 4: 1021–1029
41. Ward JM, Diwan BA, Ohshima M, Hu H, Schuller HM, Rice JM (1986) Tumor-initiating and promoting activities of di(2-ethylhexyl) phthalate in vivo and in vitro. Environ Health Perspect 65: 279–291
42. Ward JM, Lynch P, Riggs C (1988) Rapid development of hepatocellular neoplasms in aging male C3H/HeNCr mice given phenobarbital. Cancer Lett 39: 9–18
43. Ward JM, Diwan BA, Lubet RA, Henneman JR, Devor DE (1990) Liver tumor promoters and other mouse liver carcinogens. In: Stevenson D, McClain M, Popp J, Slaga T, Ward J, Pitot H (eds) Mouse liver carcinogenensis: mechanisms and species comparisons Alan R. Liss, New York, pp 85–108
44. Ward JM, Anver MR, Haines DC, Benveniste RE (1994) Chronic active hepatitis in mice caused by *Helicobacter hepaticus*. Am J Pathol 145: 959–968
45. Ward JM, Fox JG, Anver MR, Haines DC, George CV, Collins MJ Jr, Gorelick PL, Nagashima K, Gonda MA, Gilden RV, et al. (1994) Chronic active hepatitis and associated liver tumors in mice caused by a persistent bacterial infection with a novel *Helicobacter* species. J Natl Cancer Inst 86: 1222–1227
46. Yoshitomi K, Alison RH, Boorman GA (1986) Adenoma and adenocarcinoma of the gallbladder in aged laboratory mice. Vet Pathol 23: 523–527
47. Yoshitomi K, Boorman GA (1994) Tumours of the gallbladder. Pathology of tumours in laboratory animals, vol 2. Tumours of the mouse, 2nd edn. IARC Scientific Publications No. 111, Lyon, 271–279

Respiratory System and Mesothelium

D.L. Dungworth[1] · S. Rittinghausen[2,3] · L. Schwartz[2] · J.R. Harkema
Y. Hayashi · B. Kittel · D. Lewis · R.A. Miller · U. Mohr · K.T. Morgan
S. Rehm · M.V. Slayter

[1]Chair; [2]Co-chair; [3]Initial draft.

Hyperplasia, Epithelial *(H)*

Squamous cell • Transitional (nonciliated cuboidal/columnar) cell • Mucous (goblet) cell
Olfactory epithelial cell • Basal cell
With cellular atypia (dysplasia) • Mixed

Histogenesis

Respiratory, transitional, olfactory or squamous epithelium or subepithelial glandular epithelium.

Diagnostic Features

- Epithelium thickened due to an increase in number of basal, mucous, non-ciliated cuboidal/columnar, squamous or olfactory cells
- Epithelial hyperplasia can result in an undulating rugose surface of the epithelium
- Ciliated cells can be crowded or less evident due to the increase in number of non-ciliated cells
- Frequently irregular arrangement of cell layers
- Often associated with degeneration and inflammation
- Occasionally increase in the number of cells of subepithelial glands in the nasal septum, maxillo-, naso- and ethmoturbinates
- Epithelial hyperplasia can include proliferation of atypical or pleomorphic basal or undifferentiated cells (epithelial hyperplasia with cellular atypia), but without disruption of the underlying basal lamina

Squamous Cell
- The lesion occurs in the squamous epithelium of the nasal vestibule or ventral meatus
- Focal increase in the number of cells to five or more epithelial cell layers
- Normal differentiation pattern
- Occasional mitoses are observed
- The hyperplastic squamous cells have larger nuclei, more prominent nucleoli and more abundant cytoplasm

Transitional (Non-Ciliated Cuboidal/Columnar) Cell
- Epithelial hyperplasia of non-ciliated cuboidal/columnar epithelium composed of three or more cell layers

Mucous (Goblet) Cell
- In mucous cell hyperplasia, mucous cells in the surface epithelium can form intraepithelial glands
- Mucous cells are taller, with apical mucous granules and basally located nuclei

Olfactory Epithelial Cell
- Hyperplasia in olfactory epithelium is frequently characterized by an increase in the number of olfactory sensory and/or basal cells

Basal Cell
- Density and thickness of the basal-cell layer increased

With Cellular Atypia (Dysplasia)
- Increased thickness and disorganized structure of the epithelium
- Proliferation of basal and undifferentiated cells is present
- Atypia and pleomorphism of undifferentiated, polyhedral cells are present with variable numbers of ciliated and secretory columnar cells in the respiratory epithelium, or sustentacular and sensory cells in the olfactory epithelium
- Proliferation of atypical or pleomorphic basal or undifferentiated cells can extend downwards into the ducts of underlying glands forming nodular clusters in the lamina propria

Mixed
- Proliferation of several cell types is present

Differential Diagnosis

Adenoma

Adenoma mostly is characterized by an expansile nodule with occasional protrusion into the nasal or paranasal cavities. Cellular atypia is common in the case of endophytic growth. Adenoma of the subepithelial glands causes compression of the adjacent structures

Papilloma, Squamous Cell

Squamous cell papilloma is characterized by exophytic growth of squamous cells, which can be mixed with cuboidal or mucus-producing cells, resting on a vascularized stalk of connective tissue

Adenocarcinoma or Carcinoma, Adenosquamous or Carcinoma, Neuroepithelial

Epithelial hyperplasia with cellular atypia might progress to adenocarcinoma, adenosquamous carcinoma or neuroepithelial carcinoma (depending on site and principal cell involved). Diagnosis of malignant tumor is based on one or more of the following: overall size, loss of polarity of epithelium, high degree of atypia and mitotic index, and invasive behavior

Comment

Epithelial hyperplasia is frequently a reversible alteration. Squamous cell metaplasia can occur within hyperplastic areas. Areas of epithelial hyperplasia with cellular atypia should not be confused with regenerative responses which follow acute injury to this type of epithelium.

Epithelial hyperplasia of transitional epithelium lining the lateral meatus is commonly caused by inhaled irritants.

Eosinophilic globules (eosinophilic intracytoplasmic proteinaceous accumulations) can be a prominent feature associated with all types of epithelial hyperplasia. They can also be found in non-hyperplastic cell populations and are generally believed to be a degenerative change (dilated endoplasmic reticulum containing proteinaceous material).

Since the thickness of the epithelium varies slightly depending on the location within the nasal vestibule, careful comparison to controls is important for the identification of mild squamous cell hyperplasia.

References

(See 6, 7, 17, 24, 30, 41, 47, 48, 62, 63, 71)

Figs. 1-9 see pp. 90-92

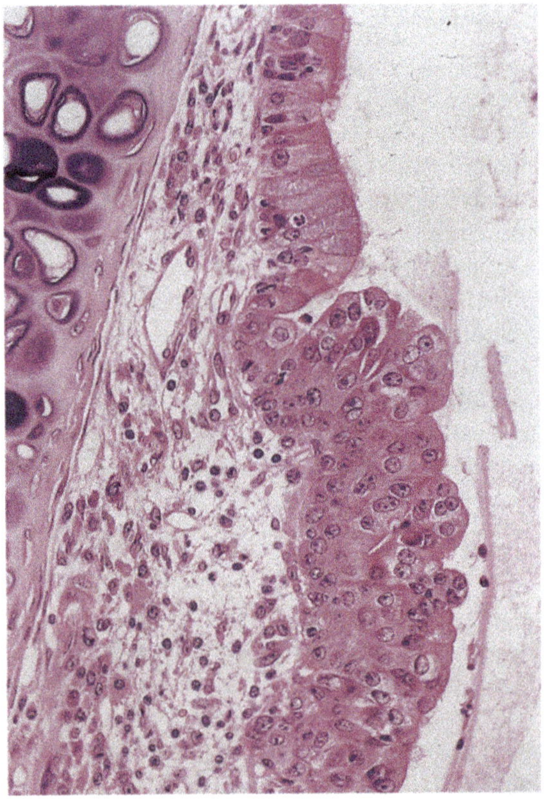

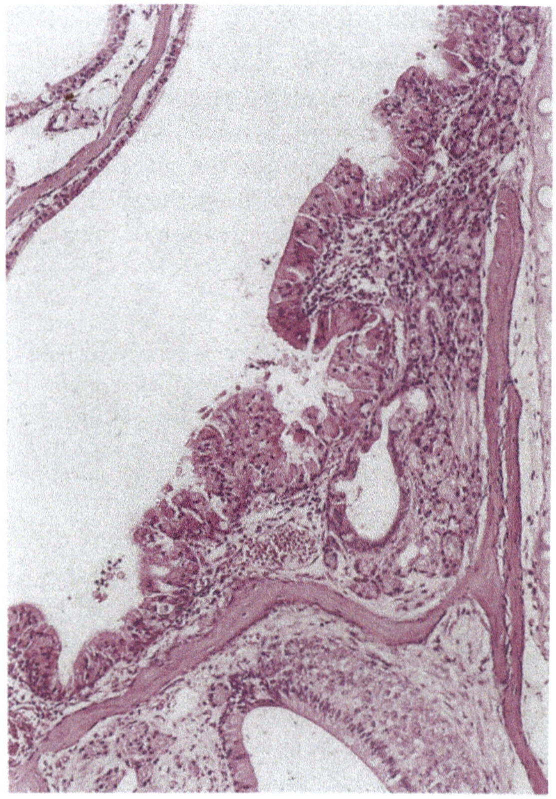

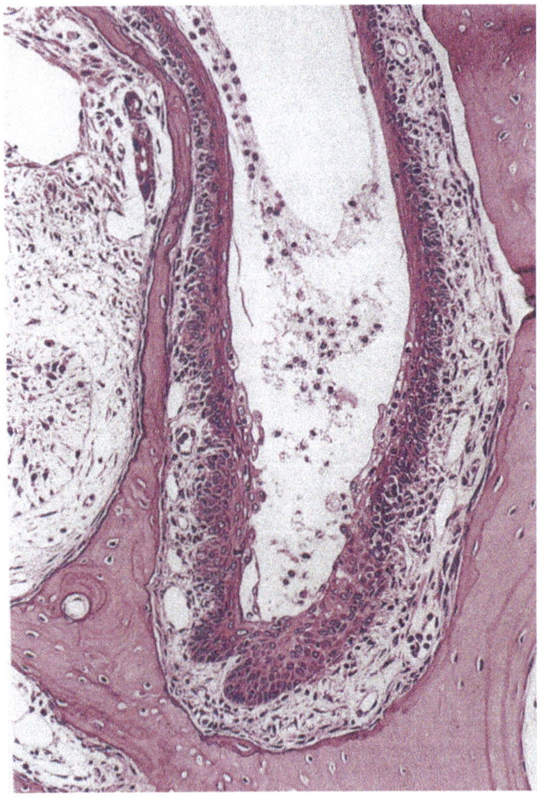

Fig. 1 *(upper left)*. Epithelial hyperplasia of the nasal cavity showing multilayering. Swiss mouse, female, treated, age 70 weeks. H&E

Fig. 2 *(upper right)*. Papillary epithelial hyperplasia of the nasal cavity. C57BL mouse, female, treated, age 66 weeks. H&E

Fig. 3 *(below)*. Epithelial hyperplasia, squamous cell, of the nasal cavity. Swiss mouse, female, treated, age 62 weeks. H&E

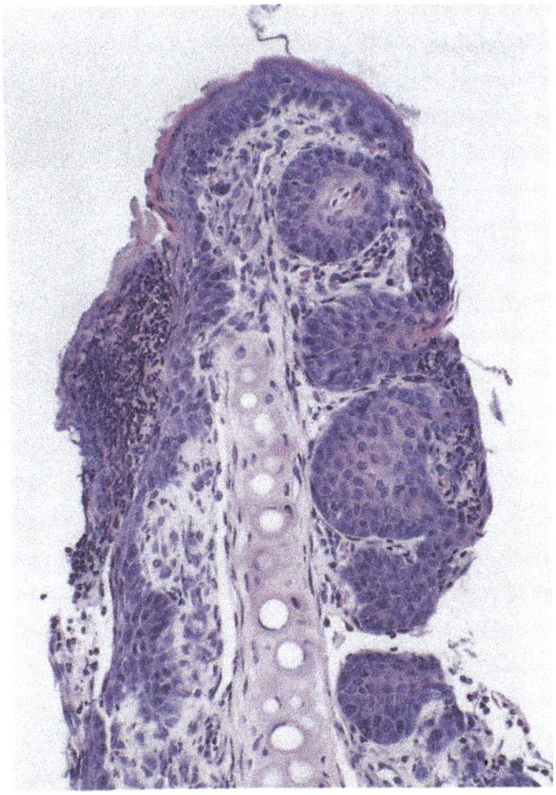

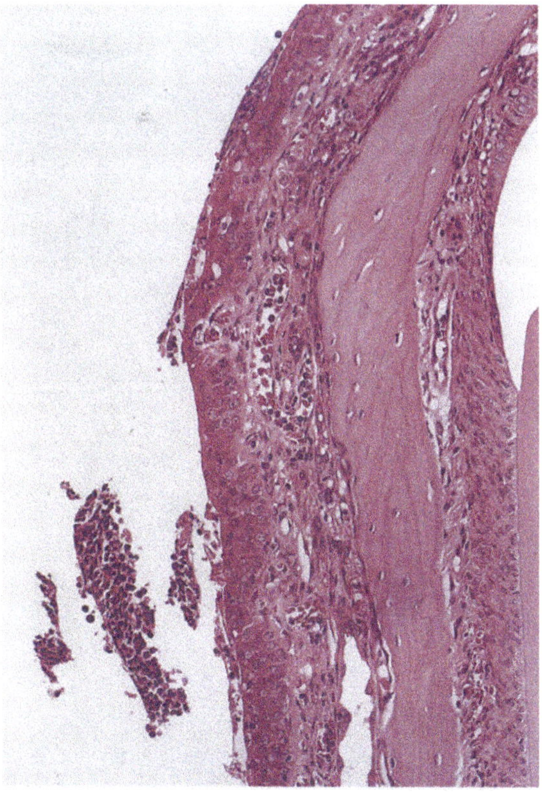

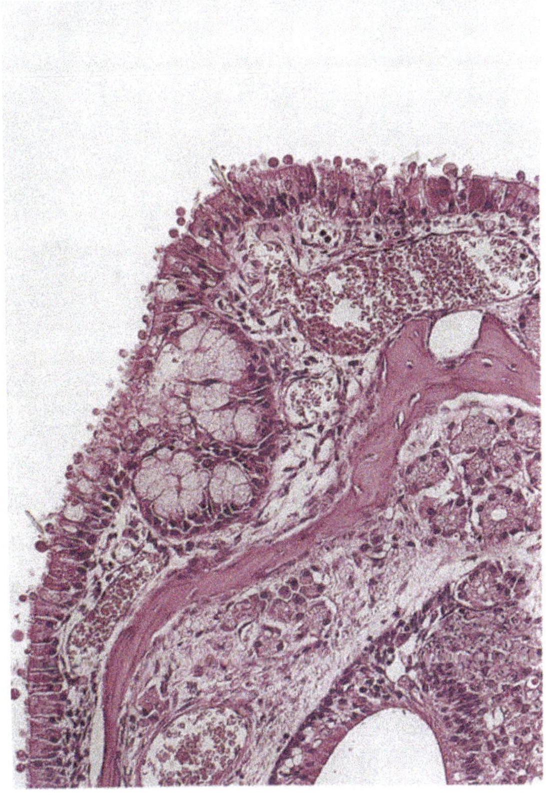

Fig. 4 *(upper left)*. Epithelial hyperplasia, squamous cell, of the dorsal atrioturbinate of the nasal cavity. B6C3F1 mouse. H&E

Fig. 5 *(upper right)*. Epithelial hyperplasia, transitional (non-ciliated cuboidal/columnar) cell, of the nasal cavity. NMRI mouse, female, treated, age 46 weeks. H&E

Fig. 6 *(below)*. Epithelial hyperplasia, mucous (goblet) cell of the nasal cavity. C57BL mouse, female, treated, age 66 weeks. H&E

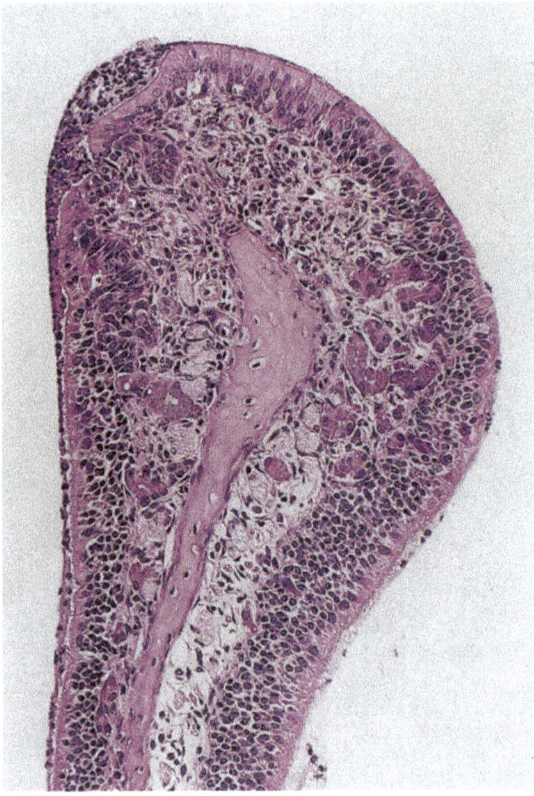

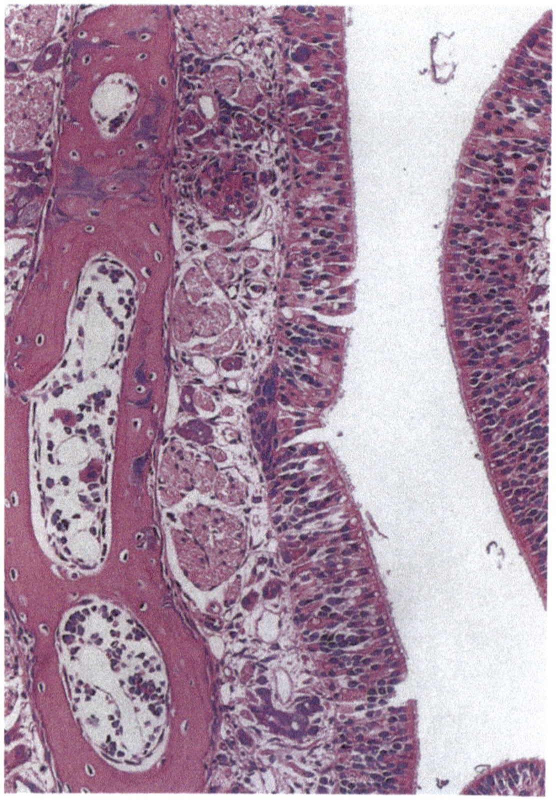

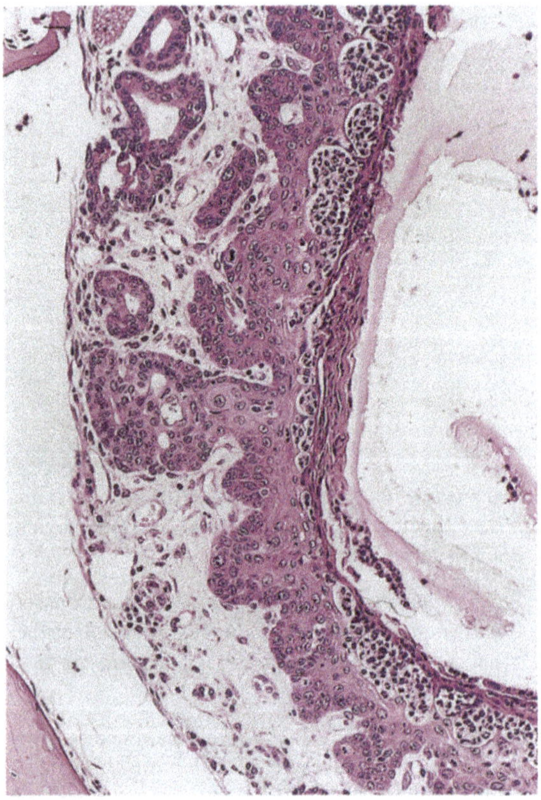

Fig. 7 *(upper left)*. Epithelial hyperplasia, olfactory epithelial cell of the nasal cavity. Swiss mouse, female, treated, age 70 weeks. H&E

Fig. 8 *(upper right)*. Epithelial hyperplasia, basal cell of the nasal cavity. Swiss mouse, female, treated, age 68 weeks. H&E

Fig. 9 *(below)*. Epithelial hyperplasia of ducts of Bowman's glands of the nasal cavity. C57BL/6NCr mouse, female, treated, age 44 weeks. H&E

Metaplasia, Squamous Cell *(H)*

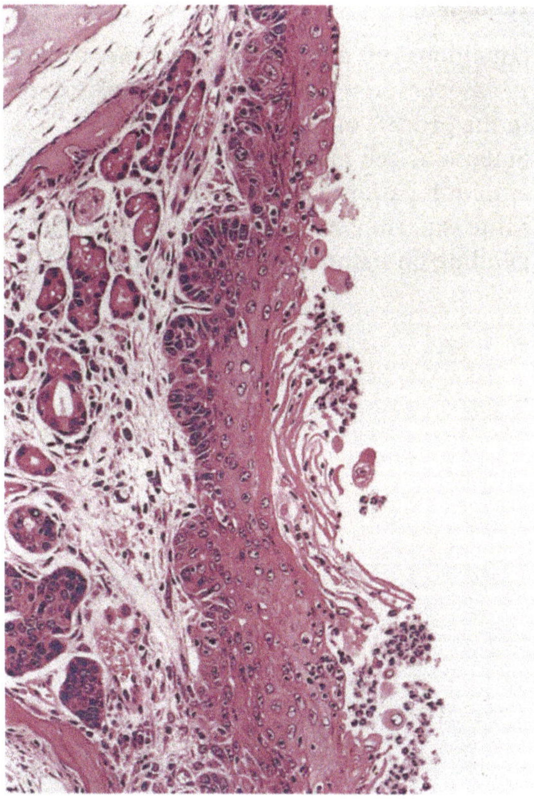

Fig. 10. Keratinizing squamous cell metaplasia of the nasal cavity associated with suppurative inflammation. Swiss mouse, female, treated, age 70 weeks. H&E

Histogenesis

Metaplasia of transitional, respiratory, or olfactory epithelium. Can also occur in the nasolacrimal ducts and ducts of glands in the lamina propria.

Diagnostic Features

- Characterized by replacement of transitional, respiratory, olfactory or ductal cells by squamous cells
- At least several layers of stratified epithelial cells, with flattening of the more superficial cells
- Compact or large epithelial cells with complete loss of cilia
- Deeper nuclei are mostly round to oval with a distinct nucleolus; superficial nuclei more flattened
- Sometimes slight nuclear polymorphism and cellular atypia
- Surface cells might contain only keratohyaline granules, or might be excessively keratinized
- Often desquamation of surface cells

Differential Diagnosis

Papilloma, Squamous Cell

Squamous cell papilloma is characterized by papillary or filiform projection above the surface of the epithelium or extension into the lumen of the ducts of the submucosal glands. There is delicate vascular mesenchymal stroma and marked proliferative thickening of the epithelium

Carcinoma, Squamous Cell

Squamous cell carcinoma is characterized by destruction of the basement membrane, cellular atypia and disorientation, and frequent mitoses. Presence of further signs of malignancy such as invasive growth and/or metastases

Epithelial Regeneration

Epithelial regeneration usually follows acute injury. Cells are one or possibly two layers thick with increased basophilia, slight karyomegaly but no horizontal layering (stratification) of flattened cells as in squamous cell metaplasia

Comment

Squamous cell metaplasia sometimes occurs in association with chronic inflammation or in the process of regeneration.
Squamous cell metaplasia with normal differentiation pattern can be reversible under some experimental circumstances (e.g. depending on nature of inhaled irritant and duration of exposure), but in other situations it might eventually give rise to squamous cell papilloma or squamous cell carcinoma.

References

(See 6, 17, 24, 30, 41, 47, 62, 71, 78)

Metaplasia, Respiratory Epithelial *(H)*

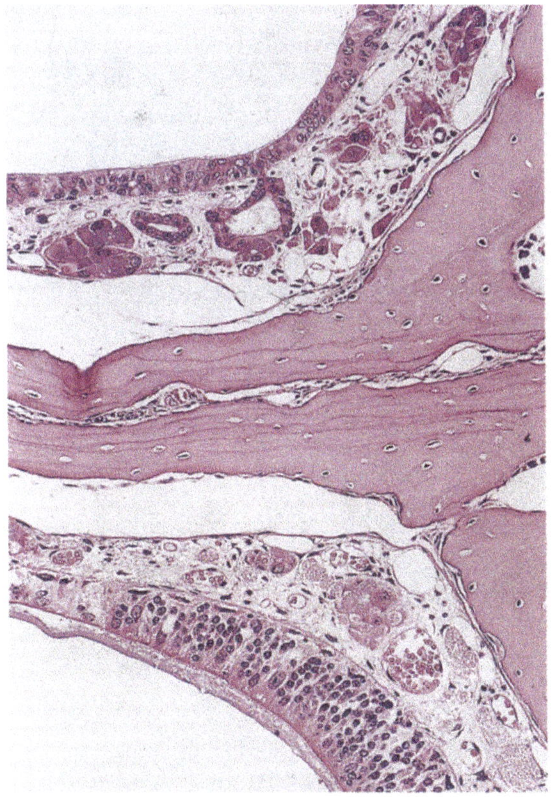

Histogenesis

Cells of olfactory epithelium and/or cells of glands in the lamina propria.

Diagnostic Features

- Respiratory epithelial metaplasia of olfactory epithelium is most common on the dorsal medial meatus
- This lesion can also arise in the submucosal glands
- Characterized by loss of sensory and sustentacular neuroepithelial cells which can be associated with focal atrophy and degeneration
- Frequently respiratory epithelial metaplasia of submucosal glands beneath the respiratory epithelium is observed
- Replacement is by simple columnar epithelium, with or without cilia, that resembles respiratory epithelium
- In the area of respiratory epithelial metaplasia of olfactory epithelium, submucosal glands are usually replaced by respiratory epithelium

Differential Diagnosis

Precise anatomic localization and comparison with controls are necessary to ensure that the site should normally be lined by olfactory epithelium

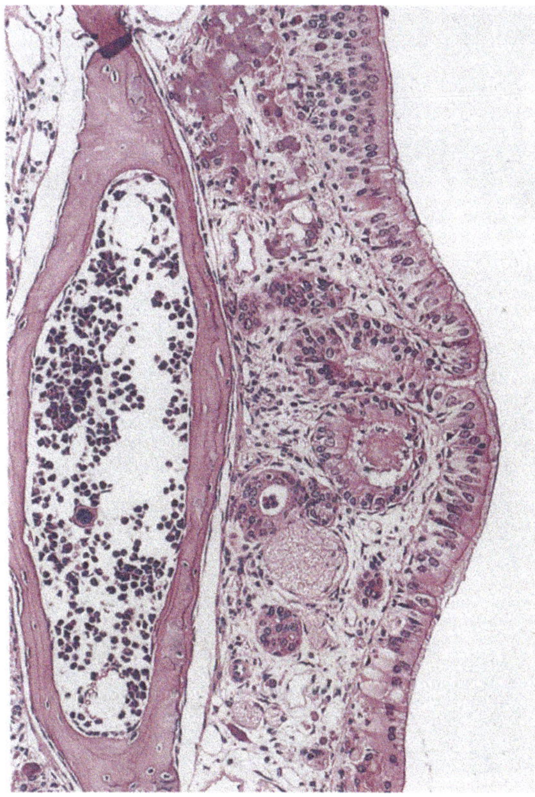

Fig. 11 *(above)*. Respiratory epithelial metaplasia of olfactory epithelium of the nasal cavity. For comparison normal olfactory epithelium in the contralateral nasal cavity. NMRI mouse, female. H&E

Fig. 12 *(below)*. Respiratory epithelial metaplasia of submucous glands of the nasal cavity. Swiss mouse, female, treated, age 68 weeks. H&E

Comment

This change is occasionally seen spontaneously in aged mice or can be a response to an irritant. The spontaneous lesion tends to affect one side more than the other and is associated with slightly dilated Bowman's glands, possibly containing neutrophils and/or eosinophilic granular debris. There is no evidence that respiratory epithelial metaplasia is pre-neoplastic.

The normal extension of olfactory epithelium into the anterior region of the dorsal meatus can vary with age and can be possibly strain-dependent.

References

(See 6, 17, 47, 48, 49)

Adenoma *(B)*

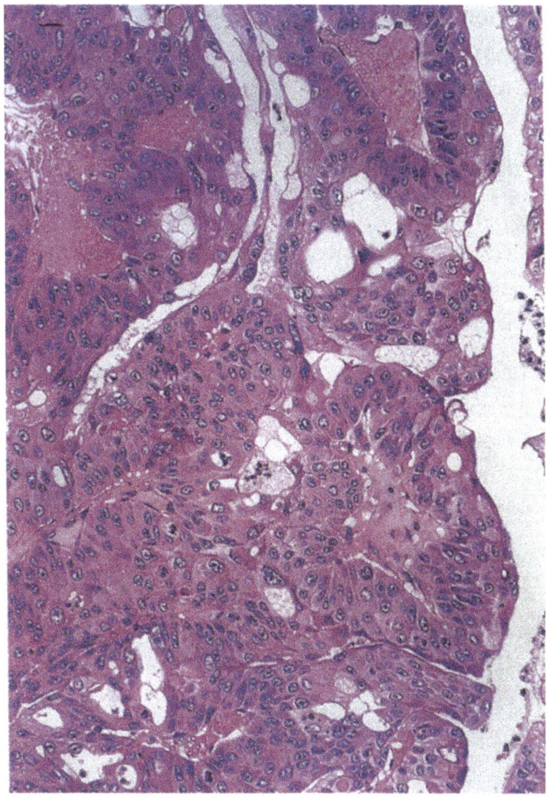

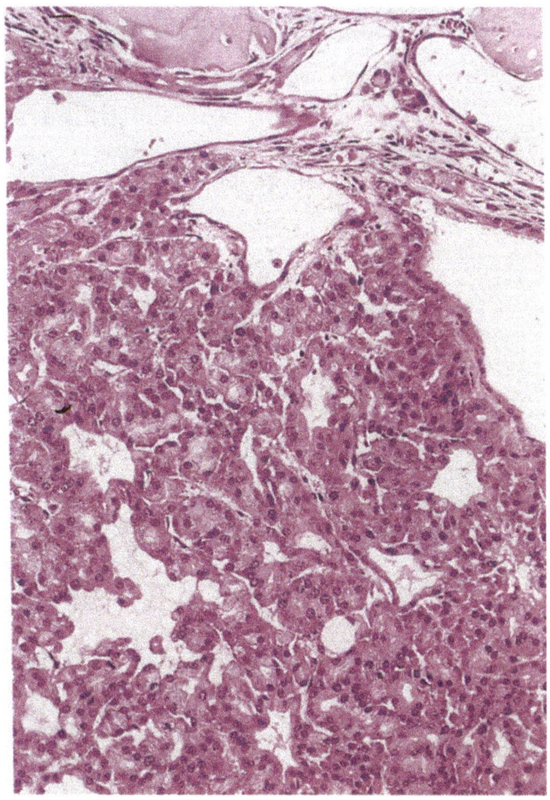

Histogenesis

Transitional, respiratory, or glandular epithelial cells.

Diagnostic Features

- Usually arises in the most anterior part of the nasal cavity, originating from the mucosa of the naso- or maxilloturbinates or from the lateral wall of the anterior nasal cavity
- Characterized by an expansile growth with occasional protrusion into the nasal or paranasal cavities, but can show endophytic growth as well
- Compression can be caused by endophytic adenomas of respiratory epithelium or adenomas of submucosal glands
- Cystic glandular structures or sheets of cells, sometimes with pseudoacinar structures due to drop-out of dead cells
- Cysts can contain PAS-positive material, sloughed epithelium, and inflammatory cells
- Secretory activity is mostly visible as mucus production
- Non-ciliated tumor cells with cuboidal to low columnar shape
- Basophilic cytoplasm and centrally located nuclei

Fig. 13 *(above)*. Adenoma of the nasal cavity (transitional type). Swiss mouse, female, treated, age 35 weeks. H&E

Fig. 14 *(below)*. High magnification of adenoma of the nasal cavity. p53 heterozygous mouse, male, untreated, age 129 weeks. H&E

- Occasionally focal areas of squamous cell metaplasia
- Epithelial cells can contain eosinophilic globules.

Differential Diagnosis

Hyperplasia, Epithelial
: No exophytic growth, no nodular lesion, absence of compression of the adjacent tissue

Papilloma, Squamous Cell
: Predominantly squamous cell proliferation. Exophytic, mostly papillary or filiform growth. Vascularized stalk of connective tissue

Adenocarcinoma
: Presence of signs of malignancy such as cellular and/or nuclear atypia, frequent mitoses, invasive growth, and/or metastases

Comment

Adenomas with exophytic growth are sometimes called polypoid or villous adenomas. Location of the tumor usually determines the type of adenoma (transitional or respiratory epithelial type). Transitional tumors are mostly found in the lateral meatus of the anterior (proximal) aspect of the nasal cavity. In many cases, it can be difficult to determine by light microscopy whether the cell of origin is of transitional, respiratory or submucosal glandular epithelium. Adenomas of the glands can be expected ultrastructurally to show periacinar myoepithelial cells and large apical secretory granules, as demonstrable in normal submucosal glands.
Adenomas might arise from glandular epithelium in the lamina propria although this has not been reported.
Eosinophilic globules (eosinophilic intracytoplasmic proteinaceous accumulations) can be a prominent feature associated with adenoma. They are generally believed to be a degenerative change (dilated endoplasmic reticulum containing proteinaceous material).

References

(See 6, 7, 17, 40, 46, 62, 63)

Papilloma, Squamous Cell *(B)*

Histogenesis

Can originate from transitional, respiratory or olfactory epithelium, or from the squamous epithelium of the nasal vestibule.

Diagnostic Features

- Usually exophytic mass of uniform, regularly arranged squamous cells with papillary or filiform structures
- Epithelial cells cover vascularized stalk of connective tissue
- Intact basement membrane
- Occasionally growth beneath the mucosal surface ("inverted" or "endophytic" papilloma)
- Lesion is covered by squamous surface cells which can contain only keratohyaline granules, or are highly keratinized

Differential Diagnosis

Hyperplasia, Epithelial (Squamous Cell) or Metaplasia, Squamous Cell
Usually multifocal and bilateral. No or minimal exophytic growth, does not form papillary projections. No vascularized stalk of connective tissue

Adenoma
No or minor squamous cell areas

Carcinoma, Squamous Cell
Destruction of the basement membrane. Cellular atypia and disorientation, frequent mitoses. Presence of further signs of malignancy such as invasive growth and/or metastases

Comment

Transitional and respiratory epithelial cells can be sometimes observed as a minor component.

References

(See 6, 7, 17, 62, 63)

Adenocarcinoma *(M)*

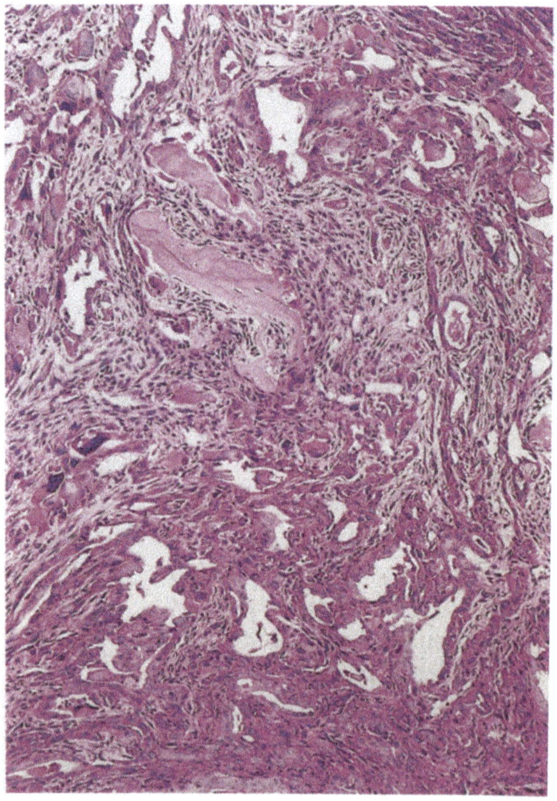

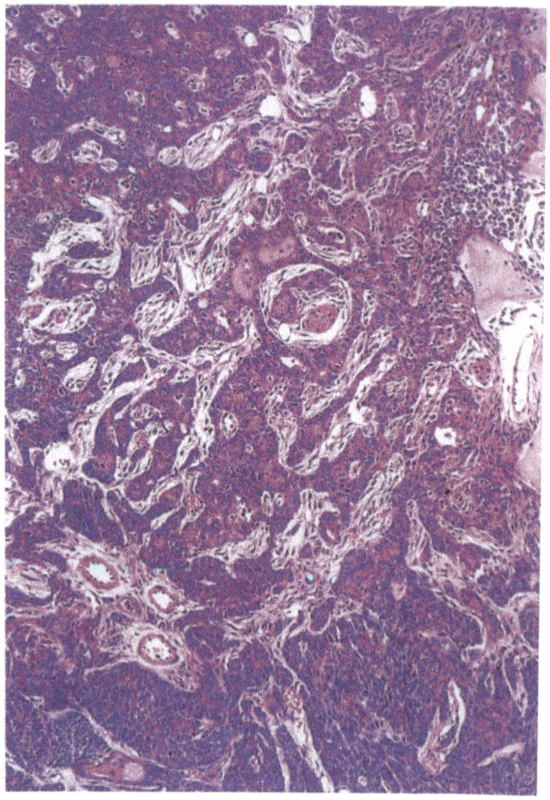

Histogenesis

Transitional, respiratory or glandular epithelium, or olfactory Bowman's gland.

Diagnostic Features

- Localized in the anterior nasal cavity (originating from subepithelial glands), in the posterior nasal cavity (often originating in the mucosa of the ethmoturbinates), or by malignant change occurring in an adenoma
- Solid, pseudoglandular, papillary or tubular formations
- Lumina can be filled with mucosubstances
- Large cuboidal to columnar, or anaplastic cells
- Loss of polarity of the epithelium
- Can show only penetration of the basement membrane or invasion of the surrounding bone, cribriform plate and/or olfactory lobes of the brain
- Invasion of the cerebrum, and/or metastases to regional lymph nodes and/or lungs can occur
- Areas of squamous differentiation can be present
- Well-differentiated adenocarcinomas have prominent glandular or cystic structures lined by more regular secretory cells

Fig. 15 *(above)*. Poorly differentiated adenocarcinoma of the nasal cavity with giant tumor cells. C57BL/6NCr mouse, female, treated, age 46 weeks. H&E

Fig. 16 *(below)*. Adenocarcinoma of Bowman's glands of the nasal cavity. C57BL/6NCr mouse, female, treated, age 44 weeks. H&E

- Poorly-differentiated adenocarcinomas have masses of anaplastic cells, inconspicuous glandular structures, a high degree of nuclear pleomorphism and often abnormal mitotic figures

Differential Diagnosis

Adenoma
Absence of signs of malignancy such as penetration of the basement membrane, invasive growth, and metastases. Glandular structures, when present, usually round and lined by well-differentiated tall columnar cells, with basally located nuclei, which do not display marked cytologic atypia

Carcinoma, Adenosquamous
The neoplastic squamous cell portions of the adenosquamous carcinoma show signs of malignancy such as frequent mitoses, cellular and/or nuclear atypia, and/or invasion into surrounding tissues, lymphatics, and/or vessels. Metaplastic squamous cell portions within an adenocarcinoma have more benign characteristics and are not as prominent

Carcinoma, Neuroepithelial
Neurogenic structures, i.e., plexiform intercellular fibrils ("neurofibrils") or rosettes, should be demonstrable

Comment

Location of the tumor usually determines the type of adenocarcinoma (transitional or respiratory epithelial). Transitional tumors are mostly found in the lateral meatus of the anterior (proximal) aspect of the nasal cavity. Occasionally, differentiation between poorly differentiated adenocarcinoma and neuroepithelial carcinoma is only possible by electron microscopic analysis. When a neurogenic component is not demonstrable, a diagnosis of adenocarcinoma should be favored over one of a neuroepithelial carcinoma.

References

(See 6, 8, 17, 21, 40, 62, 63, 72)

Carcinoma, Adenosquamous *(M)*

Histogenesis

Respiratory epithelial cells, ductal cells of submucosal glands, or basal and sustentacular cells of the olfactory epithelium, and from metaplastic areas of these epithelial cells.

Diagnostic Features

- Portions of the tumor consist of malignant squamous cells
- Other portions reveal a malignant glandular part
- Squamous components of the tumors can show typical keratin pearl formation
- Tumor can contain undifferentiated epithelial cells
- Presence of evidence of malignancy such as frequent mitoses, cellular and/or nuclear atypia, and/or invasion into surrounding tissues, lymphatics, vessels, and/or metastases

Differential Diagnosis

Papilloma, Squamous Cell
This tumor type can be composed of a mixture of squamous and mucus-producing cells (mucoepidermoid type), but evidence of malignancy such as marked cytologic atypia, penetration of the basement membrane, invasive growth, and/or metastases are absent. Squamous proportions predominate in squamous cell papilloma

Adenocarcinoma
In the case of areas of squamous cell metaplasia within adenocarcinoma, the differentiation might cause difficulties. Adenocarcinomas with squamous metaplasia have a large neoplastic glandular part and minor squamous cell portions with more regular (benign) characteristics

Carcinoma, Squamous Cell
Sometimes this tumor has enclosed respiratory or submucosal glandular epithelial cells. The glandular component is composed of normal-appearing, non-neoplastic epithelial cells entrapped and surrounded by nests of neoplastic squamous cells

References

(See 6, 17, 63)

Carcinoma, Squamous Cell *(M)*

Histogenesis

Can originate from squamous differentiation of transitional, respiratory or olfactory epithelial cells, epithelial cells of the subepithelial glands, or from the squamous epithelium of the nasal vestibule.

Diagnostic Features

- Most often in the anterior nasal cavity, arising in the lateral walls, septum, naso- and maxilloturbinates, and in the ethmoturbinates
- Composed of solid, often branching, cords or masses of cells with various degrees of anaplasia
- Shape and size of cells irregular, large and polygonal, or flattened and stratified
- Cytoplasm eosinophilic and granular to hyalinized due to the high keratin content. Cells can contain only keratohyaline granules or form keratin pearls
- Presence of signs of malignancy such as frequent mitoses, cellular and/or nuclear atypia, and/or invasion into surrounding tissues
- Well-differentiated squamous cell carcinomas are predominantly composed of cells with prominent intercellular bridges, normal keratinization, minimal nuclear atypia and a low mitotic index
- Poorly differentiated squamous cell carcinomas have cells with few intercellular bridges, abnormal keratinization (dyskeratosis), nuclear and cytoplasmic atypia and abnormal mitotic figures. Spindle-shaped cells might be prominent

Differential Diagnosis

Metaplasia, Squamous Cell or Hyperplasia, Epithelial (Squamous Cell) or Papilloma, Squamous Cell

Absence of characteristics of malignancy such as penetration of the basement membrane, invasion into surrounding tissues, lymphatics, vessels, bronchi, and metastases. Minimal cellular atypia and disorganization

Carcinoma, Adenosquamous

The differentiation can cause difficulties in the case of squamous cell carcinoma with enclosed glandular components of the underlying tissue. This glandular part is composed of normal-appearing, non-neoplastic epithelial cells, whereas the neoplastic glandular part of the adenosquamous carcinoma reveals evidence of malignancy, e.g., frequent mitoses, and cellular and/or nuclear atypia, and/or invasion into surrounding tissues, lymphatics, and/or vessels

Comment

In the case of poorly differentiated tumors, ultrastructural (tonofilaments) and/or immunohistochemical (keratins) verification might be necessary to determine the origin.
A poorly differentiated non-keratinizing squamous cell carcinoma might be misdiagnosed as a poorly differentiated adenocarcinoma.

References

(See 6, 7, 17, 38, 62, 63, 71)

Figs. 17-19 see p. 104

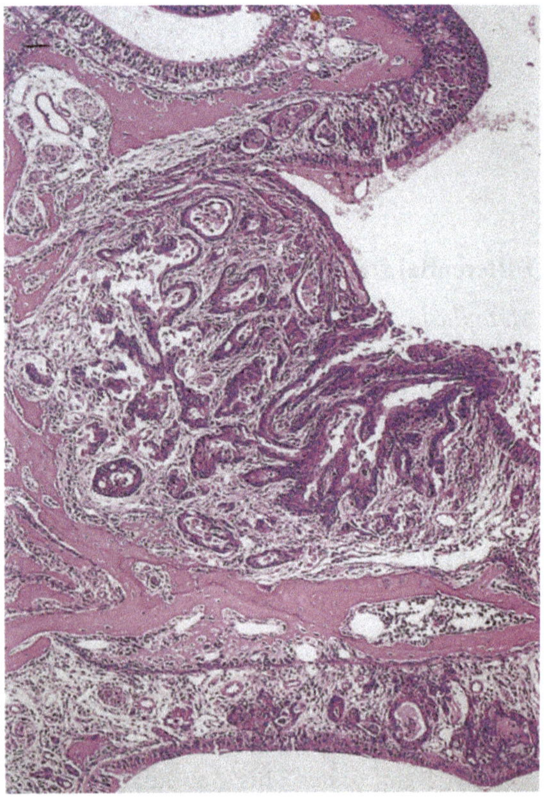

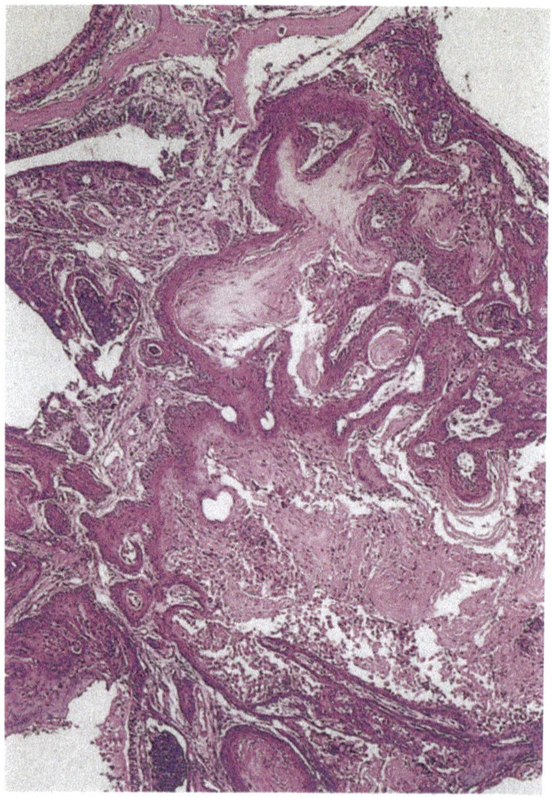

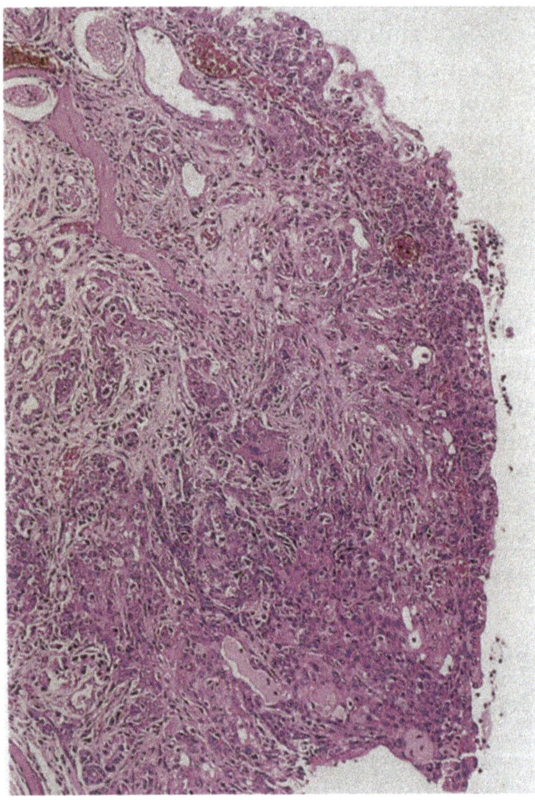

Fig. 17 *(upper left)*. Squamous cell carcinoma of the nasal cavity. Swiss mouse, female, treated, age 62 weeks. H&E

Fig. 18 *(upper right)*. Squamous cell carcinoma of the nasal cavity. C57BL/6NCr mouse, female, treated, age 43 weeks. H&E

Fig. 19 *(below)*. Squamous cell carcinoma of the nasal cavity. V-Ha-ras Oncomouse, male, untreated, age 38 weeks. H&E

Carcinoma, Neuroepithelial (*M*)

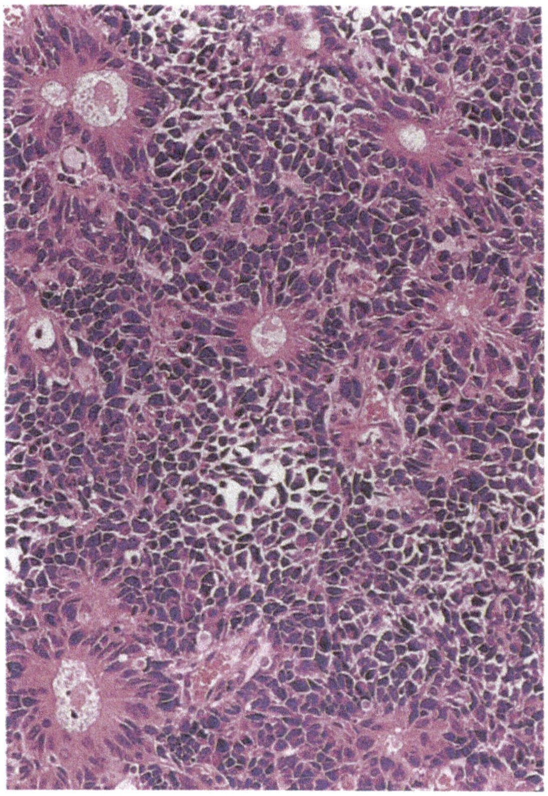

Fig. 20. Neuroepithelial carcinoma of the nasal cavity. C57BL/6NCr mouse, female, treated, age 45 weeks. H&E

Synonym(s). Neuroepithelioma, olfactory; neuroblastoma, olfactory; esthesioneuroblastoma

Histogenesis

Olfactory epithelium (sustentacular cells, basal cells, immature sensory cells and possibly ductal cells of Bowman's glands).

Diagnostic Features

- Arises from the olfactory epithelium, which covers a small dorsocranial portion of the septum and much of the posterior nasal cavity
- Frequently there is compartmentalization of sheets of neoplastic cells into lobules by fibrovascular septa; rosettes and pseudorosettes can be present
- Small round or columnar cells with poorly defined, pale-staining cytoplasm
- Round to oval basally-located nuclei which do not display marked cytologic atypia
- Distinct, sharply-defined nuclear chromatin
- True rosette (Flexner-Wintersteiner rosette): tumor cells surround an open central lumen bounded by distinct cell membranes, thus mimicking glandular structures
- Pseudorosette (Homer-Wright rosette): tumor cells arranged in a circle around a small central lumen which is filled with amorphous or tangled fibrillary material
- Plexiform intercellular fibrils
- Areas of anaplastic cells can be present

Differential Diagnosis

Adenocarcinoma

Absence of rosettes or plexiform intercellular fibrils. A useful distinguishing feature between true rosettes sometimes present in neuroepithelial carcinomas and the acinar spaces in adenocarcinomas is that in neuroepithelial carcinomas the nuclei of cells forming the rosettes blend peripherally with the uniform population forming the rest of the tumor

Comment

Often invasion of the ethmoid bone and brain. Frequency and morphology of rosette structures highly variable.
Whether tumors arising from olfactory epithelium are more appropriately classified

under the term neuroepithelial carcinoma or olfactory neuroblastoma, cannot be resolved on the basis of current information. The general term neuroepithelial carcinoma is used here because it allows for the origin of neoplasms from sensory or sustentacular cells, whereas olfactory neuroblastoma implies origin only from neurogenic components. There is clearly a need to develop procedures to distinguish neuroepithelial tumors of neurogenic origin (neuroblastoma, esthesioneuroblastoma), from those of non-neurogenic cell types, especially sustentacular cells and their precursors.

When neither rosettes (true or pseudo) nor plexiform intercellular fibrils are demonstrable, a diagnosis of adenocarcinoma should be favored over one of neuroepithelial carcinoma. Ultrastructural analysis of a neuroepithelial carcinoma should reveal some clear characteristics of olfactory epithelium such as olfactory vesicles, cilia, and microtubules.

The histogenic relationship of Bowman's glands to the other components of the nasal epithelium has not been clearly defined, but the working group believes their origin to be non-neurogenic. Therefore, carcinomas originating from Bowman's glands should be classified with adenocarcinomas of the nasal cavity.

Up to now, most cases were negative for immunohistochemical reactions with antibodies against intermediate filaments.

References

(See 6, 17, 18, 21, 26, 54, 63)

Hyperplasia, Epithelial *(H)*

Mucous (goblet) cell • Neuroendocrine cell
Respiratory epithelial • Squamous cell
With cellular atypia (dysplasia)

Histogenesis

Respiratory epithelium, or submucosal glands. Squamous cells of laryngeal epithelium. Neuroendocrine cells of bronchioles.

Diagnostic Features

Mucous (Goblet) Cell

- Replacement of respiratory epithelium by mucous cells either as single-layer or appearing pseudostratified

Neuroendocrine Cell

- Clusters of uniform small cells which protrude into the lumen of bronchioles. Cells have scant cytoplasm and nuclei with stippled chromatin
- Definitive identification of neuroendocrine cells requires immunostaining for specific markers. For the mouse the best markers appear to be protein gene product 9.5 (PGP), calcitonin, calcitonin gene-related peptide (CGRP) and helodermin

Respiratory Epithelial

- Diffuse or focal increase of epithelial cells, possibly forming luminal protrusions
- Papillary hyperplasia present when small fronds of epithelium with simple connective tissue core project into the airway lumen

Squamous Cell (Larynx only)

- Focal or diffuse increase of squamous cell layers of the laryngeal epithelium, which can be keratinized
- No or minimal exophytic growth
- Area of cells with more prominent nucleoli and more abundant cytoplasm
- Mild nuclear atypia and pleomorphism can be detected
- Medial to arytenoid cartilage can be preferential site

With Cellular Atypia (Dysplasia)

- Variation in cellular/nuclear size and shape of lining cells
- Increased nuclear basophilia
- Increased nuclear invagination
- Bi- or multinucleated cells

Differential Diagnosis

Papilloma
Characterized by expansile/compressive growth of papillary structures

Adenocarcinoma
Increased cellular atypia, destruction of the basement membrane and/or other signs of malignancy such as invasion and/or destruction of adjacent tissues

Metaplasia, Squamous Cell
Replacement of respiratory epithelium with regular squamous epithelium that frequently has superficial keratinization

Carcinoma, Squamous Cell
Growth pattern usually comprises cell clusters with central keratinization, or without overt keratinization showing distinct intercellular bridges, and with cytologic and behavioral features common to squamous cell carcinomas

Comment

Bronchi and bronchioles are both included here as anatomic locations because of differences in nomenclature of conducting airways in rodent lungs, i.e. whether to classify according to type of epithelium or to presence/absence of cartilage.

Papillary respiratory epithelial hyperplasia is frequently associated with papillomas observed in transgenic mice, or can be induced by multiple N-nitrosodiethylamine treatment, or intratracheal carcinogen applications.
Small clusters of neuroendocrine cells that protrude slightly into the airway lumen are normal histologic features and must be distinguished from the larger hyperplastic lesions.
In neuroendocrine cells of the larynx and trachea, serotonin could not be demonstrated, whereas in the lungs, serotonin containing cells were seen.
Epithelial hyperplasia of mucous cells can occur in association with subacute and chronic inflammation caused by infectious agents or due to the exposure to local carcinogenic or non-carcinogenic irritants.
Epithelial hyperplasia with cellular atypia is observed primarily following the systemic application of nitrosamines or after intratracheal carcinogen instillation. Cellular atypia can also be a common feature of neoplasia and can be present without hyperplasia.
Secretory cells of trachea, bronchi, and bronchioles can also show eosinophilic globules (eosinophilic intracytoplasmic proteinaceous accumulations), a degenerative change similar to the epithelium of the nasal cavity (dilated endoplasmic reticulum containing proteinaceous material). The eosinophilic globules do not constitute a dysplastic alteration, but can be a response to tissue irritation, and increases with age.
Further subdivisions of hyperplasia could be used, especially for experimental studies with agents targeting specific cell types (e.g. Clara cells), but are not essential for routine toxicologic pathology.

References

(See 24, 35, 37, 41, 51, 55, 71)

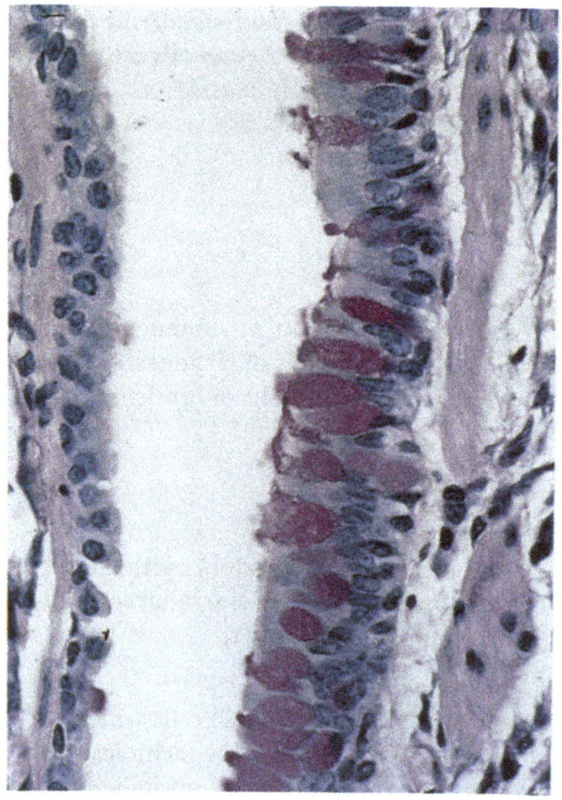

Fig. 21 *(above)*. Epithelial hyperplasia, mucous (goblet cell) cell, in a bronchiole. PAS

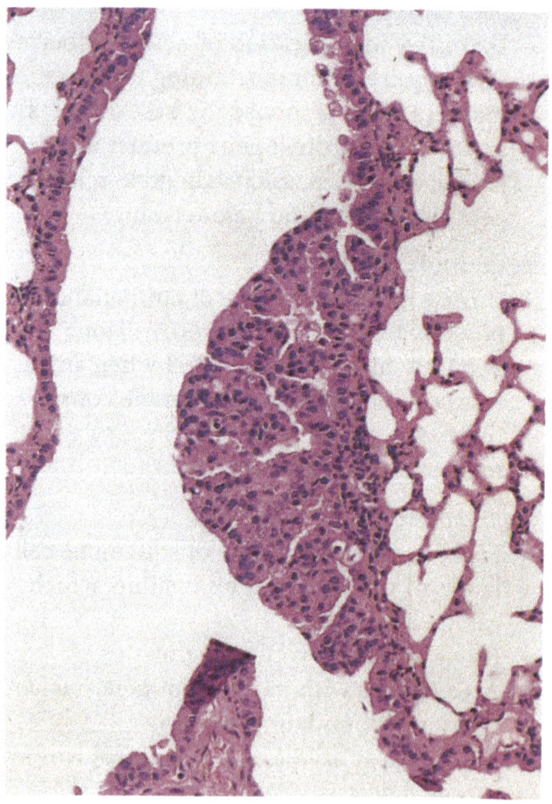

Fig. 22 *(below)*. Papillary epithelial hyperplasia of a bronchiole. NMRI mouse, male, untreated, 105 weeks. H&E

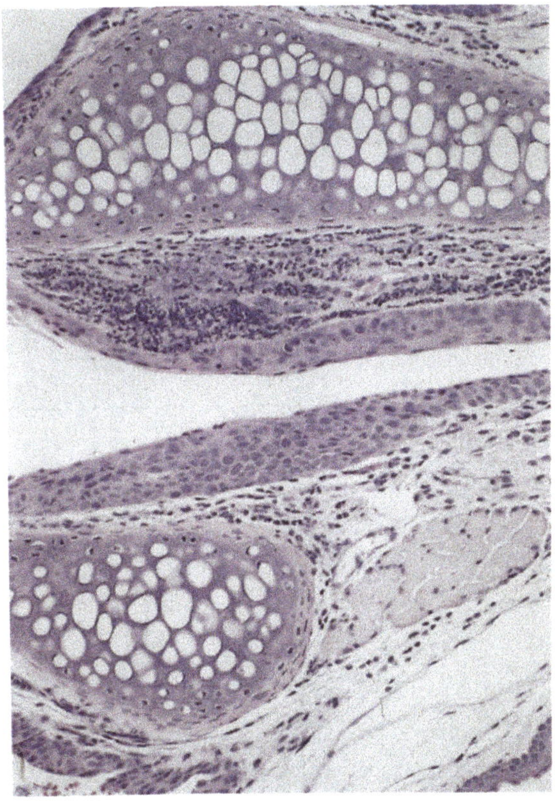

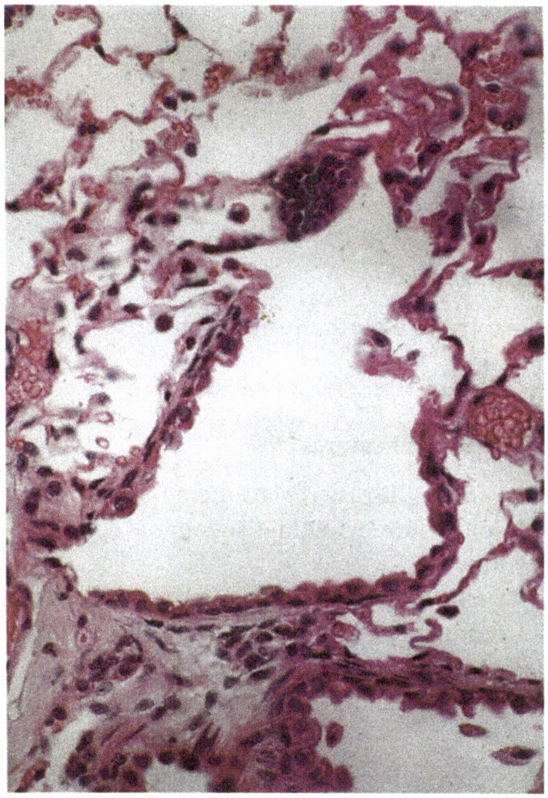

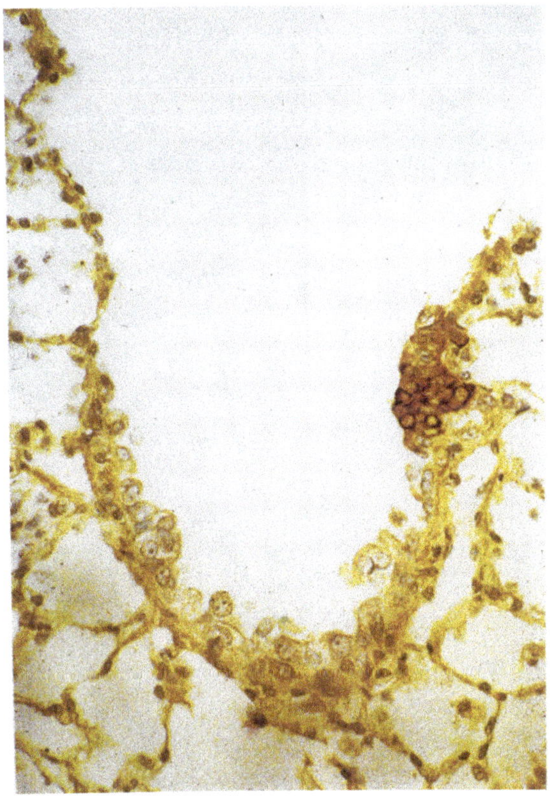

Fig. 23 *(upper left).* Squamous cell hyperplasia of the larynx at the site of the arytenoid cartilage of a treated mouse. In addition, inflammatory cell infiltration. Treated mouse. H&E (photograph by kind permission of Dr. Gary Boorman, National Toxicology Program)

Fig. 24 *(upper right).* High power view of focus of neuroendocrine cells in a bronchiole of a mouse. Treated mouse. Depending on the strain of mouse and experimental condition this might be either considered to be at the upper limit of normal size for a neuroepithelial body or mild neuroendocrine hyperplasia. H&E

Fig. 25 *(below).* High power view of focus of neuroendocrine cells in a bronchiole of a mouse. Positive chromogranin immunostaining

Metaplasia, Squamous Cell *(H)*

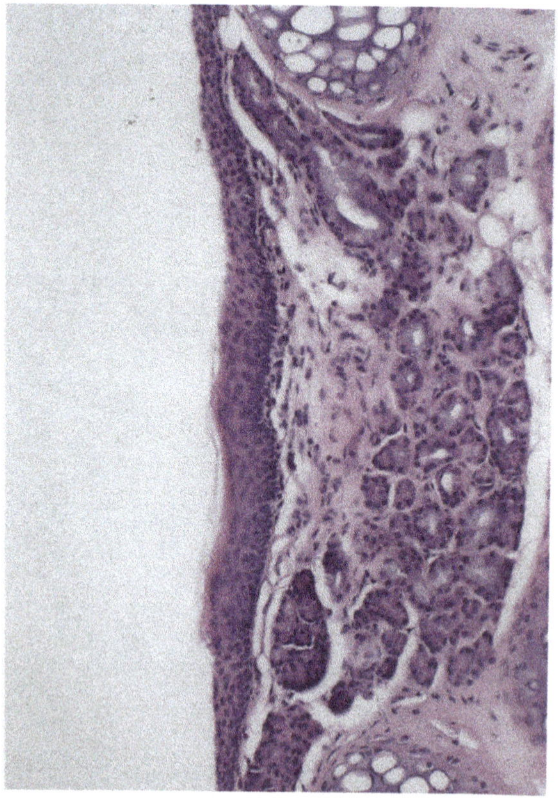

Histogenesis

Respiratory epithelium, submucosal glands.

Diagnostic Features

- Focal or diffuse replacement of respiratory epithelium by regular squamous epithelium
- Surface cells can be highly keratinized and desquamating, non-keratinized or contain keratohyaline granules
- Cells can show distinct intercellular bridges
- Frequently associated with pseudostratified/stratified hyperplasia

Differential Diagnosis

Papilloma

Projection above the surface of the respiratory epithelium or extension into the lumen of the ducts of the submucosal glands. Presence of delicate cores of vascular mesenchymal stroma covered by marked proliferative thickening of the epithelium

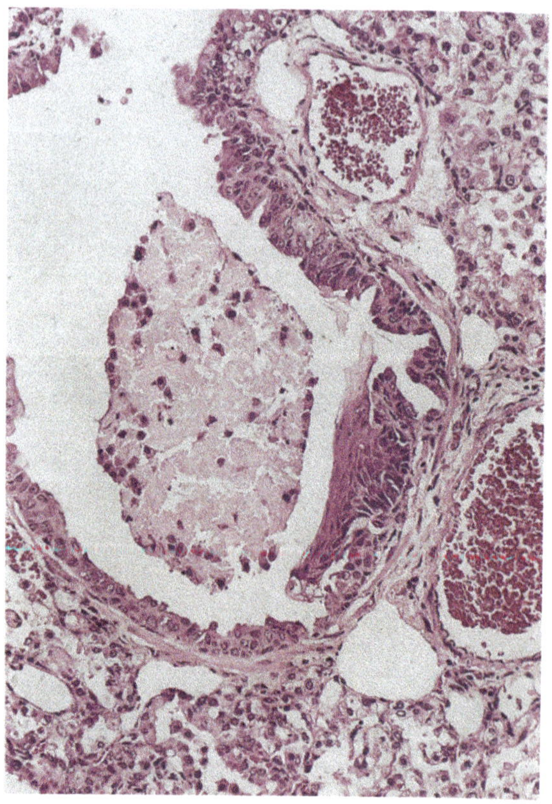

Fig. 26 *(above).* Squamous cell metaplasia at the base of epiglottis of the larynx. Treated mouse, female. H&E (photograph by kind permission of Dr. Gary Boorman, National Toxicology Program)

Fig. 27 *(below).* Squamous cell metaplasia of a bronchiole. NMRI mouse, female, treated, age 45 weeks. H&E

Carcinoma, Squamous Cell
Characterized by more pronounced cellular atypia and disorganization. Invasion of basement membrane and adjacent tissue structures

Comment

The laryngeal sites most susceptible to squamous cell metaplasia in mice and rats are at the base of the epiglottis and in the laryngeal pouch where the epithelium overlies a small cluster of submucosal glands. Accurate planes of section and precise anatomic localization are necessary for proper evaluation of whether or not squamous cell metaplasia is present.

References

(See 17, 20, 24, 35, 40, 41, 71)

Papilloma *(B)*

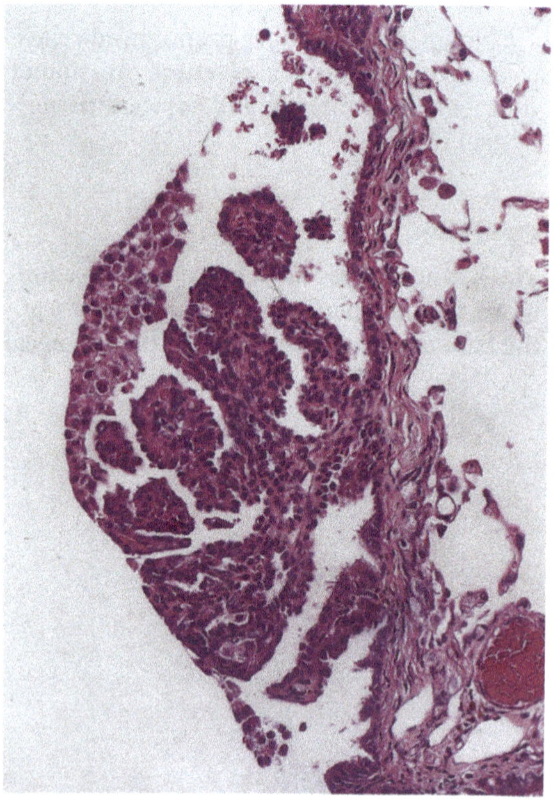

Fig. 28. Papilloma of a bronchiole. NMRI mouse, female, treated, age 42 weeks. H&E

Histogenesis

Respiratory epithelium or respiratory epithelium which has undergone squamous metaplasia.

Diagnostic Features

- Airway is expanded or distorted by growth of branching papillary structures with central connective tissue stalk and peripheral lining of cuboidal/respiratory epithelial cells
- Seen to arise from airway epithelium if plane of section appropriate
- Frequently associated with papillary hyperplasia in some experimental studies
- If papilloma originates in terminal bronchioles, there can be growth by expansion into alveolar parenchyma
- No clear evidence of invasion of adjacent structures
- The branching connective tissue stalk is usually covered by various proportions of cuboidal or columnar respiratory epithelium, but might occasionally be covered by squamous cells

Differential Diagnosis

Hyperplasia, Epithelial or Metaplasia, Squamous Cell
Absence of central connective tissue stalk or absence of luminal expansion/distortion by branching growth

Adenocarcinoma
Characterized by increased cellular pleomorphism and invasion of adjacent pulmonary structures

Carcinoma, Squamous Cell
Destruction of the basement membrane. Cellular atypia and disorientation, more frequent mitoses. Presence of further signs of malignancy such as invasive growth and/or metastases

Adenoma, Bronchiolo-Alveolar (Papillary) or Carcinoma, Bronchiolo-Alveolar of the Lung
These neoplasms arise in alveolar parenchyma. The adenoma can extend with papillary growth from an alveolar duct into a terminal bronchiole. The carcinoma is more invasive and destructive

Comment

Differentiation between a papilloma originating in a small bronchiole and a papillary bronchiolo-alveolar adenoma might not be possible (or necessary). Unless there is clear evidence that the tumor arose from airway epithelium, it should be classified in the bronchiolo-alveolar category.

Diagnosis of bronchiolar papilloma is provided for here because in some experimental studies papillomas arising from the epithelium of small airways can be detected and are therefore separable from the usual types of bronchiolo-alveolar tumors such as occur spontaneously in the lungs of mice.

Inflammatory polyps need to be kept in mind for differential diagnosis but these have not been reported for mice. They have occurred in the larynx of rats exposed to cobalt sulfate aerosols.

References

(See 11, 12, 17, 42, 61, 67)

Adenocarcinoma *(M)*

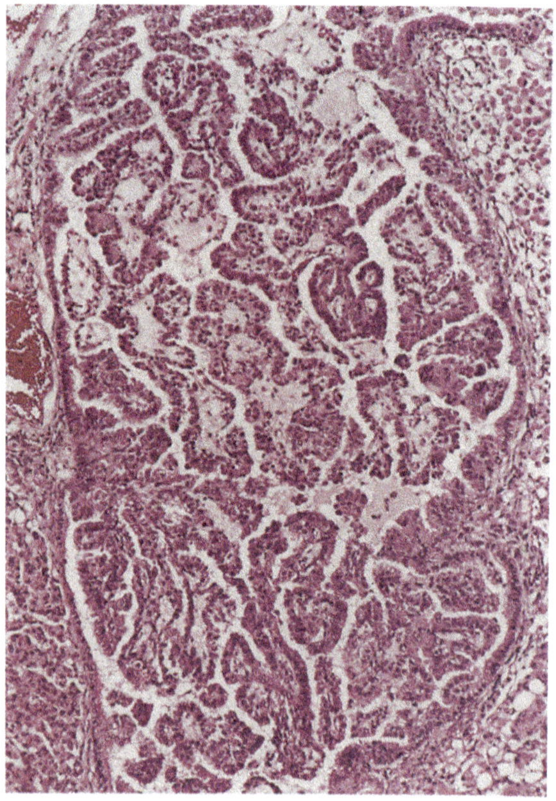

Fig. 29. Papillary adenocarcinoma of a bronchus. Treated mouse. H&E

Histogenesis

Respiratory epithelium.

Diagnostic Features

- Definitive evidence of origin from a conducting airway is required
- Shows evidence of invasion of basement membrane or adjacent pulmonary structures
- Foci of mucinous cell differentiation can be present
- Papillary growth in early stages with central connective tissue stalk lined by cuboidal to columnar or pleomorphic epithelium
- Irregular tubular/glandular structures can be present
- Cytologic features of malignancy and possible evidence of stromal invasion and/or destruction of airway wall

Differential Diagnosis

Papilloma
Characterized by fairly uniform cuboidal cells without indication of tissue invasion

Carcinoma, Squamous Cell
Growth pattern usually comprises cell clusters with central keratinization; in the absence of overt keratinization there are distinct intercellular bridges

Comment

The literature reports a few examples of these tumors, for instance in transgenic mice, or chemically induced.

References

(See 13, 17, 42, 61)

Carcinoma, Squamous Cell *(M)*

Histogenesis

Respiratory epithelium, which has undergone squamous metaplasia.

Diagnostic Features

- Growth pattern in cell clusters or irregular structures with central keratinization (keratin pearls), or without overt keratinization but forming distinct intercellular bridges Cellular debris and necrosis can be common as well as the presence of inflammatory cells, in particular neutrophils
- Cytologic features common to squamous cell carcinoma, i.e., dysplasia to anaplasia
- Cells can frequently be quite pleomorphic, including the formation of very large cells ("giant cells") 60 μm or more at the largest diameter
- Penetration of the basement membrane and invasion of adjacent tissues

Differential Diagnosis

Metaplasia, Squamous Cell
Shows no or minimal exophytic growth. Does not form papillary projections. Regular, orderly progression of keratinization with little or no dysplasia and atypia. No vascularized stalk of connective tissue stroma

Papilloma
Intact basement membrane. Absence of signs of malignancy such as frequent mitosis, cellular atypia, and invasion into surrounding tissues, lymphatics, vessels, and bronchi. Exophytic growth

Adenocarcinoma
Composed of papillary strands lined by cuboidal or pleomorphic cells, usually not exhibiting squamous cell metaplasia

Carcinoma, Adenosquamous of the Lung
Neoplasm composed of malignant squamous as well as non-squamous, glandular elements

Metastases
Metastases from squamous cell carcinoma with its primary at another organ site mostly perivascular, mostly multifocal

Comment

There are few reports of experimentally-induced squamous cell carcinomas in the airways of mice but spontaneous occurrence does not appear to have been reported.

References

(See 11, 17, 25, 61, 71)

Hyperplasia, Bronchiolo-Alveolar (*H*)

Alveolar • Bronchiolar (bronchiolization)

Synonym(s). Hyperplasia, bronchiolar/alveolar; hyperplasia, type II cell

Histogenesis

Alveolar type II cells and/or bronchiolar secretory cells.

Diagnostic Features

- Solitary or multiple, segmental (cone-shaped) foci of increased cellularity
- Lack of strongly convex or spherical border
- Background of bronchiolo-alveolar architecture still detectable
- Epithelial cells dominant cause of hypercellularity
- Epithelial cells usually single-layered
- Depending on the situation in which it arises (see comment) it is often possible to classify it into one of the following types

Alveolar

- Round to oval or cuboidal often/mostly hypertrophic alveolar type II cells with abundant eosinophilic cytoplasm prominently outlining alveolar walls. Increased basophilia can be associated with cuboidal cell shape
- Cytoplasm can be vacuolated
- Cells form a single layer that is contiguous throughout the area of hyperplasia
- Areas of alveolar hypercellularity can be single or multiple, round or cone-shaped if localized subpleurally. Peripheral margins are indistinct
- Associated with terminal airways by extension of peripheral growth
- Formation of solid cell clusters and papillary projections marks transition towards neoplasia
- Can be associated with influx of alveolar macrophages

Bronchiolar (bronchiolization)

- Alveolar walls are lined by cuboidal to tall columnar or possibly pleomorphic cells that might exhibit bronchiolar cell differentiation, e.g., formation of cilia, or Clara cell resemblance, or presence of mucous granules or eosinophilic globules
- Can be associated with squamous cell metaplasia
- Normal resident alveolar type II and type I cells can be interspersed
- Cells form a single layer or focal tufts of cells that can appear pseudostratified
- Areas of hyperplasia centered on terminal bronchioles, but in some planes of section the connection between the foci of bronchiolization and the terminal bronchiole might not be observed
- Peripheral borders indistinct as with alveolar cell hyperplasia
- Formation of increased cellular atypia, acini contiguously lined by single or stratified cells and basement membrane invasion mark neoplasia

Differential Diagnosis

Adenoma, Bronchiolo-Alveolar

More densely cellular rounded nodule. Obscuring of the normal alveolar architecture by solid to papillary tumor formations within alveoli and alveolar ducts

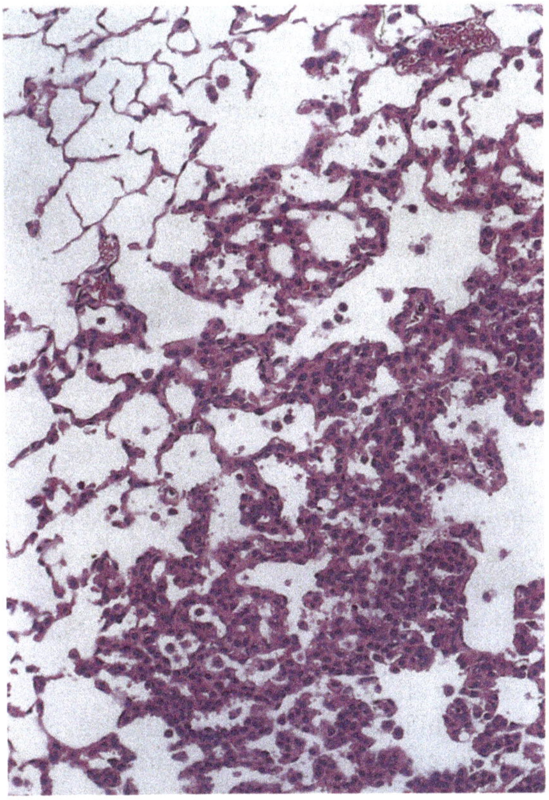

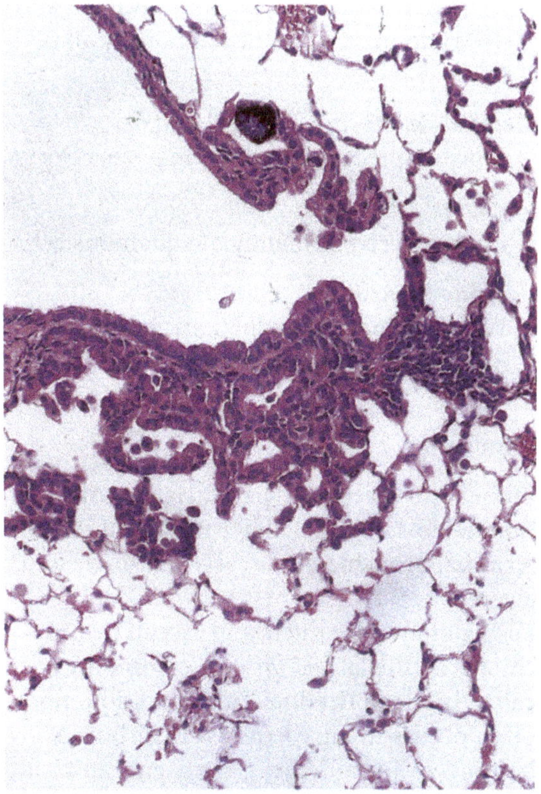

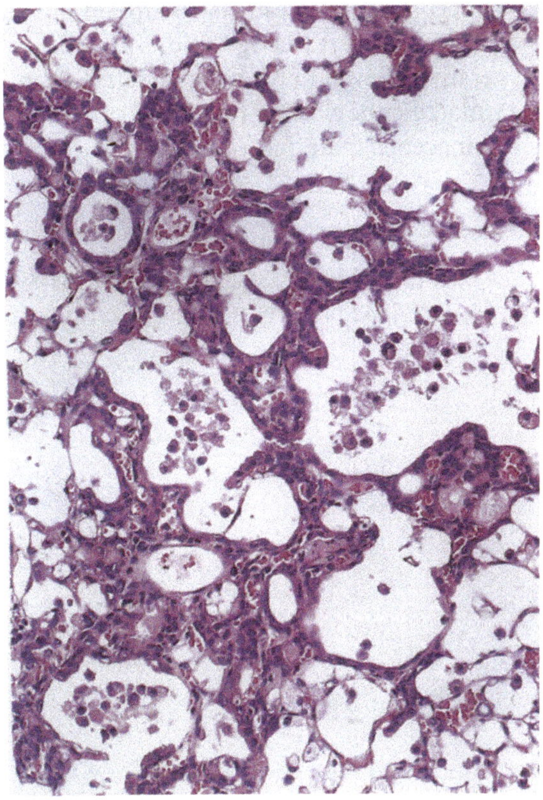

Fig. 30 *(upper left)*. Bronchiolo-alveolar hyperplasia, alveolar, of the lung. NMRI mouse, female, treated, age 38 weeks. H&E

Fig. 31 *(upper right)*. Bronchiolo-alveolar hyperplasia, bronchiolar, of the lung (bronchiolization). NMRI mouse, female, treated, age 45 weeks. H&E

Fig. 32 *(below)*. Mixed bronchiolo-alveolar hyperplasia of the lung. NMRI mouse, female, treated, age 45 weeks. H&E

Carcinoma, Bronchiolo-Alveolar
Increased cellular atypia and invasion and destruction of adjacent pulmonary structures

Metaplasia, Mucous (Goblet) Cell
Consists predominantly of mucous cells

Metaplasia, Squamous Cell
Consists predominantly of squamous cells

Carcinoma, Acinar
Well-developed glandular structures with invasion of basement membranes

Comment

Bronchiolo-alveolar hyperplasia, by convention, refers to epithelial hyperplasia distal to terminal bronchioles, i.e., alveolar ducts and alveoli.

Lining of alveolar ducts and alveoli by bronchiolar epithelial cell types is commonly called bronchiolization. This condition most often occurs in mice exposed to airborne irritants or intratracheal treatments and is a late sequelae to Sendai virus infections. The relative contribution of distal growth of preexisting bronchiolar epithelium and metaplasia (transdifferentiation) of alveolar type II cells is not clear. The relative contributions probably vary somewhat according to causal situation.

In the lungs of control mice, the separation of bronchiolization and alveolar type II cell hyperplasia is often fairly clear. In chronically inflamed lungs of mice, however, there is much more likely to be mixed bronchiolo-alveolar hyperplasia.

The differentiation of bronchiolo-alveolar hyperplasia from bronchiolo-alveolar adenoma is difficult or impossible in borderline cases. Arbitrary guidelines have been suggested, such as obliteration of three or more contiguous alveolar spaces as the criterion for an adenoma. This type of criterion is not so easy to apply in practice with the lungs as it is with solid organs.

It might not always be possible with routine light microscopic techniques to identify bronchiolar cells (especially Clara cells) and alveolar type II cells. To date, the best immunohistochemical markers are the 16-kDa-protein (CC16) for Clara cells and surfactant apoprotein A or C for type II cells. Surfactant apoprotein A has been used as a marker for type II cells although Clara cells can also be positive. Where cells are positive for surfactant apoprotein A and negative for CC16, however there is good presumptive evidence that they are of type II origin.

Eosinophilic globules (eosinophilic intracytoplasmic proteinaceous accumulations) can be a prominent feature associated with all types of hyperplasia. They can also be found in non-hyperplastic cell populations and are generally believed to be a degenerative change (dilated endoplasmic reticulum containing proteinaceous material).

References

(See 3, 4, 17, 19, 20, 22, 32, 45, 50, 51, 55, 57 59, 61, 69, 73, 80, 82)

Metaplasia, Mucous (Goblet) Cell *(H)*

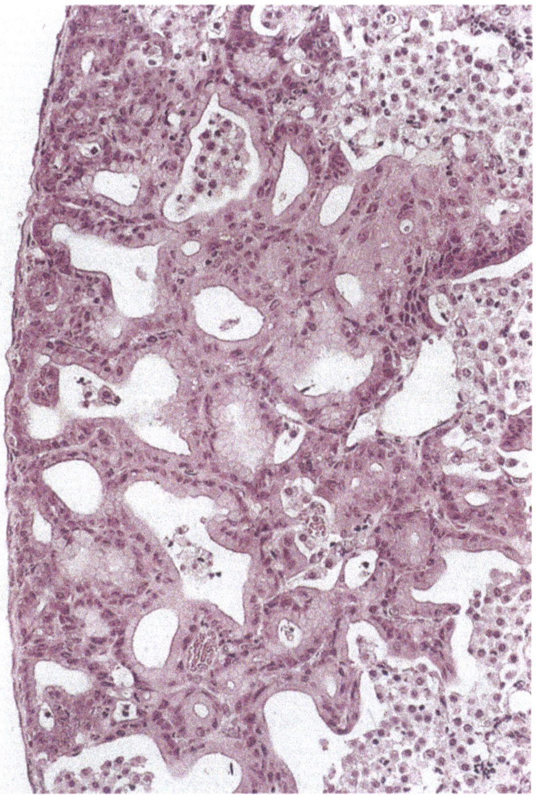

Fig. 33. Mucous cell metaplasia in the peripheral lung. NMRI mouse, female, treated, age 41 weeks. H&E

Histogenesis

Clara cells and/or alveolar type II epithelial cells which undergo mucous cell metaplasia.

Diagnostic Features

- Mucus-containing spaces lined predominantly by a single layer of mature mucous cells in the periphery
- Very low mitotic activity
- PAS/Alcian blue-staining enhances identification of mucous cells and mucus
- Usually accompanied by features of chronic inflammation and fibrosis
- Proliferation of other epithelial elements as described for bronchiolo-alveolar hyperplasia can be present
- Background of normal bronchiolo-alveolar architecture retained to various degrees, depending on amount of associated inflammation and fibrosis

Differential Diagnosis

Hyperplasia, Bronchiolo-Alveolar
Absence or insignificant amounts of mucous cells and intraluminal mucus (see comment)

Adenoma, Bronchiolo-Alveolar
More densely cellular rounded nodule. Obscuring of the normal alveolar architecture by papillary to solid tumor formations within alveoli and alveolar ducts. Inconspicuous or absent inflammatory components within tumor proper

Carcinoma, Acinar
Well developed glandular structures with invasion of basement membranes

Comment

Mucous cell metaplasia, when present, usually occurs in lungs of animals chronically exposed to airborne irritants and therefore frequently is associated with foci of bronchiolo-alveolar hyperplasia.
Small foci of mucous cell metaplasia can occur incidentally in untreated mice. Exaggerated mucous cell metaplasia has been observed as a rare lesion at the bifurcations of medium-sized airways in mice exposed to cobalt sulfate aerosols.
Mucous cell metaplasia can be associated with squamous cell metaplasia.

References

(See 17)

Metaplasia, Squamous Cell (*H*)

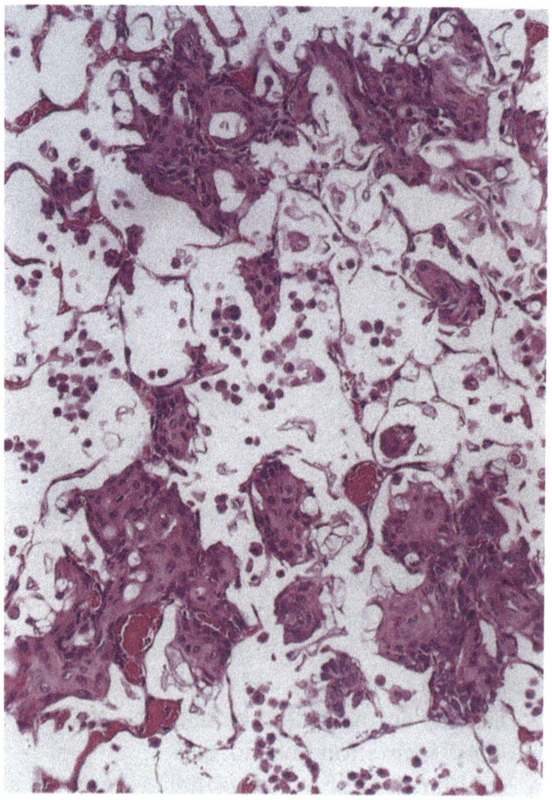

Fig. 34. Squamous cell metaplasia in the lung of a NMRI mouse. NMRI mouse, female, treated, age 45 weeks. H&E

Histogenesis

Clara cells and/or alveolar type II cells which undergo squamous cell metaplasia.

Diagnostic Features

- Replacement of alveolar epithelium by squamous cells, which can contain only keratohyaline granules, or can be highly keratinized
- More commonly multifocal
- Orderly progression from basal cells to keratinizing surface epithelium in areas of keratinization
- Rarely slight nuclear polymorphism and cellular atypia; occasional mitotic figures can be present
- Usually associated with Clara, mucous and/or ciliated cells

Differential Diagnosis

Carcinoma, Squamous Cell
Normal pulmonary architecture disturbed. Mostly singular nodule. Presence of signs of malignancy such as cellular and/or nuclear atypia, frequent mitoses, invasive growth, and/or metastases

Hyperplasia, Bronchiolo-Alveolar
Associated with insignificant extent of squamous metaplasia

Comment

Different usages of hyperplasia, metaplasia and transdifferentiation have been applied for changes in cell populations lining damaged bronchiolo-alveolar regions. The term metaplasia here is used for the presence of a mature cell type not normally resident at the junction of terminal bronchioles and alveolar ducts i.e., squamous or mucous cells.

References

(See 17, 20, 23, 59, 61)

Adenoma, Bronchiolo-Alveolar *(B)*

Solid • Papillary • Mixed

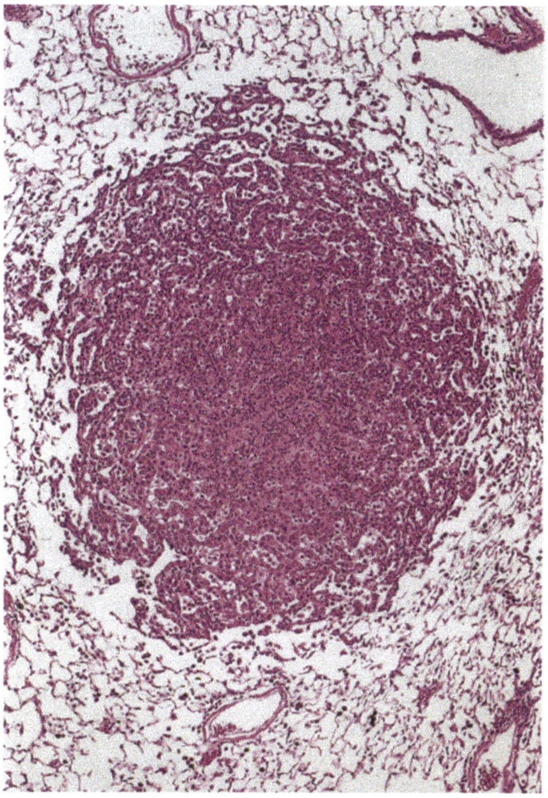

Synonym(s). Adenoma, type II cell

Histogenesis

It is generally believed that solid tumors are composed of cells expressing alveolar type II cell features and represent benign alveolar type II cell tumors. Papillary tumors, however, are considered either to be less well-differentiated type II cell tumors progressing towards a malignant phenotype or to be of Clara cell origin.

Diagnostic Features

- Frequently located at the lung periphery and usually small in size (less than 3–4 mm in diameter)

Solid

- Alveolar spaces obliterated by proliferating round to oval cells. Frequently these solid areas are surrounded by alveoli lined by hyperplastic alveolar type II cells, i.e., there might be no sharp demarcation between the tumor and normal parenchyma
- Cells usually have abundant eosinophilic cytoplasm that can appear granular or vacuolated. The nucleus usually is round to oval
- Mitotic figures are rare or absent
- Can extend into bronchiole from alveolar duct
- Compression is frequently observed

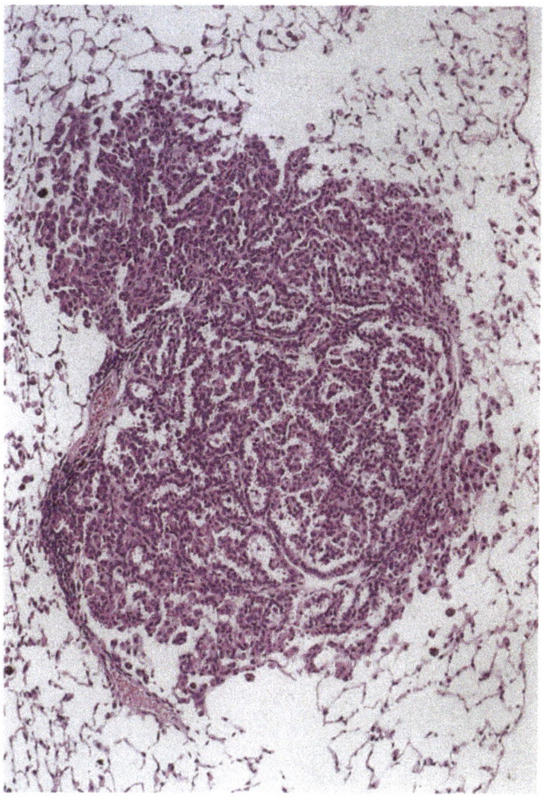

Fig. 35 *(above).* Solid bronchiolo-alveolar adenoma in the lung. NMRI mouse, female, treated, age 45 weeks. H&E

Fig. 36 *(below).* Papillary bronchiolo-alveolar adenoma in the lung. NMRI mouse, female, treated, age 44 weeks. H&E

Papillary

- Composed predominantly of delicate papillary structures lined by cuboidal to columnar cells that can be deeply basophilic
- Regular pattern (lack of distortion and focal variation in cell appearance)
- Tumors can show prominent tubular profiles, i.e., elongated lumina surrounded by cuboidal cells, depending on orientation of the histologic section
- Papillary structures sharply demarcated from surrounding alveolar parenchyma
- Can be associated with peripheral alveolar hyperplasia
- Large, sometimes foamy, macrophages can fill spaces between tumor cells
- Mitotic rate usually low. Degree of cellular pleomorphism low. There is no invasion and destruction of adjacent tissue

Mixed

- Papillary as well as solid structures in the same neoplasm

Differential Diagnosis

Hyperplasia, Bronchiolo-Alveolar (Alveolar)
Alveolar walls prominently outlined by single, continuous layer of hyperplastic alveolar type II cells. No significant obliteration of contiguous alveolar spaces by solidly packed cells or papillary fronds. No compression

Bronchiole-Papilloma
Can be morphologically identical to papillary bronchiolo-alveolar adenoma of lung parenchyma. A diagnosis of bronchiolar papilloma should only be made if there is clear evidence that the tumor arose from bronchiolar epithelium

Carcinoma, Bronchiolo-Alveolar
Increased cellular pleomorphism and/or atypia and/or mitotic figures; distortion of pulmonary parenchyma, possibly with local invasion/metastasis

Comment

It can be difficult to distinguish clearly bronchiolo-alveolar hyperplasia from bronchiolo-alveolar adenoma or bronchiolo-alveolar adenoma from bronchiolo-alveolar carcinoma and diagnosis can, therefore, at times be arbitrary.
There are no reports on malignant solid carcinomas in mice.
Compression is present as an artifact in immersion-fixed lung tissue but is not a significant feature in lungs fixed at near normal volume by intratracheal infusion of fixative.
Eosinophilic globules can be a prominent feature associated with bronchiolo-alveolar adenoma.

References

(See 1, 2, 3, 5, 13, 14, 15, 17, 22, 27, 28, 29, 31, 39, 42, 43, 45, 52, 53, 56, 57, 58, 60, 61, 65, 66, 69, 74, 75, 76, 77, 79, 80, 82)

Carcinoma, Bronchiolo-Alveolar *(M)*

Synonym(s). Adenocarcinoma, pulmonary

Histogenesis

Bronchiolo-alveolar carcinomas can originate from either alveolar type II or Clara cells. Depending on the experimental design and/or carcinogen, tumors can be induced selectively or different types can be observed in a single mouse.

Diagnostic Features

- Irregular nodular growth, moderately well to poorly circumscribed neoplasms (often greater than 3–4 mm in diameter) can occupy an entire lobe
- Composed of papillary connective tissue lined by cuboidal to columnar or pleomorphic cells, cytoplasm can contain glycogen and/or neutral lipids
- Tumors can exhibit areas of increased cytoplasmic basophilia and atypia indicating local expansion of less differentiated tumor cells
- Frequently associated with influx of macrophages into the tumor and in adjacent alveoli
- Indication of malignancy, such as destruction of parenchyma, invasion of bronchiolar walls, interstitial tissues and/or pleura, and/or dissemination through lymphatics, air spaces, and/or distant metastases
- Large tumors can exhibit necrosis, cholesterol clefts, fibrosis (especially at subpleural localization), and obliterating fibrosis of bronchioles
- Advanced stages of malignancy and invasion (e.g., to pleura) frequently associated with formation of spindle cells, round atypical cells, scirrhous changes, increased mitotic rate, hemorrhage and necrosis

Differential Diagnosis

Adenoma, Bronchiolo-Alveolar
: No invasive growth into pulmonary vessels or extrapulmonary parenchyma, no metastases. No or only slight cellular pleomorphism or atypia

Carcinoma, Acinar
: Predominantly glandular/acinar pattern representing tumor cells that utilize preexisting alveolar walls. Composed of cuboidal to columnar or pleomorphic cells without distinguishing features, or, more commonly, mixed with ciliated cells, or mucous cells. Large portions or the entire tumor can show differentiation in favor of single cell type, e.g., mucinous adenocarcinoma

Carcinoma, Adenosquamous or Carcinoma, Squamous Cell
: Tumor shows areas of malignant squamous cells or is composed entirely of characteristically squamous carcinoma

Mesothelioma, Malignant
: Differentiation of pleural spread of papillary bronchiolo-alveolar carcinoma from malignant mesothelioma of epithelioid type can be difficult. Metastases of bronchiolo-alveolar carcinomas (to pleura etc.) frequently are very anaplastic and can be misdiagnosed as malignant mesotheliomas of mesenchymal type

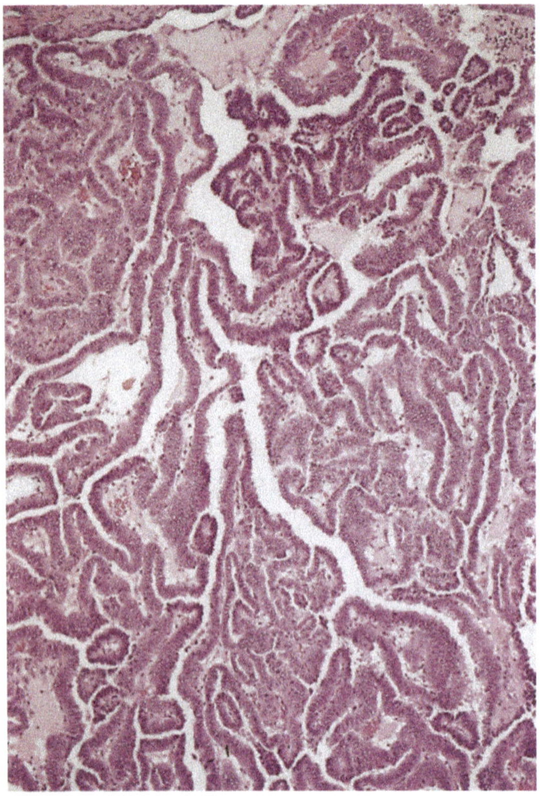

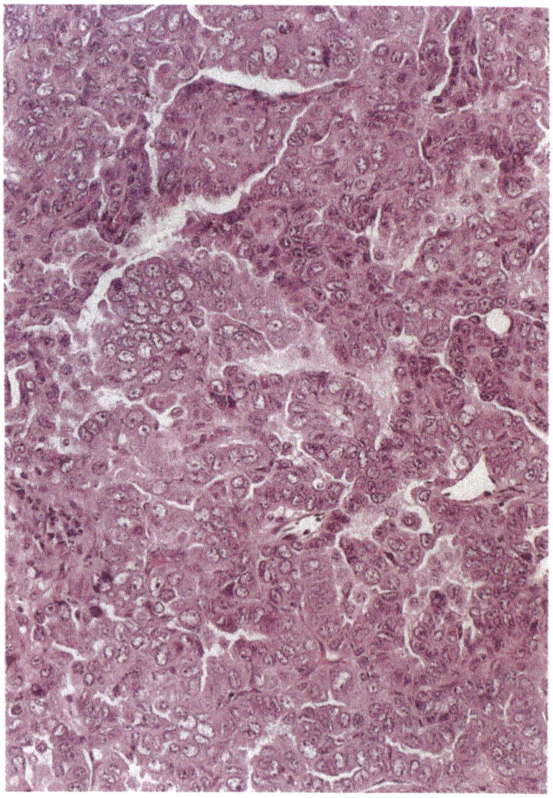

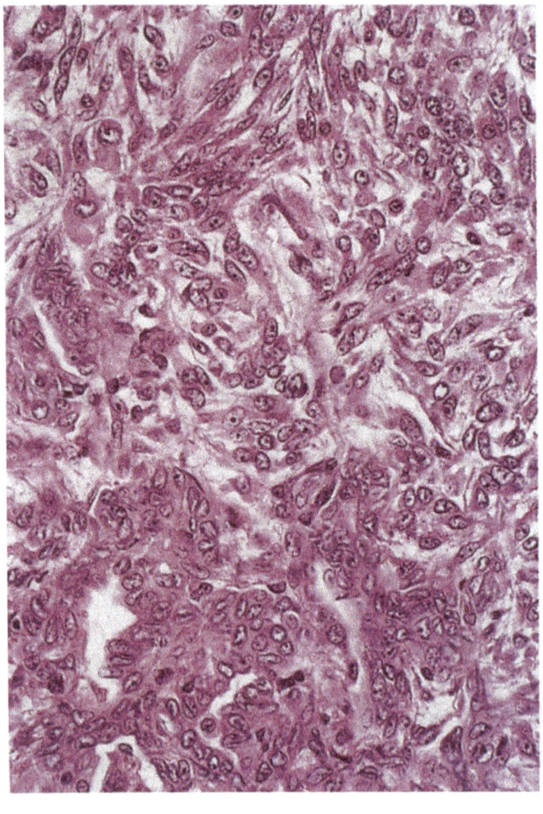

Fig. 37 *(upper left)*. Well-differentiated bronchiolo-alveolar carcinoma in the lung showing a papillary growth pattern. NMRI mouse, female, treated, age 88 weeks. H&E

Fig. 38 *(upper right)*. Papillary bronchiolo-alveolar carcinoma in the lung exhibiting area of giant multinucleated tumour cells. NMRI mouse, female, treated, age 88 weeks. H&E

Fig. 39 *(below)*. High power view of bronchiolo-alveolar carcinoma of the lung with area of spindle cells. H&E

Comment

Of the carcinomas the papillary type is the most common tumor observed spontaneously and/or induced. With regard to histogenesis, the majority of papillary bronchiolo-alveolar tumors are considered by some investigators to represent less well-differentiated type II cell tumors processing toward a malignant phenotype, whereas, few other investigators believe that papillary tumors arise from Clara cells.

References

(See 2, 3, 13, 14, 15, 17, 22, 27, 28, 29, 31, 39, 42, 43, 44, 45, 56, 57, 58, 60, 61, 65, 66, 69, 74, 75, 76, 79, 80, 81)

Carcinoma, Acinar *(M)*

Histogenesis

Acinar carcinomas are believed to originate from bronchiolar epithelial (Clara) cells.

Diagnostic Features

- Diffusely expansile with irregular margins or more circumscribed nodule
- Predominantly glandular/acinar pattern representing tumor cells that utilize preexisting alveolar walls
- Composed of cuboidal to columnar or pleomorphic cells without distinguishing features or more commonly, mixed with ciliated cells or mucous cells
- Large portions or the entire tumor can show differentiation in favor of single cell type, e.g., mucinous adenocarcinoma
- Cells can show variable extent of eosinophilic globules
- Lack of significant squamous cell metaplasia
- Neoplasms show clear indication of malignancy such as penetration of basement membranes and tissue destruction

Differential Diagnosis

Carcinoma, Bronchiolo-Alveolar
Mostly forms papillary structures causing distinct nodular growth. Composed of cuboidal to columnar cells, lacking ciliated, mucous or squamous cells

Carcinoma, Adenosquamous or Carcinoma, Squamous Cell
Tumor is composed of malignant squamous cells in part (adenosquamous carcinoma) or entirely (squamous cell carcinoma)

Comment

Naturally occurring acinar carcinomas are extremely rare. They can be induced by intratracheal instillation of methylcholanthrene or cutaneous treatment with N-nitrosobis-(2-chloroethyl)urea. The latter does not cause the induction of solid/papillary neoplasms in strains of mice with low spontaneous lung tumor incidence.

With respect to histogenesis, acinar carcinomas are believed to arise directly from airways or from Clara cells populating alveolar walls. Some investigators believe that parenchymal Clara cells have migrated from bronchioli, whereas others consider these to have arisen from metaplasia of alveolar type II cells.

Eosinophilic globules (eosinophilic intracytoplasmic proteinaceous accumulations) can be a prominent feature associated with acinar carcinoma. They can also be found in non-neoplastic and hyperplastic cell populations and are generally believed to be a degenerative change.

References

(See 59, 65, 66)

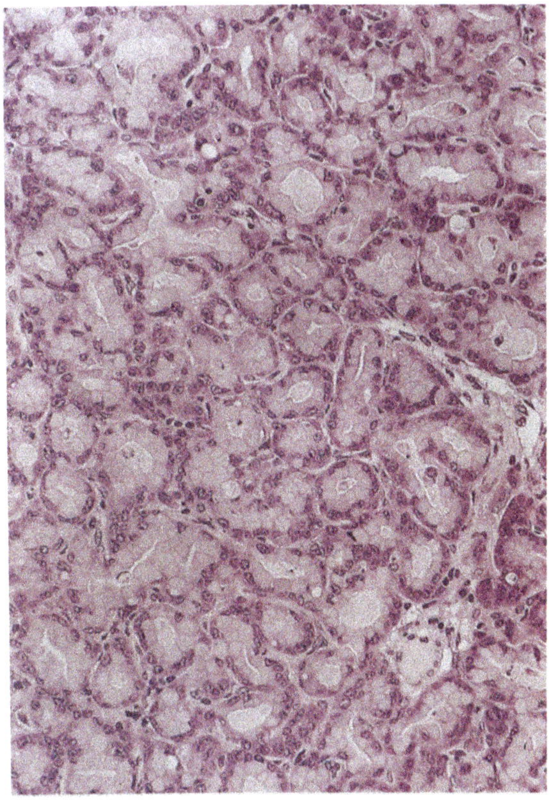

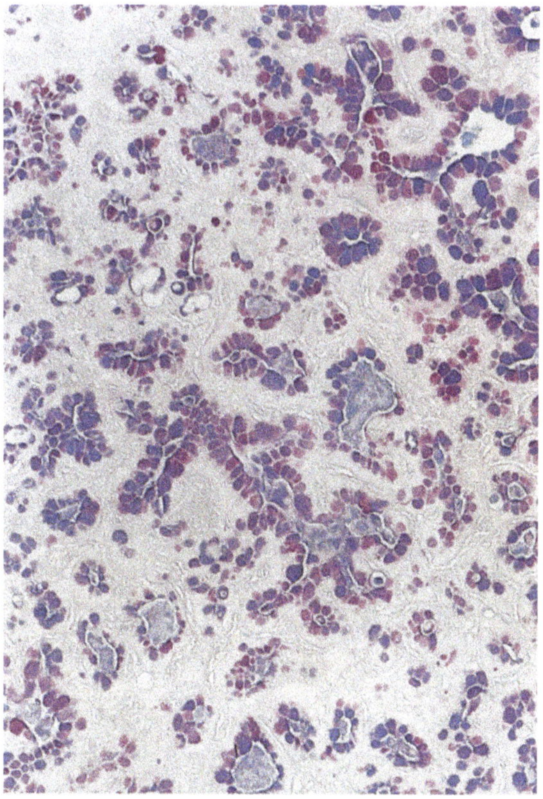

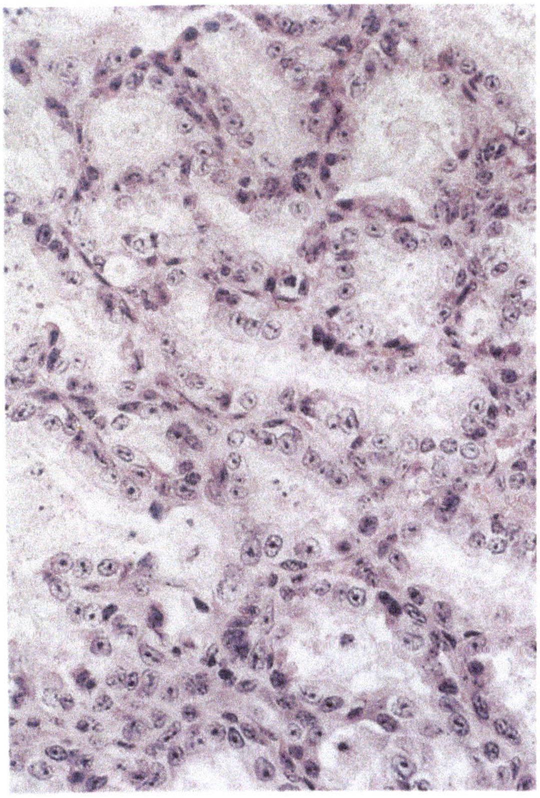

Fig. 40 *(upper left).* Acinar carcinoma of the lung in a female mouse. H&E

Fig. 41 *(upper right).* High power view of acinar carcinoma of the lung. PAS reaction

Fig. 42 *(below).* High power view of acinar carcinoma of the lung showing proliferation of ciliated cells. H&E

Carcinoma, Adenosquamous *(M)*

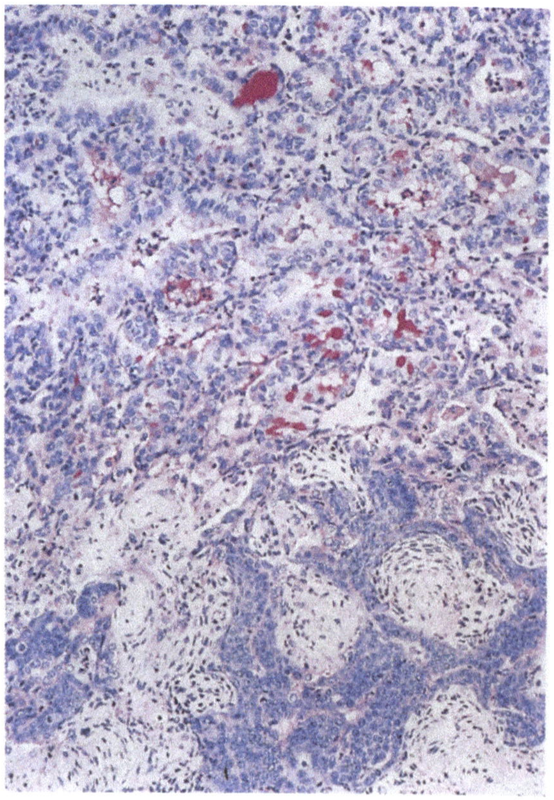

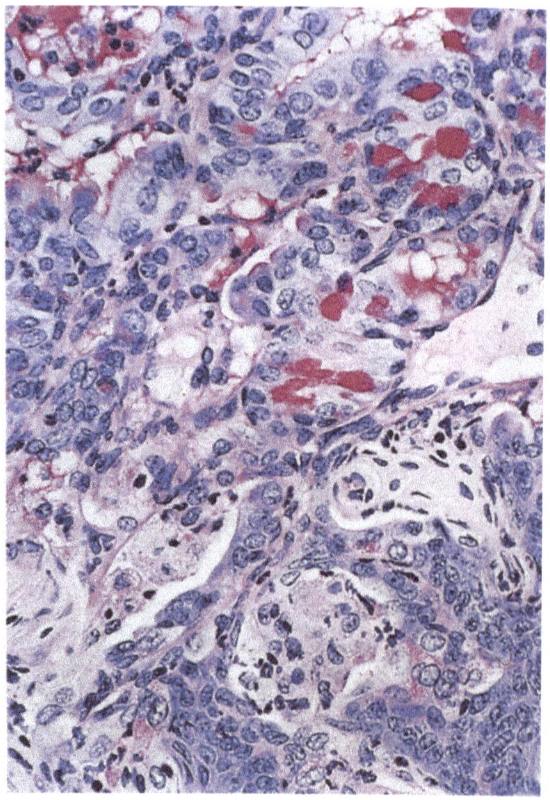

Synonym(s). Carcinoma, mucoepidermoid

Histogenesis

Adenosquamous carcinomas are believed to derive from acinar carcinomas, or possibly bronchiolo-alveolar carcinomas, with clonal shifts to malignant squamous cell phenotype.

Diagnostic Features

- Nodular, or diffusely expansile with irregular margins
- Composed of both adenocarcinomatous and malignant squamous cell components
- Squamous cells can show keratinization with formation of central keratin pearls or expansion of acini by desquamated keratinized cells
- Squamous cell differentiation can also be recognized by the formation of polygonal cells with prominent intercellular bridges lacking keratinization. Cells can also be greatly enlarged with atypical nuclei
- Neoplasms usually show clear indication of malignancy such as penetration of basement membranes and tissue destruction

Fig. 43 *(above).* Low power view of adenosquamous carcinoma of the lung. PAS reaction

Fig. 44 *(below).* Higher magnification of adenosquamous carcinoma of the lung. PAS reaction

Differential Diagnosis

Carcinoma, Bronchiolo-Alveolar or Carcinoma, Acinar
Bronchiolo-alveolar carcinoma or acinar carcinoma with squamous cell metaplasia have a large neoplastic glandular part with an insignificant portion of more regular benign looking squamous cells

Carcinoma, Squamous Cell
The neoplasm is composed entirely of malignant squamous cells, although there might be entrapped air spaces lined by hyperplastic bronchiolar or alveolar type II epithelial cells

Comment

Adenosquamous carcinomas most frequently derive from a squamous cell shift in acinar or bronchiolo-alveolar carcinomas. This is not usually seen in papillary neoplasms.

References

(See 13, 17, 59, 61, 65, 66)

Carcinoma, Squamous Cell *(M)*

Non-keratinizing • Keratinizing

Histogenesis

Squamous cell metaplasia of alveolar epithelium and/or Clara cells.

Diagnostic Features

- Growth pattern in cell clusters or irregular nests with central keratinization (keratin pearls), or without overt keratinization but forming distinct intercellular bridges
- Cellular debris and necrosis can be common as well as the presence of inflammatory cells, especially neutrophils
- Cells have cytologic features of malignancy (atypia, disorganization, and increased mitotic rate). Local invasion is more common in keratinizing than in non-keratinizing tumors
- Cells can frequently be quite pleomorphic, including the formation of very large cells ("giant cells") 60 µm or more at the largest diameter
- Can invade the adjacent lung parenchyma, pleura, vessels, and/or bronchi

Non-keratinizing

- Lack of overt keratinization but cells characteristically display distinct intercellular bridges
- Cells usually pleomorphic and can be small with scant cytoplasm resembling basal cells of the upper respiratory tract or large and eosinophilic with abundant cytoplasm

Keratinizing

- Amount of keratin ranges from small to abundant

Differential Diagnosis

Metaplasia, Squamous Cell
Preserves the normal pulmonary architecture. Small size. Tends to be multifocal. Low degree of cellular atypia

Carcinoma, Adenosquamous
Neoplasm composed of squamous as well as non-squamous, glandular elements that both show evidence of malignancy

Carcinoma, Bronchiolo-Alveolar or Carcinoma, Acinar
Composed of papillary strands or acinar structures lined by cuboidal or pleomorphic cells, with no or insignificant squamous cell metaplasia

Comment

Spontaneously occurring squamous cell carcinomas are extremely rare in lungs of mice and benign proliferations of squamous cells (keratinizing cysts and keratinizing cystic epitheliomas), as they are described for rats, have not been reported to occur in mice.

References

(See 13, 17, 20, 36, 55, 59, 61, 65, 66, 68)

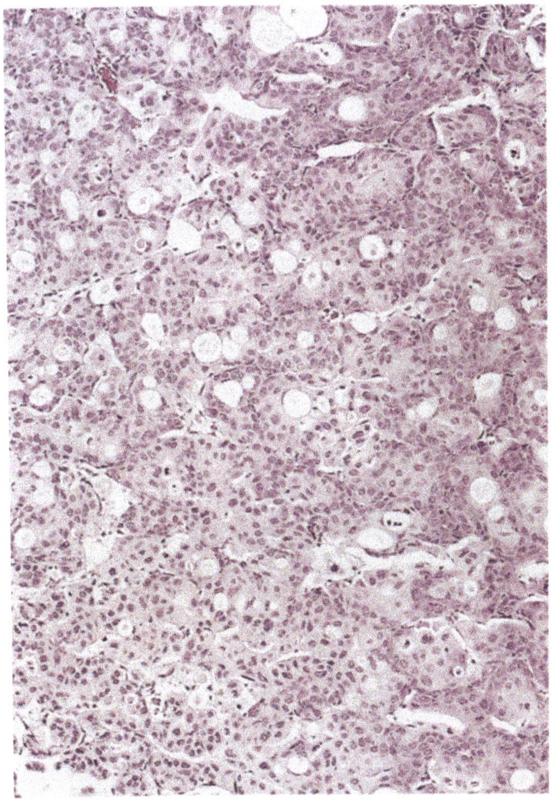

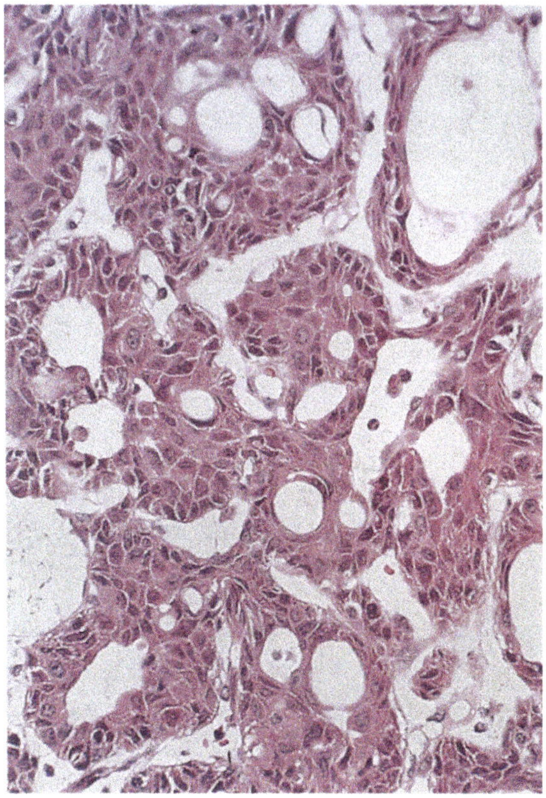

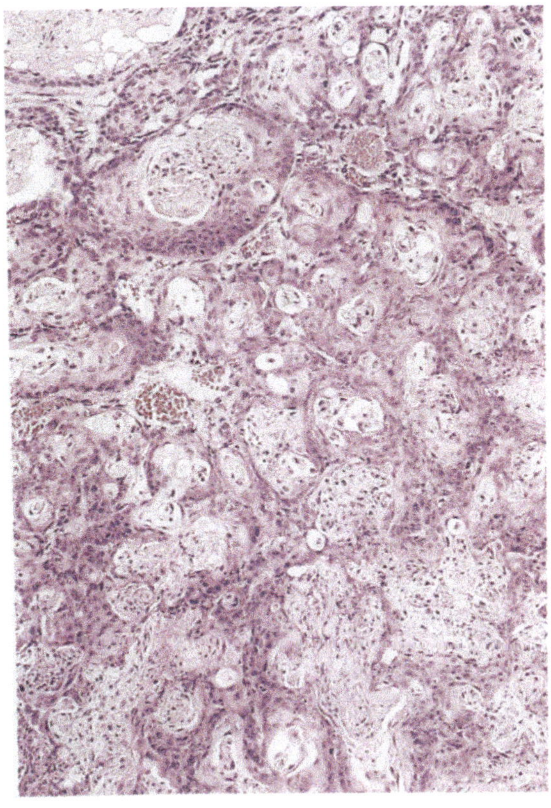

Fig. 45 *(upper left).* Non-keratinizing squamous cell carcinoma of the lung. NMRI mouse, female, treated, age 45 weeks, H&E

Fig. 46 *(upper right).* High power view of non-keratinizing squamous cell carcinoma of the lung. H&E

Fig. 47 *(below).* Keratinizing squamous cell carcinoma of the lung. H&E

Hyperplasia, Mesothelial *(H)*

Histogenesis

Mesothelial cells lining body cavities.

Diagnostic Features

- Focal or diffuse areas of large cuboidal epithelial-like cells are arranged in multiple layers without a significant amount of supporting stroma and no protrusion into the body cavity
- Cells with moderately abundant basophilic cytoplasm have large prominent dark-staining basophilic nuclei
- Cleft-like infoldings of the surface may be observed

Differential Diagnosis

Mesothelioma, Malignant (Epithelioid)
Mesothelioma is composed of papillary and tubular proliferations within a prominent stroma. A vascular connective tissue core is observed in papillary structures. Cellular pleomorphism and local tissue and vascular invasion are associated with more obvious malignant characteristics

References

(See 9, 10, 16, 20, 33, 34, 64, 70)

Mesothelioma, Malignant *(M)*

Epithelioid • Mesenchymal • Mixed

Histogenesis

Mesothelial cells lining body cavities.

Diagnostic Features

Epithelioid

- The lesion is generally composed of papillary and/or tubular structures and resembles adenocarcinoma, but areas with a solid pattern may occur
- Local invasion into underlying tissues is sometimes observed; metastases are uncommon
- Solid areas are composed of epithelial-like cells with vacuolated to foamy cytoplasm and large nuclei

Mesenchymal

- Sarcomatous component may be fibrosarcomatous, chondrosarcomatous, or osteosarcomatous
- Fibrosarcomatous pattern varies: one consists of bundles of uniform spindle-shaped cells separated by thick collagen fibers, another pattern consists of bundles of large basophilic spindle-shaped cells with vacuolated cytoplasm and large nuclei

Mixed

- In the mixed type, both epithelial and mesenchymal components are present

Differential Diagnosis

Hyperplasia, Mesothelial

Proliferated mesothelial cells do not have stroma and do not protrude into body cavities

Carcinoma, Bronchiolo-Alveolar

Differentiation between the epithelioid type and pleural spread of a papillary bronchiolo-alveolar carcinoma can be difficult. Metastases of bronchiolo-alveolar carcinomas (to pleura etc.) frequently are very anaplastic and can be misdiagnosed as malignant mesotheliomas of mesenchymal type

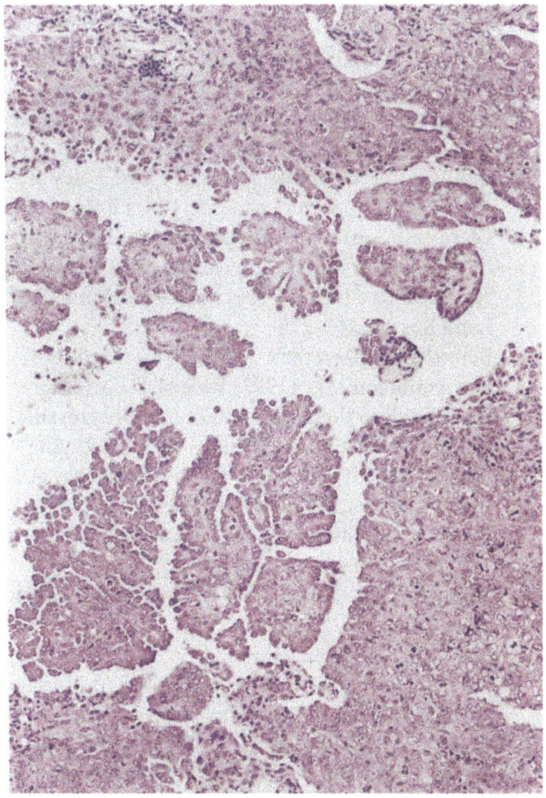

Fig. 48. Higher magnification of malignant mesothelioma, mixed type, of the peritoneum. Numerous mitotic figures. H&E (section by courtesy of Dr. Lucy Anderson, Laboratory of Comparative Carcinogenesis, Perinatal Carcinogenesis Section, Frederick Cancer Research and Development Center, National Cancer Institute)

Comment

Mesothelial proliferations are a rare spontaneous occurrence in mice although they can be induced, particularly by intrapleural or intraperitoneal injections of carcinogenic materials. The working group is not aware of benign mesotheliomas in mice having been recorded or whether, in fact, such tumors exist.

References

(See 9, 10, 16, 20, 33, 34, 64, 70)

References

1. Beer DG, Malkinson AM (1985) Genetic influence on type 2 or Clara cell origin of pulmonary adenomas in urethan-treated mice. J Natl Cancer Inst 75: 963–969
3. Belinsky SA, Devereux TR, White CM, Foley JF, Maronpot RR, Anderson MW (1991) Role of Clara cells and type II cells in the development of pulmonary tumors in rats and mice following exposure to a tobacco-specific nitrosamine. Exp Lung Res 17: 263–278
3. Belinsky SA, Devereux TR, Foley JF, Maronpot RR, Anderson MW (1992) Role of the alveolar type II cell in the development and progression of pulmonary tumors induced by 4-(methylnitrosamino)-1-(3-pyridyl)-1-butanone in the A/J mouse. Cancer Res 52: 3164–3173
4. Bernard A, Lauwerys R (1995) Low-molecular-weight proteins as markers of organ toxicity with special reference to Clara cell protein. Toxicol Lett 77: 145–151
5. Branstetter DG, Moseley PP (1991) Effect of lung development on the histological pattern of lung tumors induced by ethylnitrosourea in the C3HeB/FeJ mouse. Exp Lung Res 17: 169–179
6. Brown HR (1990) Neoplastic and potentially preneoplastic changes in the upper respiratory tract of rats and mice. Environ Health Perspect 85: 291–304
7. Brown HR, Monticello TM, Maronpot RR, Randall HW, Hotchkiss JR, Morgan KT (1991) Proliferative and neoplastic lesions in the rodent nasal cavity. Toxicol Pathol 19: 358–372
8. Chen HC, Pan IZ, Liang CT, Hong CC (1995) Nasal adenocarcinoma with myoepithelial component in a CD-1 mouse. Vet Pathol 32: 710–713
9. Davis JMG (1972) The fibrogenic effects of mineral dusts injected into the pleural cavity of mice. Br J Exp Pathol 53: 190–201
10. Davis MR, Manning LS, Whitaker D, Garlepp MJ, Robinson BW (1992) Establishment of a murine model of malignant mesothelioma. Int J Cancer 52: 881–886
11. Dickhaus S, Reznik G, Green U, Ketkar M (1977) The carcinogenic effect of beta-oxidized dipropylnitrosamine in mice. I. Dipropylnitrosamine and methyl-propylnitrosamine. Z Krebsforsch 90: 253–258
12. Dickhaus S, Reznik G, Green U, Ketkar M (1978) The carcinogenic effect of beta oxidized dipropylnitrosamine in mice. II. 2-hydroxypropyl-n-propylnitrosamine and 2-oxo-propyl-n-propylnitrosamine. Z Krebsforsch 91: 189–193
13. Dixon D, Maronpot RR (1991) Histomorphologic features of spontaneous and chemically-induced pulmonary neoplasms in B6C3F1 mice and Fischer 344 rats. Toxicol Pathol 19: 540–556
14. Dixon D, Horton J, Haseman JK, Talley F, Greenwell A, Nettesheim P, Hook GE, Maronpot RR (1991) Histomorphology and ultrastructure of spontaneous pulmonary neoplasms in strain A mice. Exp Lung Res 17: 131–155
15. Dixon D, Herbert RA, Sills RC, Boorman GA (1999) Lungs, pleura, and mediastinum. In: Maronpot RR, Boorman GA, Gaul BW (eds) Pathology of the mouse. Reference and atlas. Cache River Press, Vienna, pp 293–332
16. Donna A, Betta PG, Robutti F, Bellingeri D (1991) A one-year carcinogenicity study with 2,6-dichlorobenzonitrile (Dichlobenil) in male Swiss mice: preliminary note. Cancer Detect Prev 15: 41–44
17. Dungworth DL, Hahn FF, Hayashi Y, Keenan K, Mohr U, Rittinghausen S, Schwartz L (1992) 1. Respiratory system. In: Mohr U, Capen CC, Dungworth DL, Griesemer RA, Ito N, Turusov VS (eds) International classification of rodent tumours. Part I: The rat. IARC Scientific Publications No. 122, Lyon, pp 1–57
18. Elkon D (1983) Olfactory esthesioneuroblastoma. In: Reznik G, Stinson SF (eds) Nasal tumours in animals and man, vol II. Tumour pathology. CRC Press, Boca Raton, pp 129–147
19. Ernst H, Dungworth DL, Kamino K, Rittinghausen S, Mohr U (1996) Non-neoplastic lesions in the lungs. In: Mohr U, Dungworth DL, Capen CC, Carlton WW, Sundberg JP, Ward JM (eds) Pathobiology of the aging mouse, vol 1. ILSI Press, Washington DC, pp 281–300
20. Faccini JM, Abbott DP, Paulus GJJ (1990) Respiratory tract. Mouse histopathology. Elsevier, Amsterdam, pp 48–63
21. Feron VJ, Woutersen RA, Spit BJ (1986) Pathology of chronic nasal toxic responses including cancer. In: Barrow CS (ed) Toxicology of the nasal passages. Chemical Industry Institute of Toxicology Series. Hemisphere, Washington, pp 67–89
22. Foley JF, Anderson MW, Stoner GD, Gaul BW, Hardisty JF, Maronpot RR (1991) Proliferative lesions of the mouse lung: progression studies in strain A mice. Exp Lung Res 17: 157–168
23. Frith CH, Ward JM (1988) Respiratory system. In: (eds)Color atlas of neoplastic and non-neoplastic lesions in aging mice. Elsevier, Amsterdam, pp 27–32
24. Gopinath C, Prentice DE, Lewis DJ (1987) The respiratory system. In: (eds) Atlas of experimental pathology. MTP Press, Lancaster, pp 22–42
25. Green U, Konishi Y, Ketkar MB, Althoff J (1980) Comparative study of the carcinogenic effect of

BHP and BAP on NMRI mice. Cancer Lett 9: 257–261
26. Griciute L, Castegnaro M, Bereziat JC (1981) Influence of ethyl alcohol on carcinogenesis with N-nitrosodimethylamine. Cancer Lett 13: 345–352
27. Gunning WT, Castonguay A, Goldblatt PJ, Stoner GD (1991a) Strain A/J mouse lung adenoma growth patterns vary when induced by different carcinogens. Toxicol Pathol 19: 168–175
28. Gunning WT, Stoner GD, Goldblatt PJ (1991b) Glyceraldehyde-3-phosphate dehydrogenase and other enzymatic activity in normal mouse lung and in lung tumors. Exp Lung Res 17: 255–261
29. Gunning WT, Goldblatt PJ, Stoner GD (1992) Keratin expression in chemically induced mouse lung adenomas. Am J Pathol 140: 109–118
30. Harkema JR (1990) Comparative pathology of the nasal mucosa in laboratory animals exposed to inhaled irritants. Environ Health Perspect 85: 231–238
31. Heath JE, Frith CH, Wang PM (1982) A morphologic classification and incidence of alveolar-bronchiolar neoplasms in BALB/c female mice. Lab Anim Sci 32: 638–647
32. Hermans C, Knoops B, Wiedig M, Arsalane K, Toubeau G, Falmagne P, Bernard A (1999) Clara cell protein as a marker of Clara cell damage and bronchoalveolar blood barrier permeability. Eur Respir J 13: 1014–1021
33. Hoch-Ligeti C, Restrepo C, Stewart HL (1986) Comparative pathology of cardiac neoplasms in humans and in laboratory rodents: a review. J Natl Cancer Inst 76: 127–142
34. Kane AB (1992) Animal models of mesothelioma induced by mineral fibers: implications for human risk assessment. Prog Clin Biol Res 374: 37–50
35. Karube T, Katayama H, Takemoto K, Watanabe S (1989) Induction of squamous metaplasia, dysplasia and carcinoma in situ of the mouse tracheal mucosa by inhalation of sodium chloride mist following subcutaneous injection of 4-nitroquinoline 1-oxide. Jpn J Cancer Res 80: 698–701
36. Lijinsky W, Reuber MD (1988) Neoplasms of the skin and other organs observed in Swiss mice treated with nitrosoalkylureas. J Cancer Res Clin Oncol 114: 245–249
37. Luts A, Uddman R, Absood A, Hakanson R, Sundler F (1991) Chemical coding of endocrine cells of the airways: presence of helodermin-like peptides. Cell Tissue Res 265: 425–433
38. Maita K, Hirano M, Harada T, Mitsumori K, Yoshida A, Takahashi K, Nakashima N, Kitazawa T, Enomoto A, Inui K, Shirasu Y (1988) Mortality, major cause of moribundity, and spontaneous tumors in CD-1 mice. Toxicol Pathol 16: 340–349
39. Malkinson AM (1989) The genetic basis of susceptibility to lung tumors in mice. Toxicology 54: 241–271
40. Maronpot RR (1990) Pathology Working Group review of selected upper respiratory tract lesions in rats and mice. Environ Health Perspect 85: 331–352
41. Maronpot RR, Miller RA, Clarke WJ, Westerberg RB, Decker JR, Moss OR (1986) Toxicity of formaldehyde vapor in B6C3F1 mice exposed for 13 weeks. Toxicology 41: 253–266
42. Maronpot RR, Palmiter RD, Brinster RL, Sandgren EP (1991) Pulmonary carcinogenesis in transgenic mice. Exp Lung Res 17: 305–320
43. Mason RJ, Kalina M, Nielsen LD, Malkinson AM, Shannon JM (2000) Surfactant protein C expression in urethane-induced murine pulmonary tumors. Am J Pathol 156: 175–182
44. Matsuzaki O (1975) Histogenesis and growing patterns of lung tumors induced by potassium 1-methyl-1,4-dihydro-7-(2-(5-nitrofuryl)vinyl)-4-oxo-1,8-naphthyridine-3-carboxylate in ICR mice. GANN 66: 259–267
45. Miller RA, Boorman GA (1990) Morphology of neoplastic lesions induced by 1,3-butadiene in B6C3F1 mice. Environ Health Perspect 86: 37–48
46. Miller RR, Young JT, Kociba RJ, Keyes DG, Bodner KM, Calhoun LL, Ayres JA (1985) Chronic toxicity and oncogenicity bioassay of inhaled ethyl acrylate in Fischer 344 rats and B6C3F1 mice. Drug Chem Toxicol 8: 1–42
47. Monticello TM, Morgan KT, Uraih L (1990) Non-neoplastic nasal lesions in rats and mice. Environ Health Perspect 85: 249–274
48. Morgan KT (1991) Approaches to the identification and recording of nasal lesions in toxicology studies. Toxicol Pathol 19: 337–351
49. Nagano K, Enomoto M, Yamanouchi K, Aiso S, Katagiri T (1988) Toxicologic pathology of upper respiratory tract. Jap J Toxicol Pathol 1: 115–127
50. Nettesheim P, Szakal AK (1972) Morphogenesis of alveolar bronchiolization. Lab Invest 26: 210–219
51. Pack RJ, Al-Ugaily LH, Morris G (1981) The cells of the tracheobronchial epithelium of the mouse: a quantitative light and electron microscope study. J Anat 132: 71–84
52. Palmer KC (1985) Clara cell adenomas of the mouse lung. Interaction with alveolar type 2 cells. Am J Pathol 120: 455–463
53. Palmer KC, Grammas P (1987) Beta-adrenergic regulation of secretion from Clara cell adenomas of the mouse lung. Lab Invest 56: 329–334

54. Rabstein LS, Peters RL (1973) Tumors of the kidneys, synovia, exocrine pancreas and nasal cavity in BALB-cf-Cd mice. J Natl Cancer Inst 51: 999–1006
55. Rehm S, Kelloff GJ (1991) Histologic characterization of mouse bronchiolar cell hyperplasia, metaplasia, and neoplasia induced intratracheally by 3-methylcholanthrene. Exp Lung Res 17: 229–244
56. Rehm S, Ward JM (1989) Quantitative analysis of alveolar type II cell tumors in mice by whole lung serial and step sections. Toxicol Pathol 17: 737–742
57. Rehm S, Ward JM, ten Have-Opbroek AAW, Anderson LM, Singh G, Katyal SL, Rice JM (1988) Mouse papillary lung tumors transplacentally induced by N-nitrosoethylurea: evidence for alveolar type II cell origin by comparative light microscopic, ultrastructural, and immunohistochemical studies. Cancer Res 48: 148–160
58. Rehm S, Devor DE, Henneman JR, Ward JM (1991a) Origin of spontaneous and transplacentally induced mouse lung tumors from alveolar type II cells. Exp Lung Res 17: 181–195
59. Rehm S, Lijinsky W, Singh G, Katyal SL (1991b) Mouse bronchiolar cell carcinogenesis. Histologic characterization and expression of Clara cell antigen in lesions induced by N-nitrosobis-(2-chloroethyl) ureas. Am J Pathol 139: 413–422
60. Rehm S, Ward JM, Anderson LM, Riggs CW, Rice JM (1991c) Transplacental induction of mouse lung tumors: stage of fetal organogenesis in relation to frequency, morphology, size, and neoplastic progression of N-nitrosoethylurea-induced tumors. Toxicol Pathol 19: 35–46
61. Rehm S, Ward JM, Sass B (1994) Tumours of the lungs. In: Mohr U, Turusov VS (eds) Pathology of tumours in laboratory animals, vol. 2. Tumours of the mouse, 2nd edn. IARC Scientific Publications No. 111, Lyon, pp 325–339
62. Renne RA, Giddens WE, Boorman GA, Kovatch R, Haseman JE, Clarke WJ (1986) Nasal cavity neoplasia in F344/N rats and (C57BL/6 x C3H)F1 mice inhaling propylene oxide for up to two years. J Natl Cancer Inst 77: 573–582
63. Reznik GK, Schuller HM, Stinson SF (1994) Tumours of the nasal cavities. In: Mohr U, Turusov VS (eds) Pathology of tumours in laboratory animals, vol. 2. Tumours of the mouse, 2nd edn. IARC Scientific Publications No. 111, Lyon, pp 305–324
64. Rice JM, Kovatch RM, Anderson LM (1989) Intraperitoneal mesotheliomas induced in mice by a polycyclic aromatic hydrocarbon. J Toxicol Environ Health 27: 153–160
65. Rittinghausen S, Dungworth DL, Ernst H, Mohr U (1996a) Naturally occurring pulmonary tumors in rodents. In: Jones TC, Mohr U, Hunt RD (eds) Monographs on pathology of laboratory animals. Respiratory system, 2nd edn. Springer, Berlin Heidelberg New York Tokyo, pp 183–206
66. Rittinghausen S, Dungworth DL, Ernst H, Mohr U (1996b) Primary pulmonary tumors. In: Mohr U, Dungworth DL, Capen CC, Carlton WW, Sundberg JP, Ward JM (eds) Pathobiology of the aging mouse, vol 1. ILSI Press, Washington DC, pp 301–314
67. Schueller HM (1987) Experimental carcinogenesis in the peripheral lung. In: McDowell EM (ed) Lung carcinomas. Churchill Livingstone, Edinburgh, pp 243–254
68. Schulte A, Ernst H, Peters L, Heinrich U (1994) Induction of squamous cell carcinomas in the mouse lung after long-term inhalation of polycyclic aromatic hydrocarbon-rich exhausts. Exp Toxicol Pathol 45: 415–421
69. Singh G, Katyal SL, Ward JM, Gottron SA, Wong-Chong ML, Riley EJ (1985) Secretory proteins of the lung in rodents: immunocytochemistry. J Histochem Cytochem 33: 564–568
70. Suzuki Y, Kohyama N (1984) Malignant mesothelioma induced by asbestos and zeolite in the mouse peritoneal cavity. Environ Res 35: 277–292
71. Takahashi A, Iwasaki I, Ide G (1985) Effects of minute amounts of cigarette smoke with or without nebulized N-nitroso-N-methylurethane on the respiratory tract of mice. Jpn J Cancer Res 76: 324–330
72. Tamano S, Hagiwara A, Shibata MA, Kurata Y, Fukushima S, Ito N (1988) Spontaneous tumors in aging (C57BL/6 N x C3H/HeN)F1 (B6C3F1) mice. Toxicol Pathol 16: 321–326
73. Ten Have-Opbroek AAW (1986) The structural composition of the pulmonary acinus in the mouse. A scanning electron microscopical and developmental-biological analysis. Anat Embryol (Berl) 174: 49–57
74. Thaete LG, Malkinson AM (1990) Differential staining of normal and neoplastic mouse lung epithelia by succinate dehydrogenase histochemistry. Cancer Lett 52: 219–227
75. Thaete LG, Malkinson AM (1991) Cells of origin of primary pulmonary neoplasms in mice: morphologic and histochemical studies. Exp Lung Res 17: 219–228
76. Thaete LG, Gunning WT, Stoner GD, Malkinson AM (1987) Cellular derivation of lung tumors in sensitive and resistant strains of mice: results at 28 and 56 weeks after urethan treatment. J Natl Cancer Inst 78: 743–749

77. Thaete LG, Nesbitt MN, Malkinson AM (1991) Lung adenoma structure among inbred strains of mice: the pulmonary adenoma histologic type (Pah) genes. Cancer Lett 61: 15–20
78. Turk MAM, Henk WG, Flory W (1987) 3-Methylindole-induced nasal mucosal damage in mice. Vet Pathol 24: 400–403
79. Ward JM, Rehm S (1990) Applications of immunohistochemistry in rodent tumor pathology. Exp Pathol 40: 301–312
80. Ward JM, Singh G, Katyal SL, Anderson LM, Kovatch RM (1985) Immunocytochemical localization of the surfactant apoprotein and Clara cell antigen in chemically induced and naturally occurring pulmonary neoplasms of mice. Am J Pathol 118: 493–499
81. Witschi HP (1985) Enhancement of lung tumor formation in mice. Carcinog Compr Surv 8: 147–158
82. Witschi HP (1986) Separation of early diffuse alveolar cell proliferation from enhanced tumor development in mouse lung. Cancer Res 46: 2675–2679

Urinary System

G.C. Hard[1] · B. Durchfeld-Meyer[2,3] · B. Short[2] · A. Bube · K. Krieg[3]
D. Creasey · J. Everitt · C.H. Frith · J. Glaister · J.C. Seely · H. Tsuda
V.S. Turusov

[1]Chair; [2]Co-chair; [3]Initial draft.

Hyperplasia, Renal Tubule *(H)*

Simple • Atypical

Synonym(s). Simple tubule hyperplasia; atypical tubule hyperplasia; focal tubule dysplasia

Histogenesis

Arising from renal tubule epithelium.

Diagnostic Features

- Unilateral or bilateral
- Focal or multifocal increase in the number of lining cells
- Solitary or multiple tubules are affected
- No evidence of compression of adjacent parenchyma
- Integrity of the single tubule structure is essentially maintained
- Lumen may be dilated or cystic
- Cells are smaller than, similar to, or slightly larger than regular epithelial cells
- Cytoplasm is eosinophilic, basophilic or clear
- Nuclei are crowded
- Nuclei are uniform round or slightly oval
- Mitotic figures may be present

Simple
- Multifocal
- Cytoplasm is basophilic, but may be unchanged
- Nuclei may be crowded
- Tubule dimension is normal, decreased, or increased due to lumen dilation or increased cell number

Atypical
- Solitary
- Cytoplasm is eosinophilic, basophilic or clear
- Cellular atypia is present
- Cellular and nuclear pleomorphism is present
- Lining cells exhibit multilayered or solid growth, partially or completely obliterating tubule lumen
- Tubule dimensions are consistently increased through increased cell number and sometimes lumen dilatation
- No evidence of vascular ingrowth

Differential Diagnosis

Hypertrophy, Renal Tubule
Cells are enlarged and
Cell number is not increased

Regeneration, Renal Tubule
Cells are basophilic and
Cell number is not usually increased and
Tubule size is not increased

Adenoma, Renal Tubule
Growth extends beyond integrity of the single tubule structure or
Vascular ingrowth is present

Fig. 1 *(upper left).* Renal tubule hyperplasia, simple, of the kidney. The tubule lumen is dilated. NMRI mouse. PAS ▶

Fig. 2 *(upper right).* Renal tubule hyperplasia, simple, of the kidney. Cells are crowded together but the single cell layer is retained. NMRI mouse. PAS

Fig. 3 *(lower left).* Renal tubule hyperplasia, atypical, of the kidney. Multicellular, papillary projections from tubule lining into the dilated lumen. NMRI mouse. H&E

Fig. 4 *(lower right).* Renal tubule hyperplasia, atypical, of the kidney. Multilayered complex proliferation occludes lumen to produce a solid tubule profile. H&E

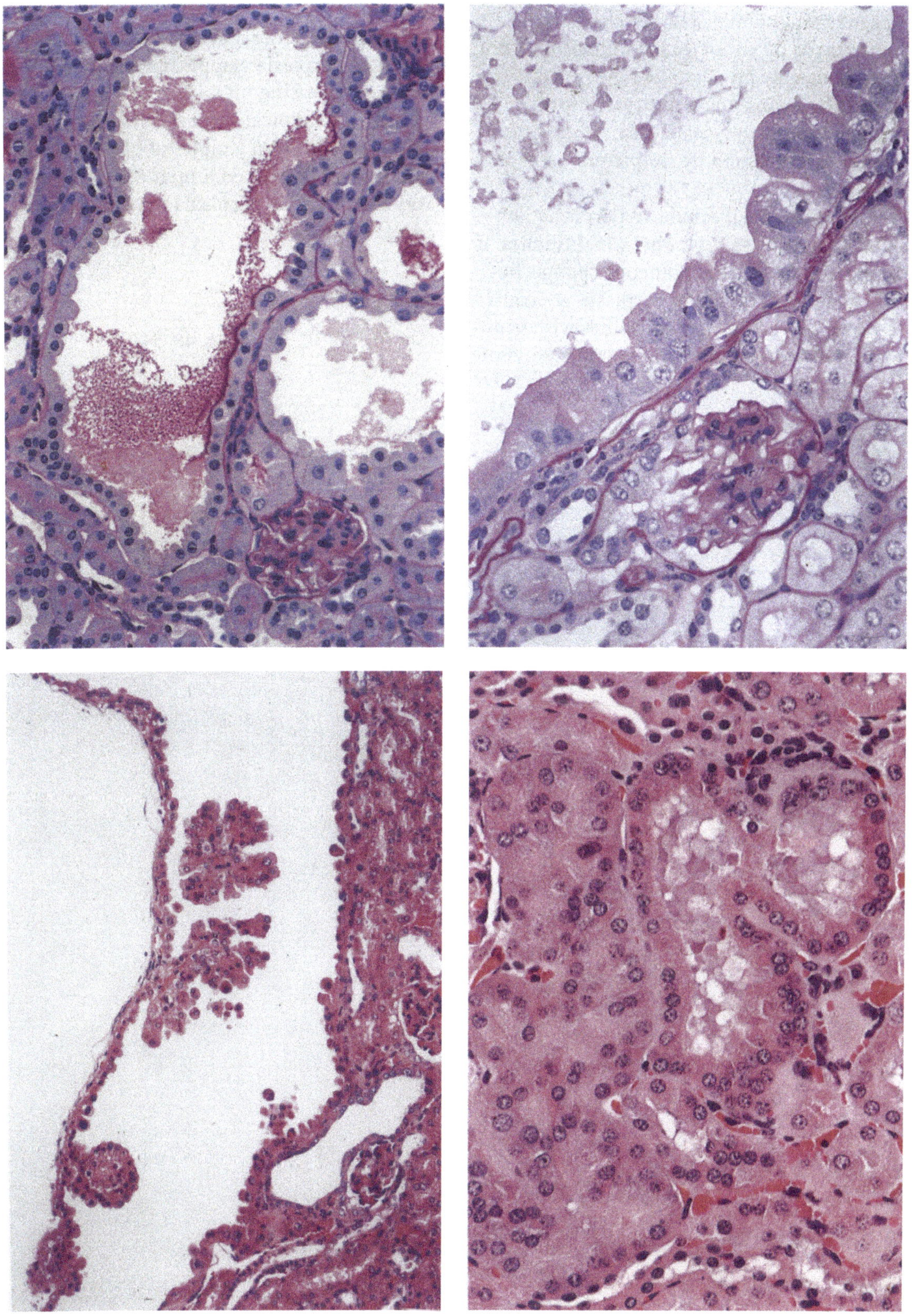

Comment

Regenerating tubule cells are characterized temporarily by an increased mitotic rate, but there is usually no persistent overproduction of cells following acute injury. However, if the injury is chronically sustained, regeneration can be accompanied by simple tubule hyperplasia.

In the mouse renal tubule, hyperplasia has been observed mainly in proximal tubules. It is uncommon as a spontaneous change, but foci of atypical tubule hyperplasia are consistently observed in chronic studies with genotoxic renal carcinogens, for example ethylnitrosourea, streptozotocin and the flame retardant, Tris, and following X-irradiation. There is now general acceptance that atypical tubule hyperplasia can be a preneoplastic lesion progressing into adenoma, and consequently the atypical form should always be distinguished from simple tubule hyperplasia. When there is marked lumen dilation the atypical tubule hyperplasia may be designated as of cystic type.

References

(See 16, 26, 30, 38, 44, 46, 48, 54)

Adenoma, Renal Tubule *(B)*

Solid • Papillary • Cystic • Cystopapillary • Mixed

Synonym(s). Renal cell adenoma; renal tubule cell adenoma; benign renal cell tumor; benign renal epithelioma

Histogenesis

Arising from the renal tubular epithelium.

Diagnostic Features

- Solitary, but occasionally multiple and bilateral
- Usually well circumscribed and discrete
- Compression of adjacent parenchyma is present
- Growth pattern is solid, papillary, cystic, cystopapillary or mixed
- Extends beyond confines of original tubule structure
- No evidence of areas of necrosis or hemorrhage
- Early vascular ingrowth is present
- Cells are well-differentiated
- Cytoplasm is basophilic, eosinophilic, clear or mixed
- Nuclei are regular and vesicular
- Mitotic figures are uncommon

Differential Diagnosis

Hyperplasia, Renal Tubule
Growth is limited to original tubule structure and
No evidence of neovascularization

Carcinoma, Renal Tubule
Local invasion is present or
Metastasis is present or
Areas of necrosis
or hemorrhage are present and
Cellular pleomorphism is present

Comment

The morphology and histogenesis of renal adenomas in the mouse have not been studied as closely as in the rat, although like the rat, adenomas of mouse kidneys probably develop from various segments of the nephron. In the mouse, adenomas are often mildly cystic (representing very dilated lumen), combined with a papillary pattern of proliferation, but solid adenomas also occur. The lack of areas of necrosis and hemorrhage is a useful (and readily recognizable) criterion for distinction from renal tubule carcinoma.

References

(See 11, 13, 14, 16, 17, 21, 23, 26, 30, 31, 35, 38, 47, 48, 49, 54, 58)

Figs. 5-7 see p. 144.

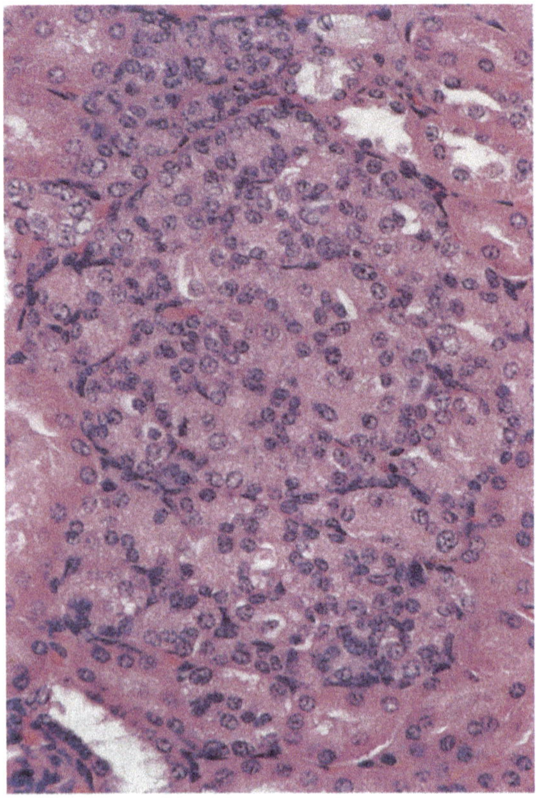

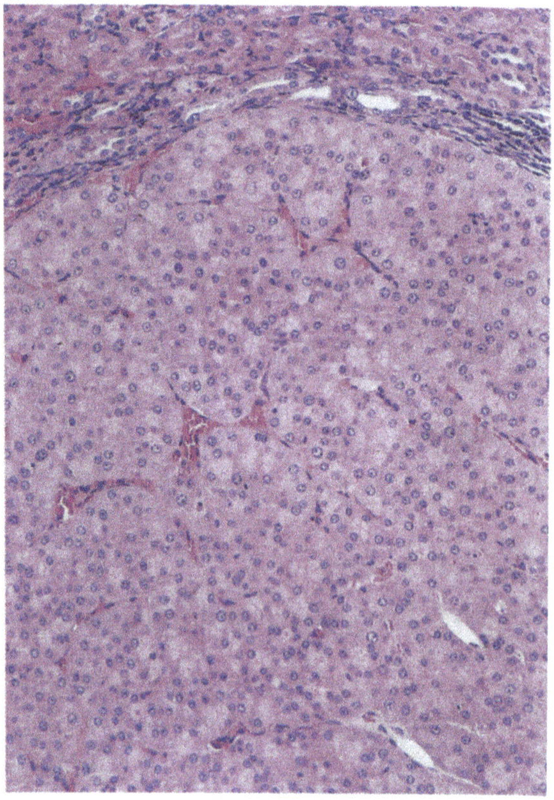

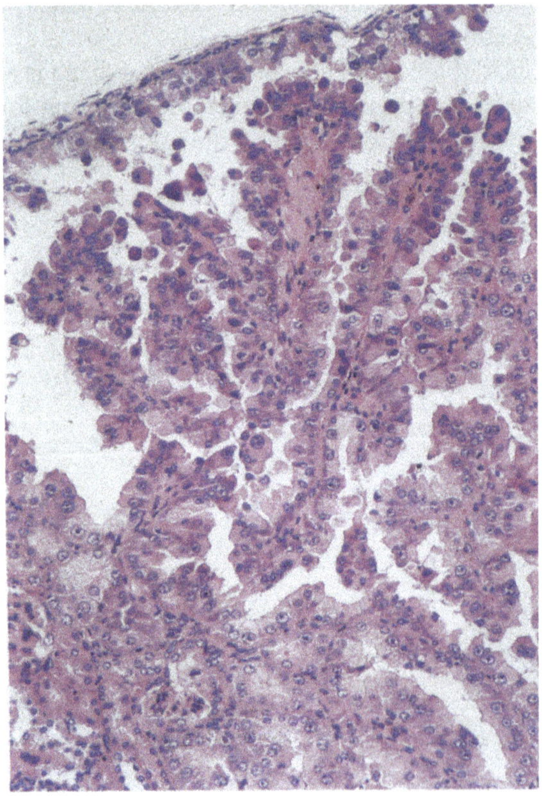

Fig. 5 *(upper left)*. Renal tubule adenoma of the kidney. There is solid proliferation beyond the integrity of a single tubule. H&E

Fig. 6 *(upper right)*. Renal tubule adenoma of the kidney. Basophilic cells with solid growth pattern. NMRI mouse. H&E

Fig. 7 *(below)*. Renal tubule adenoma of the kidney. Basophilic cells with cystopapillary growth pattern. NMRI mouse. H&E

Carcinoma, Renal Tubule *(M)*

Solid • Papillary • Anaplastic • Mixed

Synonym(s). Renal cell carcinoma; renal adenocarcinoma; malignant renal epithelioma

Histogenesis

Arising from renal tubule epithelium.

Diagnostic Features

- Solitary, but occasionally multiple and bilateral
- Compression of adjacent parenchyma is present
- Growth pattern is solid, papillary, anaplastic, or mixed
- Growth is expansive
- Areas of necrosis or hemorrhage are present
- Vascular ingrowth is prominent
- Cells vary from small and uniform, to large and pleomorphic
- Cytoplasm is basophilic, eosinophilic, clear or mixed
- Nuclei vary from small, round or oval, to large, pleomorphic, vesicular
- Mitotic figures may be numerous
- Local invasion into adjacent parenchyma may be present
- Metastasis can occur but is rare. Main site are the lungs

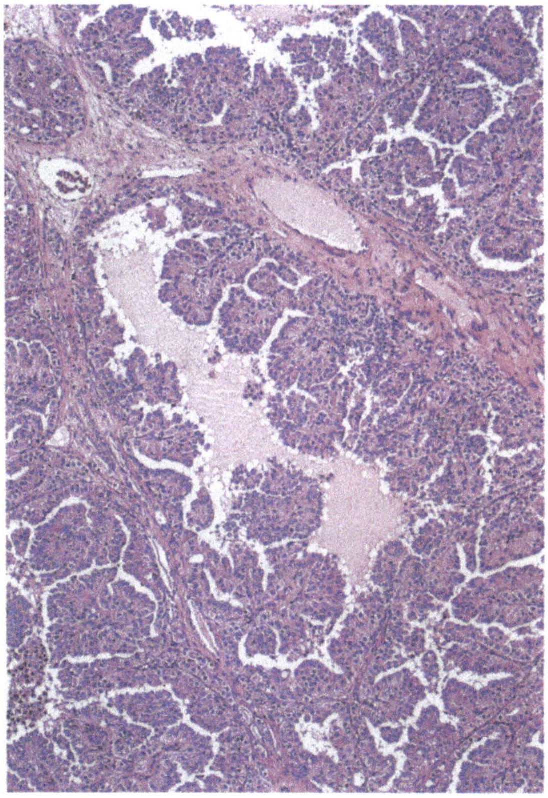

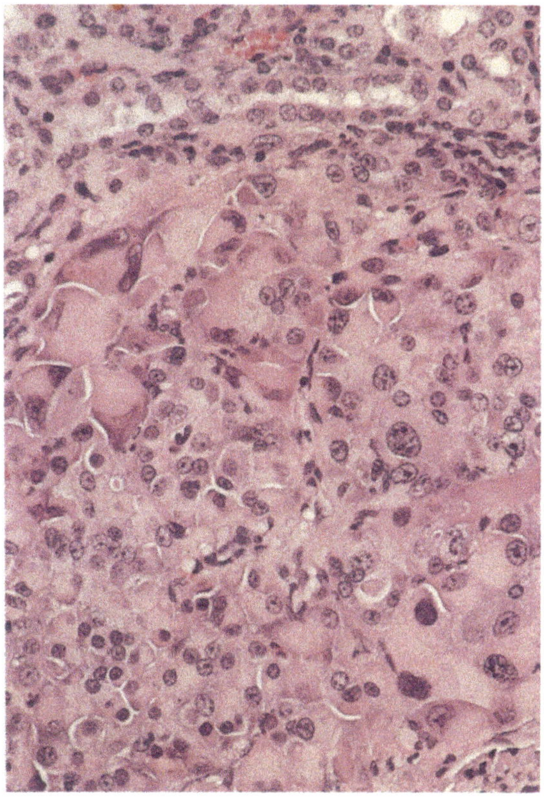

Fig. 8 *(above)*. Renal tubule carcinoma of the kidney. Basophilic cells with papillary growth pattern. Note the well-developed fibrovascular stroma. NMRI mouse. H&E

Fig. 9 *(below)*. Renal tubule carcinoma of the kidney. Basophilic cells with anaplastic growth pattern. H&E

Differential Diagnosis

Adenoma, Renal Tubule
No evidence of areas of necrosis and hemorrhage and
Mitotic figures are uncommon and
No evidence of cellular pleomorphism and
No evidence of invasion and
No evidence of metastasis is present

Carcinoma, Transitional Cell
Great majority of cells are transitional cells

Comment

Spontaneously occurring adenomas and carcinomas in the mouse are rare but they are the most frequent kidney tumors induced experimentally by carcinogenic chemicals. It is generally accepted that carcinomas develop from, and represent progressive expansive growth of adenomas. The most useful distinction from adenoma is the presence of areas of necrosis or hemorrhage which implies a stage of autonomous growth that outstrips neovascularization. Studies with genotoxic chemicals, for example streptozotocin and ethylnitrosourea, demonstrate that larger renal tubule carcinomas in the mouse can metastasize to the lungs.

References

(See 14, 16, 17, 21, 23, 26, 30, 38, 47, 48, 49, 54, 59, 60)

Nephroblastoma *(M)*

Synonym(s). Embryonal nephroma; Wilms' tumor

Histogenesis

Presumed to originate from the metanephric blastema.

Diagnostic Features

- Unilateral and solitary
- May be well demarcated
- Compression of surrounding tissue is present
- Growth is expansive
- Cells are closely packed or arranged in reticular sheets
- Various stages of differentiation may be present, into primitive or more mature tubular profiles, and structures resembling glomeruloid bodies
- Tubular structures usually have lumens
- Glomeruloid bodies consist of invaginated avascular tufts
- Cysts are sometimes present
- Tubular structures and cysts are lined by single to multiple cuboidal cell layers
- Stroma consists of non-neoplastic fibroblast-like cells
- Cells are primitive basophilic blastema cells

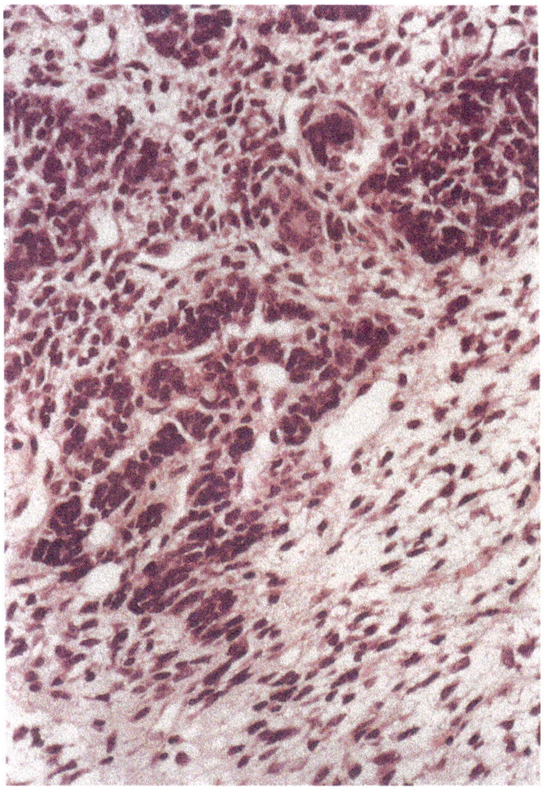

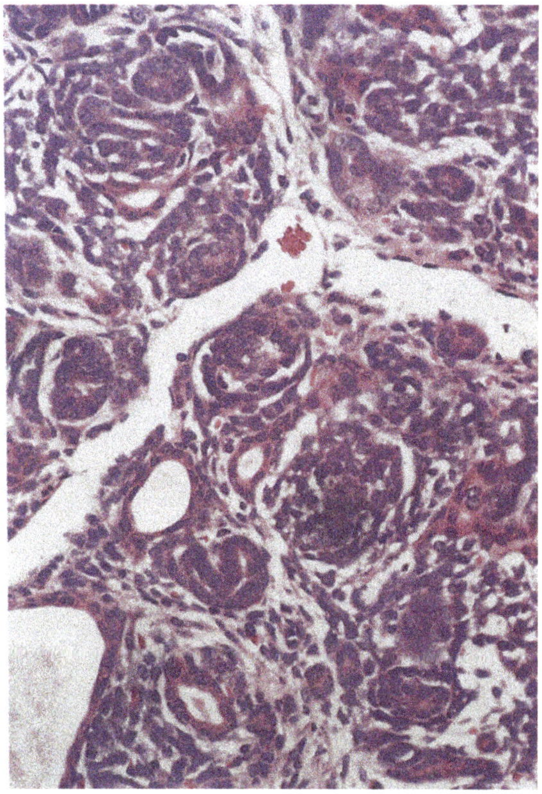

Fig. 10 *(above).* Nephroblastoma of the kidney. Clusters of highly basophilic blast cells, primitive tubules, and mature ducts. H&E

Fig. 11 *(below).* Nephroblastoma of the kidney. Dense clusters of highly basophilic cells beside a loose reticular sheet of dispersed blast cells. H&E

- Cells in tubular and cystic structures are cuboidal to columnar
- Cells in glomeruloid bodies are very small cuboidal with basophilic nuclei
- Mitotic figures are rare or may be more numerous
- Inflammatory cells may be present
- Invasive growth may be present

Differential Diagnosis

Sarcoma, Renal

Demarcation is poor and
Growth is infiltrative and
Fibroblast-like spindle cells are present and
No evidence of blastema and organoid epithelial structures, e.g., primitive tubules or glomeruloid bodies

Comment

Nephroblastoma in the mouse is exceedingly rare and to date only four cases appear to have been reported in the literature. This tumor appears to be morphologically similar to that observed in the rat, but some cases have shown a predominance of loose, reticular sheets of blastemal cells with little organoid differentiation. Organoid differentiation into structures resembling primitive avascular glomeruli has been described in one case. In contrast to experience with the rat, there is no unequivocal evidence of chemical induction of this tumor in mice. The human term of Wilms' tumor is used as a synonym for human nephroblastoma, but should not be used for rodent nephroblastoma.

References

(See 23, 27, 37, 43, 48, 57)

Sarcoma, Renal *(M)*

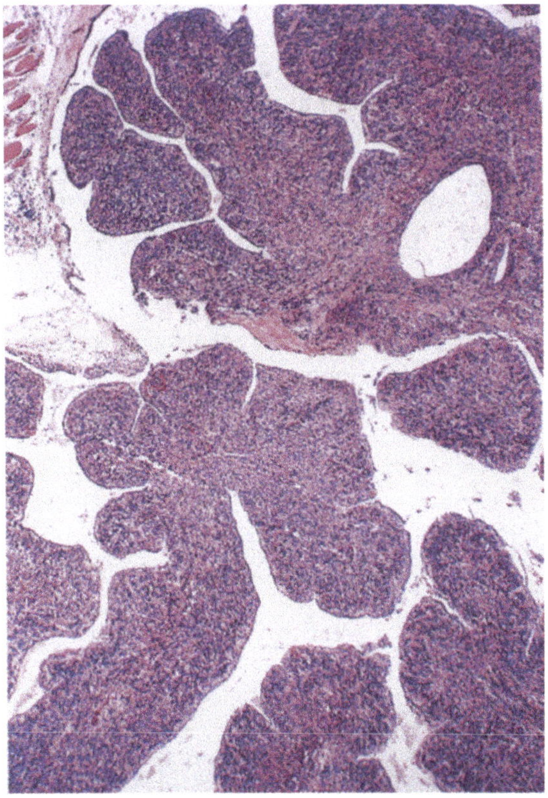

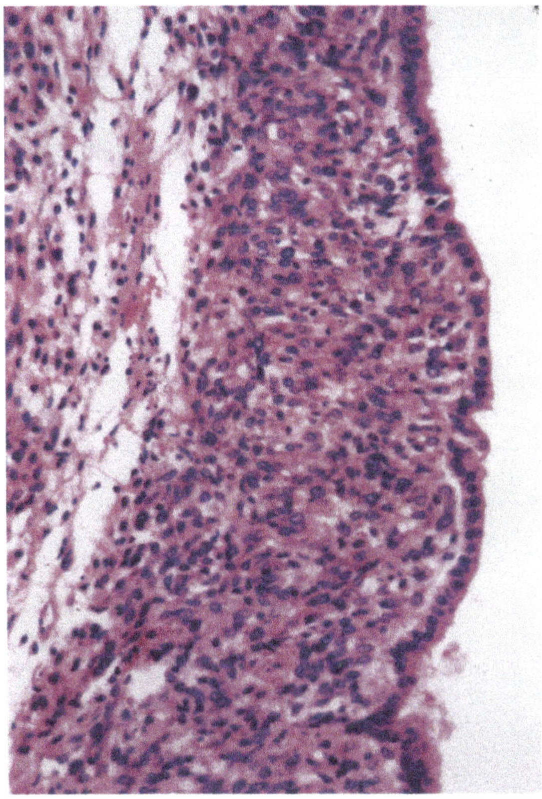

Synonym(s). Renal fibrosarcoma; renal mesenchymal tumor

Histogenesis

Probably arising from interstitial mesenchymal cells in the cortex or medulla.

Diagnostic Features

- Unilateral or bilateral
- Demarcation is poor
- Growth is infiltrative, replacing parenchyma
- Entrapped pre-existing tubules are present
- Areas of hemorrhage and necrosis may be present
- Cells are arranged in dense sheets or forming interlacing bundles, associated with fine collagen fibers
- Cells are fibroblast-like and spindle-shaped
- Nuclei are large, fusiform, often possessing several nucleoli
- Mitotic figures are present

Fig. 12 *(above).* Renal sarcoma of the kidney. Densely cellular polypoid projections of tumor tissue into renal pelvis. Induced by infection with polyoma virus. H&E (by courtesy of Dr. C. Dawe)

Fig. 13 *(below).* Renal sarcoma of the kidney. Dense sheets of highly basophilic fusiform cells. Note the intact urothelial lining. Induced by infection of polyoma virus. H&E (by courtesy of Dr. C. Dawe)

Differential Diagnosis

Nephroblastoma
Growth is expansive and
Predominant component is blastema, with some organoid epithelial differentiation, e.g., primitive tubules or glomeruloid bodies

Comment

Spontaneously occurring renal sarcomas are very rare in the mouse. They have most often been induced experimentally using the infective agent, polyoma virus. They have also been induced by intrarenal administration of 20-methylcholanthrene. Virus-induced renal sarcomas develop rapidly after inoculation, and progressively invade and replace the renal tissue.

References

(See 12, 29, 38, 50, 51, 52, 54, 56)

Hyperplasia, Transitional Cell *(H)*

Without atypia • Atypical

Synonym(s). Hyperplasia, urothelial

Histogenesis

Arising from transitional cell epithelium.

Diagnostic Features

- Multicellular thickening of the transitional cell lining extends beyond the normal number of cell layers
- Growth pattern is simple, papillary or nodular
- Simple: linear thickening of the lining lacks prominent outward or inward focal growth
- Papillary: exophytic projection of lining into the lumen is supported by a simple branching fibrovascular core
- Nodular: solid nest or nests of cells are protruding outwards into the lumen, or inwards as a downgrowth
- Chronic cystitis or calculi may be present

Focal

- Cells may be of normal size, smaller or larger than normal
- Cytoplasm may be basophilic
- Increase in number of cells containing PAS-positive granules may be present

Diffuse

- Cells are uniform retaining the differentiated and relatively organized character of regular urothelium
- Cells may be of normal size, smaller or larger than normal
- Cytoplasm may be basophilic
- Increase in number of cells containing PAS-positive granules may be present

Without Atypia

- Cells are uniform retaining the differentiated and relatively organized character of regular urothelium

Atypical

- Focal or multifocal
- Cellular and nuclear pleomorphism, hyperchromatism, and enlarged nucleoli are present
- Irregular arrangement of cells is present
- Growth pattern is nodular, but may be papillary
- Cytoplasm is basophilic
- Mitotic figures may be present

Figs. 14-16 see p. 152

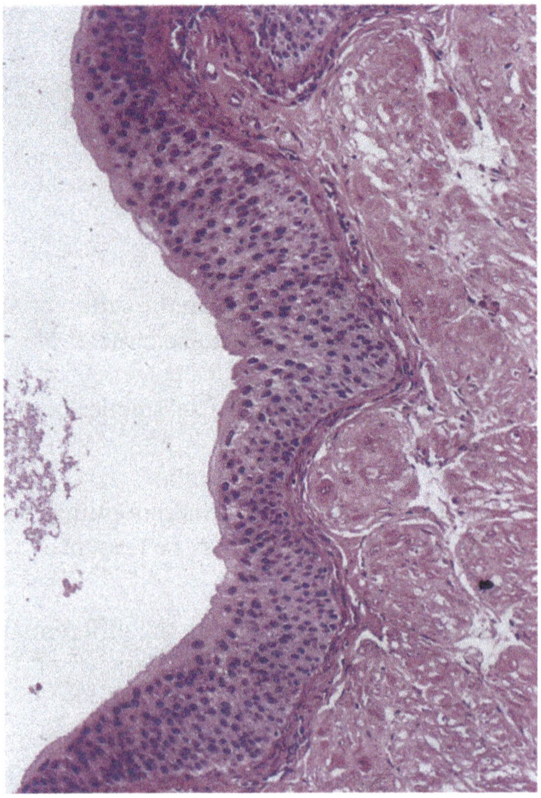

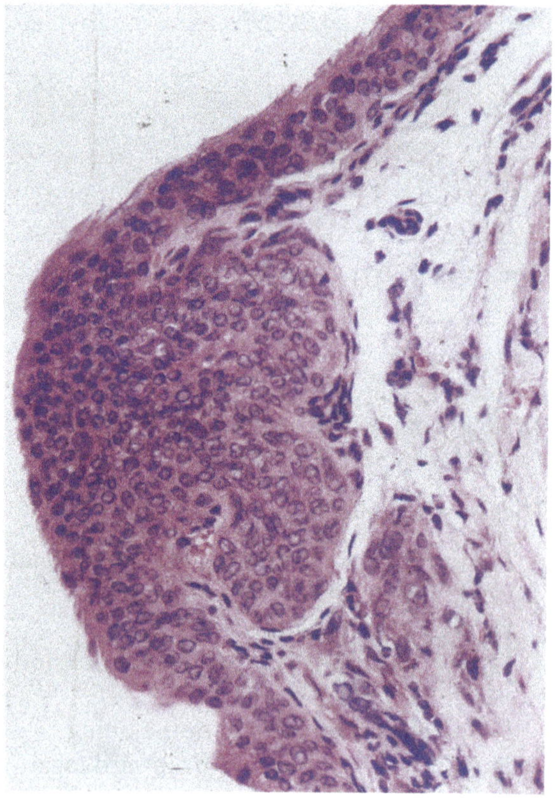

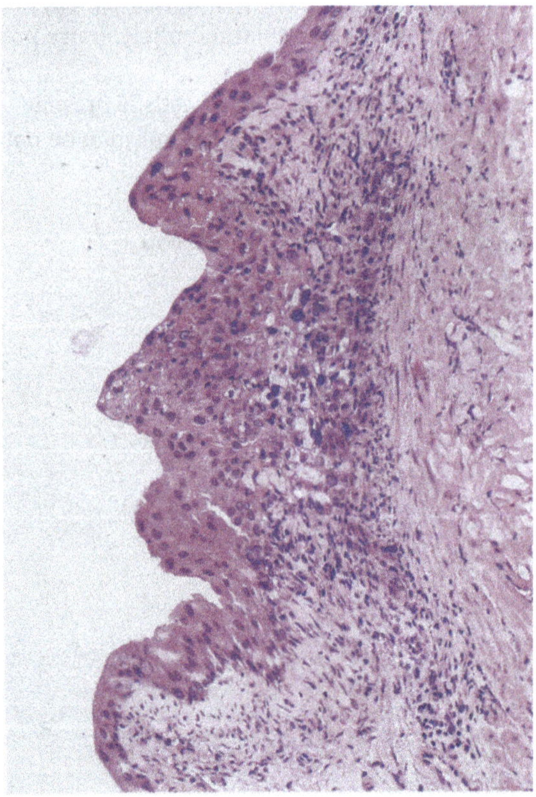

Fig. 14 *(upper left)*. Diffuse transitional cell hyperplasia of the urinary bladder. Uniform increase in thickness of urothelial lining. NMRI-BR mouse, male. H&E

Fig. 15 *(upper right)*. Focal nodular transitional cell hyperplasia of the urinary bladder. Nodular focus of regular transitional cells forming a downgrowth of the lining (endophytic). NMRI mouse, female. H&E

Fig. 16 *(below)*. Focal atypical transitional cell hyperplasia of the urinary bladder. Focus of irregular and disorganized transitional epithelium with cell pleomorphism. NMRI mouse, male. H&E

Differential Diagnosis

Papilloma, Transitional Cell
- Cells are uniformly transitional and
- Projections may have a complex branching slender stalk and
- Lesion is usually solitary, but may be multiple

Comment

Spontaneous transitional cell hyperplasia in the mouse is rare but there is a potential for its induction by chemicals causing injury to these cells. In addition, it could be a response to infective agents causing pyelonephritis, and may be secondary to lower urinary tract obstruction.

Atypical hyperplasia should always be distinguished from transitional cell hyperplasia without atypia because, when associated with bladder carcinogens, the atypical form is considered to be a precursor of epithelial bladder cancer.

Atypical transitional cell hyperplasia has been associated with genotoxic urothelial carcinogens such as N-butyl-N-(4-hydroxybutyl)nitrosamine (OH-BBN).

References

(See 2, 8, 11, 17, 19, 20, 21, 23, 25, 36, 42, 54, 61)

Papilloma, Transitional Cell *(B)*

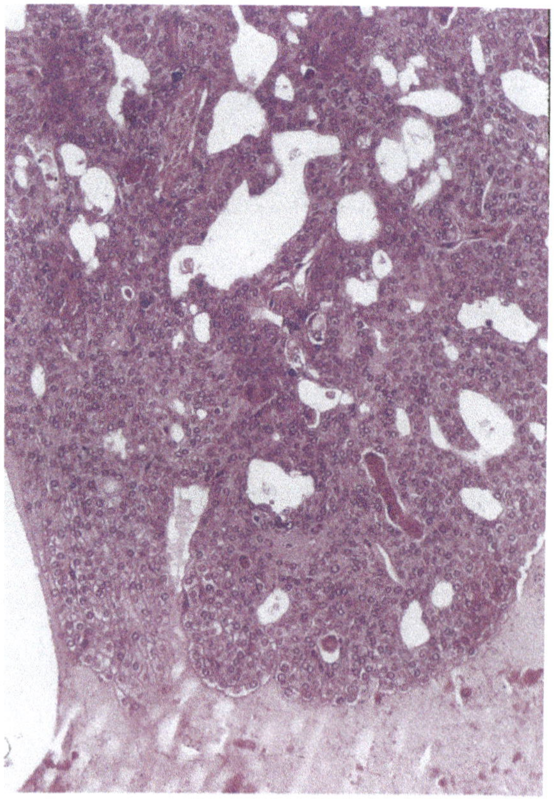

Fig. 17. Transitional cell papilloma of the urinary bladder. Cells are uniform and well differentiated. NMRI-BR mouse, male. H&E

Synonym(s). Papilloma, urothelial

Histogenesis

Arising from transitional cell epithelium.

Diagnostic Features

- Solitary, but may be multiple
- Growth pattern is exophytic papillary or polypoid projecting into the lumen from a wide base, or from a slender stalk
- Cells are well organized
- A complex, branching, fibrovascular core is present
- Glandular metaplasia or squamous metaplasia may be present
- Squamous metaplasia may exhibit keratin formation
- Cells are uniform and well differentiated
- Cytoplasm may be slightly more basophilic than in normal cells
- Mitotic figures are rare or absent
- No evidence of invasion is present

Differential Diagnosis

Hyperplasia, Transitional Cell
Multifocal or diffuse and
Growth pattern is broad-based, nodular or exophytic with simple branching projections and less complex

Carcinoma, Transitional Cell
Metastasis is present or
Invasion of bladder wall is present and
Demarcation is poor and
Cellular atypia is present and
Hemorrhage or necrosis may be present

Comment

A diagnosis of transitional cell papilloma of mouse renal pelvis should not be used because small lesions found in the restricted confines of the renal pelvis represent either urothelial hyperplasia or early transitional cell carcinoma. Chemically-induced papillomas of the urinary bladder usually consist of a solid mass of transitional epithelial cells with scant connective tissue, while spontaneously occurring papillomas tend to have a thin cover of transitional epithelium over a fibrovascular stromal core. Chemically induced bladder papilloma may be completely squamous. Spontaneously occurring squamous cell papilloma has not been reported.

References

(See 2, 8, 11, 21, 25, 61)

Carcinoma, Transitional Cell *(M)*

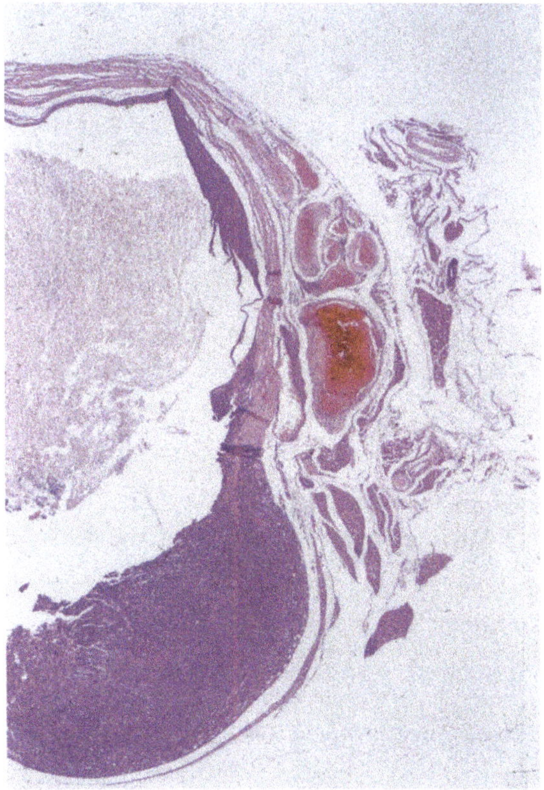

Synonym(s). Carcinoma, urothelial

Histogenesis

Arising from transitional cell epithelium.

Diagnostic Features

- Solitary or multiple
- Demarcation is poor
- Growth pattern may be papillary projecting into the lumen from a slender stalk, or solid originating from a broad base as a sessile lesion
- Cords, solid sheets, and solid or hollow nests of transitional cells are separated by delicate strands of connective tissue
- Squamous metaplasia may be present
- Vascularization may be prominent
- Hemorrhage or necrosis may be present
- Great majority of cells are transitional cells
- Cell density may be increased
- Cellular polarity is present in solid or hollow nests of transitional cells
- In sheets of transitional cells cellular polarity is lost
- Cellular pleomorphism may be present
- Nuclei may be bizarre or spindle-shaped
- Mitotic figures may be numerous

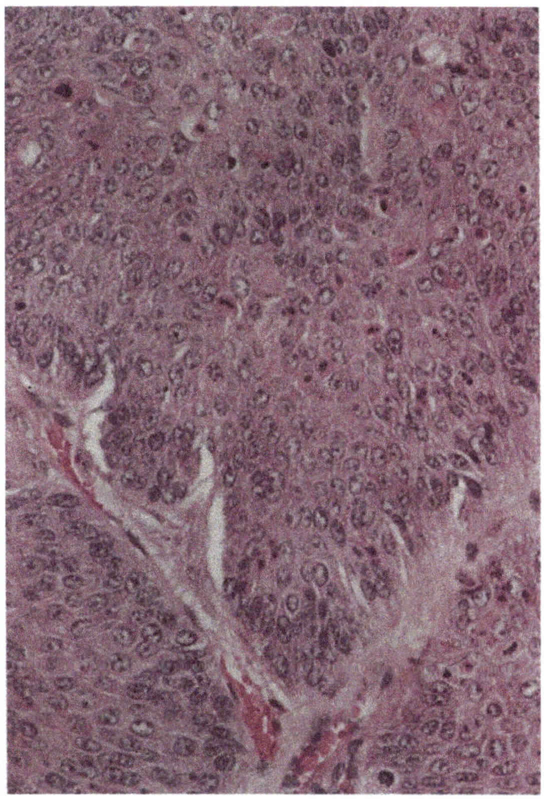

Fig. 18 *(above).* Transitional cell carcinoma of the urinary bladder. Solid growth in the urinary bladder wall. NMRI mouse, female. H&E

Fig. 19 *(below).* Transitional cell carcinoma of the urinary bladder. Sheets of relatively well-differentiated transitional cell epithelium separated by connective tissue stroma. H&E

- Inflammatory cells, especially lymphocytes and mast cells, may be present
- Invasion of the kidney parenchyma, muscle wall or lymphatics is present
- Metastasis may be present

Differential Diagnosis

Papilloma, Transitional Cell
No evidence of invasion and
No evidence of cellular pleomorphism and
Fibrovascular core is complex branching

Carcinoma, Squamous Cell
Great majority of cells are squamous cells

Carcinoma, Renal Tubule
No evidence of non-tubule-derived cells

Comment

Spontaneous transitional cell carcinomas of the renal pelvis have not been reported in control mice, but they have been induced by genotoxic chemicals such as dimethylnitrosamine and OH-BBN. The OH-BBN studies in NON/Shi mice illustrate the potential of this tumor to metastasize to the lungs.

Induced transitional cell carcinomas of the ureter are extremely rare in the mouse. Spontaneous transitional carcinomas of the mouse urinary bladder or urethra are rare. They have been induced experimentally by a few genotoxic chemicals, notably N[4-5-nitro-2-furyl)-2-thiazolyl]formamide (FANFT) and OH-BBN. Bladder tumors induced by chemical carcinogens in mice are usually flat, poorly differentiated, invasive carcinomas in contrast to the exophytic papillary bladder tumors induced in rats. OH-BBN is a particularly potent bladder carcinogen for the mouse, rapidly inducing highly aggressive transitional cell carcinomas that can metastasize to the lungs.
Transitional cell carcinomas appear to metastasize less frequently in the mouse than do the squamous cell type.

References

(See 2, 4, 6, 8, 11, 15, 18, 20, 21, 22, 23, 25, 40, 41, 42, 50, 55, 61)

Carcinoma, Squamous Cell *(M)*

Synonym(s). Carcinoma, epidermoid

Histogenesis

Arising from transitional cell epithelium.

Diagnostic Features

- Solitary or multiple
- Demarcation is poor
- Growth pattern is irregular
- Squamous cells are arranged in cords, sheets, or irregular nests
- Vascularization is very prominent
- Great majority of cells are squamous cells
- Cellular atypia and pleomorphism is pronounced
- Keratinization and pearl formation is usually present
- Mitotic figures are usually present
- Inflammatory cell infiltrate is often present
- Invasion of the surrounding tissue is present
- Metastasis is common.

Differential Diagnosis

Carcinoma, Transitional Cell
Great majority of cells are transitional cells and
Areas of squamous differentiation are small

Comment

Areas of squamous differentiation may be present in transitional cell tumors. The diagnosis squamous cell carcinoma should be reserved for tumors where the great majority of cells are of the squamous type.

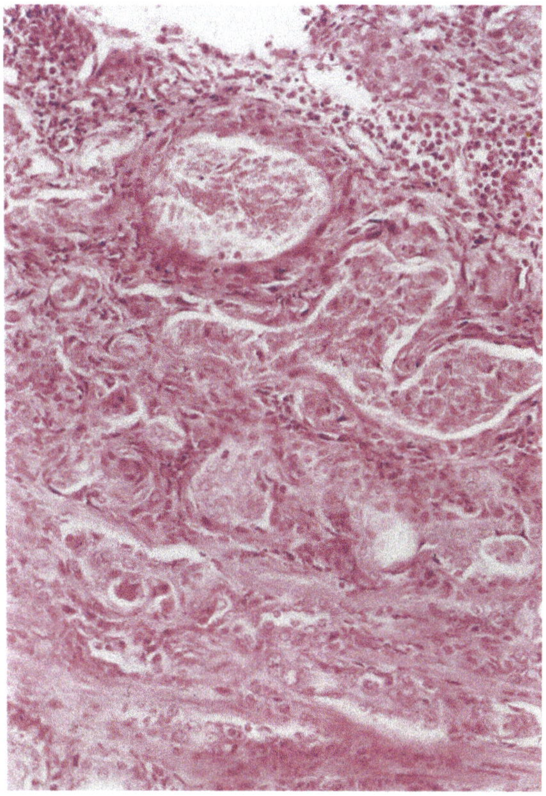

Fig. 20. Squamous cell carcinoma of the urinary bladder. Clusters of squamous cells with some keratinization. Note the prominent inflammatory cell infiltrate. NMRI mouse, female. H&E

Spontaneous squamous carcinomas are extremely rare, but this tumor can be induced experimentally in the renal pelvis following the introduction of polynuclear hydrocarbons into the kidney or renal pelvis, and in hydronephrotic NON/Shi mice after treatment with OH-BBN. In the urinary bladder it has been induced by dibutylnitrosamine and OH-BBN.
However, squamous cell carcinomas are encountered less frequently in association with carcinogens than the transitional cell counterpart, but appear to metastasize more frequently.

References

(See 1, 3, 5, 8, 11, 20, 22, 23, 24, 25, 32, 41, 42, 50, 53, 54)

Mesenchymal Proliferative Lesion (*B*)

Solid • Polypoid

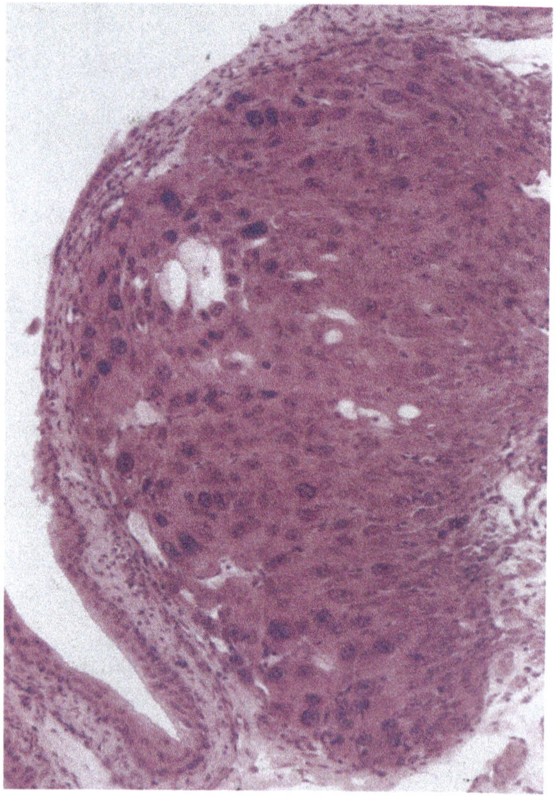

Fig. 21. Mesenchymal proliferative lesion of the urinary bladder. Solid growth located between mucosa and muscle layer. NMRI mouse, female. H&E

Synonym(s). Vegetative lesion; mesenchymal tumor; granular cell tumor; undifferentiated sarcoma; hemangioma/sarcoma; leiomyosarcoma; leiomyoblastoma

Histogenesis

Unknown.

Diagnostic Features

- Located in the submucosa, usually in caudal half of bladder, often near trigone
- Solitary, may be multiple
- Is distinctly demarcated
- Non-encapsulated
- The growth pattern is solid or polypoid
- Vascular channels are well developed
- Areas of coagulative necrosis may be present
- Cells of two types are present, large epithelioid cells in compact islands, and spindle cells located more peripherally
- Epithelioid cells have distinct borders, eosinophilic, homogeneous or fibrillar cytoplasm, and eosinophilic granules may be present
- Nuclei of epithelioid cells are eccentric, often with pleomorphic or bizarre outline, and prominent nucleoli
- Spindle cells are fibrocyte-like or smooth muscle-like
- Mitotic figures may be present
- Mononuclear inflammatory cell infiltrate is invariably present, usually peripherally
- Hemosiderin-laden macrophages are often present
- Does not invade surface urothelium or go beyond serosa
- Local invasion of smooth muscle and submucosa by spindle cell component may be present

Solid

- Single or multiple submucosal proliferations

Polypoid

- Polypoid mass is bulging or protruding into the lumen
- Mass has a fibrovascular stroma covered by urothelium
- Close to the surface, suburothelial foci of solid mesenchymal proliferative lesions are present

Differential Diagnosis

None

Comment

The definite nature of this lesion is still under debate. One view regards some of these lesions as mesenchymal cell proliferations and others as benign tumors (B) based on the potential to be locally invasive. The other considers the lesion to be a decidual-like reaction because of the morphological similarities to the uterine decidual reaction, which also can appear to be locally invasive, but is still considered to be a hyperplastic lesion (H).

This lesion is primarily spontaneous and strain-specific, being found most frequently in Swiss-derived mice. Thus, it has been seen predominantly in Swiss Webster, NMRI and CD1 strains, but not as yet in B6C3F1 mice. Identical bladder lesions have been described in association with surgically implanted glass or paraffin pellets, particularly in the vicinity of sutures used to close the bladder after implantation. Whether these lesions can be induced by chemical agents is not presently known, although an increased incidence has been noted with various combinations of endogenously administered estrogens and progesterone. Primary lesions of similar morphology have been identified in reproductive organs of both male and female mice, namely the prostate, seminal vesicles and uterus.

References

(See 7, 9, 10, 25, 28, 33, 34, 39, 45)

References

1. Akagi G, Akagi A, Kimura M, Otsuka H (1973) Comparison of bladder tumors induced in rats and mice with N-butyl-N-(4-hydroxybutyl)-nitrosoamine. GANN 64: 331–336
2. Armstrong EC, Bonser GM (1945) Epithelial tumours of the urinary bladder in mice induced by 2-acetylaminofluorene. J Pathol Bacteriol 57: 517–522
3. Becci PJ, Thompson HJ, Grubbs CJ, Squire RA, Brown CC, Sporn MB, Moon RC (1978) Inhibitory effect of 13-cis-retinoic acid on urinary bladder carcinogenesis induced in C57BL/6 mice by N-butyl-N-(4-hydroxybutyl)-nitrosamine. Cancer Res 38: 4463–4466
4. Becci PJ, Thompson HJ, Strum JM, Brown CC, Sporn MB, Moon RC (1981) N-Butyl-N-(4-hydroxybutyl)nitrosamine-induced urinary bladder cancer in C57BL/6 x DBA/2 F1 mice as a useful model for study of chemoprevention of cancer with retinoids. Cancer Res 41: 927–932
5. Bertram JS, Craig AW (1970) Induction of bladder tumours in mice with dibutylnitrosamine. Br J Cancer 24: 352–359
6. Bertram JS, Craig AW (1972) Specific induction of bladder cancer in mice by butyl-(4-hydroxybutyl)-nitrosamine and the effects of hormonal modifications on the sex difference in response. Eur J Cancer 8: 587–594
7. Bonser GM, Jull JW (1957) The histopathological changes in the mouse urinary bladder following surgical implantation of paraffin wax pellets containing various chemicals. J Pathol Bacteriol 72: 499–505
8. Boorman GA, Wood M, Fukushima S (1994) Tumours of the urinary bladder. Pathology of tumours in laboratory animals, vol 2. Tumours of the mouse, 2nd edn. IARC Scientific Publications No. 111, Lyon, pp 383–406
9. Butler WH, Cohen SH, Squire RA (1997) Mesenchymal tumors of the mouse urinary bladder with vascular and smooth muscle differentiation. Toxicol Pathol 25: 268–274
10. Chandra M, Frith CH (1991) Spontaneously occurring leiomyosarcomas of the mouse urinary bladder. Toxicol Pathol 19: 164–167
11. Cohen SM, Friedell GH (1982) Neoplasms of the urinary system. In: Foster HL, Small JD, Fox JG (eds) The mouse in biomedical research. Academic Press, New York, pp 449–473
12. Dawe CJ, Law LW, Dunn TB (1960) Studies of parotid-tumor agent in cultures of leukemic tissues of mice. J Natl Cancer Inst 23: 717–798
13. Delahunt B, Cartwright PR, Thornton A, Dady PJ (1995) Proliferation kinetics of streptozotocin-induced renal tumours in mice. Virchows Arch 425: 577–582
14. Dunn TB (1967) Renal disease of the mouse. In: Cotchin E, Roe FJC (eds) Pathology of laboratory rats and mice. Blackwell Scientific Publications, Oxford, pp 150–176
15. Ertürk E, Cohen SM, Bryan GT (1970) Urinary bladder carcinogenicity of N-[4-(5-nitro-2-furyl)-2-thiazolyl] formamide in female Swiss mice. Cancer Res 30: 1309–1311
16. Eustis SL, Hailey JR, Boorman GA, Haseman JK (1994) The utility of multiple-section sampling in the histopathological evaluation of the kidney for carcinogenicity studies. Toxicol Pathol 22: 457–472
17. Faccini JM, Abbott DP, Paulus GJJ (1990) Mouse histopathology. A glossary for use in toxicity and carcinogenicity studies. Elsevier, Amsterdam
18. Frith CH (1978) Report of a workshop on urothelial lesions in mice. J Environ Pathol Toxicol 1: 617–626
19. Frith CH (1979) Morphologic classification of inflammatory, nonspecific, and proliferative lesions of the urinary bladder of mice. Invest Urol 16: 435–444
20. Frith CH (1986) Transitional cell carcinoma, urinary tract, mouse. In: Jones TC, Mohr U, Hunt RD (eds) Monographs on pathology of laboratory animals. Urinary system. Springer, Berlin Heidelberg New York Tokyo, pp 341–348
21. Frith CH, Ward JM (1988) Urinary system. In: (eds) Color atlas of neoplastic and non-neoplastic lesions in aging mice. Elsevier, Amsterdam, pp 67–76
22. Frith CH, Farmer JH, Greenman DL, Shaw GW (1980) Biologic and morphologic characteristics of urinary bladder neoplasms induced in BALB/c female mice with 2-acetylaminofluorene. J Environ Pathol Toxicol 3: 103–119
23. Frith CH, Terracini B, Turusov VS (1994) Tumours of the kidney, renal pelvis and ureter. Pathology of tumours in laboratory animals, vol 2. Tumours of the mouse, 2nd edn. IARC Scientific Publications No. 111, Lyon, pp 357–381
24. Gagne F, Roulin M, Lacour F, Guerin M (1955) Cancerisation du bassinet par benzopyrene provoquee apres hydronephrose chez le rat et la souris. Bull Cancer 42: 270–279
25. Gaillard ET (1999) Ureter, urinary bladder, and urethra. In: Maronpot RR, Boorman GA, Gaul BW (eds) Pathology of the mouse. Reference and atlas. Cache River Press, Vienna, pp 235–258
26. Greaves P (1990) Histopathology of preclinical toxicity studies: interpretation and relevance in drug safety evaluation. Elsevier, Amsterdam

27. Guerin M, Chouroulinkov I, Riviere MR (1969) Experimental kidney tumors. In: Rouiller C, Muller AF (eds) The kidney: morphology, biochemistry, physiology, vol II. Academic Press, New York, pp 199–268
28. Halliwell WH (1998) Submucosal mesenchymal tumors of the mouse urinary bladder. Toxicol Pathol 26: 128–136
29. Ham AW, Siminovich L (1961) Viral carcinogenesis with particular reference to in vivo and in vitro studies with the polyoma virus. Prog Exp Tumor Res 2: 67–89
30. Hard GC (1985) Identification of a high-frequency model for renal carcinoma by the induction of renal tumors in the mouse with a single dose of streptozotocin. Cancer Res 45: 703–708
31. Hard GC, Alden CL, Stula EF, Trump BF (1995) Proliferative lesions of the kidney in rats. In: (eds) Guides for toxicologic pathology. STP/ARP/AFIP, Washington DC
32. Ilfeld FW (1937) The experimental production of visceral tumors with hydrocarbons. Am J Cancer 26: 744–754
33. Jacobs JB, Cohen SM, Arai M, Friedell GH, Bulay O, Urman HK (1976) Chemically induced smooth muscle tumors of the mouse urinary bladder. Cancer Res 36: 2396–2398
34. Karbe E, Hartmann E, George C, Wadsworth P, Harleman J, Geiss V (1998) Similarities between the uterine decidual reaction and the "mesenchymal lesion" of the urinary bladder in aging mice. Exp Toxicol Pathol 50: 330–340
35. Kondalenko VF, Turusov VS, Chemeris GY (1997) Renal cell tumours in mice: electron microscopy versus immunohistochemistry. Exp Toxicol Pathol 49: 129–134
36. Levi PE, Knowles JC, Cowen DM, Wood M, Cooper EH (1971) Disorganization of mouse bladder epithelium induced by 2-acetylaminofluorene and 4-ethylsulfonylnaphthalene-1-sulfonamide. J Natl Cancer Inst 46: 337–352
37. Liebelt AG, Sass B, Sobel HJ, Werner RM (1989) Spontaneous nephroblastoma in a strain CE/J mouse. A case report. Toxicol Pathol 17: 57–61
38. Lombard LS, Rice JM, Vesselinovitch SD (1974) Renal tumors in mice: light microscopic observations of epithelial tumors induced by ethylnitrosourea. J Natl Cancer Inst 53: 1677–1685
39. McConnell RF (1989) General observations on the effects of sex steroids in rodents with emphasis on long-term oral contraceptive studies. In: Michael F (ed) Safety requirements for contraceptive steroids. Cambridge University Press, Cambridge, pp 211–229
40. Murai T, Mori S, Hosono M, Takeuchi Y, Ohara T, Makino S, Takeda R, Hayashi Y, Fukushima S (1991) Renal pelvic carcinoma which shows metastatic potential to distant organs, induced by N-butyl-N-(4-hydroxybutyl)nitrosamine in NON/Shi mice. Jpn J Cancer Res 82: 1371–1377
41. Murai T, Mori S, Hosono M, Takeuchi Y, Ohara T, Makino S, Hayashi Y, Takeda R, Fukushima S (1994a) Influences of aging and sex on renal pelvic carcinogenesis by N-butyl-N-(4-hydroxybutyl)nitrosamine in NON/Shi mice. Cancer Lett 76: 147–153
42. Murai T, Mori S, Hosono M, Takeuchi Y, Ohara T, Makino S, Takeda R, Fukushima S (1994b) Effect of phenacetin pretreatment on renal pelvic carcinogenesis by N-butyl-N-(4-hydroxybutyl)nitrosamine in NON/Shi mice of both sexes. Teratog Carcinog Mutagen 14: 193–201
43. Prejean JD, Peckham JC, Casey AE, Griswold DP, Weisburger EK, Weisburger JH (1973) Spontaneous tumors in Sprague-Dawley rats and Swiss mice. Cancer Res 33: 2768–2773
44. Reznik G, Ward JM, Hardisty JF, Russfield A (1979) Renal carcinogenic and nephrotoxic effects of the flame retardant tris(2,3-dibromopropyl) phosphate in F344 rats and (C57BL/6 N X C3H/HeN)F1 mice. J Natl Cancer Inst 63: 205–212
45. Roe FJC (1964) An illustrated classification of the proliferative and neoplastic changes in mouse bladder epithelium in response to prolonged irritation. Br J Urol 37: 239–254
46. Rosen VJ, Cole LJ (1962) Accelerated induction of kidney neoplasms in mice after X-radiation (690 rad) and unilateral nephrectomy. J Natl Cancer Inst 28: 1031–1042
47. Sass B (1986) Adenoma, adenocarcinoma, kidney, mouse. In: Jones TC, Mohr U, Hunt RD (eds) Monographs on pathology of laboratory animals. Urinary system. Springer, Berlin Heidelberg New York Tokyo, pp 87–101
48. Seely JC (1999) Kidney. In: Maronpot RR, Boorman GA, Gaul BW (eds) Pathology of the mouse. Reference and atlas. Cache River Press, Vienna, pp 207–234
49. Shinohara Y, Frith CH (1980) Morphologic characteristics of benign and malignant renal cell tumors in control and 2-acetylaminofluorene-treated BALB/c female mice. Am J Pathol 100: 455–468
50. Stevenson JL, von Haam E (1962) Induction of kidney tumors in mice by the use of 20-methylcholanthrene-impregnated strings. Cancer Res 22: 1177–1179
51. Stewart SE (1963) Polyoma virus carcinogenesis. Acta Un Int Cancer 19: 256–262
52. Takayama S, Oota K (1965) Induction of malignant tumors in various strains of mice by oral

administration of N-nitrosodimethylamine and N-nitrosodiethylamine. GANN 57: 189–199
53. Tamano S, Hagiwara A, Suzuki E, Okada M, Shirai T, Fukushima S (1991) Time- and dose-dependent induction of invasive urinary bladder cancers by N-ethyl-N-(4-hydroxybutyl)nitrosamine in B6C3F1 mice. Jpn J Cancer Res 82: 650–656
54. Terracini B, Campobasso O (1979) Tumours of the kidney, renal pelvis and ureter. In: Turusov VS (ed) Pathology of tumours in laboratory animals, vol II. Tumours of the mouse. IARC Scientific Publications No. 23, Lyon, pp 289–300
55. Terracini B, Palestro G, Gigliardi MR, Montesano G, Montesano R (1966) Carcinogenicity of dimethylnitrosamine in Swiss mice. Br J Cancer 20: 871–876
56. Toth B, Rappaport H, Shubik P (1963) Influence of dose and age on the induction of malignant lymphomas and other tumors by 7,12-dimethylbenzanthracene in Swiss mice. J Natl Cancer Inst 30: 723–742
57. Turusov VS (1992) Nephroblastoma in a strain CBA male mouse treated with 1,2-dimethylhydrazine. Exp Oncol 14: 20–22
58. Turusov VS, Chemeris GY (1992) Renal cell tumors induced in CBA male mice by 1,2-dimethylhydrazine. Toxicol Pathol 20: 570–575
59. Turusov VS, Day NE, Tomatis L, Gati E, Charles RT (1973) Tumors in CF-1 mice exposed for six consecutive generations to DDT. J Natl Cancer Inst 51: 983–997
60. Turusov VS, Lanko NS, Parfenov YD, Gordon WP, Nelson SD, Hillery PS, Keefer LK (1988) Carcinogenicity of deuterium-labeled 1,2-dimethylhydrazine in mice. Cancer Res 48: 2162–2167
61. Wood M, Bonser GM (1979) Tumours of the urinary bladder. In: Turusov VS, Davis V (eds) Pathology of tumours in laboratory animals, vol II. Tumours of the mouse. IARC Scientific Publications No. 23, Lyon, pp 301–324

Male Genital System

S. Rehm[1] · J.H. Harleman[2,3] · M. Cary · D. Creasy · R.A. Ettlin
S.L. Eustis · G.L. Foley · J.-L. LeNet · A. Maekawa · K. Mitsumori
R.F. McConnell · G. Reznik[3](†)

[1]Chair; [2]Co-chair; [3]Initial draft.

Hyperplasia, Leydig Cell *(H)*

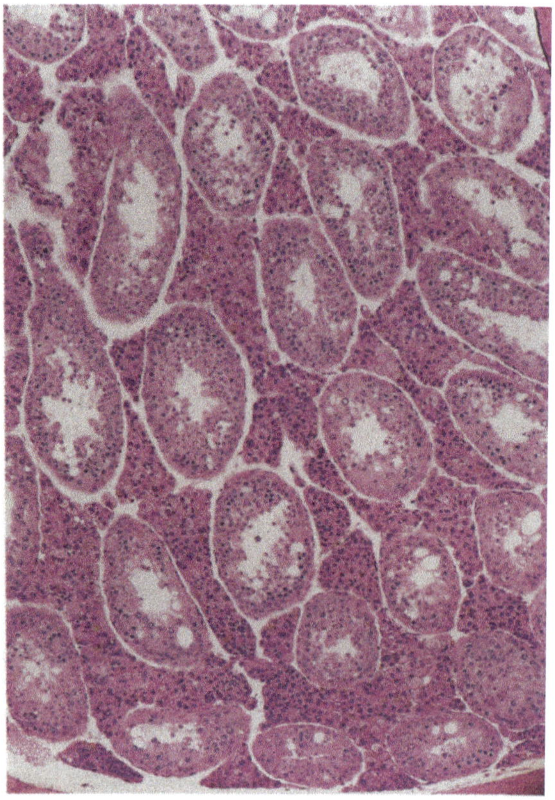

Fig. 1. Diffuse Leydig cell hyperplasia of the testis. CD-1 mouse, 2-year study, age >12 months. H&E

Synonym(s). Hyperplasia, interstitial cell

Histogenesis

Leydig cells.

Diagnostic Features

- Focal, multifocal or diffuse collections of Leydig cells between seminiferous tubules
- Cells with central nucleus and often prominent nucleolus, abundant eosinophilic cytoplasm, sometimes vacuolated

Focal/Multifocal
- Triangular or rounded Leydig cell aggregation between seminiferous tubules
- For consistent diagnosis, dimensions are selected that range from cell aggregates occupying 25% of a seminiferous tubule diameter up to an area of approximately 3 seminiferous tubules
- No or only minimal compression of the surrounding tissue is seen

Diffuse
- Hyperplasia forming a diffuse connecting network of interstitial cells between the seminiferous tubules
- Lesion may be locally extensive involving a large portion of the testis

Differential Diagnosis

Adenoma, Leydig Cell
Usually symmetrical peripheral compression of adjacent tubules; larger than three seminiferous tubules; cells and nuclei generally larger or pleomorphic, may have some nuclear atypia; presence of dilated vascular network

Carcinoma, Leydig Cell
Polymorphism, invasion into adjacent tissue (vessels, capsule) or formation of metastases

Comment

Diffuse hyperplasia is the most common type in mice. Both diffuse and focal lesions should be recorded in a study since most Leydig cell tumors begin as hyperplasias. The diagnostic criteria for separating hyperplasias and tumors are arbitrary. Leydig cell hyperplasia and neoplasia are particularly common in some mouse strains such as the NMRI, but not so common in other strains. Diffuse hyperplasia precedes tumor formation in mice with testicular feminization and is associated with estrogenic compounds or 5α-reductase inhibitors.

References

(See 7, 9, 12, 15, 16, 24, 33, 35)

Adenoma, Leydig Cell *(B)*

Synonym(s). Tumor, Leydig cell, benign; interstitial cell tumor, benign

Histogenesis

Leydig cells.

Diagnostic Features

- Mass with often circumferential peripheral compression of adjacent seminiferous tubules (may be uni- or bilateral)
- Composed predominantly of uniform polyhedral cells with abundant eosinophilic, finely granular or vacuolated cytoplasm
- Nucleus generally central, round with evenly distributed chromatin and single prominent nucleolus. Low nuclear/cytoplasmic ratio. Polyploidy and larger nuclei may be found
- Large tumors may have areas of less well-differentiated basophilic cells with scanty cytoplasm or elongated spindle-shaped cells
- Mitotic rate varies; no atypical mitotic figures
- Some tumors moderately vascular containing dilated thin-walled vessels and hemorrhage; focal areas of necrosis occasionally present
- Cystic areas contain proteinaceous material or blood
- Stroma usually scanty, areas of hyalinization and fibrosis of entrapped seminiferous tubules may be present
- Generally not encapsulated; adjacent seminiferous tubules often show variable degrees of atrophy
- Diameter of proliferative lesion is larger than three seminiferous tubules

Differential Diagnosis

Hyperplasia, Leydig Cell
Usually no compression of adjacent seminiferous tubules, nuclei smaller with more infoldings and without atypia. Diameter does not exceed three seminiferous tubules

Carcinoma, Leydig Cell
Polymorphism, invasion into adjacent tissue (vessels, capsule) or formation of metastases

Tumor, Sertoli Cell, Malignant
Cells are elongated with pale or vacuolated cytoplasm and are generally in palisades forming tortuous tubular structures

Tumor, Granulosa Cell, Benign
Composed of follicle-like nests or cords filled with small round cells with scant cytoplasm

Seminoma, Malignant
Cells are large, clear or eosinophilic with distinct cell boundaries and have a high nuclear/cytoplasmic ratio

Comment

The diagnosis of a benign Leydig cell tumor is mainly based on the size of the lesion, exceeding three normal seminiferous tubules and two or more distinct neoplastic nodules separated by tubular tissue are recorded as multifocal. Since these lesions appear to represent a continuous spectrum from small collections of hyperplastic cells to large tumors, the diagnostic criteria for separating hyperplasias and tumors are arbitrary. Leydig cell adenomas can be very large, replacing most or all of the original tissue and two or more distinct neoplastic nodules, separated by tubular tissue, are recorded as multifocal. Leydig cell tumors occur in mice with testicular feminization and can be induced with estrogenic compounds, 5α-reductase inhibitors or cadmium. Leydig cell tumors have also been reported in the epididymides.

References

(See 7, 8, 9, 12, 15, 16, 24, 26, 27, 33, 34, 35, 38)

Fig. 2 *(upper left).* Leydig cell adenoma of the testis. Neoplasm displaced most of normal architecture. Numerous distended vascular spaces evident. CD-1 mouse, 2-year study, age >12 months. H&E

Fig. 3 *(upper right).* Leydig cell adenoma of the testis. Cells with granular, eosinophilic cytoplasm and variable size. CD-1 mouse, 2-year study, age >12 months. H&E

Fig. 4 *(lower left).* Leydig cell adenoma of the testis. Cells with pale vacuolated cytoplasm surrounding rete testis tubule. CD-1 mouse, 2-year study, age >12 months. H&E ▶

Fig. 5 *(lower right).* Leydig cell adenoma of the testis. Small basophilic, undifferentiated cells surrounding atrophic seminiferous tubules. B6C3F1 mouse, 2-year study, age >12 months. H&E

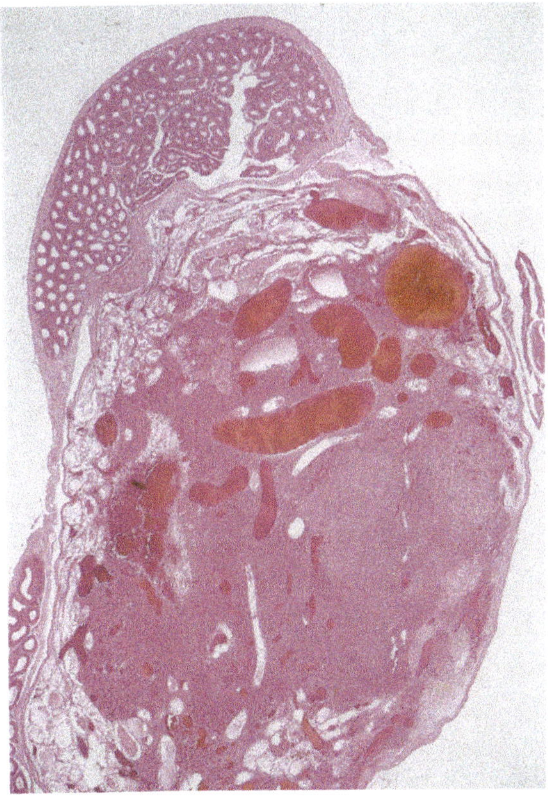

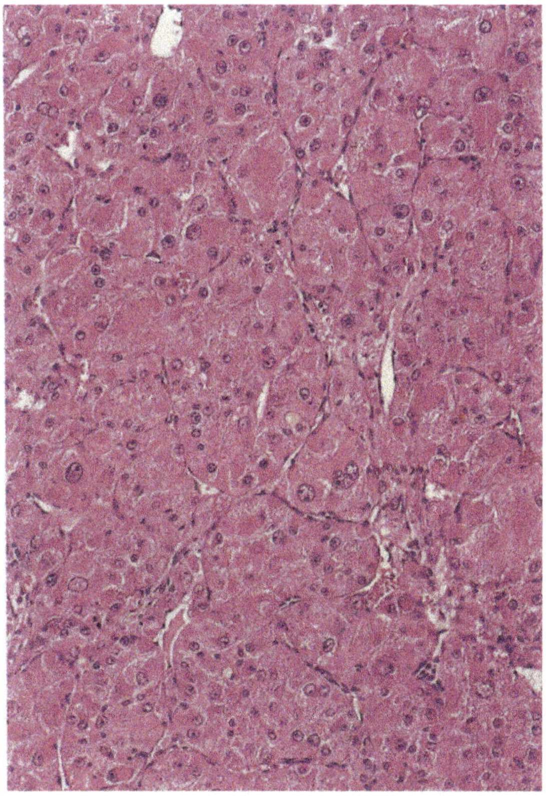

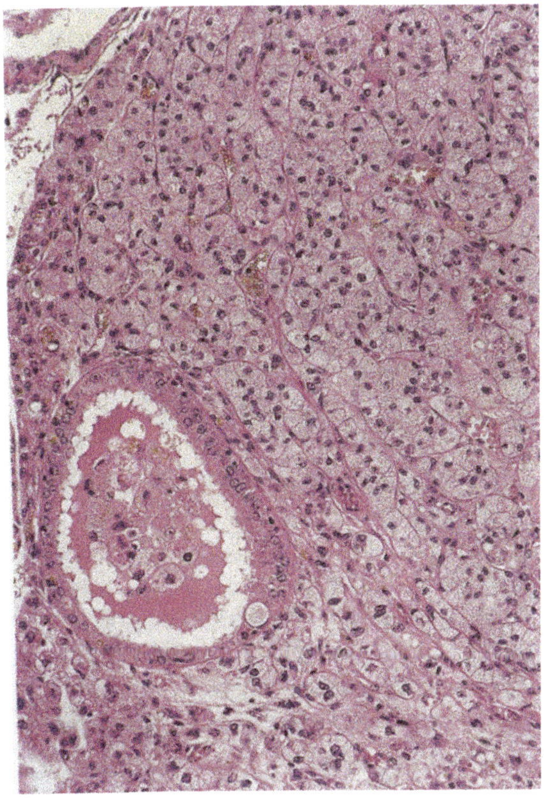

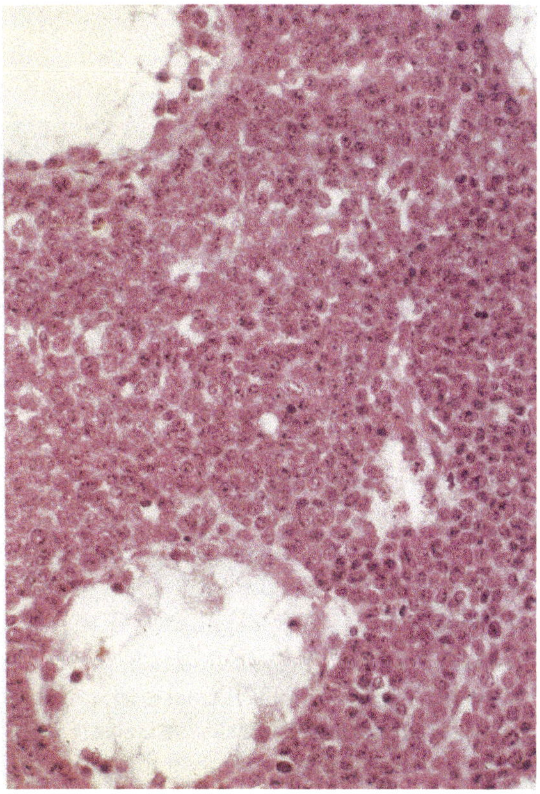

Carcinoma, Leydig Cell *(M)*

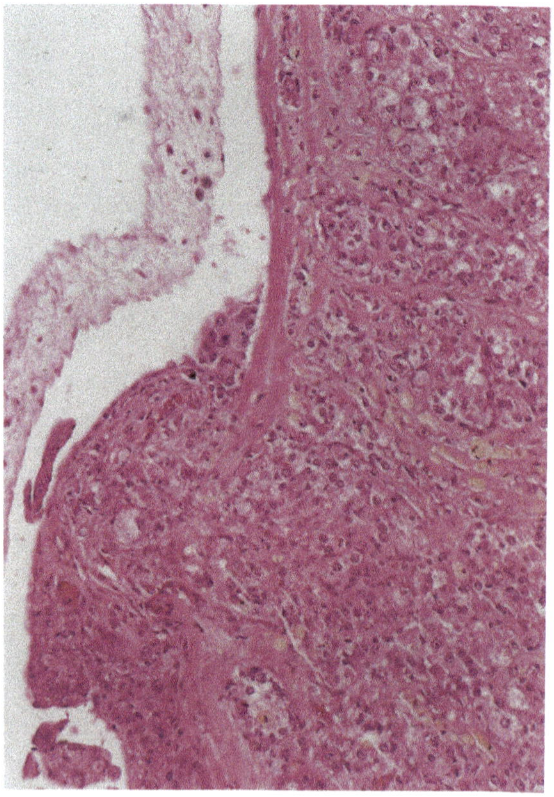

Fig. 6. Leydig cell carcinoma of the testis. Pleomorphic cells invading through the tunica vaginalis. CD-1 mouse, 2-year study, age >12 months. H&E

Synonym(s). Tumor, Leydig cell, malignant; interstitial cell tumor, malignant

Histogenesis

Leydig cells.

Diagnostic Features

- Mass showing invasion into the capsule or adjacent tissue or distant metastases
- Cellular pleomorphism is present such as poorly differentiated basophilic cells or spindle-shaped cells with scanty cytoplasm
- Mitoses are infrequent; mitotic figures are sometimes atypical
- Entrapped seminiferous tubules often seen
- Areas of necrosis and/or hemorrhage often present

Differential Diagnosis

Adenoma, Leydig Cell
: Although areas of spindle-shaped Leydig cells may occur, adenomas lack pronounced cellular pleomorphism and invasion of adjacent tissues or metastases

Seminoma, Malignant
: Cells are large, clear or eosinophilic with distinct cell boundaries and have a high nuclear/cytoplasmic ratio

Tumor, Sertoli Cell, Malignant
: Cells are elongated with pale or vacuolated cytoplasm and generally arranged in palisades forming tortuous tubular structures

Tumor, Granulosa Cell, Benign
: Composed of follicle-like nests or cords filled with small round cells with scant cytoplasm. Lack invasion. Due to the rarity of both granulosa cell tumors and Leydig cell carcinomas composed of small cells with scant cytoplasm distinguishing features have not been delineated

Comment

Invasion into the testicular tunica, the surrounding tissue or the spermatic cord or metastases are the most important criteria for the distinction between carcinoma and adenoma. In man, infiltration of the testicular tunica albuginea by Leydig cells can occur without neoplastic transformation. Identification of invasion into blood vessels and lymphatics is problematic. False positive diagnoses are easily made because of the blood and protein-filled spaces, which occur in Leydig cell tumors. Leydig cell carcinomas were seen in CD-1 mice exposed to diethylstilbestrol in utero.

References

(See 7, 9, 12, 14, 15, 16, 24, 27, 34, 38)

Tumor, Sertoli Cell, Malignant *(M)*

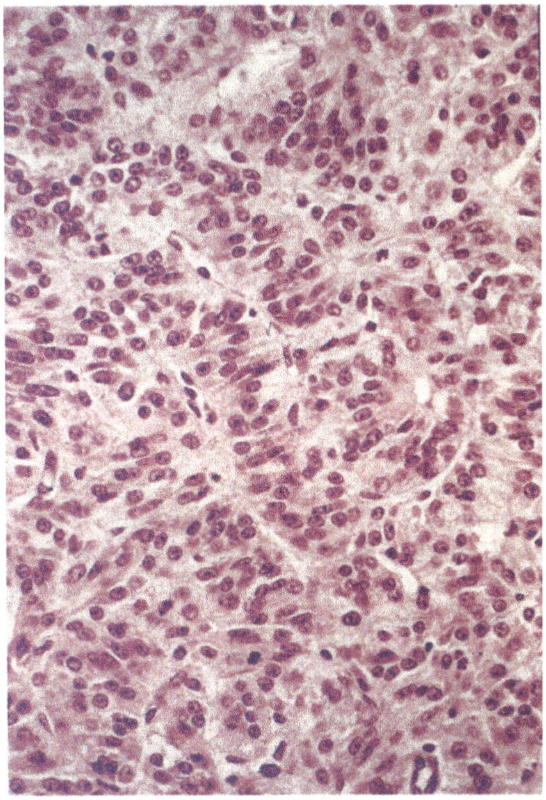

Fig. 7. Malignant Sertoli cell tumor of the testis. Nuclear palisading array of elongated cells with abundant pale cytoplasm. B6C3F1 mouse, 2-year study, age >12 months. H&E

Synonym(s). Androblastoma; arrhenoblastoma; sustentacular cell tumor, malignant; tumor, gonadal stromal, malignant; tumor, sex cord stromal, malignant

Histogenesis

Sex cord/stromal cells; Sertoli cells.

Diagnostic Features

- Cells generally arranged in palisades on a delicate fibrovascular stroma resulting in the formation of tortuous tubular structures without distinct lumina
- Occasionally less highly organized diffuse pattern, but tubular structures generally present
- Tumor cells elongated with generally pale and vacuolated cytoplasm and finely stippled nuclear chromatin, nucleoli not prominent, cell borders indistinct
- Mitotic figures quite common especially in less well-differentiated areas

Differential Diagnosis

Adenoma, Leydig Cell
or Carcinoma, Leydig Cell
May also contain elongated and spindle-shaped cells but not arranged in a palisading/tubular pattern as found in Sertoli cell tumor

Tumor, Granulosa Cell, Benign
Composed of follicle-like nests or cords filled with small round cells with scant cytoplasm

Comment

These tumors are extremely rare in mice. The diagnosis is based mainly on the cellular arrangement and the nuclear orientation. Because of the extreme rarity of these tumors, no distinction is made between benign and malignant Sertoli cell tumors in this classification. Malignant tumor must have invasion outside capsule, vasculature or distant metastasis.

References

(See 8, 12, 24, 27, 34)

Tumor, Granulosa Cell, Benign *(B)*

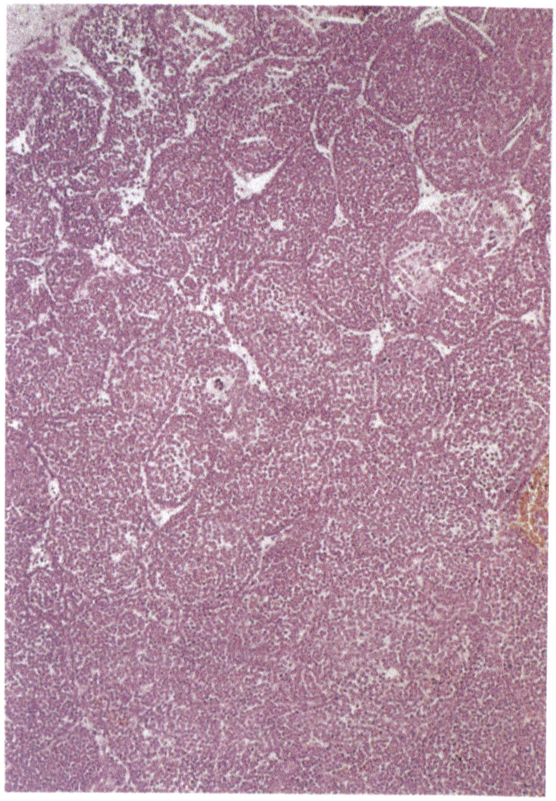

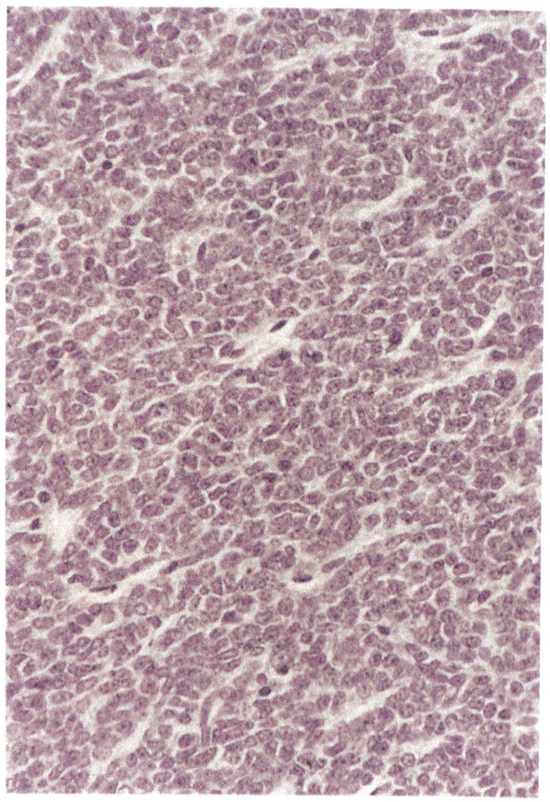

Histogenesis

Sex cord/stromal cells; Sertoli cells.

Diagnostic Features

- Growth pattern reminiscent of ovarian granulosa cell tumors, i.e., forming follicle-like nests and nodules, or cords and sheets replacing seminiferous tubules
- Small cells similar to those of ovarian granulosa cell tumors with scant eosinophilic cytoplasm and nuclei with a single nucleolus and stippled chromatin
- Cells frequently round but may be spindle-shaped or pleomorphic; mitotic figures are occasionally present

Differential Diagnosis

Tumor, Sertoli Cell, Malignant
Forms seminiferous tubule-like structures lined by highly elongated cells with abundant cytoplasm

Adenoma, Leydig Cell
Spindle-shaped cells usually are part of tumor composed of typical round cells

Fig. 8 *(above)*. Benign granulosa cell tumor of the testis. Small basophilic cells forming follicle-like structures. B6C3F1 mouse, 2-year study, age >12 months. H&E

Fig. 9 *(below)* Benign granulosa cell tumor of the testis. Basophilic cells with scant cytoplasm. CD-1 mouse, 2-year study, age >12 months. H&E

Carcinoma, Leydig Cell
Invasion into adjacent tissue (vessels, capsule) or formation of metastasis. Due to the rarity of both granulosa cell tumors and Leydig cell carcinomas composed of small cells with scant cytoplasm distinguishing features have not been delineated

Seminoma, Malignant
Large, clear or eosinophilic cells with distinct cell boundaries, high nuclear to cytoplasmic ratio

Comment

Because of the extreme rarity of these tumors no distinction is made between benign and malignant granulosa cell tumors in this classification. Malignant tumor must have invasion outside capsule, vasculature or distant metastasis.

References

(See 1, 24, 27, 34, 36, 38)

Seminoma, Malignant *(M)*

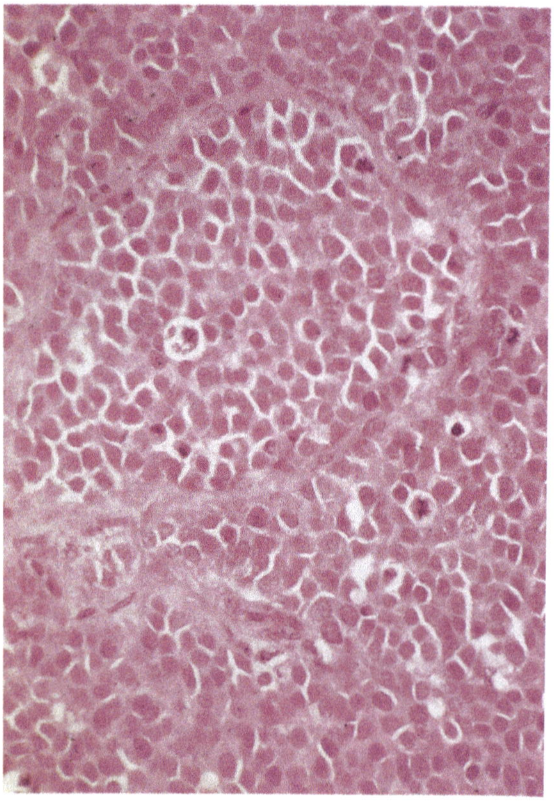

Fig. 10. Seminoma of the testis. Tubular structures filled with sheets of round to polyhedral cells with distinct cell borders. B6C3F1 mouse, 2-year study, age >12 months. H&E

Synonym(s). Germinoma, malignant; spermatoblastoma; spermatocytoma, malignant

Histogenesis

Spermatogenic cells (seminiferous tubules).

Diagnostic Features

- Cells arranged in sheets and lobules, may fill tubules and diffusely infiltrate stroma
- Large round to polyhedral cells with distinct cell boundaries
- Cells of different size, resembling primitive germ cells/spermatogonia
- Cytoplasm usually clear or eosinophilic, usually containing glycogen, surrounding a large centrally located spherical nucleus with 1–2 nucleoli
- Mitotic figures can be numerous and may be atypical
- The tumor may have lymphocytic or granulomatous reaction in the stroma

Differential Diagnosis

Adenoma, Leydig Cell
or Carcinoma, Leydig Cell
Uniform cells with abundant eosinophilic cytoplasm and sometimes also areas with spindle-shaped cells and small round nuclei and the pattern of growth is primarily intertubular

Tumor, Granulosa Cell, Benign
Composed of follicle-like nests or cords filled with small round cells with scant cytoplasm

Comment

Seminomas are extremely rare tumors in mice and not well described. The diagnosis of a malignant seminoma is mainly based on the cytological features of the tumor cells resembling spermatogonia, or at times spermatocytes. Because of the rarity of these tumors, no morphological distinction is made between benign and malignant seminomas in this classification.

References

(See 3, 7, 12, 24, 27, 34, 38)

Carcinoma, Yolk Sac *(M)*

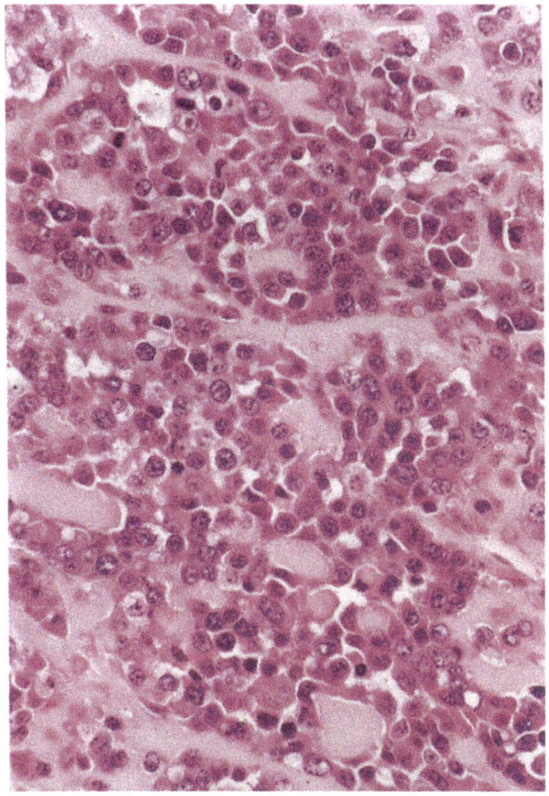

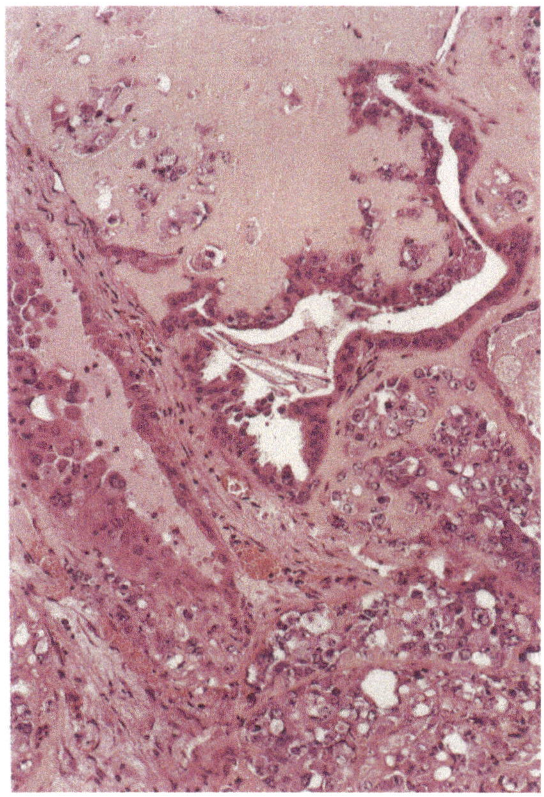

Histogenesis

Germ cell tumor variant.

Diagnostic Features

- Most characteristic features are tumor cells producing abundant eosinophilic, PAS-positive matrix in which neoplastic cells are embedded
- Tumors may have cell types and patterns mimicking two layers of fetal membranes, i.e., the parietal and visceral yolk sac
- Parietal yolk sac foci are composed of polygonal or cuboidal endodermal cells with amphophilic cytoplasm and pleomorphic nuclei with one or more nucleoli. Cytoplasm contains PAS-positive droplets or granules
- Parietal yolk cells form sheets, rosettes, cords or papillary structures. Occasionally small cysts and glomerular bodies found
- Visceral yolk sac endodermal cells do not contain eosinophilic, PAS-positive droplets and may be columnar or large polygonal cells with pale, vacuolated cytoplasm and giant nuclei
- Visceral yolk sac cells may form papillary structures surrounding capillaries, form cell clusters, or mingle with parietal cells

Fig. 11 *(above)*. Yolk sac carcinoma of the testis. Small basophilic cells producing eosinophilic, basement membrane-like material. CD-1 mouse, 2-year study, age >12 months. H&E

Fig. 12 *(below)*. Yolk sac carcinoma of the testis. Invasion of epididymis. Vacuolated large pale basophilic cells and abundant eosinophilic, basement membrane-like material. CD-1 mouse, 2-year study, age >12 months. H&E

- Metastatic spread of all tumor components may occur along the abdominal serosa, or be found at distant sites such as lymph nodes and lung

Differential Diagnosis

Carcinoma of the Rete Testis
or Adenocarcinoma of the Seminal Vesicle
or Adenocarcinoma of the Prostate
Lack abundant eosinophilic PAS-positive matrix but may have scirrhous component

Comment

Yolk sac carcinomas are rare tumors in the mouse. Currently no published reports are available on mouse testicular yolk sac carcinomas. Changes described in this classification system are based on a case found during a 2-year-carcinogenicity study with CD-1 mice.

References

(See 3, 21, 36)

Teratoma, Benign *(B)*

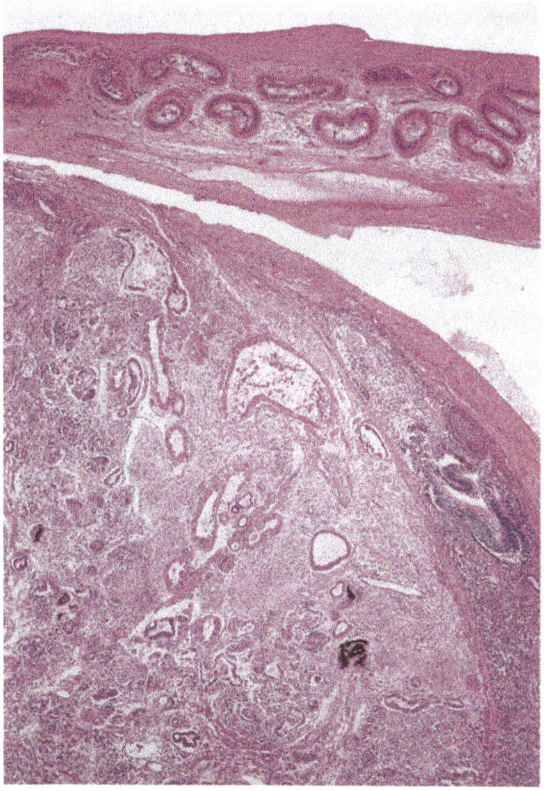

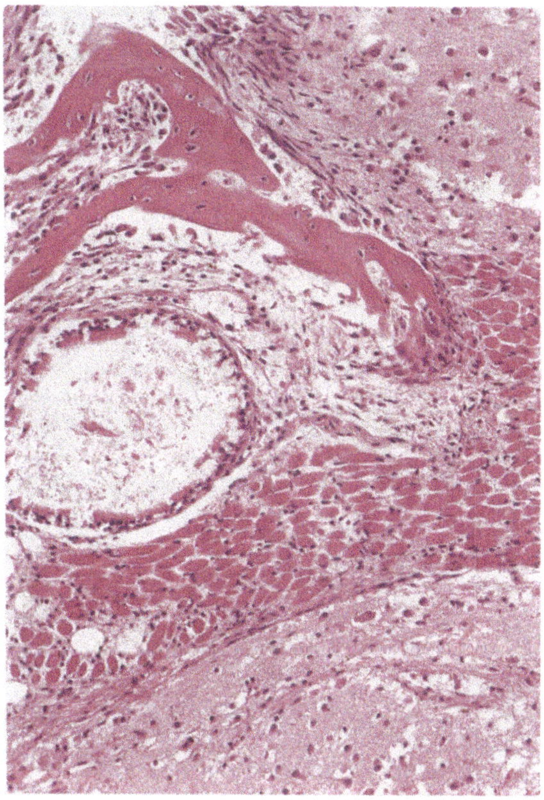

Histogenesis

Multipotential embryonic tissue which has escaped the influence of primary organizers.

Diagnostic Features

- Teratoma must contain tissue derived from the three germ layers
- Tissues components are generally well-differentiated
- Benign teratomas generally contain cysts which may be lined by cuboidal, enteric or respiratory epithelium. Smooth muscle may surround these cysts
- Other elements may consist of pancreatic tissue, gastric epithelium, thyroid, well-differentiated neuronal tissue, cartilage, bone and/or skeletal muscle

Differential Diagnosis

Teratoma, Malignant

Malignant teratomas are composed of poorly differentiated tissue elements and often show large areas with necrosis and/or hemorrhage or other malignant features (invasion and metastases)

Fig. 13 *(above)*. Benign teratoma of the testis. Cystic epithelial structures in association with neurogenic tissues; epididymis is present at one side. HuCD4/Tg mouse, age 6 months. H&E

Fig. 14 *(below)*. Benign teratoma of the testis. Cyst lined by epithelium (mucous cells) and surrounded by bony spicules, muscles and nervous tissue. HuCD4/Tg mouse, age 6 months. H&E

Comment

Teratomas are rare tumors but have been observed more commonly in substrains of strain 129. They may occur in any tissue but are most often encountered in the genital system.

References

(See 2, 27, 36, 39)

Teratoma, Malignant *(M)*

Histogenesis

Multipotential embryonic tissue which has escaped the influence of primary organizers.

Diagnostic Features

- Teratomas must contain tissue derived from the three germ layers
- In malignant teratomas, nervous, epithelial, and mesenchymal tissue are poorly differentiated resembling embryonic tissue
- These tumors are invasive and may metastasize
- Areas of necrosis and hemorrhage, indicative of a rapid growth, may be present

Differential Diagnosis

Teratoma, Benign
Benign teratomas have a higher degree of tissue maturation. Large areas with necrosis and/or hemorrhage and other malignant features such as invasion or metastasis are absent

Sex Cord Stromal Tumors
Sex cord stromal tumors do not have tissues from the three germ layers as an integral part of the tumor

Comment

Teratomas are rare tumors, which are most often encountered in the genital system but may occur in any tissue.

References

(See 2, 25, 27, 36, 39)

Hyperplasia, Rete Testis *(H)*

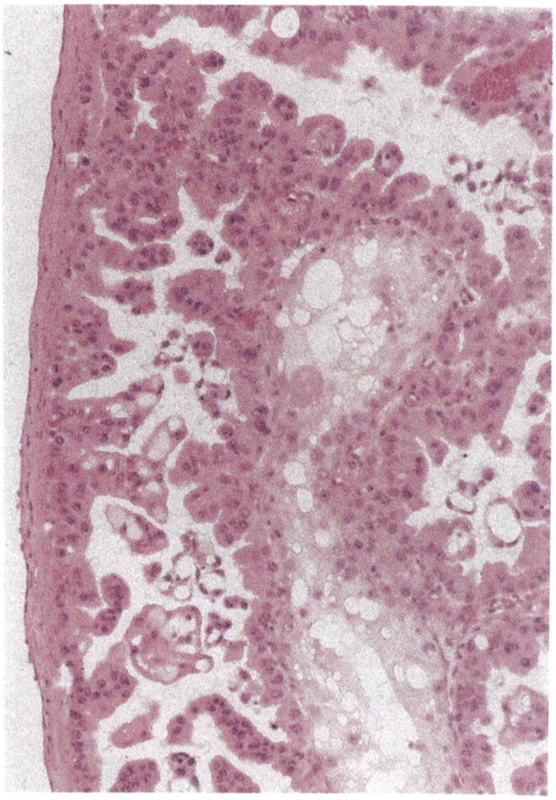

Fig. 15. Hyperplasia of the rete testis. Tubules lined by hyperplastic, papillary projections. CD-1 mouse, 2-year study, age >12 months. H&E

Histogenesis

Rete testis epithelium.

Diagnostic Features

- Focal/multifocal proliferations of up to three cell layers, frequently forming papillary knob/finger-like projections of rete testis tubules localized in the mediastinum testis
- No distortion of the normal architecture and absence of compression
- Cells are cuboidal to tall columnar with abundant eosinophilic cytoplasm and round to oval vesicular nuclei with a single nucleolus
- Cross-sections of papillary projections into lumen resemble rosettes
- Mitotic activity is low
- Cystic structures filled with spermatozoa may be present

Differential Diagnosis

Adenoma, Rete Testis

Composed of extensive papillary structures with supportive stroma and causes compression of adjacent tissue due to increased tissue mass

References

(See 5, 9, 12, 31, 44)

Adenoma, Rete Testis *(B)*

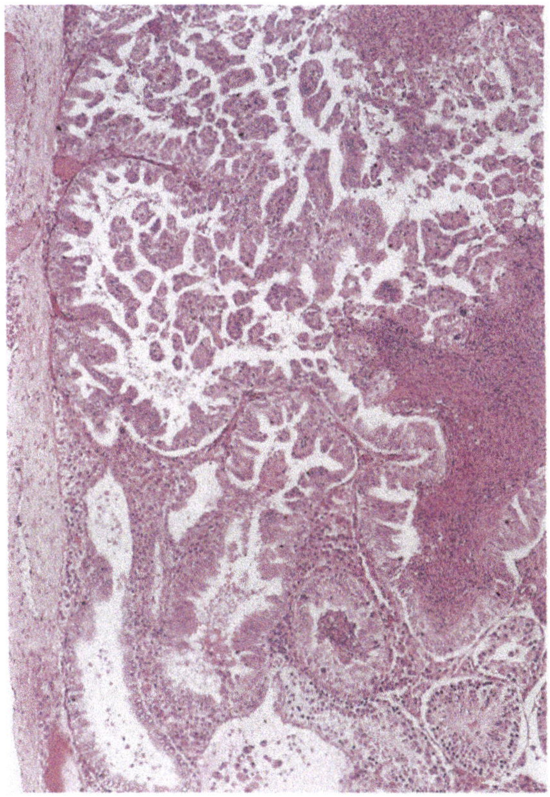

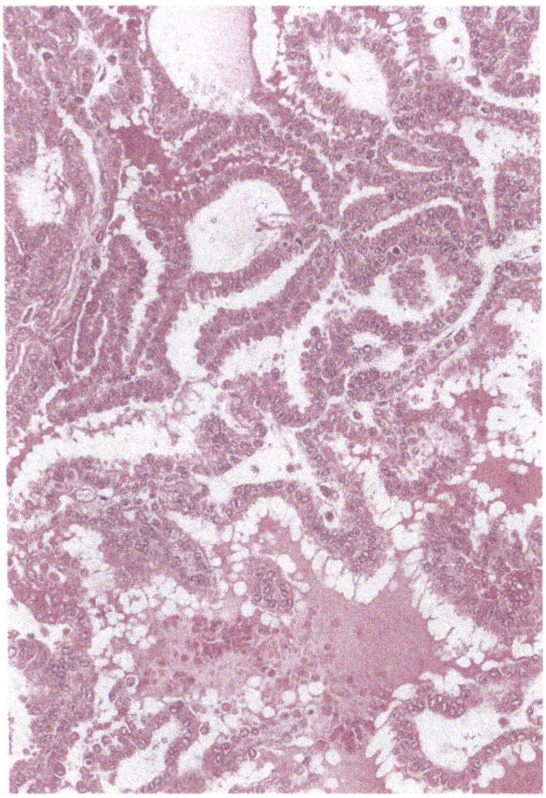

Histogenesis

Rete testis epithelium.

Diagnostic Features

- Tubulo-papillary tumor of rete testis structures localized in the mediastinum testis showing compression of the adjacent tissue
- Papillary structures and tubules generally lined with single or multiple layers of epithelial cells
- Focal crowding and distension of rete testis tubules is common
- Cells are cuboidal to tall columnar or pleomorphic with abundant eosinophilic cytoplasm and round to oval vesicular nuclei with a single nucleolus
- Mitotic activity is low
- Cystic structures filled with spermatozoa may be present
- Hyperplasia of epithelia lining tubules in tunica albuginea may be present

Differential Diagnosis

Hyperplasia, Rete Testis
Limited epithelial proliferation with little stroma and no compression of adjacent structures, atypia or invasion

Fig. 16 *(above)*. Adenoma of the rete testis. Tubulo-papillary structures displacing seminiferous tubules and causing focal accumulation of sperm. CD-1 mouse, 2-year study, age >12 months. H&E

Fig. 17 *(below)*. Adenoma of the rete testis. H&E

Carcinoma, Rete Testis
Invasion, marked atypia, solid areas and/or high mitotic index are usually present

Carcinoma/Adenocarcinoma, Metastatic
Exclude primary malignant epithelial tumor from other site

References

(See 5, 7, 12, 31, 32, 44)

Carcinoma, Rete Testis *(M)*

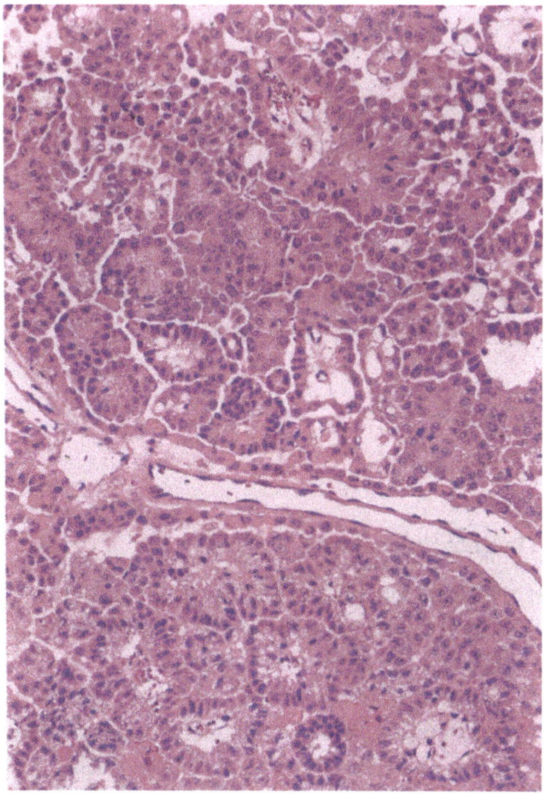

Fig. 18. Carcinoma of the rete testis. Tubular structures replaced by atypical, irregular papillary growths. CD-1 mouse, age 82 weeks. H&E

Histogenesis

Rete testis epithelium.

Diagnostic Features

- Characterized by coalescence of rete testis tubules localized in the mediastinum testis expanded by masses of irregular papillary growths
- Invasion of adjacent structures and tunica may be present
- Cells are pleomorphic, frequently tall columnar with eosinophilic cytoplasm and atypical, vesicular or basophilic nuclei
- Number of mitotic figures vary; mitoses can be atypical
- Solid areas with scirrhous stroma may be present

Differential Diagnosis

Adenoma, Rete Testis
Lacks invasive growth characteristic and is composed of well-differentiated cells

Carcinoma/Adenocarcinoma, Metastatic
Exclude primary malignant epithelial tumor from other site

References

(See 5, 7, 12, 31, 32, 44)

Adenoma, Leydig Cell *(B)*

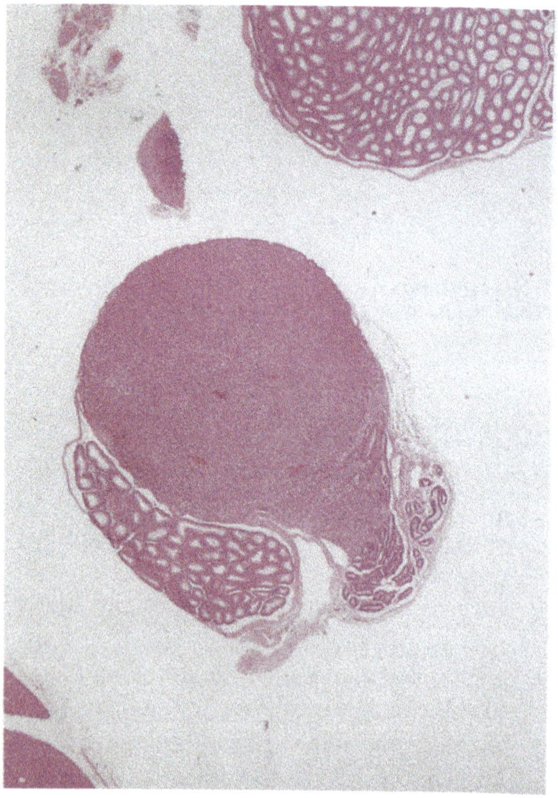

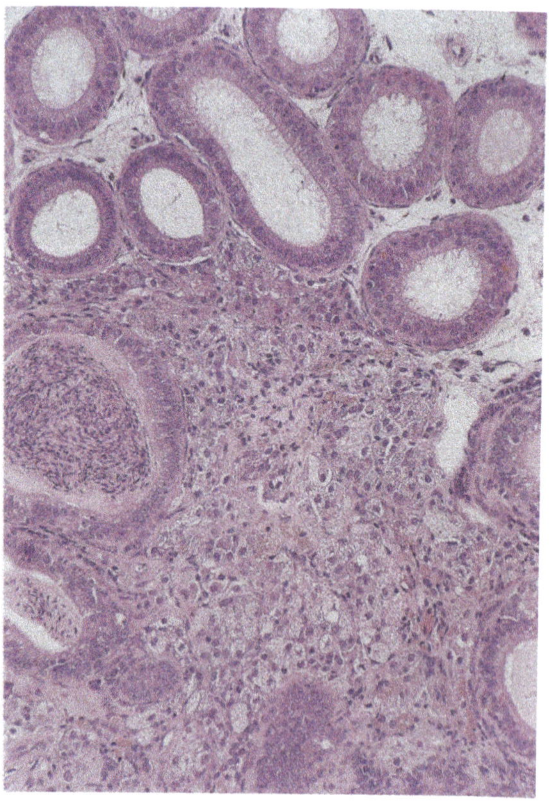

Synonym(s). Tumor, Leydig cell, benign; tumor, interstitial cell, benign

Histogenesis

Leydig cells.

Diagnostic Features

- Nodular or diffuse mass with peripheral compression, displacement or replacement of adjacent epididymal tubules
- Composed of polygonal cells with abundant eosinophilic or vacuolated cytoplasm
- Nucleus generally central, round with evenly distributed chromatin and single prominent nucleolus. Low nuclear/cytoplasmic ratio
- Smaller hyperchromatic cells with scant basophilic cytoplasm and cells with yellow-brown pigment (lipofuscin) are less common
- Mitoses infrequent with no atypical mitotic figures

Fig. 19 *(above).* Leydig cell adenoma of the epididymis. Neoplastic Leydig cell replacing epididymal structures. B6C3F1 mouse, 2-year study, age >12 months. H&E

Fig. 20 *(below).* Leydig cell adenoma of the epididymis. Vacuolated neoplastic Leydig cells displacing epididymal tubules. CD-1 mouse, 2-year study, age 86 weeks. H&E

Differential Diagnosis

Granuloma
Composed of vacuolated macrophages and other inflammatory cells; often accumulation of sperm

Comment

Rare tumor described in B6C3F1 mice. There are no reports on epididymal Leydig cell hyperplasia or carcinomas. Leydig cells normally are not found in the epididymis.

References

(See 7, 12, 24, 26, 34)

Hyperplasia *(H)*

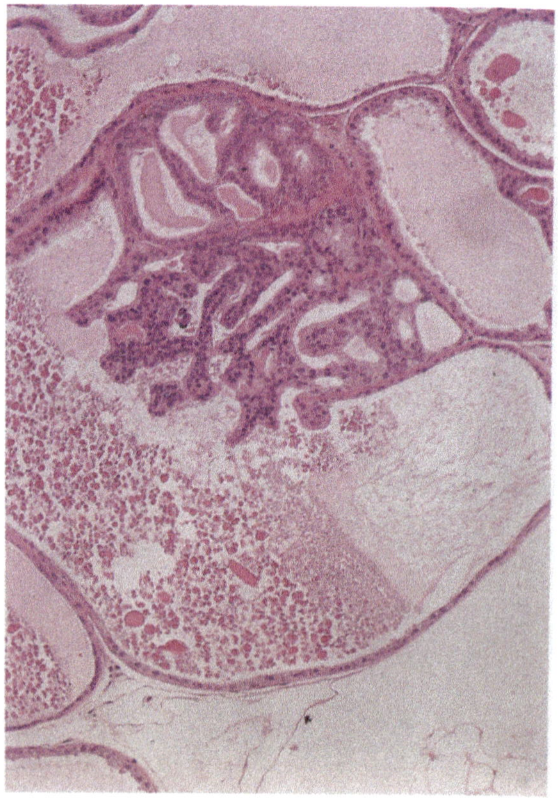

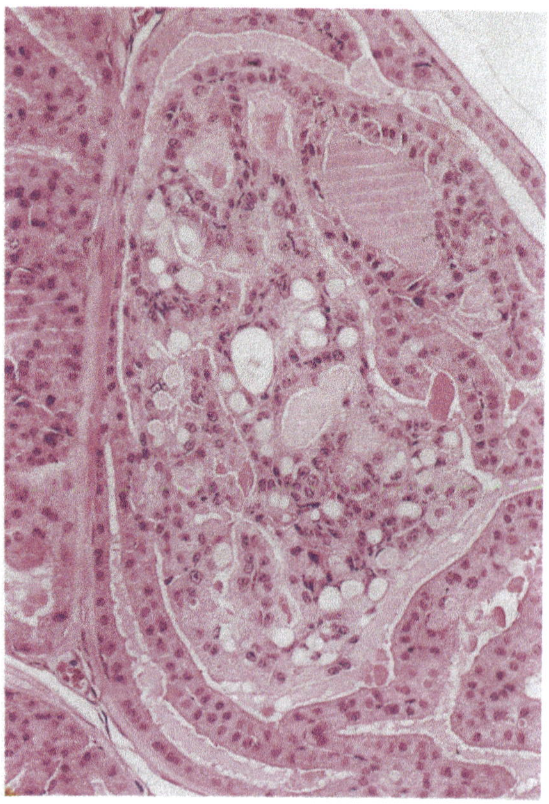

Synonym(s). Hyperplasia, atypical

Histogenesis

Prostatic and coagulating gland acinar/ductal epithelium.

Diagnostic Features

- Focal/multifocal lesion involving single or a few adjacent alveoli and occasionally excretory ducts, in ventral or dorsolateral lobe, rarely in the anterior lobe (coagulating gland)
- Proliferative lesion is not obliterating the acinar lumen and there is no disturbance of normal alveolar architecture; supporting stroma minimal or absent
- Epithelial proliferation of up to three and more cell layers, usually with cribriform or pseudostratified pattern sometimes forming papillary structures, occasionally loss of cellular polarity
- Cells are cuboidal to columnar with increased cytoplasmic basophilia
- Cellular atypia generally minimal, areas with increased cellular atypia and squamous metaplasia may be present
- Gradual transition from normal epithelium to focal hyperplastic areas or, at times, more abrupt changes
- In transition areas occasional occurrence of tall columnar cells with hypereosinophilic cytoplasm and basally located hyperchromatic round to oval nuclei

Fig. 21 *(above).* Hyperplasia of the prostate. H&E

Fig. 22 *(below).* Hyperplasia of the coagulating gland. H&E

- Decreased or high nuclear/cytoplasmic ratio, mitotic figures infrequent
- No capsule formation or compression of adjacent tissue
- Usually not accompanied by inflammation

Differential Diagnosis

Hyperplasia, Functional
Generally diffuse not focal, "crowded" but not truly multilayered epithelial cells of increased height and basophilia, forming folds into the alveolar lumina containing less secretion

Hyperplasia, Reactive
Almost always combined with inflammation, consists of a simple thickening of the epithelium to 2–6 or more cell layers, occasionally with pseudoglandular formation

Adenoma
Distortion of normal glandular architecture and compression of adjacent tissue, and/or proliferative lesion obliterating one acinar lumen. Ingrowth of vascular supply and supporting stroma and a mild degree of pleomorphism is present

Comment

The prostate is a multilobulated organ consisting of ventral, dorsal, lateral and anterior lobes. The anterior lobes are frequently referred to as coagulating gland and the dorsal and lateral lobes are often referred to as the dorsolateral lobe. Contrary to the rat, a microscopic distinction of the lobes is difficult and is best achieved by separate embedding. Spontaneous proliferative lesions of prostate lobes in mice are rare. Some transgenic mice show a high incidence up to 100%.

References

(See 7, 9, 10, 11, 12, 22, 25, 34, 37, 41)

Adenoma *(B)*

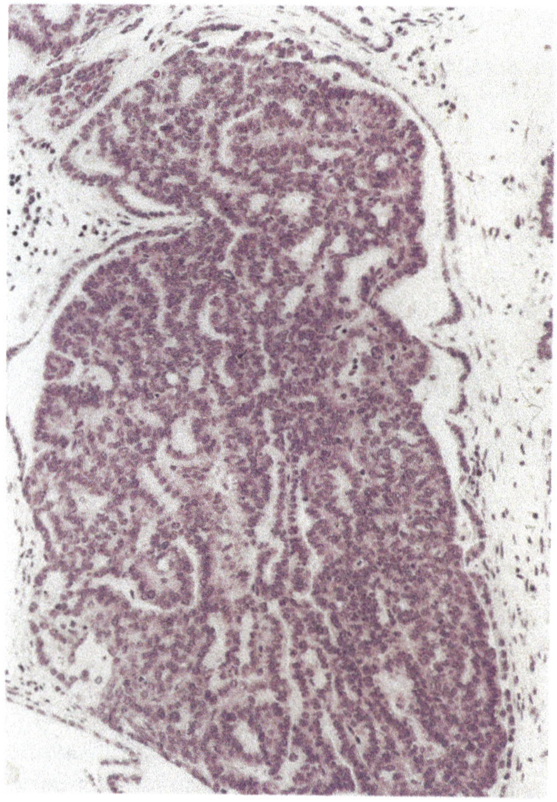

Histogenesis

Prostatic and coagulating gland acinar/ductal epithelium.

Diagnostic Features

- Intra-alveolar cribriform and/or papillary proliferations partially or completely obliterating the lumina of generally several acini
- Distortion of normal glandular architecture and compression of surrounding tissue
- Sometimes enclosed completely or partially by a fine fibrous capsule
- Ingrowth of vascular supply and supporting stroma
- Occasionally outgrowths into adjacent alveoli/ducts (no invasion)
- Cells usually enlarged with round to oval hypo- or hyperchromatic nuclei, decreased or increased nuclear/cytoplasmic ratio, loss of polarity, slightly eosinophilic or amphophilic cytoplasm
- Mild degree of pleomorphism, some areas of dysplasia and squamous metaplasia
- Mitotic figures may be seen
- Usually no inflammation
- Coagulating gland adenomas may be grossly distorted by trapped secretions

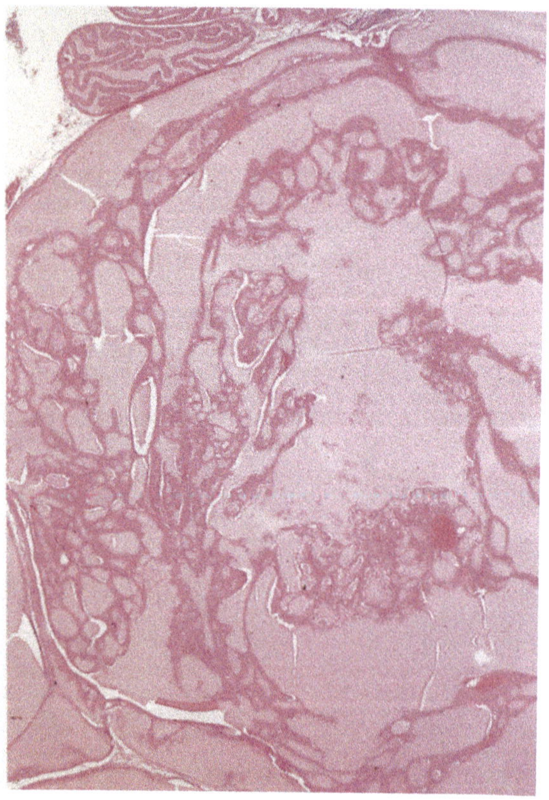

Fig. 23 *(above)*. Adenoma of the prostate. Papillary structures greatly expand acinus. CD-1 mouse, 2-year study, age >12 months. H&E

Fig. 24 *(below)*. Adenoma of the coagulating gland. Papillary and glandular formations causing accumulation of secretions and distension of gland. B6C3F1 mouse, 2-year study, age >12 months. H&E

Differential Diagnosis

Hyperplasia
No compression of surrounding tissue or capsule formation. Proliferative lesion does not obliterate one acinar lumen. Lacks supportive stroma and glandular architecture is maintained

Adenocarcinoma
Invasive growth, e.g., through the capsule, and metastases, often gross disruption of the architecture, central necrosis and hemorrhage sometimes found, increased cellular atypia

Tumor, Granular Cell, Benign
Composed of large epitheloid or fusiform eosinophilic cells with PAS-positive cytoplasmic granules

Comment

The prostate is a multilobulated organ consisting of ventral, dorsal, lateral and anterior lobes. The anterior lobes are frequently referred to as coagulating gland and the dorsal and lateral lobes are often referred to as the dorsolateral lobe. Contrary to the rat, a microscopic distinction of the lobes is difficult and is best achieved by separate embedding. Spontaneous proliferative lesions of prostate lobes in mice are rare. Some transgenic mice show a high incidence up to 100%.

References

(See 7, 9, 10, 11, 12, 22, 25, 34, 37, 41)

Adenocarcinoma *(M)*

Histogenesis

Prostatic and coagulating gland acinar/ductal epithelium.

Diagnostic Features

- Range in size from microscopically small to grossly visible masses; often gross distortion of the architecture
- Cribriform, papillary/tubular, and/or solid anaplastic type of growth pattern with invasion of adjacent tissues
- Columnar or pleomorphic/anaplastic tumor cells with low nuclear/cytoplasmic ratio, eosinophilic or basophilic cytoplasm, large vesicular or hyperchromatic nuclei with multiple nucleoli and frequent mitotic figures
- Scirrhous stroma, focal necrosis, hemorrhage and a mixed inflammatory cell infiltrate often present
- Show distinct fibrous capsule partly demarcating the tumor from the surrounding tissue
- Metastases to regional lymph nodes, lung and kidney may occur

Differential Diagnosis

Adenoma
Absence of invasive growth and less disruption of architecture and only mild pleomorphism present

Tumor, Granular Cell, Benign
Composed of large or variably-sized epitheloid or fusiform eosinophilic cells with PAS-positive granules

Proliferative Lesion, Submucosal
Composed of spindle cells and nests of large round epitheloid cells with highly atypical nuclei

Comment

The prostate is a multilobulated organ consisting of ventral, dorsal, lateral and anterior lobes. The anterior lobes are frequently referred to as coagulating gland and the dorsal and lateral lobes are often referred to as the dorsolateral lobe. Contrary to the rat, a microscopic distinction of the lobes is difficult and is best achieved by separate embedding. Spontaneous proliferative lesions of prostate lobes in mice are rare. Some transgenic mice show a high incidence.
For large invasive neoplasms of the male accessory sex glands it may be often impossible to exactly define the gland of origin. These may therefore have to be classified as adenocarcinomas, origin unknown.

References

(See 7, 9, 10, 11, 12, 22, 25, 34, 37, 41)

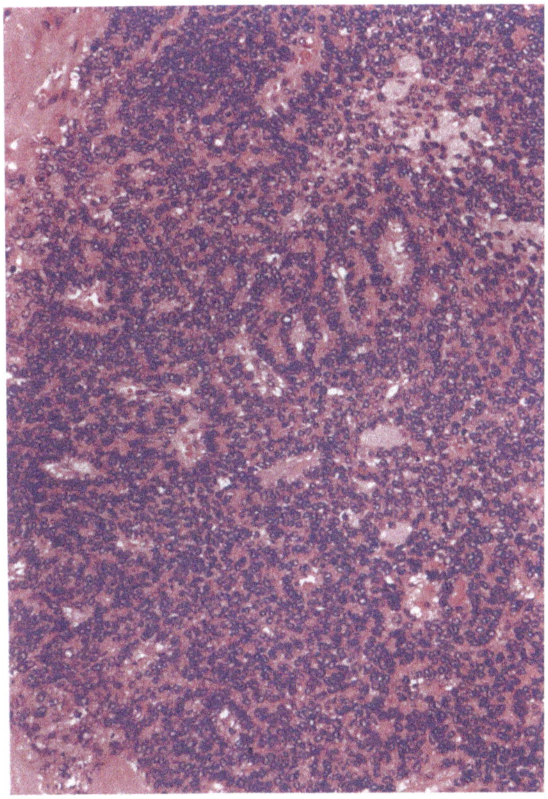

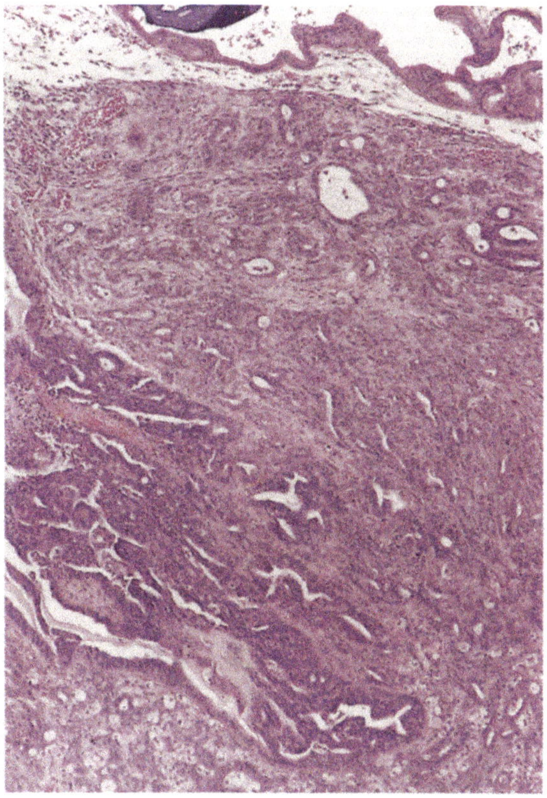

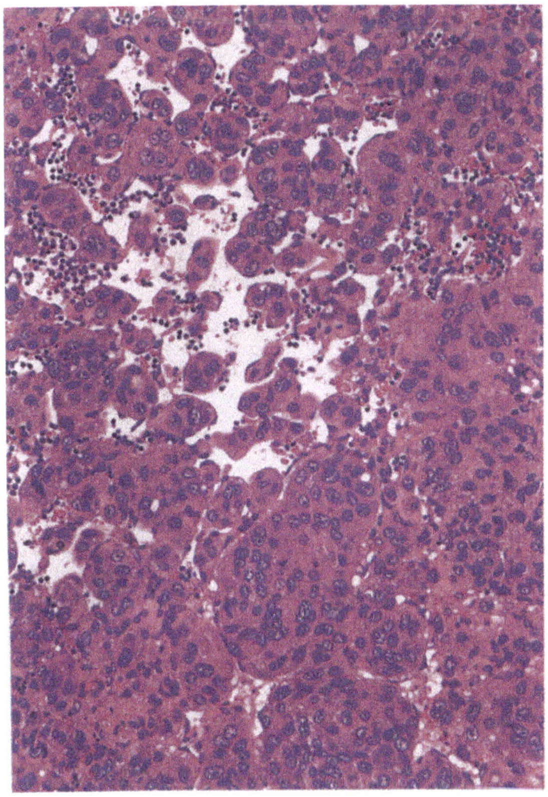

Fig. 25 *(upper left)*. Adenocarcinoma of the prostate. Atypical, variably-sized acini composed of pleomorphic cells, replacing normal architecture. CD-1 mouse, age 82 weeks. H&E

Fig. 26 *(upper right)*. Adenocarcinoma of the prostate. Neoplastic acini and cells associated with a scirrhous response invading adjacent stroma. C3(1)/SV40 Large T Antigen transgenic mouse, age 11 months. H&E

Fig. 27 *(below)*. Adenocarcinoma of the coagulating gland. Solid nests composed of eosinophilic, atypical epithelial cells. CD-1 mouse, age 79 weeks. H&E

Hyperplasia *(H)*

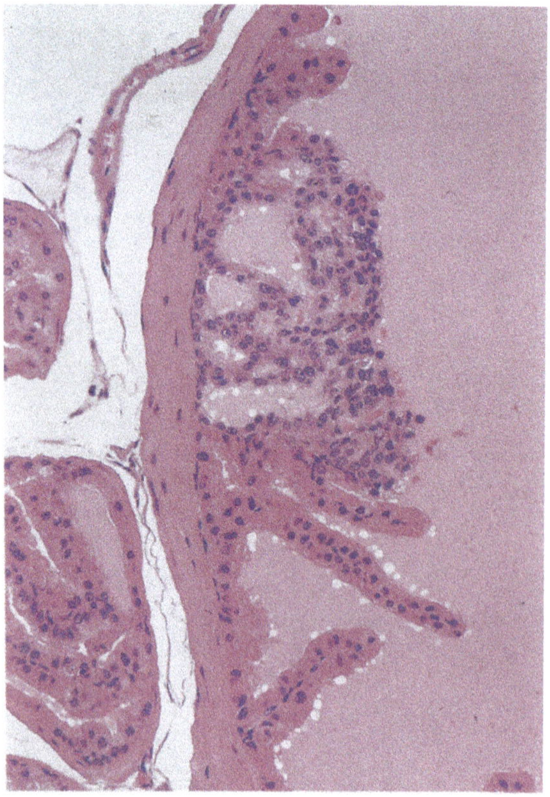

Fig. 28. Hyperplasia of the seminal vesicle. Hyperplastic papillary structures lined by pale basophilic, hypertrophic cells. CD-1 mouse, age 85 weeks. H&E

Histogenesis

Seminal vesicular epithelium.

Diagnostic Features

- Entirely intramucosal, usually focal lesion, lumen not obliterated
- Cells usually follow normal glandular contours, no piling-up, multilayered epithelium, sometimes cribriform pattern, normal glandular architecture preserved
- Enlarged, pale amphophilic cells with vesicular nuclei, normal cellular polarity, no atypia
- Sometimes mitotic figures seen
- No capsule formation or compression of the adjacent tissue
- No ingrowth of vascular supply and supporting stroma

Differential Diagnosis

Adenoma

Partly encapsulated lesion with compression of the surrounding tissue, distortion of normal architecture, (partial) loss of normal cellular polarity, arrangement of cells in cribriform, comedo or microglandular pattern

References

(See 7, 9, 12, 20, 24, 25, 27)

Adenoma *(B)*

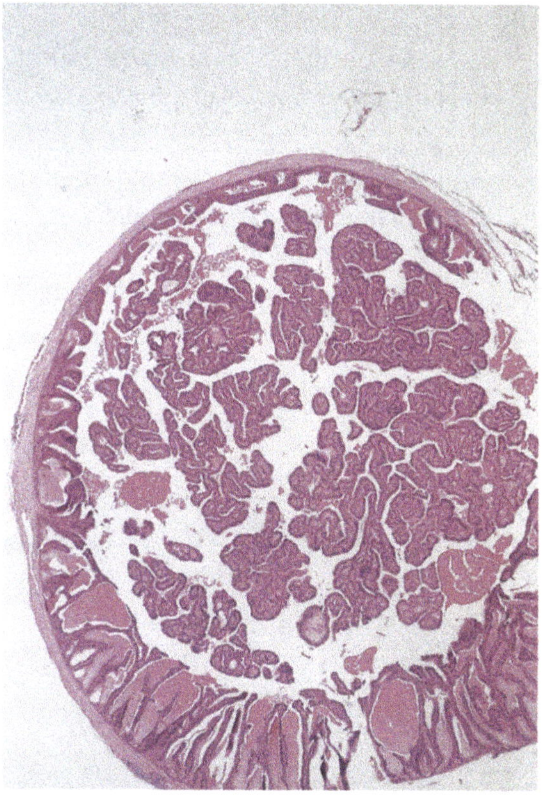

Histogenesis

Seminal vesicular epithelium.

Diagnostic Features

- Well-delineated nodule with complete or partial obliteration of the glandular lumen
- Distortion of normal glandular architecture and compression of adjacent tissue
- Cells arranged in papillary, cribriform or microglandular pattern with some loss of cellular polarity
- Some pleomorphism and nuclear crowding may be found, but low mitotic index
- Round to oval hyperchromatic nuclei, eosinophilic cytoplasm; connective tissue stroma may be increased
- May be encapsulated with fibrous septa dividing tumor into pseudolobules
- May be distorted by trapped secretions

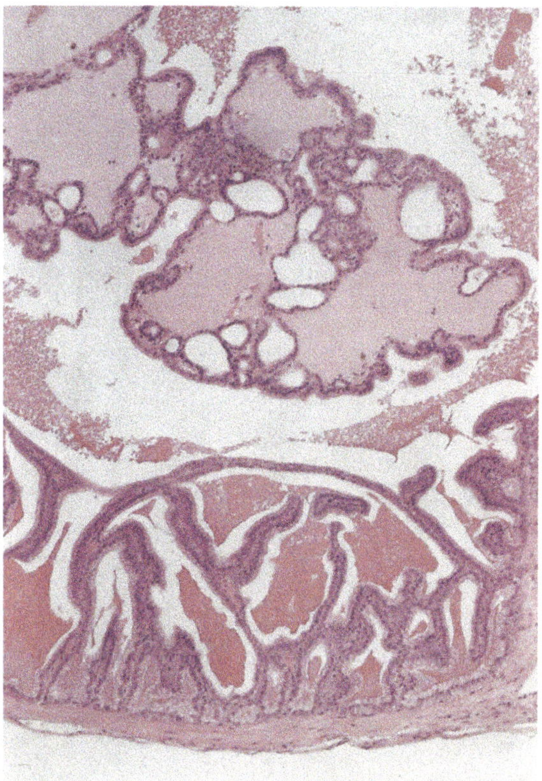

Fig. 29 *(above)*. Adenoma of the seminal vesicle. Intraluminal papillary growth pattern. B6C3F1 mouse, 2-year study, age >12 months. H&E

Fig. 30 *(below)*. Adenoma of the seminal vesicle. H&E

Differential Diagnosis

Hyperplasia

No distortion of normal architecture, cells generally follow normal glandular contours

Adenocarcinoma

Local invasion and capsule infiltration or metastases and marked cellular and nuclear pleomorphism

Tumor, Granular Cell, Benign

Solid growth of round to fusiform eosinophilic cells with PAS-positive granules

References

(See 7, 9, 12, 17, 20, 24, 25, 27, 34)

Adenocarcinoma *(M)*

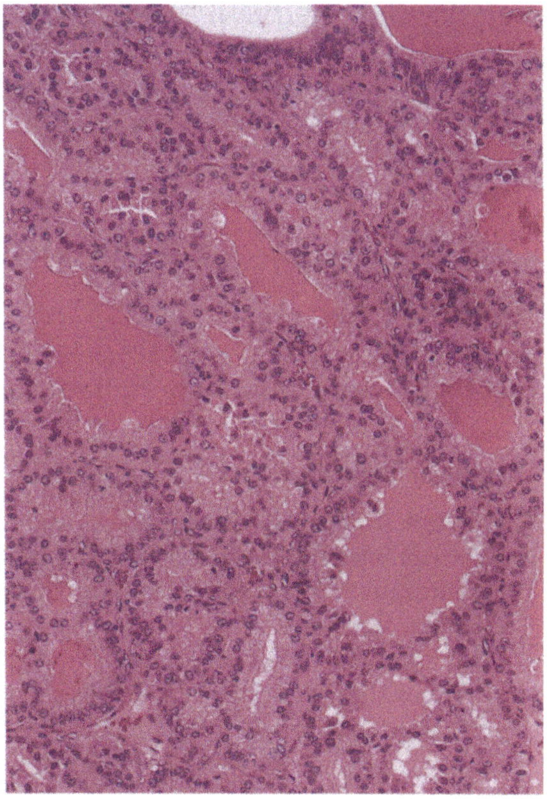

Fig. 31. Adenocarcinoma of the seminal vesicle. Irregular neoplastic acini. CD-1 mouse, age 82 weeks. H&E

Histogenesis

Seminal vesicular epithelium.

Diagnostic Features

- Growth pattern usually tubular, papillary or microglandular; less common are anaplastic solid nests
- Considerable disruption of normal glandular architecture
- Invasive growth into glandular capsule and surrounding tissue
- Cells may be well-differentiated with basophilic nuclei and abundant eosinophilic cytoplasm, anaplastic solid neoplasms are composed of cells that vary greatly in size, shape and staining
- Necrosis, hemorrhages and distant metastases may be present

Differential Diagnosis

Adenoma
No local invasion or distant metastases, minimal pleomorphism

Tumor, Granular Cell, Benign
Composed of large or variably-sized epitheloid eosinophilic cells with PAS-positive cytoplasmic granules

Proliferative Lesion, Submucosal
Composed of spindle cells and nests of large round epitheloid cells with highly atypical nuclei

Comment

Adenocarcinomas of the seminal vesicle are rare. The site of tumor origin may be difficult to establish when the tumor is highly invasive.

References

(See 7, 9, 12, 17, 20, 24, 25, 27, 34)

Carcinoma, Adenosquamous *(M)*

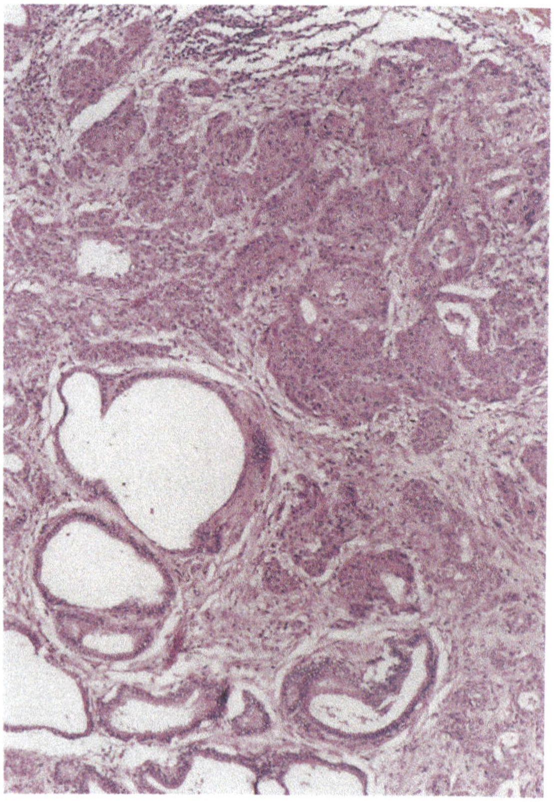

Fig. 32. Adenosquamous carcinoma, arising in the area of coagulating gland and seminal vesicles and invading prostate. Significant squamous metaplasia of neoplastic acini associated with a scirrhous response. Swiss mouse, long-term study, age >12 months. H&E

Histogenesis

Prostatic and coagulating gland acinar/ductal epithelium. Seminal vesicular epithelium.

Diagnostic Features

- Widespread squamous metaplasia of neoplastic tubules and ducts with and without keratinization in association with tubular or acinar structures
- Often large, highly invasive neoplasm with scirrhous stroma, necrosis, hemorrhage and mixed inflammatory cell infiltration

Differential Diagnosis

Adenocarcinoma
  Lacks significant squamous differentiation

Comment

The prostate is a multilobulated organ consisting of ventral, dorsal, lateral and anterior lobes. The anterior lobes are frequently referred to as coagulating gland and the dorsal and lateral lobes are often referred to as the dorsolateral lobe. Contrary to the rat, a microscopic distinction of the lobes is difficult and is best achieved by separate embedding. For large invasive neoplasms of the male accessory sex glands it may be often impossible to exactly define the gland of origin. These may therefore have to be classified as adenocarcinomas, origin unknown.

References

(See 7, 9, 12, 20, 24, 25, 27)

Tumor, Granular Cell, Benign *(B)*

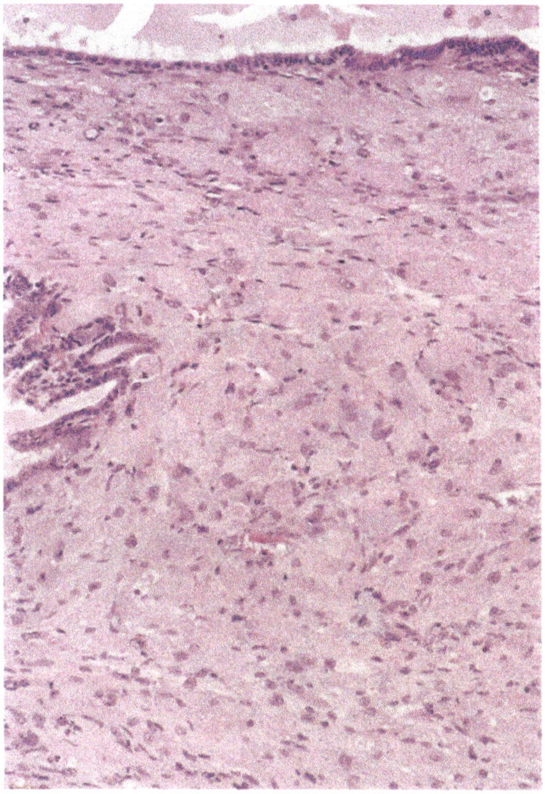

Synonym(s). Abrikossoff's tumor, benign; myoblastoma

Histogenesis

Not established; an origin from Schwann cells or mesenchymal cells has been proposed.

Diagnostic Features

- Often circumscribed and demarcated solid mass with little stroma; is composed of large epitheloid round to oval cells with large pale nuclei and abundant eosinophilic granular cytoplasm and smaller more fusiform cells with small dark uniform nuclei
- Small to medium-sized cytoplasmic granules are positive with PAS stain, are diastase-resistant, and are considered to be various stages of lysosomes present in almost every cell
- Expansive growth causes compression and atrophy of adjacent tissue
- Some tumors show a more infiltrative pattern of growth. These tumors tend to have a more fusiform morphology. In the absence of metastases or invasion into other organs these tumors are considered to be benign

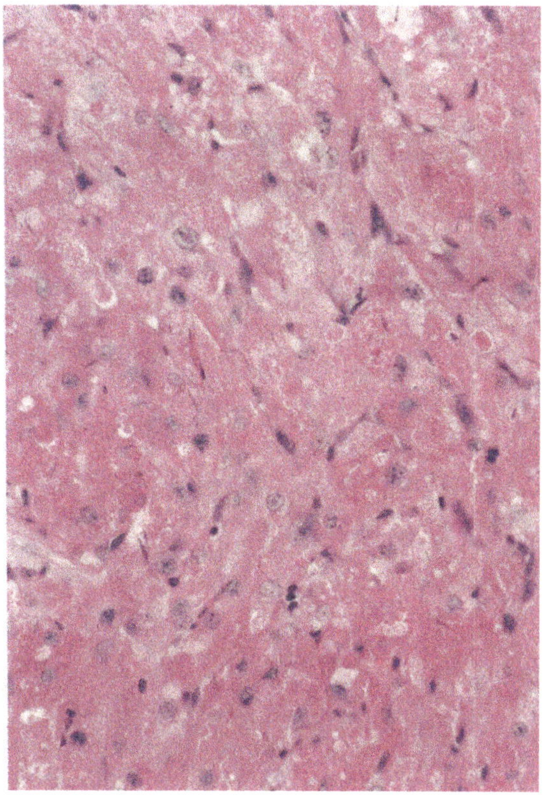

Fig. 33 *(above).* Benign granular cell tumor of the seminal vesicle. Composed of large cells with abundant finely granular cytoplasm with infiltrative growth. Swiss mouse, long-term study, age >12 months. H&E

Fig. 34 *(below).* Benign granular cell tumor of the seminal vesicle. Same specimen as in Fig. 33 showing PAS-positive staining reaction of cytoplasmic granules. Swiss mouse, long-term study, age >12 months. PAS

Differential Diagnosis

Leiomyoma
Lacks PAS-positive granules or epitheloid cells

Proliferative Lesion, Submucosal
Composed of anaplastic mixture of spindle cells without PAS-positive granules and large round, sometimes binucleated or multinucleated epitheloid cells with PAS-positive granules that are globular and intensely eosinophilic

Adenocarcinoma
Formation of glands, epithelium often amphophilic or basophilic, lack abundant PAS-positive granules

Comment

Granular cell tumors generally occur as a single pink to pale yellow-gray mass. These tumors are frequently seen in the male and female genital tract, but may occur in other organs. Granular cell tumors react positively in immunocytochemical stains for NSE, S-100 protein and peripheral myelin proteins. A malignant counterpart of the benign granular cell tumor has not been described but cannot be excluded. In this classification evidence of distant metastasis or marked invasion beyond the normal organ structures should be present before such a diagnosis is warranted. The term hyperplasia does not apply even to very small accumulations/nodules since equivalent normal structures are unknown.

References

(See 4, 17, 25, 27, 34)

Proliferative Lesion, Submucosal *(H)*

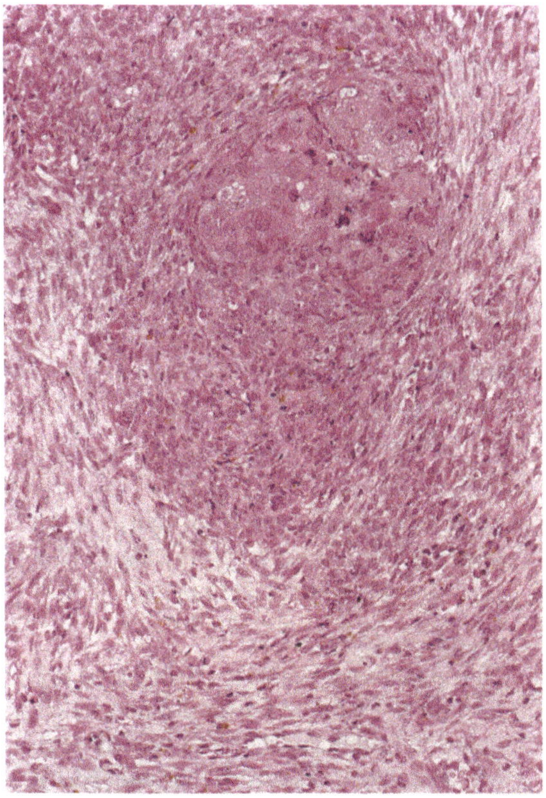

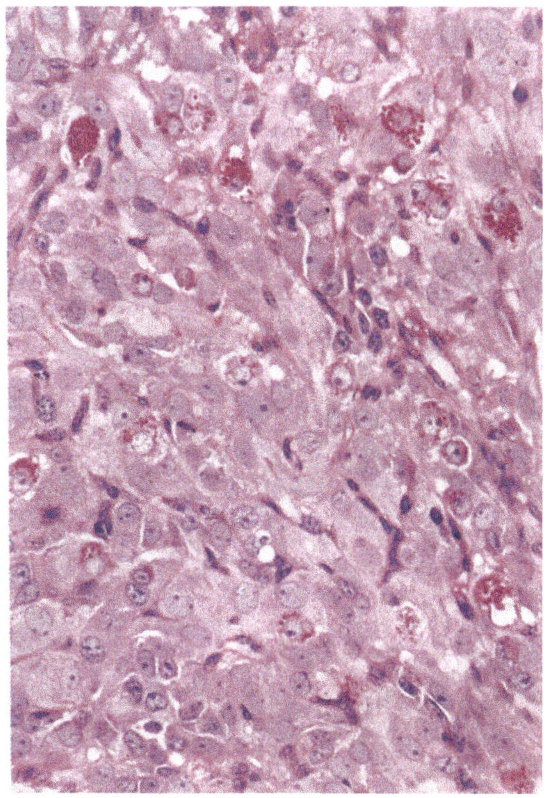

Synonym(s). Leiomyosarcoma; leiomyoblastoma; carcinosarcoma; sarcoma, undifferentiated

Histogenesis

Not established; possible origin from smooth muscle cells or mesenchymal remnants possibly of Müllarian duct origin.

Diagnostic Features

- Range from microscopically small to grossly visible masses
- May show up to three cellular components:
  - Submucosal aggregates of large round or pleomorphic cells with atypical nuclei
  - Nests and/or sheets composed of spindle cells
  - Scattered cells with large eosinophilic, PAS-positive granules
- Cells may infiltrate and expand submucosal tissues mimicking malignant invasive growth
- Spindle cells may closely resemble smooth muscle cells with abundant pale eosinophilic cytoplasm or be reminiscent of fibroblasts or immature mesenchymal cells with scant cytoplasm. Individual apoptotic cells or regressing foci may be present

Fig. 35 *(above).* Submucosal proliferative lesion of the seminal vesicle. Invasive neoplastic spindle cells and nest of round, undifferentiated cells. Swiss mouse, long-term study, age >12 months. H&E

Fig. 36 *(below).* Submucosal proliferative lesion of the seminal vesicle. Undifferentiated round cells containing large PAS positive granules. Swiss mouse, long-term study, age >12 months. PAS

- Clusters of large pleomorphic cells have abundant eosinophilic cytoplasm, may be multinucleated and show regressive (apoptotic) changes
- Mitotic rate is usually low
- Usually associated with congested secretions and inflammation
- May also be associated with scirrhous proliferative response

Differential Diagnosis

Leiomyoma or Leiomyosarcoma
Lack PAS-positive granules and large pleomorphic cells

Adenocarcinoma
Shows areas with acinar formation associated with scirrhous response. Lacks PAS-positive cytoplasmic globules

Tumor, Granular Cell, Benign
Composed of large or variably-sized epitheloid to fusiform eosinophilic cells with small to medium-sized PAS-positive cytoplasmic granules. Large decidual cell-like binucleated or multinucleated epithelial cells are not present

Comment

The pathobiology of this rare lesion is poorly understood and is probably similar to the proliferative lesion/decidual-like reaction of the submucosa in the mouse urinary bladder. Terminology still remains unresolved. It has only been documented in the mouse, apparently does not progress, shows regressive changes and is considered to be non-neoplastic. (For review see, E. Karbe, Toxicol Pathol 27: 354–362, 1999).

References

(See 6, 13, 17, 18, 19, 20, 25)

Hyperplasia, Atypical *(H)*

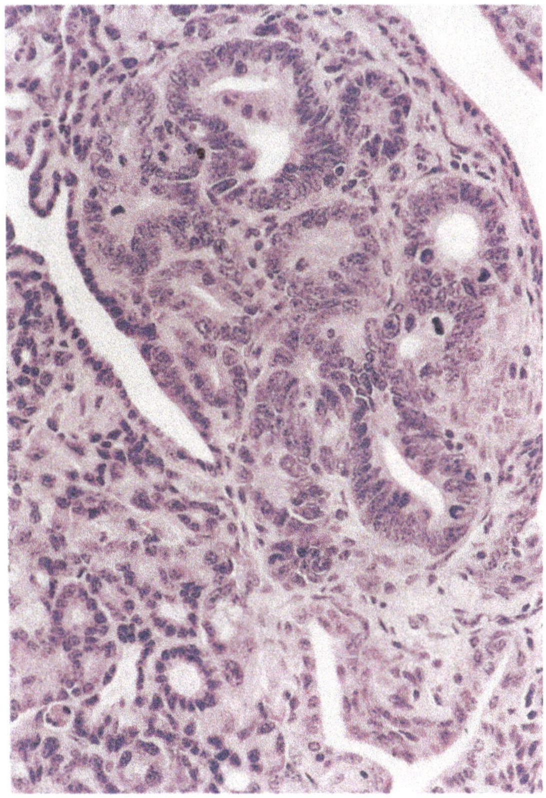

Fig. 37. Atypical hyperplasia of the periurethral gland. Focal increase in number of acini lined by hypertrophic and atypical epithelium. Numerous mitotic figures. C3(1)/SV40 Large T Antigen transgenic mouse, age 11 months. H&E

Histogenesis

Acinar epithelium.

Diagnostic Features

- Focal or multifocal glandular proliferations without stromal invasion
- Loss of secretory activity
- Irregular distorted glands are lined by atypical cells
- Cells are pleomorphic, enlarged and pale basophilic
- Mitotic figures may be frequent and atypical

Differential Diagnosis

Adenocarcinoma
Invasion of adjacent tissues

Comment

There are no reports on spontaneous periurethral gland lesions. Changes described in this classification system are observed in SV40 TAG transgenic mice.

References

(See 25, 37)

Adenocarcinoma *(M)*

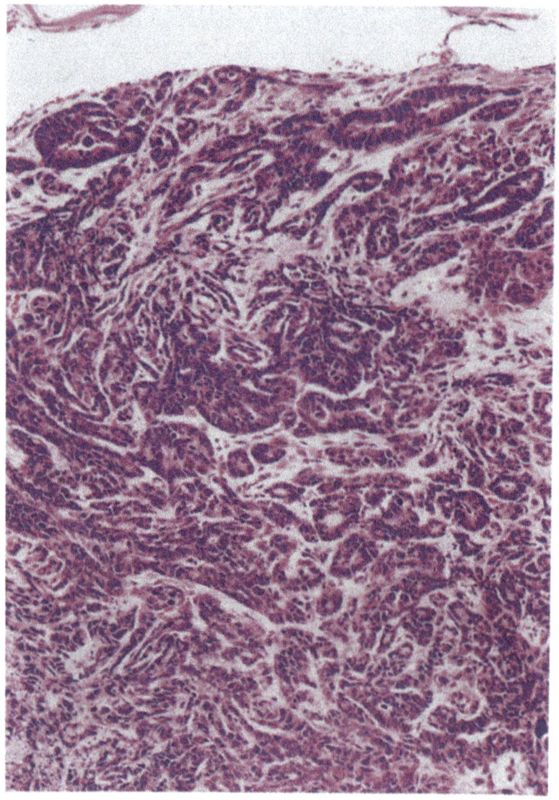

Fig. 38. Adenocarcinoma of the periurethral gland showing highly irregular glands. C3(1)/SV40 Large T Antigen transgenic mouse, age 11 months. H&E

Histogenesis

Acinar epithelium.

Diagnostic Features

- Focal or multifocal masses that may cause urethral obstruction
- Composed of dense glandular proliferations with invasion of adjacent tissues
- Neoplastic growths frequently involve periurethral glands located in the bulbourethral glands
- Irregular distorted glands are lined by atypical cells with loss of polarity and pseudostratification
- Cells are pleomorphic, hyperchromatic and have scant cytoplasm
- Mitotic figures are frequent and atypical

Differential Diagnosis

Hyperplasia, Atypical
Lack invasion of adjacent tissues and lesions are microscopically small although a high degree of cellular atypia are present

Comment

There are no reports on spontaneous periurethral gland lesions. Changes described in this classification system are observed in SV40 TAG transgenic mice.

References

(See 25, 37)

Hyperplasia, Acinar Cell *(H)*

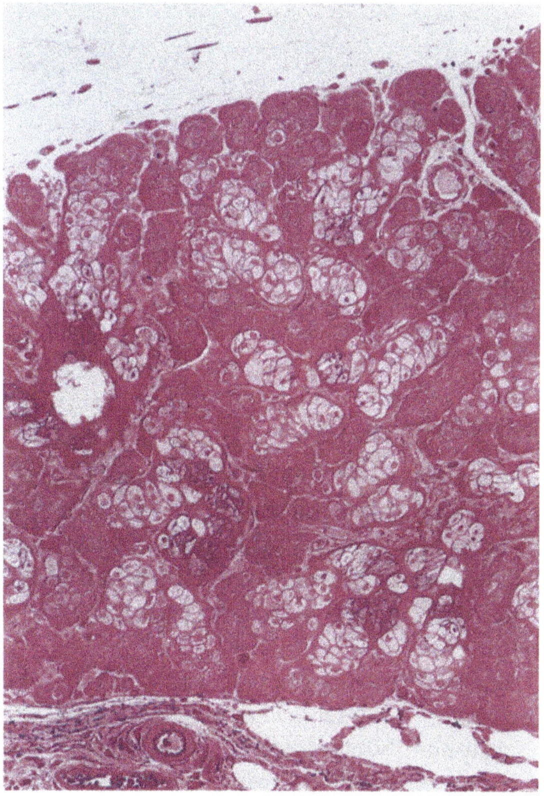

Fig. 39. Acinar cell hyperplasia of the preputial gland. Hyperplastic acinar epithelium forming typical nests. CD-1 mouse, long-term study, age 37 weeks. H&E

Histogenesis

Acinar epithelium.

Diagnostic Features

- Focal or multifocal increase of eosinophilic cells surrounded by dark flat basal cells
- Foci show lobulation similar to normal glands
- Morphology of hyperplastic acinar cells is similar to normal glandular tissue
- Central round large nucleus with one or rarely two large round nucleoli
- Condensation of heterochromatin on the nuclear envelope
- Cytoplasm with lipid-filled vesicles and vacuoles depending on age, containing typical pink granules

Differential Diagnosis

Hyperplasia, Reactive
Associated with chronic inflammation; very common in group-housed male mice. Normal architecture maintained

Adenoma, Acinar Cell
Adenoma of acinar cells has compression and loss of normal structure

Comment

This lesion occurs rarely spontaneously, mainly in older animals (2 years old), but can be induced by several carcinogens.

References

(See 12, 25, 28, 40)

Adenoma, Acinar Cell *(B)*

Histogenesis

Acinar epithelium.

Diagnostic Features

- Growth is usually solid with irregular nodular formations that extend peripherally, mimicking acini of the normal gland; excretory ducts are infrequent and an ill-defined connective tissue capsule may be present
- Neoplastic acini are composed centrally of pale, foamy sebaceous type cells, and peripherally, of basophilic basal cells
- No invasive growth or high degree of atypia and mitosis
- Architecture not maintained

Differential Diagnosis

Hyperplasia, Acinar Cell
Smaller, no tissue capsule, regular acini, uniformity of cells. Excretory ducts commonly present

Papilloma, Squamous Cell
Lacks glandular components

Carcinoma, Acinar Cell
Invasion of surrounding connective tissue. Variably sized acini. Tumor cells show cellular atypia and resemble immature acinar cells

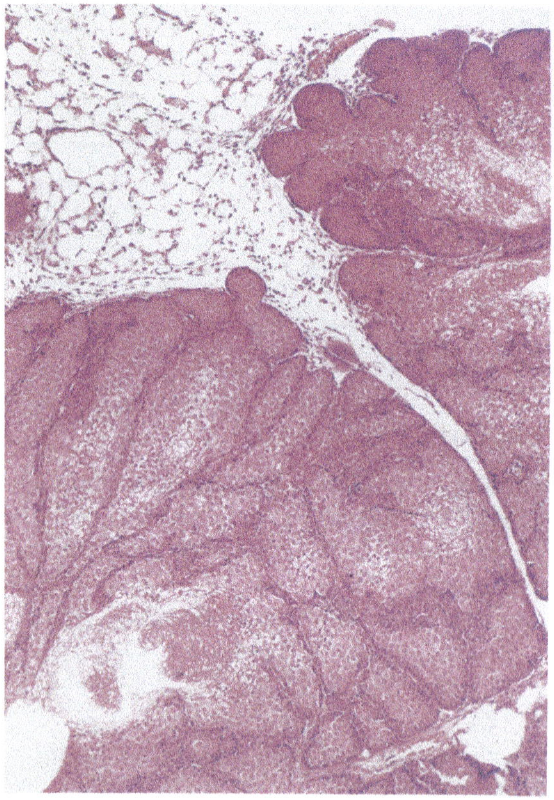

Fig. 40. Acinar cell adenoma of the preputial gland. Expansive growth of acinar epithelium showing basal cell and foamy, sebaceous cell differentiation. B6C3F1 mouse, 2-year study, age >12 months. H&E

Comment

Spontaneous adenomas are extremely rare, but they can be induced by several compounds. Cystic and papillary growth variants as seen in the rat have not been described in the mouse.

References

(See 12, 23, 25, 28, 29, 30, 40, 42, 43)

Carcinoma, Acinar Cell *(M)*

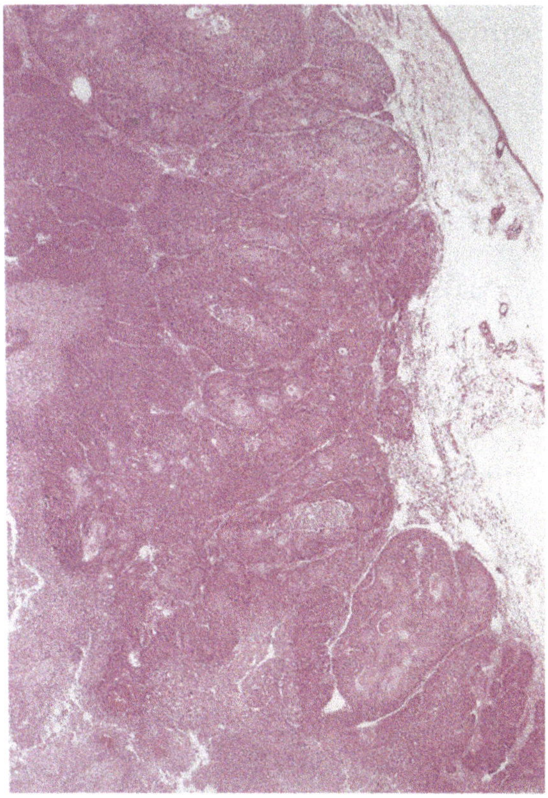

Synonym(s). Adenocarcinoma, acinar cell

Histogenesis

Acinar cells.

Diagnostic Features

- Consists of variably-sized nests and nodules morphologically distinct from the normal gland with frank invasion of adjacent structures
- Cells in the center of the nests and nodules may show sebaceous differentiation, cells at the periphery are smaller, pleomorphic and dark-staining
- Squamous metaplasia may be present and mitotic figures are common
- Poorly differentiated tumors are composed of smaller nodules and epithelial cords, separated by dense connective tissue stroma
- Sebaceous differentiation and squamous metaplasia may be present
- Tumors may show cystic areas, necrosis and inflammation

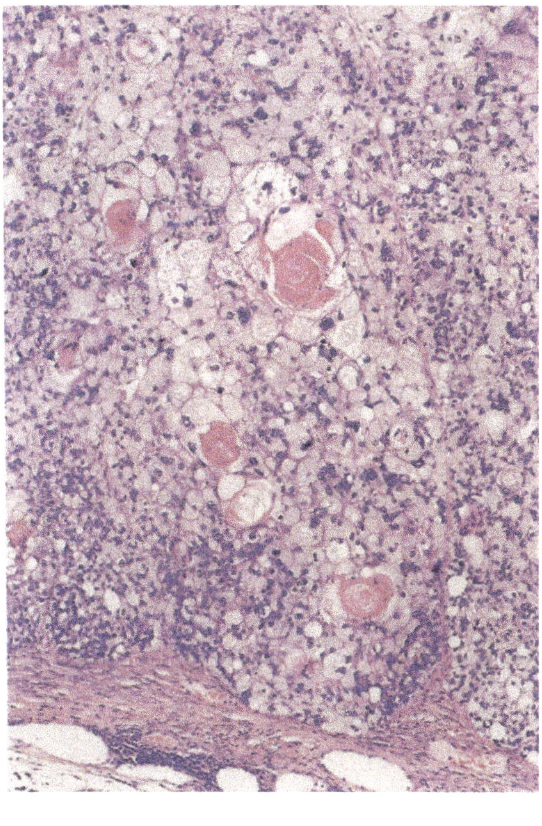

Fig. 41 *(above).* Acinar cell carcinoma of the preputial gland. Variably-sized acini extending into surrounding tissue. Tumor cells resemble immature basal cells but sebaceous differentiation is present in some acini. B6C3F1 mouse, treated with benzene, age 110 weeks. H&E

Fig. 42 *(below).* Acinar cell carcinoma of the preputial gland. Sheets composed of highly pleomorphic cells with foamy sebaceous differentiation. B6C3F1 mouse, treated with benzene, age 110 weeks. H&E

Differential Diagnosis

Hyperplasia, Acinar Cell or Adenoma, Acinar Cell
No cellular atypia and invasion of adjacent tissues present

Carcinoma, Squamous Cell
Lack neoplastic sebaceous cells, composed of non-keratinizing or keratinizing squamous cells

Adenocarcinoma of the Mammary Gland
Extremely rare tumors in males, glandular structures are present. Differentiation can be difficult, anatomic localization may be helpful. Sebaceous differentiation is not often seen in mammary carcinomas

Adenoma, Acinar Cell
Irregular distorted glands with low numbers of excretory ducts and increased numbers of basophilic basal cells

Comment

Carcinoma is considered the preferred term since the preputial gland is a skin appendage composed of modified sebaceous glands.

References

(See 12, 23, 25, 28, 29, 30, 40, 42, 43)

Hyperplasia, Squamous Cell *(H)*

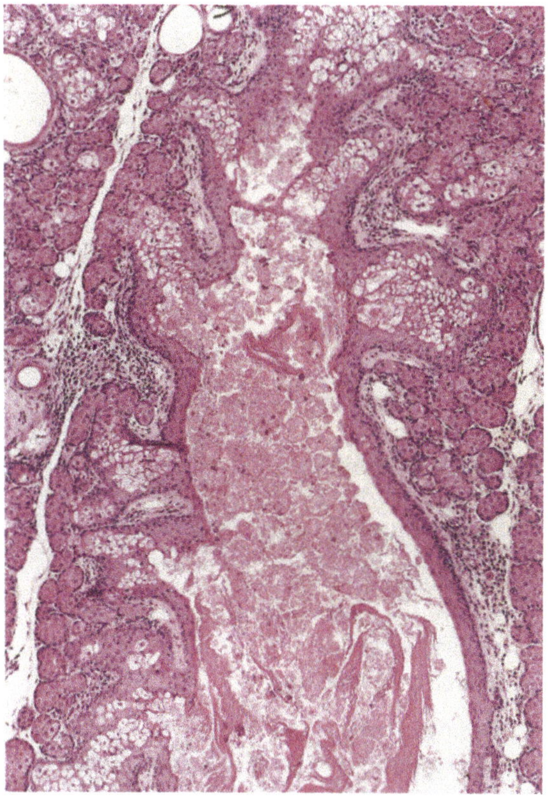

Fig. 43. Squamous cell hyperplasia of the preputial gland. Ducts lined by multiple layers of keratinizing squamous epithelium; associated dilation of duct lumen and tissue inflammation. Swiss mouse, long-term study, age >12 months. H&E

Histogenesis

Squamous epithelial cells of central and excretory ducts.

Diagnostic Features

- May occur focal, multifocal or generalized
- Epithelium consists of 3–7 flat layers of squamous epithelium
- Squamous epithelium contains keratohyaline granules and is keratinized

Differential Diagnosis

Papilloma, Squamous Cell
Compression of surrounding duct structures, focal papillary growth into the duct lumina

Comment

This lesion is seen spontaneously in older mice with dilatation of ducts and chronic inflammation. The lesion may be uni- or bilateral. It is not considered a pre-neoplastic change in these older mice but rather a reactive hyperplasia and there is no evidence of progression. Squamous cell hyperplasia is also associated with squamous cell papilloma but has not been studied as precursor for neoplasia.

References

(See 25, 40, 42, 43)

Metaplasia, Squamous Cell *(M)*

Keratinizing • *Non-keratinizing*

Histogenesis

Basal cells of central and excretory ducts–origin, basal cells.

Diagnostic Features

- Focal or multifocal

Keratinizing
- Squamous epithelium contains kerato-hyaline granules
- Is highly keratinized

Non-keratinizing
- The squamous cells may be non-keratinized

Differential Diagnosis

Hyperplasia, Squamous Cell
Mostly papillary

Comment

Very often in older animals; glands should be cut in longitudinal direction to identify the main excretory duct and several side branches.

References

(See 40)

Papilloma, Squamous Cell *(B)*

Histogenesis

Squamous cell epithelium of ducts and basal cells.

Diagnostic Features

- Papillary structures with central connective tissue stalk lined by squamous epithelium with or without keratinization
- Mitotic figures are rare
- Central and excretory ducts may also show squamous cell hyperplasia
- Papillomas may be single or multiple and can obstruct the ducts

Differential Diagnosis

Carcinoma, Squamous Cell
Destruction of basal membrane and invasion into surrounding acinar tissue. Cellular atypia and disorientation and frequent mitoses

Hyperplasia, Squamous Cell
Lacks central connective tissue stalk

Adenoma, Acinar Cell
Shows sebaceous differentiation

References

(See 12, 23, 25, 28, 29, 40, 42, 43)

Carcinoma, Squamous Cell *(M)*

Synonym(s). Carcinoma, epidermoid

Histogenesis

Duct epithelial cells. Acinar cells undergoing squamous cell differentiation.

Diagnostic Features

- Consists of irregular papillary fronds or nodules of pleomorphic poorly differentiated squamous cells with frank invasion of adjacent tissues
- Keratinization varies from keratohyalin granules, keratin pearls to highly keratinized material
- Cells may be spindle-shaped, show distinct intercellular bridges, have eosinophilic or amphophilic cytoplasm, or sebaceous cell differentiation
- Mitotic figures, necrosis and inflammation are common

Differential Diagnosis

Carcinoma, Acinar Cell
Extensive sebaceous cell differentiation with smaller squamous component, lacks prominent intercellular bridges; poorly differentiated types show scirrhous interstitium

Papilloma, Squamous Cell
Lacks invasion of surrounding structures

Carcinoma, Squamous Cell of the Skin
Differentiation may be impossible if invading preputial gland; non-preputial gland duct origin

Adenoacanthoma, Malignant of the Mammary Gland
Extremely rare in male mice; molluscoid radiating growth pattern, presence of mammary gland acini

Comment

Non-tumorous acini may be atrophic or cystic. Distinction between skin, mammary and preputial glands may be very difficult if characteristic features are not present.

References

(See 12, 23, 25, 28, 29, 30, 40, 42, 43)

References

1. Abdi MM (1995) Granulosa cell tumor of the testis in a CD-1 mouse. Vet Pathol 32: 91–92
2. Alison RH, Morgan KT (1987) Teratoma, ovary, mouse. In: Jones TC, Mohr U, Hunt RD (eds) Monographs on pathology of laboratory animals. Genital system. Springer, Berlin Heidelberg New York Tokyo, pp 46–52
3. Ashley DJB (1978) Evan's histological appearances of tumors. Churchill Livingstone, Edinburgh, pp 663–845
4. Ashley DJB (1990) Evan's histological appearances of tumors. Churchill Livingstone, Edinburgh, pp 59–62
5. Bullock BC, Newbold RR, McLachlan JA (1988) Lesions of testis and epididymis associated with prenatal diethylstilbestrol exposure. Environ Health Perspect 77: 29–31
6. Chandra M, Frith CH (1991) Spontaneously occurring leiomyosarcomas of the mouse urinary bladder. Toxicol Pathol 19: 164–167
7. Faccini JM, Abbott DP, Paulus GJJ (1990) Mouse histopathology. A glossary for use in toxicity and carcinogenicity studies. Elsevier, Amsterdam
8. Franks LM (1968) Spontaneous interstitial and Sertoli cell tumors of a testis in a C3H mouse. Cancer Res 28: 125–127
9. Frith CH, Ward JM (1988) Color atlas of neoplastic and non-neoplastic lesions in aging mice. Elsevier, Amsterdam
10. Gingrich JR, Greenberg NM (1996) A transgenic mouse prostate cancer model. Toxicol Pathol 24: 502–504
11. Gingrich JR, Barrios RJ, Morton RA, Boyce BF, DeMayo FJ, Finegold MJ, Angelopoulou R, Rosen JM, Greenberg NM (1996) Metastatic prostate cancer in a transgenic mouse. Cancer Res 56: 4096–4102
12. Gordon LR, Majka JA, Boorman GA (1996) Spontaneous nonneoplastic and neoplastic lesions and experimentally induced neoplasms of the testes and accessory sex glands. In: Mohr U, Dungworth DL, Capen CC, Carlton WW, Sundberg JP, Ward JM (eds) Pathobiology of the aging mouse, vol 1. ILSI Press, Washington DC, pp 421–441
13. Halliwell WH (1998) Submucosal mesenchymal tumors of the mouse urinary bladder. Toxicol Pathol 26: 128–136
14. Honma K, Yamada T (1984) Zur Bedeutung der "Kapselinfiltration" der testikulären Leydigzellen – eine Obduktionsstudie. Zentralbl Allg Pathol 129: 403–411
15. Huseby RA (1980) Demonstration of a direct carcinogenic effect of estradiol on Leydig cells of the mouse. Cancer Res 40: 1006–1013
16. Juriansz RL, Huseby RA, Wilcox RB (1988) Interactions of putative estrogens with the intracellular receptor complex in mouse Leydig cells: relationship to preneoplastic hyperplasia. Cancer Res 48: 14–18
17. Karbe E (1987) Granular cell tumors of genital organs, mice. In: Jones TC, Mohr U, Hunt RD (eds) Genital system. Springer, Berlin Heidelberg New York, pp 282–286
18. Karbe E (1999) "Mesenchymal tumor" or "decidual-like reaction"? Toxicol Pathol 27: 354–362
19. Karbe E, Hartmann E, George C, Wadsworth P, Harleman J, Geiss V (1998) Similarities between the uterine decidual reaction and the "mesenchymal lesion" of the urinary bladder in aging mice. Exp Toxicol Pathol 50: 330–340
20. Kaspareit J, Deerberg F (1987) Spontaneous tumours of the seminal vesicles in male Han:NMRI mice. Z Versuchstierkd 29: 277–281
21. Majeed SK, Alison RH, Boorman GA, Gopinath C (1986) Ovarian yolk sac carcinoma in mice. Vet Pathol 23: 776–778
22. Maroulakou IG, Anver M, Garrett L, Green JE (1994) Prostate and mammary adenocarcinoma in transgenic mice carrying a rat C3(1) simian virus 40 large tumor antigen fusion gene. Proc Natl Acad Sci USA 91: 11236–11240
23. Mehlman MA, Legator MS (1991) Dangerous and cancer-causing properties of products and chemicals in the oil refining and petrochemical industry – Part II: Carcinogenicity, mutagenicity, and developmental toxicity of 1,3-butadiene. Toxicol Ind Health 7: 207–220
24. Mitsumori K, Elwell MR (1988) Proliferative lesions in the male reproductive system of F344 rats and B6C3F1 mice: incidence and classification. Environ Health Perspect 77: 11–21
25. Mitsumori K, Elwell MR (1994) Tumours of the male accessory sex glands. In: Turusov VS, Mohr U (eds) Pathology of tumours in laboratory animals, vol 2. Tumours of the mouse, 2nd edn. IARC Scientific Publications No. 111, Lyon, pp 431–449
26. Mitsumori K, Talley FA, Elwell MR (1989) Epididymal interstitial (Leydig) cell tumors in B6C3F1 mice. Vet Pathol 26: 65–69
27. Mostofi FK, Davis J, Rehm S (1994) Tumours of the testis. In: Turusov VS, Mohr U (eds) Pathology of tumours in laboratory animals, vol 2. Tumours of the mouse, 2nd edn. IARC Scientific Publications No. 111, Lyon, pp 407–429

28. National Toxicology Program. Technical report series No. 289 (1986) Toxicology and carcinogenesis studies of benzene (Case No. 71–43–2) in F344/N rats and B6C3F1 mice (gavage studies). U.S. Department of Health and Human Services. National Institutes of Health
29. National Toxicology Program. Technical report series No. 316 (1986) Toxicology and carcinogenesis studies of dimethylvinyl chloride (1-chloro-2-methylpropene) (Case No. 513–37–1) in F344/N rats and B6C3F1 mice (gavage studies). U.S. Department of Health and Human Services. National Institutes of Health
30. National Toxicology Program. Technical report series No. 434 (1993) Toxicology and carcinogenesis studies of 1,3-Butadiene (Case No. 106–99–0) in B6C3F1 mice (inhalation studies). U.S. Department of Health and Human Services. National Institutes of Health
31. Newbold RR, Bullock BC, McLachlan JA (1985) Lesions of the rete testis in mice exposed prenatally to diethylstilbestrol. Cancer Res 45: 5145–5150
32. Newbold RR, Bullock BC, McLachlan JA (1986) Adenocarcinoma of the rete testis. Diethylstilbestrol-induced lesions of the mouse rete testis. Am J Pathol 125: 625–628
33. Prahalada S, Majka JA, Soper KA, Nett TM, Bagdon WJ, Peter CP, Burek JD, MacDonald JS, van Zwieten MJ (1994) Leydig cell hyperplasia and adenomas in mice treated with finasteride, a 5 alpha-reductase inhibitor: a possible mechanism. Fundam Appl Toxicol 22: 211–219
34. Radovsky A, Mitsumori K, Chapin RC (1999) Male reproductive tract. In: Maronpot RR, Boorman GA, Gaul BW (eds) Pathology of the mouse. Reference and atlas. Cache River Press, Vienna, pp 381–407
35. Reddy JK, Rao MS (1987) Testicular feminization, testes, and testicular tumors, rat, mouse. In: Jones TC, Mohr U, Hunt RD (eds), Monographs on pathology of laboratory animals. Genital system. Springer, Berlin Heidelberg New York Tokyo, pp 204–212
36. Sass B, Rehm S (1994) Tumours of the ovary. In: Turusov VS, Mohr U (eds) Pathology of tumours in laboratory animals, vol 2. Tumours of the mouse, 2nd edn. IARC Scientific Publications No. 111, Lyon, pp 493–526
37. Shibata MA, Ward JM, Devor DE, Liu ML, Green JE (1996) Progression of prostatic intraepithelial neoplasia to invasive carcinoma in C3(1)/SV40 large T antigen transgenic mice: histopathological and molecular biological alterations. Cancer Res 56: 4894–4903
38. Squire RA, Goodman DG, Valerio MG, Fredrickson T, Strandberg JD, Levitt MH, Lingeman CH, Harshbarger JC, Dawe CJ (1978) Tumors. Male reproductive system. In: Benirschke K, Garner FM, Jones TC (eds) Pathology of laboratory animals, vol II. Springer, Berlin, pp 1213–1225
39. Stevens LC (1973) A new inbred subline of mice (129-terSv) with a high incidence of spontaneous congenital testicular teratomas. J Natl Cancer Inst 50: 235–242
40. Stolte M (1993) Histomorphological age changes and ultrastructural characteristics of the preputial and clitoral glands of mice. J Exp Anim Sci 35: 166–176
41. Tehranian A, Morris DW, Min BH, Bird DJ, Cardiff RD, Barry PA (1996) Neoplastic transformation of prostatic and urogenital epithelium by the polyoma virus middle T gene. Am J Pathol 149: 1177–1191
42. Toth B, Nagel D (1980) N-Ethyl-N-formylhydrazine tumorigenesis in mice. Carcinogenesis 1: 61–65
43. Toth B, Taylor J, Gannett P (1991) Tumor induction with hexanal methylformylhydrazone of *Gyromitra esculenta*. Mycopathologia 115: 65–71
44. Yoshitomi K, Morii S (1984) Benign and malignant epithelial tumors of the rete testis in mice. Vet Pathol 21: 300–303

Female Genital System

B. Davis[1] · J.H. Harleman[3] · M. Heinrichs · A. Maekawa[2] · R.F. McConnell
G. Reznik[3](†) · M. Tucker

[1]Chair; [2]Co-chair; [3]Initial draft.

Hyperplasia, Tubulostromal *(H)*

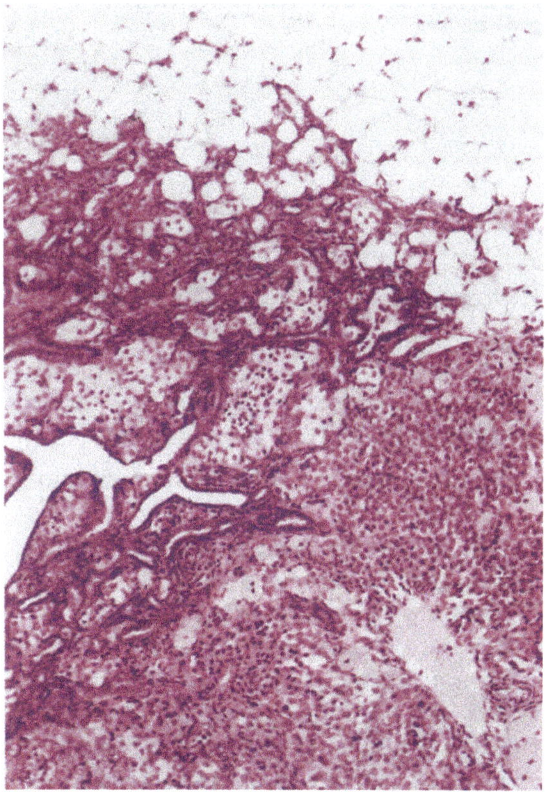

Fig. 1. Tubulostromal hyperplasia of the ovary. H&E

Histogenesis

Derived from the surface epithelium of the ovary.

Diagnostic Features

- Hyperplasia is generally diffuse and not a discrete focal lesion
- Infiltration of surface epithelial cells in the ovary proper associated with a variable proliferation of stromal cells
- If nodular, lesions without atypia and in diameter smaller than or equal size of corpus luteum are considered to represent hyperplasia
- Lesion often forms a ring of hyperplasia around the ovary
- At the hilus minimal extension on the bursa surface may occur

Differential Diagnosis

Adenoma, Tubulostromal
Tubulostromal adenomas are discrete often nodular lesions, which without atypia are larger than a corpus luteum in diameter

Comment

Extremely common lesion in old mice. Differentiation from tubulostromal adenoma sometimes arbitrary.

References

(See 2, 6, 7, 14, 20, 21, 22, 23, 24)

Adenoma, Tubulostromal *(B)*

Histogenesis

Surface epithelium of the ovary.

Diagnostic Features

- Nodular structure, which may show compression
- Delicate tubules lined by cuboidal epithelium that resemble or are continuous with surface mesothelium of the ovary
- Tubules are separated by packets of cells of probable sex cord stromal origin, which may show various degrees of luteinization
- Ratio of tubular to non-tubular components, degree of tubular dilatation and amount of vacuolation/luteinization highly variable and/or
- Diameter of proliferative lesion is larger than the size of a corpus luteum
- Some tumors may have testis-like tubules lined by sertoli-form cells and glomerular-like structures, but these structures rarely predominate
- Some tumors have areas with cystic dilation of the tubular structures
- Some extension in the ovarian bursa, especially near the hilus may be present

Differential Diagnosis

Adenocarcinoma, Tubulostromal
The differentiation between tubulostromal adenoma and tubulostromal adenocarcinoma is made on the basis of usual criteria of malignancy such as cellular atypia, local invasion, metastasis, etc.

Cystadenoma or Cystadenocarcinoma
Absence of the interstitial cell component and the presence of a generally higher cuboidal to column epithelium forming cystic spaces. Infolding of epithelium forming papillary structures. Presence of cystic spaces

Hyperplasia, Tubulostromal
Does not form discrete masses, small nodules without atypia; diameter is smaller than or equal size of a corpus luteum. May form a rim of hyperplastic tissue around the ovary

Comment

Frequent in some strains of mice but rare in rats.

References

(See 2, 4, 6, 7, 14, 19, 20, 21, 22, 23, 24)

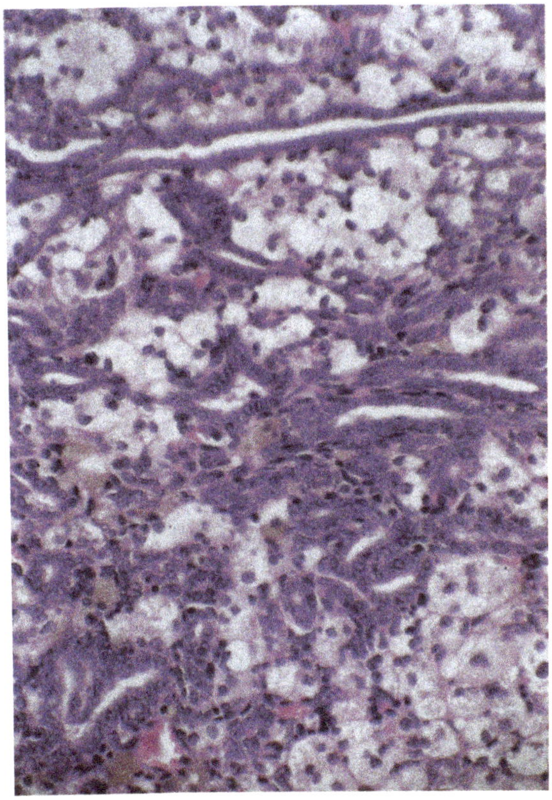

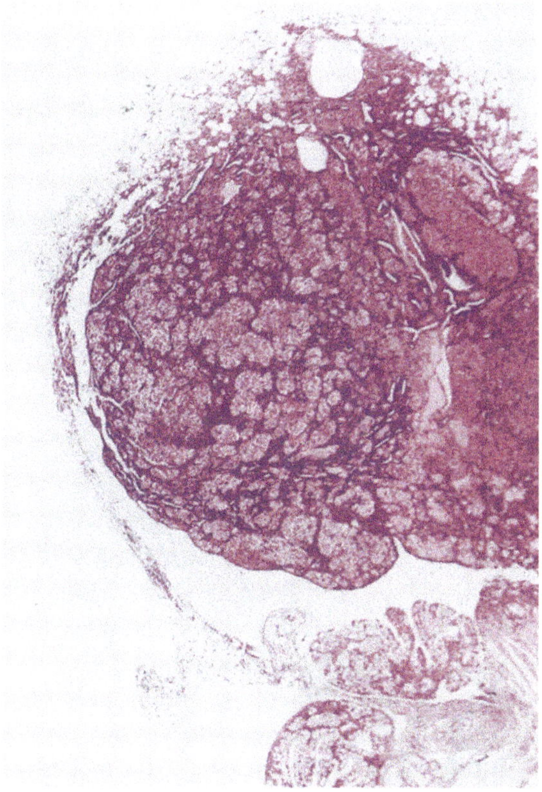

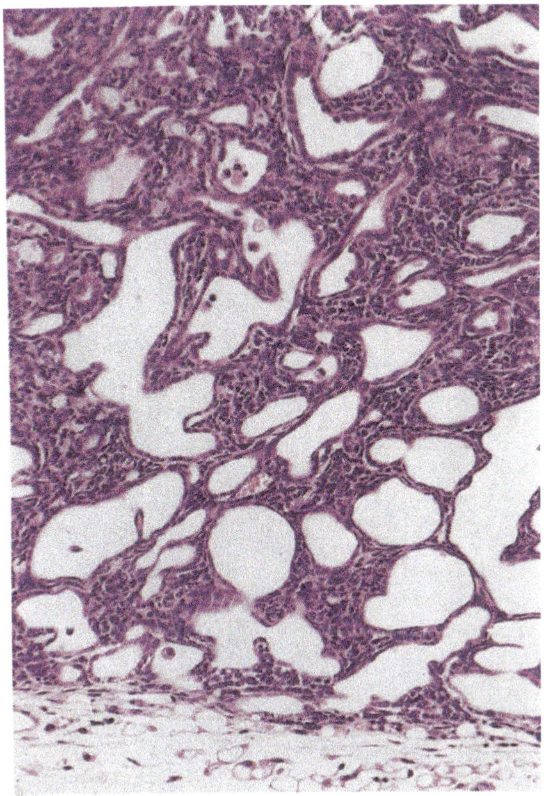

Fig. 2 *(upper left)*. Tubulostromal adenoma of the ovary demonstrating well-differentiated tubules separated by packets of cells of sex cord stromal origin. H&E

Fig. 3 *(upper right)*. Tubulostromal adenoma of the ovary. H&E

Fig. 4 *(below)*. Tubulostromal adenoma of the ovary. H&E

Adenocarcinoma, Tubulostromal *(M)*

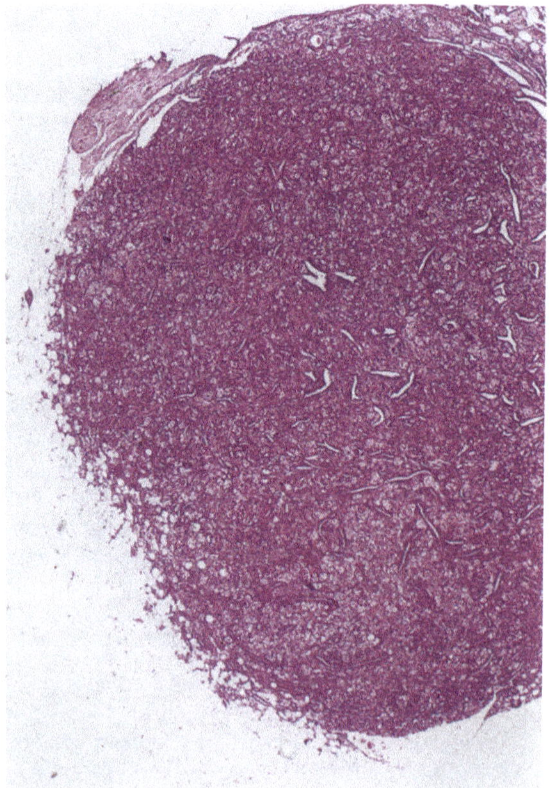

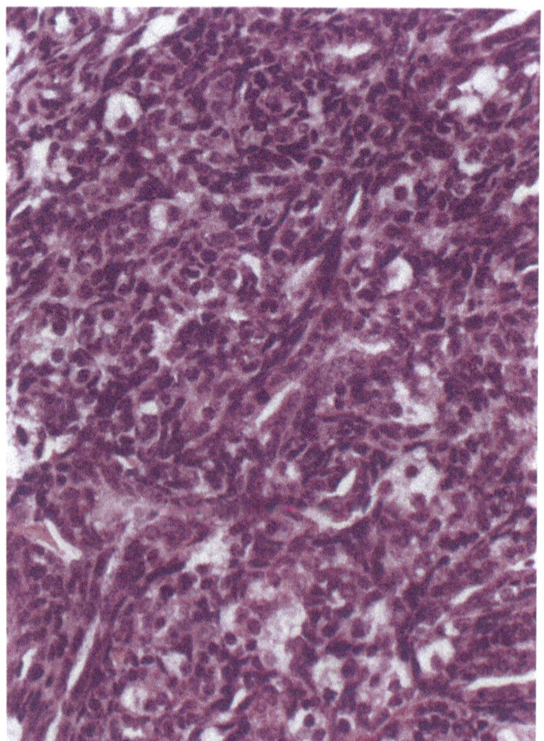

Histogenesis

Surface epithelium of the ovary.

Diagnostic Features

- Delicate tubules lined by cuboidal epithelium that resemble or are continuous with surface mesothelium of the ovary
- Tubules are separated by packets of cells of probable sex cord stromal origin, which may show various degrees of luteinization
- Ratio of tubular to non-tubular components, degree of dilatation and amount of vacuolation/luteinization highly variable
- Tumor shows a high degree of pleomorphism, atypia and infiltrative growth pattern and/or metastasis outside of the ovary
- Usual criteria for malignancy such as cellular atypia, local invasion, metastasis etc.

Differential Diagnosis

Adenoma, Tubulostromal

The differentiation between tubulostromal adenoma and tubulostromal adenocarcinoma is made on the basis of the usual criteria of malignancy such as cellular atypia, local invasion, mitotic index, hemorrhage, metastasis etc. Invasion should be beyond the ovary/ovarian bursa

Fig. 5 *(above)*. Tubulostromal adenocarcinoma, effacing entire ovary with local invasion into adjacent adipose tissue. H&E

Fig. 6 *(below)*. Tubulostromal adenocarcinoma of the ovary. Higher magnification demonstrating cellular atypia, pleomorphism and high mitotic index. H&E

Cystadenoma or Cystadenocarcinoma
Cystadenoma/cystadenocarcinoma can be differentiated from tubulostromal neoplasms by the absence of the interstitial cell component and the presence of generally higher cuboidal to column epithelium forming cystic structures. Cystadenoma/cystadenocarcinoma generally show infolding of epithelium forming papillary structures

Comment

Rare in aging mice.

References

(See 2, 4, 6, 7, 14, 20, 21, 22, 23, 24)

Hyperplasia, Cystic/Papillary *(H)*

Histogenesis

Derived from the surface epithelium of the ovary or rete anlage.

Diagnostic Features

- Small focal lesions often located at the hilus of the ovary
- Characterized by small cystic lesions which are covered by a single layer of cuboidal to columnar epithelium and often show folds or papillary projections. Within the papillary projections there may be up to three layers of well-differentiated epithelial cells present
- Occasionally papillary proliferation of the surface epithelium are noted in the ovary. These lesions mimic papillary cystadenomas without cysts are therefore included in this category
- Diameter is smaller than or equal size of a corpus luteum

Differential Diagnosis

Cystadenoma
Cystadenomas are generally lesions which may show some atypia (polymorphism, crowding) and in the absence of atypia are larger than size of one corpus luteum

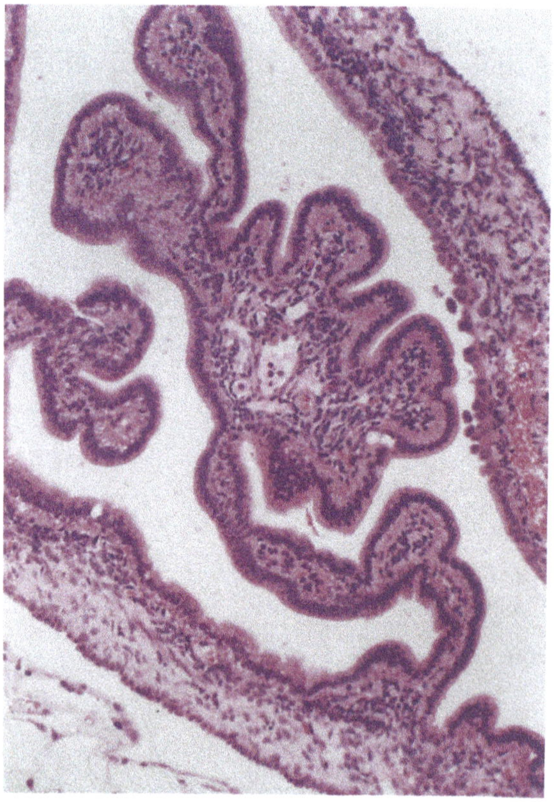

Fig. 7. Cystic/papillary hyperplasia of the ovary. H&E

Comment

These lesions are common in the hilus area of the ovary in certain strains. Only high grade lesions are recorded in the data base (larger than half of a corpus luteum).

References

(See 2, 6, 7, 14, 20, 21, 22, 23, 24)

Cystadenoma *(B)*

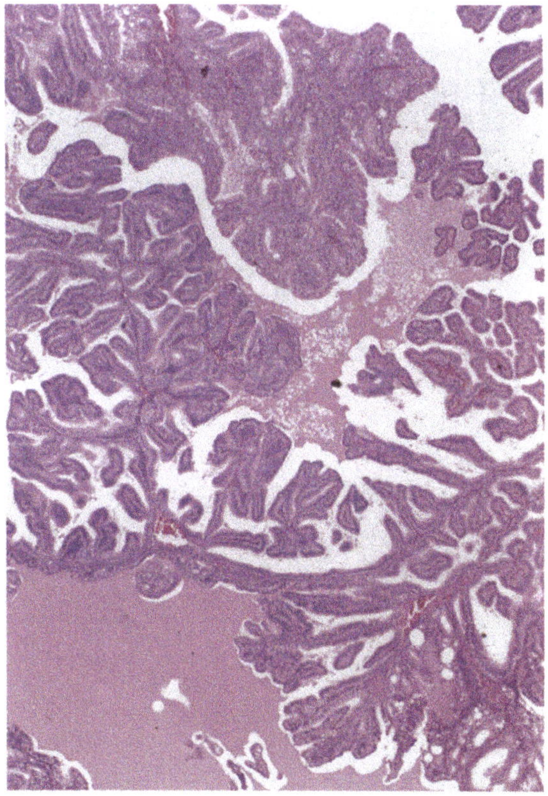

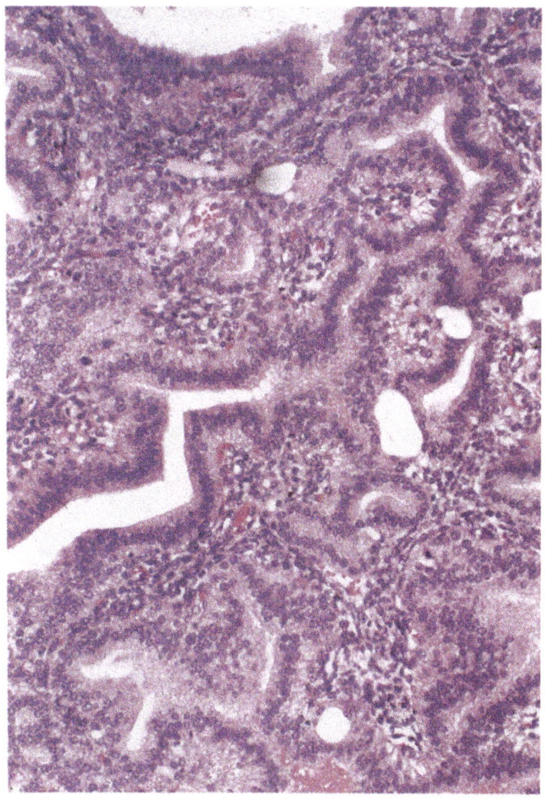

Histogenesis

Derived from the surface epithelium of the ovary.

Diagnostic Features

- Tumor often within a single or multiple cysts lined by cuboidal or columnar epithelium which is generally not ciliated
- Folds or papillary projection into the lumen may be present
- There is often compression of the adjacent ovarian stroma but little or no invasion or cellular atypia
- The cyst may contain serous fluid or blood
- There is normal delicate stromal separating the cystic structures
- Occasionally small focal papillary-like nodules apparently derived from surface epithelium. Although these are not cystic or present within a cyst, these lesions are included in this category and/or
- Diameter of proliferative lesion is larger than a corpus luteum

Differential Diagnosis

Hyperplasia, Cystic/Papillary
Well differentiated with a single layer of epithelial cells and no atypia and diameter is smaller than or equal in size to a corpus luteum

Fig. 8 *(above)*. Cystadenoma of the ovary, consisting of multiple branching papillary projections into a lumen filled with serous fluid. H&E

Fig. 9 *(below)*. Cystadenoma of the ovary, higher magnification demonstrating well-differentiated cuboidal to columnar cells lining papillary projections. H&E

Adenoma, Tubulostromal or Adenocarcinoma, Tubulostromal
Cystadenomas do not have the (luteinized) stromal compartment of the tubulostromal tumor. The stromal compartment is delicate and not an inherent part of the tumor

Cystadenocarcinoma
Cystadenocarcinomas show atypia and invasive growth pattern, and high mitotic activity

Comment

Cystadenomas are rare tumors in the rat in contrast to the mouse.

References

(See 2, 4, 6, 7, 14, 18, 20, 21, 22, 23, 24)

Cystadenocarcinoma *(M)*

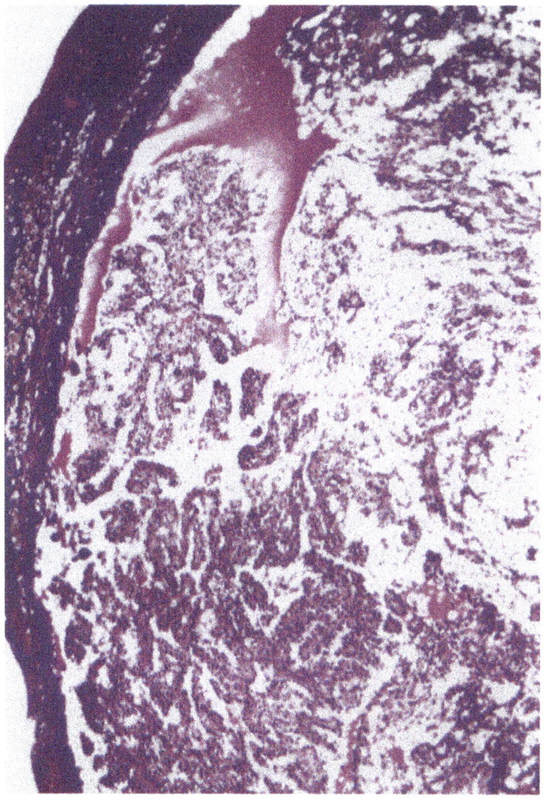

Fig. 10. Cystadenocarcinoma, effacing the ovary and invading into adjacent tissue. H&E

Histogenesis

Derived from the surface epithelium of the ovary.

Diagnostic Features

- Tumor consists of solid to cystic mass lined by cuboidal pleomorphic epithelium, which may be ciliated
- Folds or papillary projections may be present
- The tumor has a high mitotic index and infiltration of the adjacent tissue
- One or more of the following optional qualifiers may be used depending upon the morphological characteristics of the tumor: papillary, cystic, serous, mucinous

Differential Diagnosis

Cystadenoma

Cystadenomas do not have atypia, invasive growth pattern or high mitotic rate

Adenoma, Tubulostromal or Adenocarcinoma, Tubulostromal

Cystadenoma/carcinomas do not have the (luteinized) stromal compartment of the tubulostromal tumor. The stromal compartment is delicate and not an inherent part of the tumor

Cystadenomas form cystic spaces

References

(See 2, 4, 6, 7, 14, 18, 20, 21, 22, 23, 24)

Hyperplasia, Sex Cord Stromal *(H)*

Granulosa cell • Mixed type
Theca cell • Sertoli cell

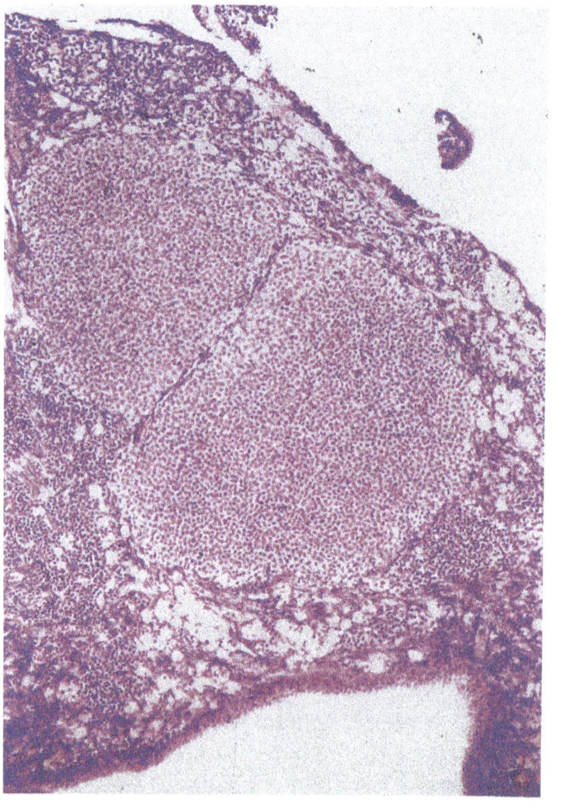

Fig. 11. Granulosa cell hyperplasia of the ovary, focally extensive lesion. H&E

Histogenesis

Sex cord stromal cells.

Diagnostic Features

- Small focal distinct lesion which may show variable morphological spectrum including granulosa cells, theca cells, sertoli cells, luteal cells, etc., often in varying quantities
- Focal lesions are discrete lesions, which are well-demarcated from the adjacent tissue
- Lesions show often the predomination (>70%) of one cell type, i.e., granulosa, theca, or sertoli. Mixed lesions are lesions in which more than one cell type is present and in which no one cell type predominates
- Diameter is smaller than or of equal size to a corpus luteum
- Focal discrete lesions larger than a large corpus luteum are, in the absence of any other morphological criteria, such as atypia, pleomorphism, high mitotic rate and invasiveness, considered to be a tumor

Differential Diagnosis

Tumor, Granulosa Cell, Benign or Tumor, Sertoli Cell, Benign or Thecoma, Benign
Discrete nodule and compression and diameter of proliferative lesion is larger than size of a corpus luteum. One cell type generally predominates

References

(See 2, 6, 7, 14, 20, 21, 22, 23, 24)

Tumor, Granulosa Cell, Benign *(B)*

Synonym(s). Tumor, sex cord stromal, benign, granulosa cell type

Histogenesis

Sex cord stromal cells.

Diagnostic Features

- Cellular morphology resembles that of normal granulosa cells
- Nuclei round to oval with coarsely stippled chromatin
- Cytoplasm varies from scanty to moderate depending on degree of luteinization and is faintly eosinophilic and vacuolated
- Call-Exner bodies uncommon
- Several patterns such as follicular, solid and trabecular identified
- Some large tumors contain areas of hemorrhage and necrosis with lipofuscin granules
- Occasionally, larger granulosa cell tumors show areas or are partially composed of fusiform theca-like cells. The predominant component, however is the granulosa cell
- Tumor may often show variable degrees of luteinization and/or
- Diameter of proliferative lesion is larger than size of a corpus luteum

Differential Diagnosis

Hyperplasia, Sex Cord Stromal (Granulosa Cell), Focal
No discrete nature and no distinct compression and displacement of surrounding tissue. In the absence of other morphological characteristics: lesion of predominantly one cell type smaller than or equal to a large corpus luteum, should be called a hyperplasia and larger than that a tumor

Tumor, Granulosa Cell, Malignant
The distinction between benign and malignant granulosa cell tumor is made on the degree of atypia, infiltrative growth pattern, presence of metastasis, and areas of necrosis and hemorrhage indicative of a high growth rate

Luteoma, Benign or Thecoma, Benign or Tumor, Sertoli Cell, Benign
The differentiation of luteoma/thecoma/sertoli cell tumor is made on the predominant characteristics of the tumor. In the case of granulosa cell tumors the granulosa cell type should be the predominant cell type. Luteomas do not have areas of non-luteinized granulosa cells, but have a very high degree of luteinization throughout the tumor

Comment

Incidence of granulosa cell tumors varies markedly between the various strains. Granulosa cell tumors occur with a high frequency in the SWR strain. In most other strains relatively uncommon. Tumor may secrete estrogen.

References

(See 2, 4, 6, 7, 14, 20, 21, 22, 23, 24)

Figs. 12-14 see p. 224.

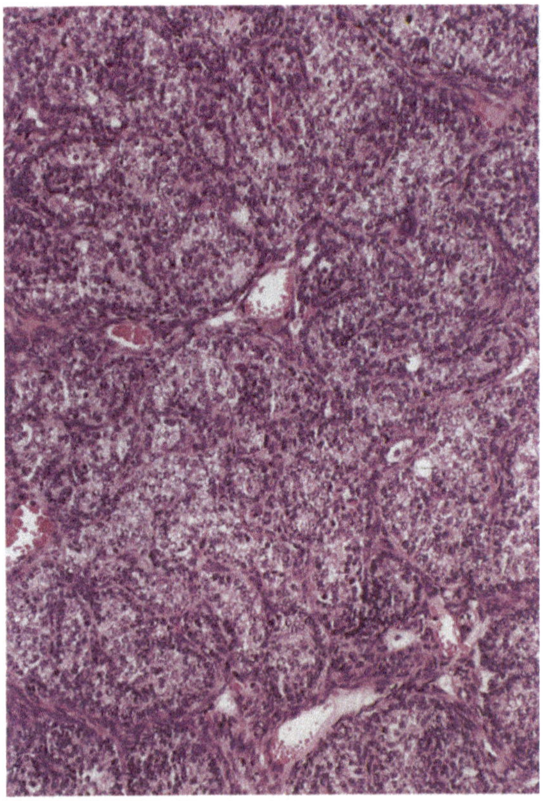

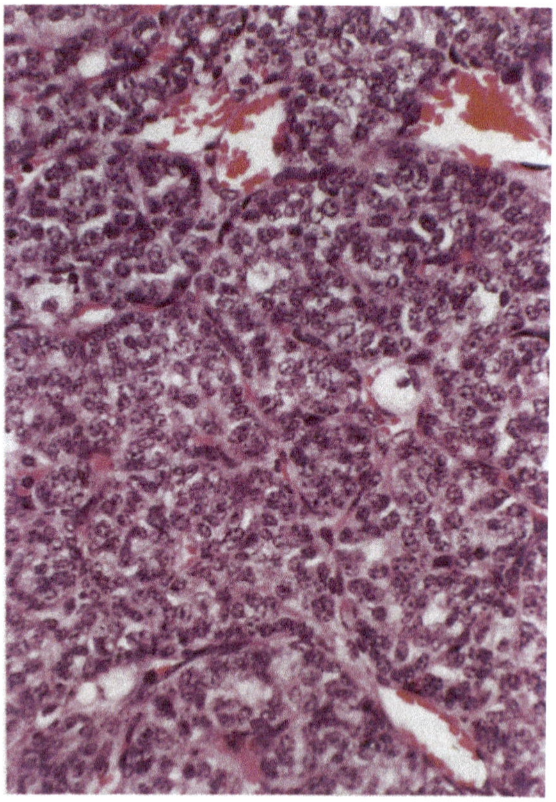

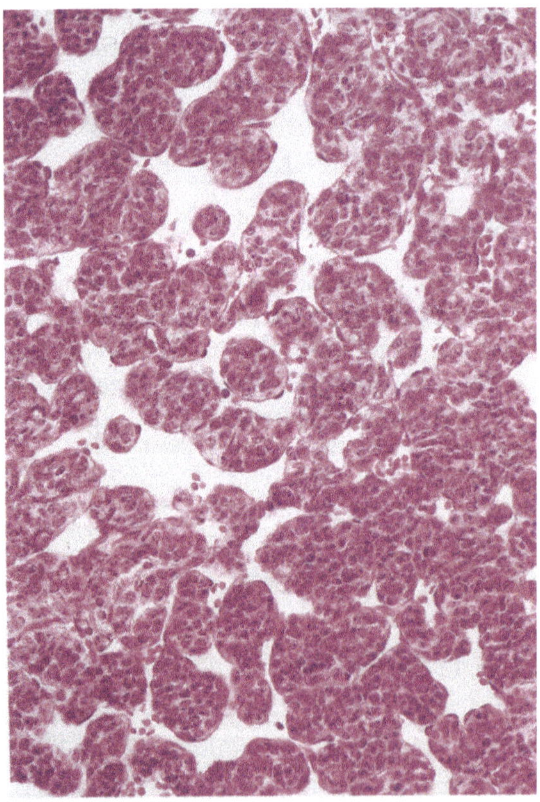

Fig. 12 *(upper left)*. Benign granulosa cell tumor of the ovary showing follicular pattern. H&E

Fig. 13 *(upper right)*. Benign granulosa cell tumor of the ovary. Higher magnification demonstrating stippled nuclei of neoplastic granulosa cells. H&E

Fig. 14 *(below)*. Benign granulosa cell tumor of the ovary. H&E

Tumor, Granulosa Cell, Malignant *(M)*

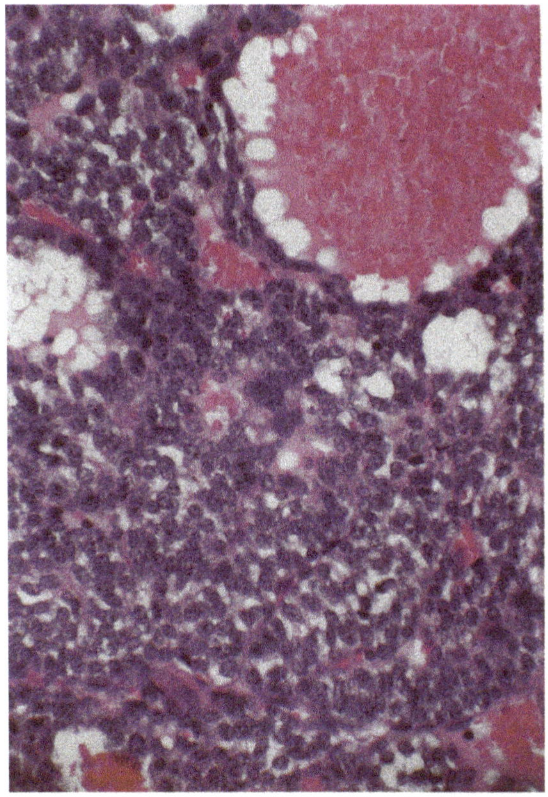

Synonym(s). Gynoblastoma; tumor, sex cord stromal, malignant, granulosa cell type

Histogenesis

Sex cord stromal cells.

Diagnostic Features

- Malignant granulosa cell tumors have the evidence of local invasion, a high degree of cellular pleomorphism and a high mitotic rate
- Focal areas of necrosis and hemorrhages are frequently present
- Tumor may show distant metastases to kidneys, lungs, and lymph nodes
- Cellular morphology resembles that of normal granulosa cells
- Nuclei round to oval with coarsely stippled chromatin
- Cytoplasm varies from scanty to moderate depending on degree of luteinization and is faintly eosinophilic and vacuolated
- Call-Exner bodies uncommon
- Several patterns such as follicular, solid and trabecular identified
- Occasionally, granulosa cell tumors show areas of, or are partially composed of fusiform theca-like cells

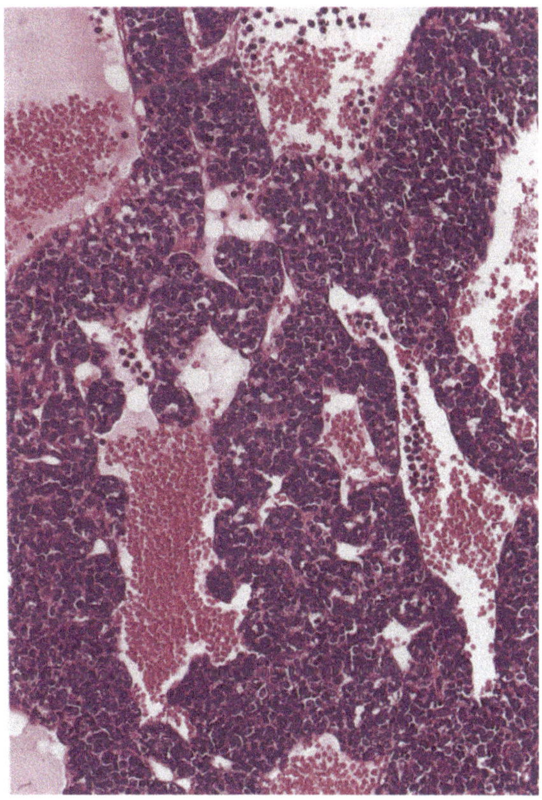

Fig. 15 *(above).* Malignant granulosa cell tumor of the ovary. Malignant variant based on cellular atypia and high mitotic index. H&E

Fig. 16 *(below).* Malignant granulosa cell tumor of the ovary. H&E

Differential Diagnosis

Tumor, Granulosa Cell, Benign
The distinction between benign and malignant granulosa cell tumor is made on the degree of atypia, infiltrative growth pattern, presence of metastasis, and areas of necrosis and hemorrhage indicative of a high growth rate

Luteoma, Benign or Thecoma, Malignant or Tumor, Sertoli Cell, Malignant
The differentiation between luteoma/thecoma/sertoli cell is made on the predominant cell type present. In the case of granulosa cell tumors the latter cell type should be the predominant cell type present

References

(See 2, 4, 6, 7, 14, 20, 21, 22, 23, 24)

Thecoma, Benign *(B)*

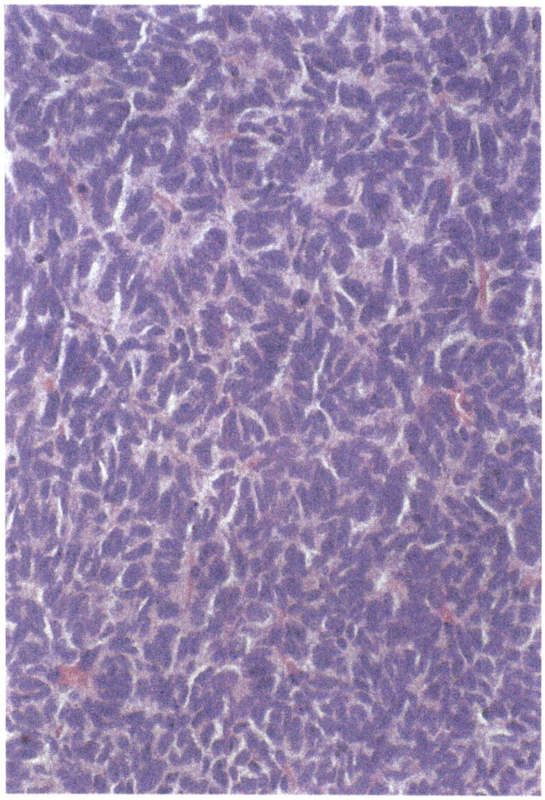

Fig. 17. Benign thecoma of the ovary. H&E

Synonym(s). Tumor, sex cord stromal, benign, thecoma type; tumor, theca cell, benign

Histogenesis

Sex cord stromal cells.

Diagnostic Features

- Composed of densely packed fusiform cells, usually arranged in whorls giving a nodular appearance
- Spindle-shaped cells arranged in interlacing bundles and whorled patterns
- Variable amount of lipid and collagen present. Collagen is mainly between bundles of cells
- Extensive necrosis in large tumors with only perivascular persistence of viable tissue
- Focal areas of mineralization and hyalinization may be present and/or
- Diameter of proliferative lesion is larger than size of a corpus luteum

Differential Diagnosis

Fibroma
: Fibromas lack lipid-laden cells and have collagen arranged around individual cells

Hyperplasia, Sex Cord Stromal
: Sex cord stromal hyperplasia may be focal, and smaller in size than or equal to a large corpus luteum, if of one cell type

Tumor, Granulosa Cell, Benign or Luteoma, Benign
: The differentiation between thecomas and the other sex cord stromal tumors is made on the predominant cell type present. In thecoma the major cell type is spindle-shaped theca cell

Thecoma, Malignant
: The differentiation between thecoma benign and malignant is made on the presence of an infiltrative growth pattern, an invasion of the adjacent tissue, degree of atypia, and metastasis in the malignant thecoma

References

(See 2, 4, 6, 7, 14, 20, 21, 22, 23, 24)

Thecoma, Malignant *(M)*

Synonym(s). Tumor, sex cord stromal, malignant, thecoma type; tumor, theca cell, malignant

Histogenesis

Sex cord stromal cells.

Diagnostic Features

- Composed of densely packed fusiform cells, usually arranged in whorls giving a nodular appearance
- Spindle-shaped cells arranged in interlacing bundles and whorled patterns
- Variable amount of lipid and collagen present. Collagen arranged between bundles of cells
- Extensive necrosis in large tumors with only perivascular persistence of viable tissue
- Focal areas of mineralization and hyalinization may occur
- Malignant thecomas also show pleomorphism and multiple areas of necrosis suggesting rapid growth, infiltration of adjacent tissue and/or metastasis

Differential Diagnosis

Fibroma or Fibrosarcoma
Fibromas/fibrosarcomas lack lipid-laden cells and have collagen arranged around individual cells

Tumor, Granulosa Cell, Malignant or Luteoma, Benign
The differentiation between thecomas and the other sex cord stromal tumors is made on the predominant cell type present. In the case of thecoma the major cell type is spindle-shaped theca cell

Thecoma, Benign
The differentiation between thecoma benign and thecoma malignant is made on the presence of an infiltrative growth pattern, the degree of atypia, invasion of the adjacent tissue, and metastasis in the malignant thecoma

References

(See 2, 4, 6, 7, 14, 20, 21, 22, 23, 24)

Luteoma, Benign *(B)*

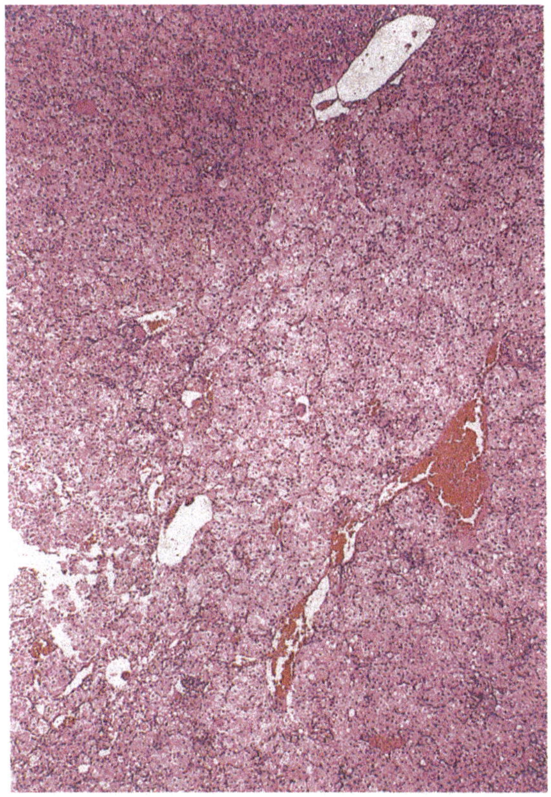

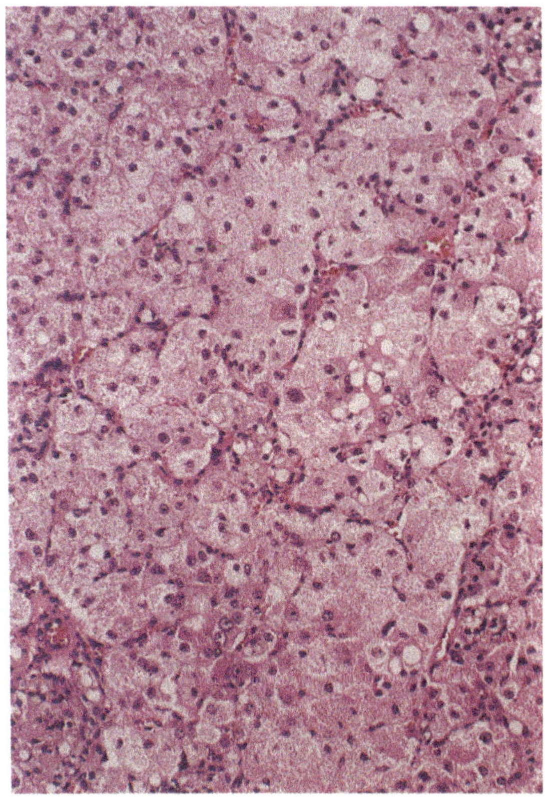

Synonym(s). Gonadal sex cord stromal tumor, benign, luteoma type; luteinized granulosa cell tumor, benign

Histogenesis

Sex cord stromal cells.

Diagnostic Features

- Composed of highly luteinized cells with extensive pale granular cytoplasm. Nuclei are round to oval without significant stippling
- Intranuclear cytoplasmic invaginations and mast cells occasionally present
- Tumor shows generally a slight degree of cellular pleomorphism
- Stromal compartment may be more irregular dividing the tumor in small lobules compared to normal corpus luteum and/or
- Diameter of proliferative lesion is larger than three corpora lutea in size

Differential Diagnosis

Tumor, Granulosa Cell, Benign or Thecoma, Benign

The differentiation between luteomas and other sex cord stromal tumors is made upon the predominant cell type present. Granulosa cell tumors and thecomas do not have such a high uniform degree of luteinization

Fig. 18 *(above).* Benign luteoma, tumor effacing entire ovary. H&E

Fig. 19 *(below).* Benign luteoma of the ovary demonstrating abundant eosinophilic cytoplasm of neoplastic cells. H&E

Comment

Luteoma, malignant has not been reported. Luteinized cells sometimes occur in malignant and benign granulosa cell tumors. In luteomas almost every cell has a high degree of luteinization and virtually no or very few smaller less luteinized cells are present (the degree of luteinization is less uniform in luteinized granulosa cell tumors).

References

(See 2, 4, 6, 7, 14, 20, 21, 22, 23, 24)

Tumor, Sertoli Cell, Benign *(B)*

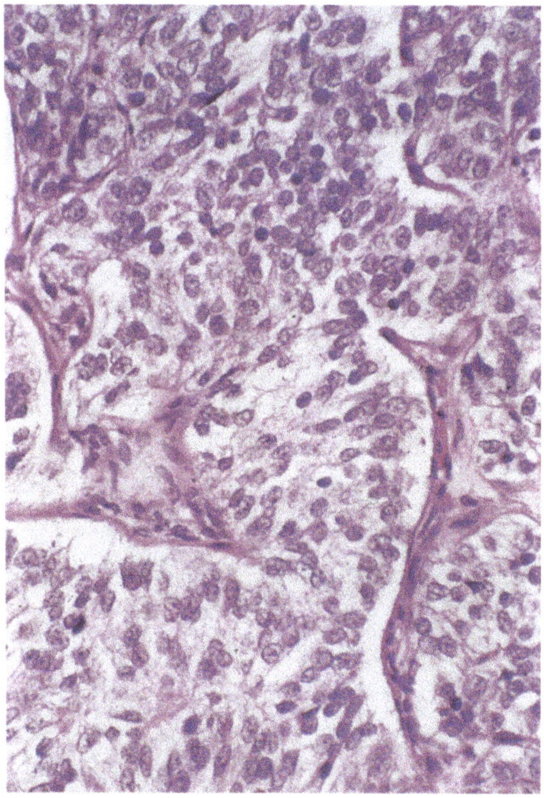

Fig. 20. Benign Sertoli cell tumor of the ovary. Tumor is composed of cords arranged in seminiferous tubule-like pattern. H&E

Synonym(s). Tumor, gonadal stromal, benign; tumor, sex cord stromal, benign; tumor, sex cord stromal, benign, sertoli type; tumor, sustentacular, benign

Histogenesis

Sex cord stromal cells.

Diagnostic Features

- Sertoli cell tumors resemble their testicular counterpart
- Characterized by seminiferous-like tubules separated by fibrovascular stroma, lined by cells with basally located nuclei and abundant faintly eosinophilic cytoplasm extending into the lumen
- No spermatogenic cells seen
- These tumors frequently have areas with nests of vacuolated cells/sertoli cells without obvious tubular structures
- These tumors frequently have a relatively high proportion of fusiform theca-like cells and/or
- Diameter of proliferative lesion is larger than size of a corpus luteum

Differential Diagnosis

Hyperplasia, Sex Cord Stromal (Sertoli Cell)
Diameter is smaller than or equal in size to a corpus luteum

Tumor, Granulosa Cell, Benign or Thecoma, Benign
Sertoli tumors are differentiated from granulosa cell tumors and other sex cord stromal tumors by the tubular growth pattern and cellular characteristics (predominant cell type present)

Tumor, Sertoli Cell, Malignant
The malignant sertoli cell tumor is differentiated from the benign sertoli cell tumor by a poorer differentiation, infiltrative growth pattern and areas of hemorrhage and necrosis

References

(See 2, 6, 7, 20, 21, 22, 23, 24)

Tumor, Sertoli Cell, Malignant *(M)*

Synonym(s). Androblastoma; arrhenoblastoma; tumor, gonadal stromal, malignant; tumor, sex cord stromal, malignant; tumor, sex cord stromal, malignant, sertoli type; tumor, sustentacular, malignant

Histogenesis

Sex cord stromal cells.

Diagnostic Features

- Sertoli cell tumors resemble their testicular counterpart
- Characterized by seminiferous-like tubules separated by fibrovascular stroma, lined by cells with basally located nuclei and abundant faintly eosinophilic cytoplasm extending into the lumen
- No spermatogenic cells seen
- Malignancy is diagnosed on the basis of focal necrosis, hemorrhage, and local invasion and/or metastasis
- Malignant sertoli cell tumors are usually less well-differentiated, exhibiting areas of rounded pleomorphic cells
- Focal areas of necroses and hemorrhages may be present
- Areas of granulosa-like cells may be present but are not the main characteristic

Differential Diagnosis

Tumor, Granulosa Cell, Malignant or Thecoma, Malignant
Sertoli tumors are differentiated from granulosa cell tumor and other sex cord stromal tumors by their tubular growth pattern and cellular characteristics

Tumor, Sertoli Cell, Benign
The malignant sertoli cell tumor is differentiated from the benign sertoli cell tumor by a poorer differentiated infiltrative growth pattern and areas of hemorrhage and necrosis

References

(See 2, 6, 7, 14, 19, 20, 21, 22, 23, 24)

Tumor, Sex Cord Stromal, Mixed, Benign *(B)*

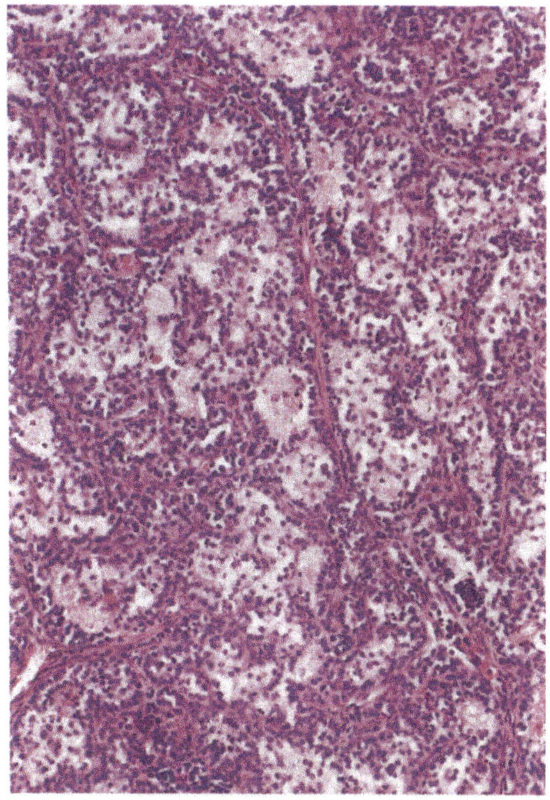

Fig. 21. Benign mixed sex cord stromal tumor of the ovary, composed of an admixture of granulosa cell, luteinized granulosa cell, and thecomatous patterns. H&E

Histogenesis

Sex cord stromal cells.

Diagnostic Features

- Tumor consists of a mixture of granulosa, luteal, theca, sertoli and stromal cells, which may show various degrees of differentiation. No one cell type dominates (>70%)
- Discrete well-demarcated focal lesions, which are bigger than one large corpus luteum

Differential Diagnosis

Tumor, Granulosa Cell, Benign or Tumor, Sertoli Cell, Benign or Thecoma, Benign or Tumor, Granulosa Cell, Malignant or Tumor, Sertoli Cell, Malignant or Thecoma, Malignant
One differentiated cell type dominates (sertoli, granulosa, theca, >70%) and is often composed of rather undifferentiated stromal cells

Hyperplasia, Sex Cord Stromal (Focal)
Little or no compression and diameter is smaller than or equal size of a corpus luteum. Discrete lesion with clear demarcation

References

(See 2, 6, 7, 14, 20, 21, 22, 23, 24)

Teratoma, Benign *(B)*

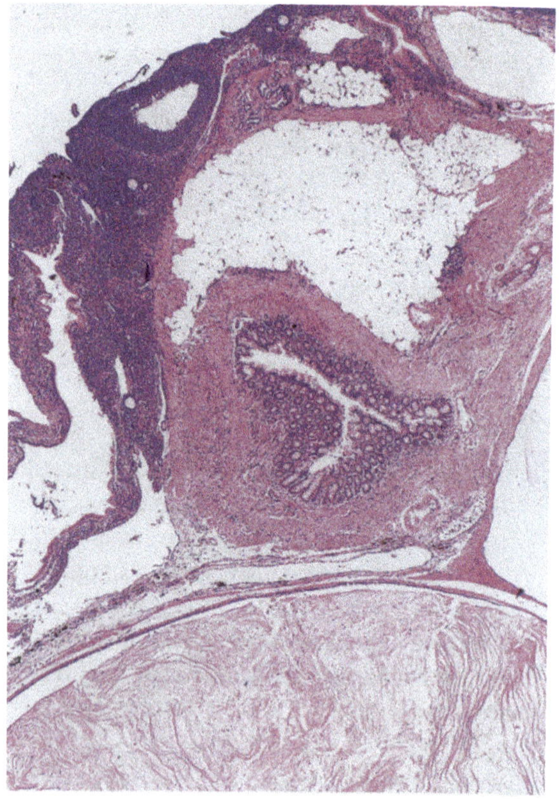

Fig. 22. Benign teratoma of the ovary demonstrating ectodermal, neuroectodermal and endodermal-derived tissues. H&E

Histogenesis

Multipotential embryonic tissue which has escaped the influence of primary organizers.

Diagnostic Features

- Teratoma must contain tissue derived from the three germ layers
- Tissues components are generally well-differentiated
- Benign teratomas generally contain cysts which may be lined by cuboidal, enteric or respiratory epithelium. Smooth muscle may surround these cysts
- Other elements may consist of pancreatic tissue, gastric epithelium, thyroid, well-differentiated neuronal tissue, cartilage, bone and/or skeletal muscle

Differential Diagnosis

Teratoma, Malignant
Malignant teratomas are composed of poorly differentiated tissue elements and often show large areas with necrosis and/or hemorrhage or other malignant features (invasion and metastases)

Comment

Teratomas are rare tumors but have been observed more commonly in substrains of strain 129. They may occur in any tissue but are most often encountered in genital system.

References

(See 1, 22)

Teratoma, Malignant *(M)*

Histogenesis

Multipotential embryonic tissue which has escaped the influence of primary organizers.

Diagnostic Features

- Teratomas must contain tissue derived from the three germ layers
- In malignant teratomas, nervous, epithelial, and mesenchymal tissue are poorly differentiated resembling embryonic tissue
- These tumors are invasive and may metastasize
- Areas of necrosis and hemorrhage, indicative of a rapid growth, may be present

Differential Diagnosis

Teratoma, Benign
Benign teratomas have a higher degree of tissue maturation. Large areas with necrosis and/or hemorrhage and other malignant features such as invasion or metastasis are absent

Sex Cord Stromal Tumors
Sex cord stromal tumors do not have tissues from the three germ layers as an integral part of the tumor

Comment

Teratomas are rare tumors, which are most often encountered in the genital system but may occur in any tissue.

References

(See 1, 22)

Carcinoma, Yolk Sac *(M)*

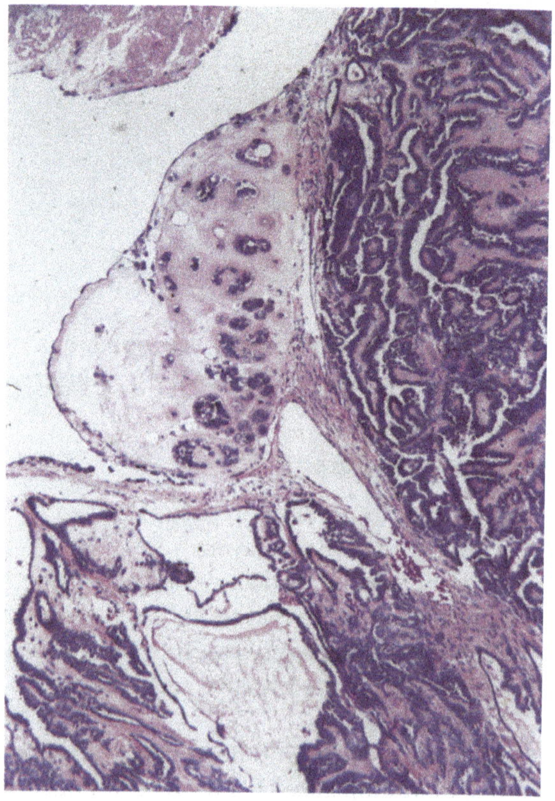

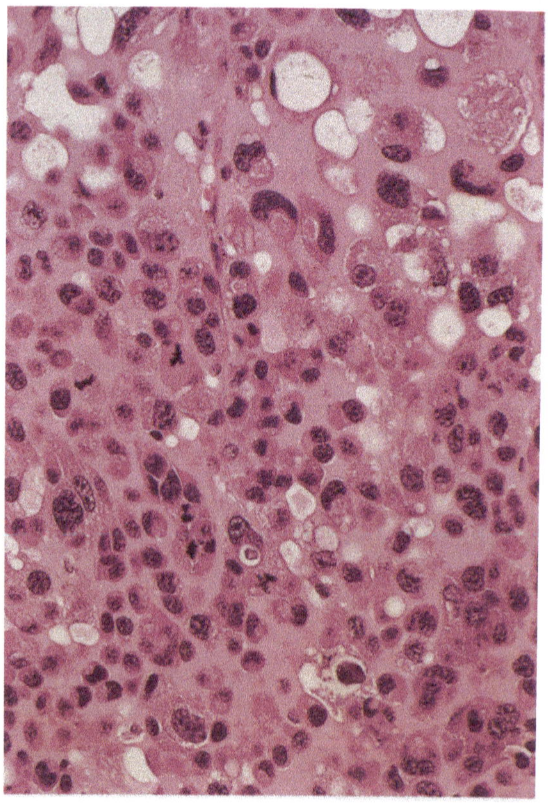

Synonym(s). Tumor, endodermal sinus, malignant; tumor, yolk sac, malignant

Histogenesis

Probably a variant of germ-cell tumor.

Diagnostic Features

- Most tumors have patterns mimicking two layers of fetal membranes, i.e., the parietal and the visceral yolk sac
- A typical characteristic of this tumor is that the tumor cells produce an abundant eosinophilic PAS-positive matrix in which nests and cords of neoplastic cells are embedded
- Parietal yolk sac foci composed of polygonal or cuboidal endodermal cells with amphiphilic cytoplasm and pleomorphic nuclei with one or more nucleoli. Cytoplasm contains PAS (+) droplets or granules
- Occasionally small cysts and glomerular bodies found
- Visceral yolk sac endodermal cells are cylindriform and do not contain PAS (+) droplets. Sometimes a papillary pattern composed of a central capillary embedded in mesenchymal tissue and surrounded by visceral endoderm is observed. In some tumors, large cells with weakly staining cytoplasm and big nuclei can be seen; occasionally giant cells surrounding blood-filled spaces present

Fig. 23 *(above)*. Yolk sac carcinoma of the ovary, demonstrating patterns from both parietal and visceral yolk sac membranes. H&E

Fig. 24 *(below)*. Yolk sac carcinoma of the ovary. H&E

- Cells form rosettes, cords or papillary structures
- Metastases mostly composed of both parietal and visceral components infiltrating lymph nodes, ovaries, kidneys, liver and spleen

Differential Diagnosis

Adenocarcinoma
Uterine adenocarcinoma is composed of cylindrical cells forming glandular structures without the abundant eosinophilic PAS(+)matrix

Comment

Spontaneous yolk sac carcinomas are rare tumors in the mouse.

References

(See 2, 4, 6, 7, 14, 17, 20, 21, 22, 23, 24)

Carcinoma, Embryonal *(M)*

Synonym(s). Teratoma, undifferentiated

Histogenesis

Probably of germ-cell origin.

Diagnostic Features

- Tumor composed of round or fusiform, poorly differentiated cells infiltrating the myometrium; cells have large nuclei with prominent nucleoli; numerous mitotic figures
- Foci of yolk sac carcinoma sometimes set in areas of embryonal carcinoma; occasionally well-differentiated tissues such as cartilage, bone, skin may be present
- Metastases mostly contain cells of embryonal carcinoma but sometimes areas of yolk sac carcinoma also found

Differential Diagnosis

Teratoma, Malignant

The embryonal carcinoma is differentiated from a teratoma in the absence of tissues from the three germ cell layers, i.e., mesenchymal, epithelial, and neuronal tissues

Comment

Spontaneous embryonal carcinoma has not been described in the mouse but can be induced in the rat by injection of mouse sarcoma virus.

References

(See 2, 3, 6, 7, 14, 20, 21, 22, 23, 24)

Choriocarcinoma *(M)*

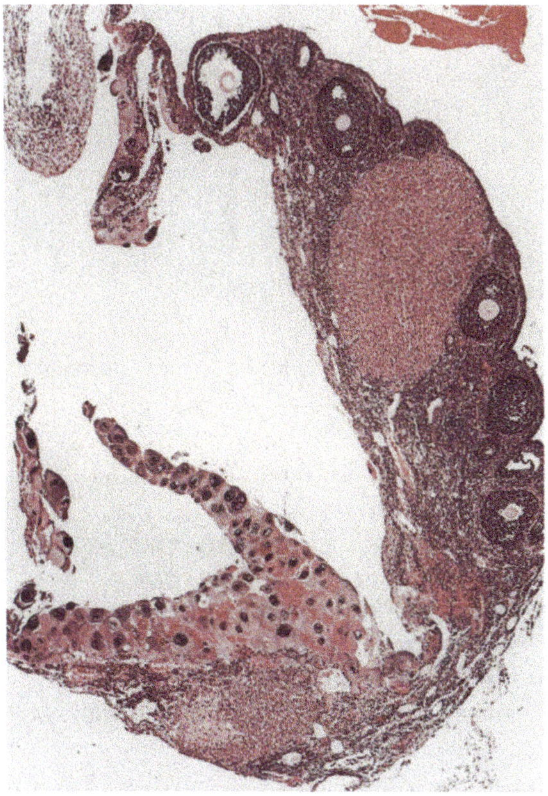

Synonym(s). Choriocarcinoma, malignant; chorioepithelioma, malignant

Histogenesis

Trophoblastic cells.

Diagnostic Features

- Tumor composed of two distinct cell types: small basophilic cells similar to placental cytotrophoblasts, and large giant cells with big nuclei and often prominent nucleoli (syncytiotrophoblasts)
- Both kinds of cells infiltrate the myometrium and produce extensive necrosis and hemorrhage
- Morphology of metastatic tumors similar to that of the primary tumor

Differential Diagnosis

Carcinoma, Yolk Sac

Choriocarcinoma does not have the abundant PAS(+)eosinophilic matrix

Hemangiosarcoma

Tumor does not form vascular spaces, although hematocysts may be present

Comment

Rare tumor in mice, only described in the uterus.

References

(See 2, 3, 6, 7, 14, 15, 20, 21, 22, 23, 24)

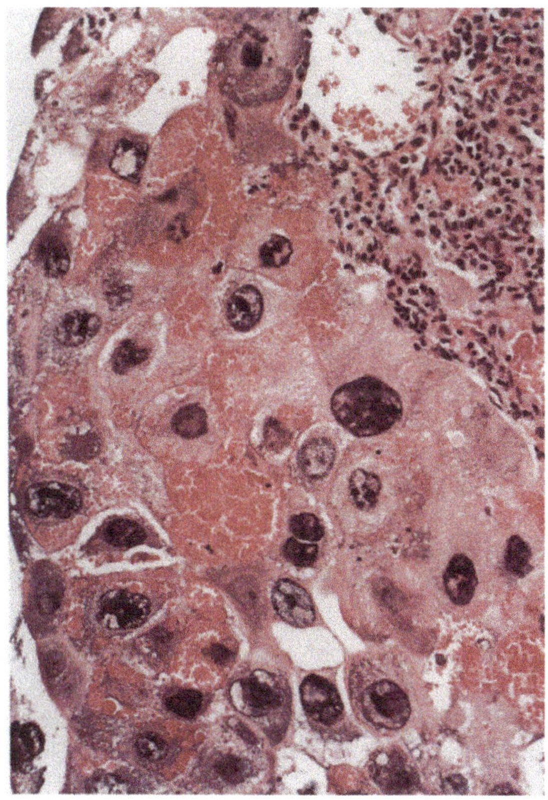

Fig. 25 *(above).* Choriocarcinoma of the ovary. H&E

Fig. 26 *(below).* Choriocarcinoma of the ovary. Higher magnification of the neoplastic syncytiotrophoblasts separated by areas of hemorrhage. H&E

Hyperplasia, Glandular *(H)*

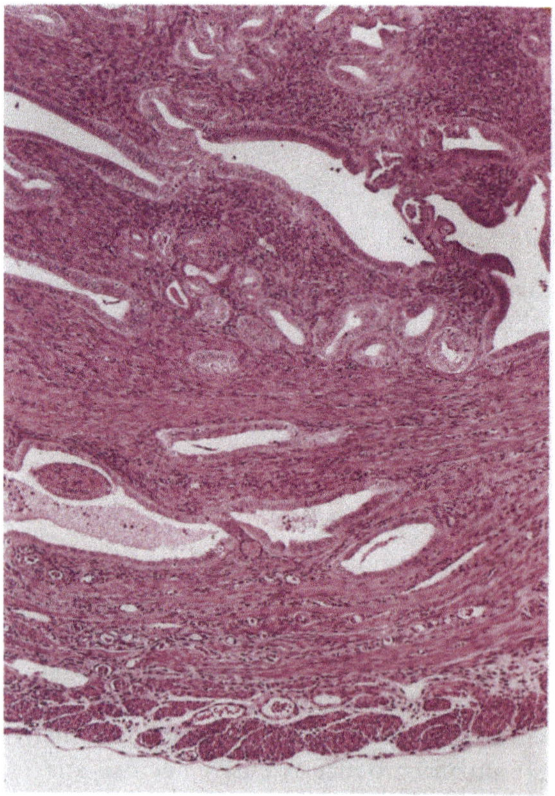

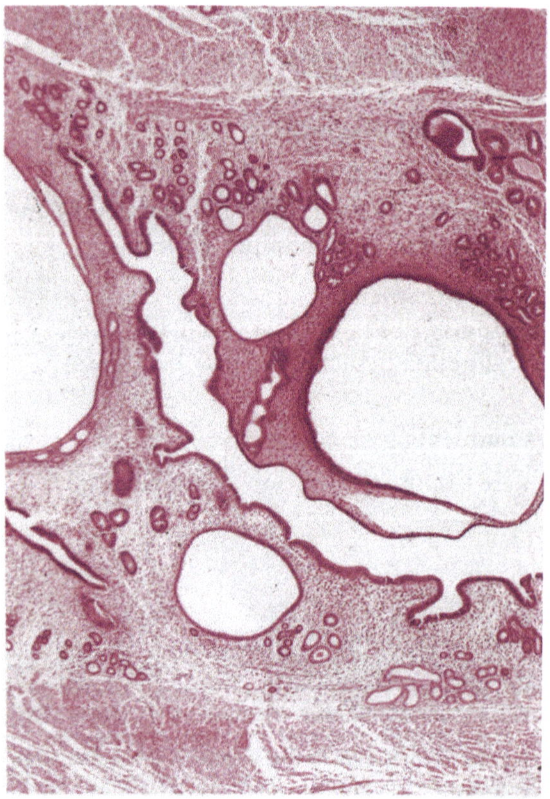

Histogenesis

Uterine epithelium.

Diagnostic Features

- Endometrial gland become tortuous, dilated and usually cystic
- Cysts lined with a single layer of epithelium, which may attain large sizes
- Lesion may be focal or diffuse
- Adenomyosis regularly seen in severe cases
- Generally results from excessive or prolonged stimulation by ovarian hormones

Differential Diagnosis

Polyp, Glandular
Polyps generally have a more prominent stroma and fibrovascular components and protrude into the lumen

Comment

Very common lesion in mice.

References

(See 5, 7, 16, 24, 27)

Fig. 27 *(above).* Glandular hyperplasia of the uterus. H&E

Fig. 28 *(below).* Uterine cystic endometrial glandular hyperplasia. H&E

Polyp, Glandular *(P)*

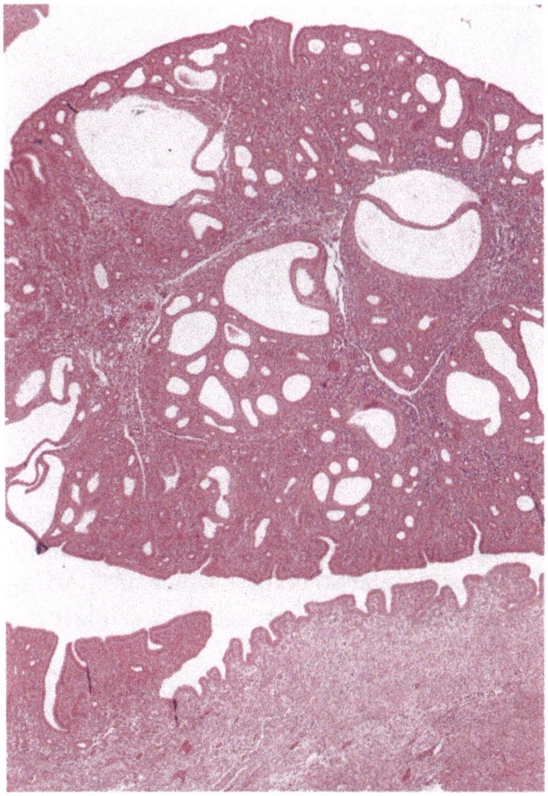

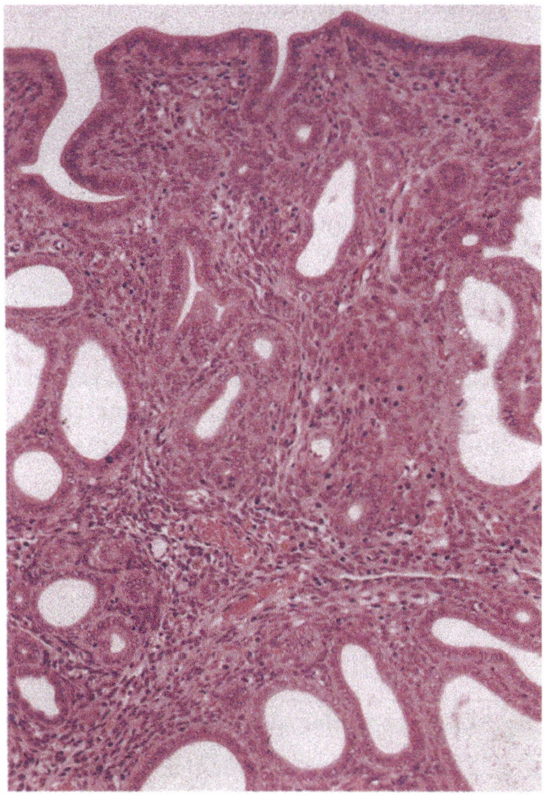

Histogenesis

Uterine glandular epithelium and fibrovascular stroma.

Diagnostic Features

- Polypoid mass protruding into the uterine lumen
- Polyp covered by cuboidal to columnar epithelium, which is continuous with and similar in appearance to the endometrial lining epithelium
- Stroma composed of spindle-shaped or stellate cells with variable amounts of collagen and endothelial-lined vascular spaces
- Endometrial glands, often cystic are seen in the stroma
- Glandular elements are prominent and often hyperplastic

Differential Diagnosis

Hyperplasia, Endometrial Stromal
Cystic endometrial hyperplasia has more compact stroma without prominent fibrovascular component

Polyp, Endometrial Stromal
Differs from stromal polyp which generally contains predominantly stromal as compared to glandular elements

Adenoma (Papillary)
Papillary adenomas have none or little proliferation of endometrial stroma which is delicate

References

(See 5, 7, 16, 24, 27)

Fig. 29 *(above)*. Glandular polyp. H&E

Fig. 30 *(below)*. Glandular polyp. H&E

Adenoma *(B)*

Papillary

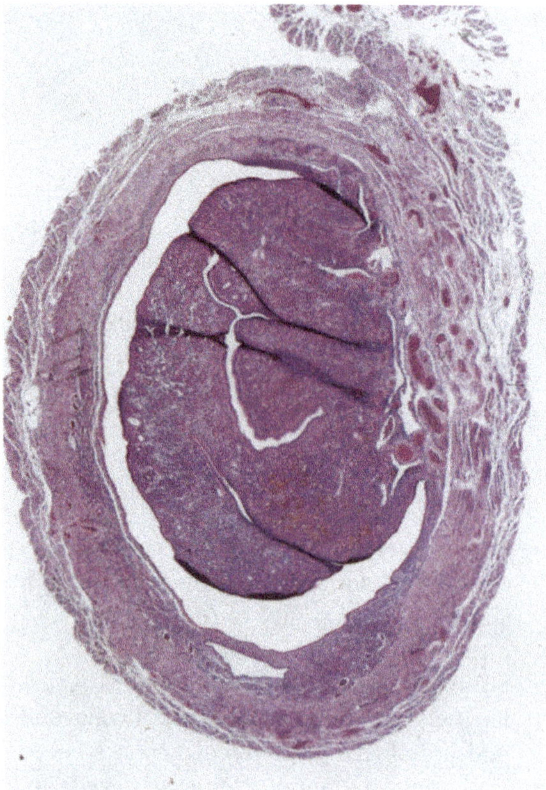

Histogenesis

Uterine surface epithelium.

Diagnostic Features

- Arises from surface epithelium and may have either a broad base or a delicate stalk
- Epithelium arranged in papillary, glandular or tubular structures which are lined by cuboidal to columnar cells 1–2 cell layers thick
- Cells have moderate to abundant eosinophilic cytoplasm with round to oval nuclei, finely stippled chromatin and 1–2 nucleoli
- Some glandular structures may contain cysts filled with exudate
- Supporting stroma generally delicate
- The modifier "papillary" is used if the predominant growth pattern of the tumor is papillary

Differential Diagnosis

Polyp, Endometrial Stromal

May be confused with endometrial stromal polyp in which the glandular component is prominent. Endometrial stroma is generally a constant feature of the endometrial polyp, however in papillary adenoma, there is little or no stromal proliferation

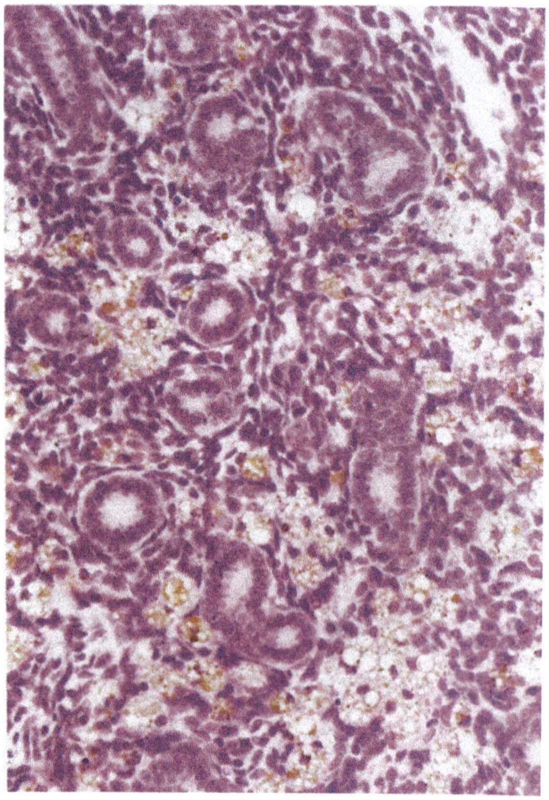

Fig. 31 *(above)*. Uterine adenoma. H&E

Fig. 32 *(below)*. Higher magnification of adenoma of the uterus, demonstrating well-differentiated glands. H&E

Hyperplasia, Endometrial Stromal
Cystic endometrial hyperplasia is generally a diffuse condition involving endometrium and adenomas are focal proliferative lesions

Hyperplasia, Glandular (Focal)
Hyperplasia is a lesion which is characterized by an increase of the epithelial component within the normal morphologic boundaries

Adenocarcinoma (Papillary)
Papillary adenocarcinomas show a higher degree of anaplasia and evidence of direct invasion of glandular elements into the underlying stroma

References

(See 3, 4, 5, 7, 16, 24, 27)

Adenocarcinoma *(M)*

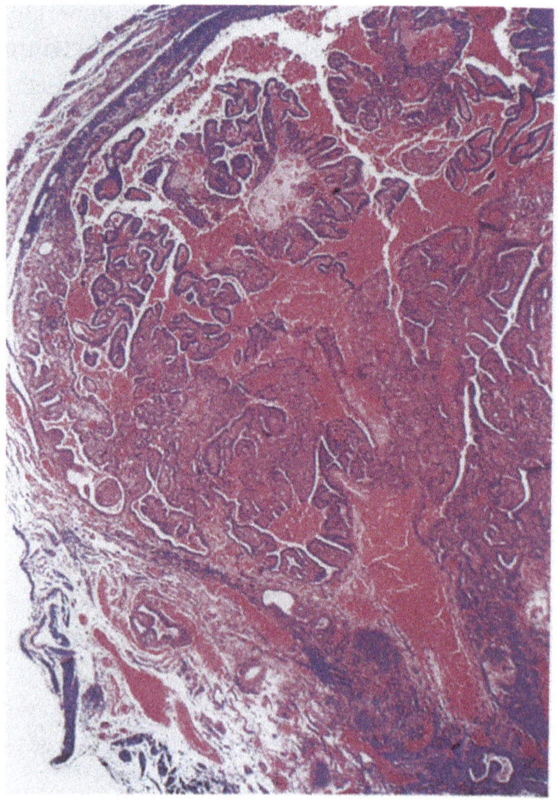

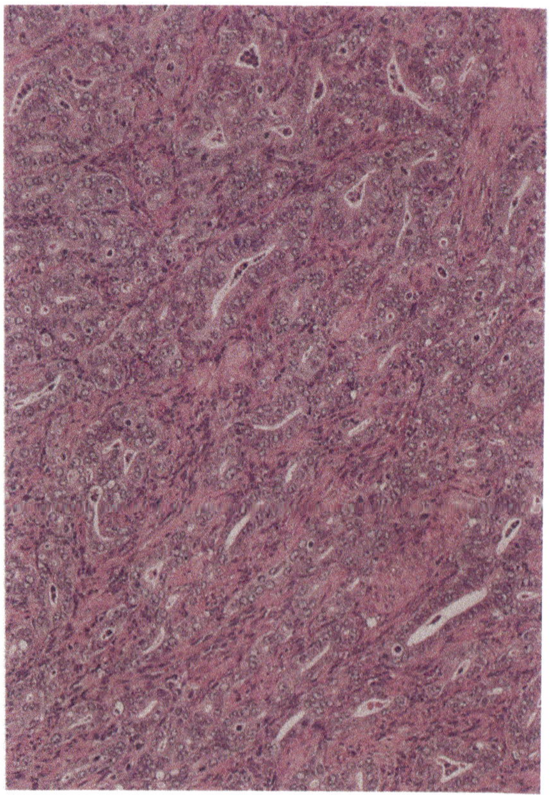

Histogenesis

Deep uterine glands and uterine surface epithelium.

Diagnostic Features

- Composed of epithelial cells with eosinophilic or amphiphilic cytoplasm and large vesicular nuclei with prominent nucleoli
- Cells arranged in acini, glandular structures and in solid nests and cords
- The neoplastic glands may show varying degrees of pleomorphism and atypia. The mitotic rate can be very high
- More anaplastic neoplasms have smaller cells with basophilic cytoplasm and numerous mitotic figures
- Neoplastic glands separated by a delicate fibrovascular stroma which may be abundant resulting in a scirrhous adenocarcinoma
- Tumor is generally highly invasive locally and distant metastasis may occur especially to the lungs
- Hemorrhages and areas of necrosis may be present

Fig. 33 *(above)*. Adenocarcinoma, effacing and infiltrating normal uterus. H&E

Fig. 34 *(below)*. Adenocarcinoma of the uterus. B6C3F1 mouse, age 89 weeks (courtesy of Dr. Shinji Yamamoto). H&E

Differential Diagnosis

Adenoma or Polyp, Endometrial Stromal or Hyperplasia, Endometrial Stromal
All of these lesions are restricted to the endometrium and lack characteristics of malignancy such as atypia, invasive growth pattern, and high mitotic rate

Carcinoma, Squamous Cell
The uterine adenocarcinoma has no or only minimal focal squamous metaplasia. Squamoid differentiation is not an integral part of this tumor

Adenomyosis
Adenomyosis lacks the malignant morphological characteristics even though islets of well-differentiated epithelial cells forming glandular structures may reach the serosal surface

References

(See 3, 4, 5, 7, 16, 24, 25)

Papilloma, Squamous Cell *(B)*

Keratinizing • Non-keratinizing

Histogenesis

Surface epithelium.

Diagnostic Features

- Marked focal papillary proliferation of the mucosa with marked squamous metaplasia
- Marked hyperkeratinization distending the endocervix
- No invasion of the underlying stroma or any cellular atypia seen
- Almost always associated with chronic suppurative inflammation
- Cells are well-differentiated

Differential Diagnosis

Metaplasia, Squamous Cell
Squamous cell metaplasia is differentiated from squamous cell papilloma in that the hyperplasia is restricted to the normal mucosal surface and does not protrude papillary-like into the lumen

Carcinoma, Squamous Cell
Squamous cell carcinomas are generally well-differentiated tumors with cells arranged in nests and cords showing various degrees of atypia infiltrating deep into the myometrium, serosa, and adjacent organs

References

(See 3, 5, 7, 16, 24, 27)

Carcinoma, Squamous Cell *(M)*

Keratinizing • Non-keratinizing

Synonym(s). Carcinoma, epidermoid

Histogenesis

Surface epithelium.

Diagnostic Features

- Usually well-differentiated and heavily infiltrated with leukocytes
- Tumor cells large and polygonal with prominent vesicular nuclei containing one or more nucleoli
- Cells arranged in cords and nests and invading deep into myometrium, serosa and contiguous organs
- Epithelium on the luminal surface of the uterus markedly thickened, dysplastic and keratinized ("keratin pearls")
- Stroma may be scant or abundant giving a scirrhous structure

Differential Diagnosis

Papilloma, Squamous Cell
The differentiation between squamous cell papilloma and squamous cell carcinoma is made on the growth pattern of the tumor and the absence of marked atypia and invasion of the adjacent tissue in the papilloma

Adenocarcinoma
The adenocarcinoma lacks the squamoid differentiation which is inherent to the squamous tumors

Comment

Rare spontaneous tumor, but may be readily induced by estrogens.

References

(See 3, 5, 7, 16, 24, 27)

Hyperplasia, Endometrial Stromal *(H)*

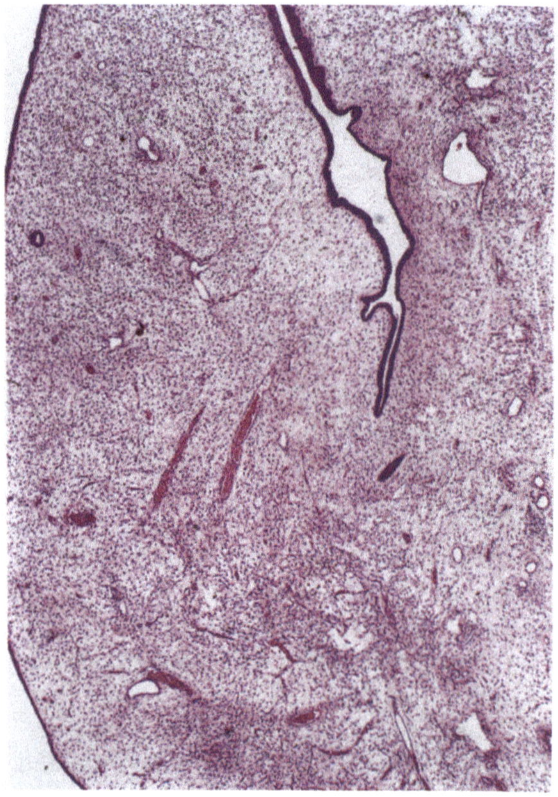

Fig. 35. Endometrial stromal hyperplasia of the uterus as a diffuse lesion. H&E

Histogenesis

Uterine glandular epithelium and fibrovascular stroma.

Diagnostic Features

- Lesion frequently found in the uterine cervix/uterus
- Proliferation of stromal cells, spindle-shape or stellate cells with variable amounts of collagen
- Lesion generally oriented along the normal anatomic structures, e.g., circular around cervix and do not cause gross anatomic abnormalities or distortions
- Growth characteristics are non-invasive
- Mitotic index and vascularization may be prominent especially in early lesions. Older lesions are characterized by a dense collagen stroma/fibrosis

Differential Diagnosis

Polyp, Endometrial Stromal
Nodular proliferation of endometrial stromal cells often protruding in the lumen covered by cuboidal to columnar epithelium

Sarcoma, Endometrial Stromal
Malignant lesion characterized by infiltrative growth, high mitotic index. Growth pattern does not follow existing anatomic structures leading to gross distortions

Comment

Common diffuse lesion in old mice in most mouse strains. Especially seen in cervix and to a lesser degree in the uterus. May be macroscopically noted as a dense firmly enlarged white cervix.

References

(See 3, 5, 7, 16, 24, 27)

Polyp, Endometrial Stromal *(P)*

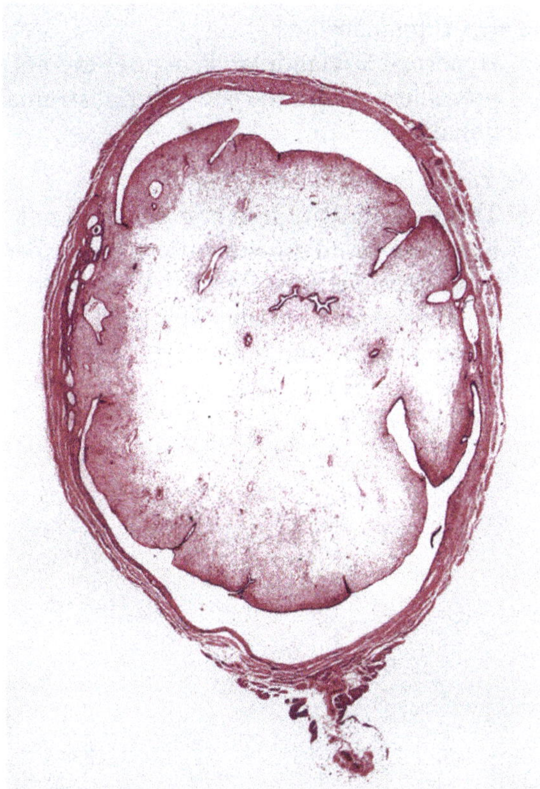

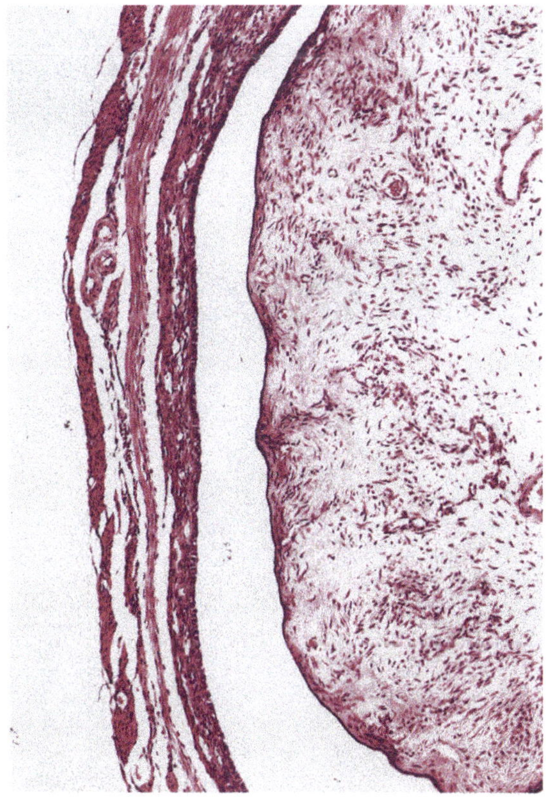

Histogenesis

Uterine epithelium and fibrovascular stroma.

Diagnostic Features

- Polypoid mass protruding into the uterine lumen
- Polyp covered by cuboidal to columnar epithelium, which continues with and is similar in appearance to the endometrial lining epithelium
- The predominant bulk of the mass is composed of stromal spindle-shaped or stellate cells with variable amounts of collagen and endothelial-lined vascular spaces
- The tumor may be solitary or multiple
- Tumor may be edematous or infarcted
- Tumors may protrude into the vagina, and may be ulcerated or inflamed and are mostly composed of predominantly loosely organized endometrial stromal cells with blood vessels and a few endometrial glands
- The diagnosis stromal polyp is also used in cases of small lesions, which do not have polypoid protrusion into the uterine lumen but are restricted to the endometrial wall. In such cases the endometrial stromal cells should be well-differentiated, otherwise such lesions can be classified as early endometrial stromal sarcomas

Fig. 36 *(above)*. Endometrial stromal polyp of the uterus, projecting into the uterine lumen. H&E

Fig. 37 *(below)*. Endometrial stromal polyp of the uterus. H&E

Differential Diagnosis

Polyp, Glandular
Hyperplastic glandular structures are not prominent within the endometrial stromal polyp

Sarcoma, Endometrial Stromal
This is an invasive lesion, which does not have a polypoid growth pattern but rather an invasive pattern of growth and other characteristics of malignancy such as pleomorphism and high mitotic rate. These tumors may originate within an endometrial stromal polyp

Comment

Occasionally endometrial stromal sarcomas arise from polyps which generally show malignant features such as rapid growth, invasiveness, metastases to distant organs.

References

(See 4, 5, 7, 8, 10, 13, 16, 24, 27)

Sarcoma, Endometrial Stromal *(M)*

Histogenesis

Uterine epithelium and fibrovascular stroma.

Diagnostic Features

- The predominant bulk of the mass is composed of stromal spindle-shaped cells with variable amounts of collagen and endothelial-lined vascular spaces
- Invasion of adjacent tissue is present
- May be present within polypoid mass
- Cells are poorly differentiated spindle cells
- Cellular pleomorphism may be present
- Cell borders are indistinct
- Cytoplasm is scant to moderate, pale and eosinophilic
- Nuclei are elliptical, elongated and hyperchromatic. They may appear oval or round when cut in cross-section
- Mitotic figures are numerous
- Areas of hemorrhage and necrosis are present
- Metastasis is rare
- May be positive for S-100 and vimentin, negative for desmin and actin

Differential Diagnosis

Polyp, Glandular
Hyperplastic glandular structures are not prominent within the endometrial stromal polyp

Polyp, Endometrial Stromal
Stromal elements are well-differentiated and protrusion into the lumen is present and no evidence of invasion of adjacent tissue/endometrial wall

Comment

Occasionally endometrial stromal sarcomas arise from polyps which generally show malignant features such as rapid growth, invasiveness, metastases to distant organs.

References

(See 4, 5, 7, 8, 10, 13, 16, 24, 27)

Tumor, Granular Cell, Benign *(B)*

Synonym(s). Abrikossoff's tumor, benign; myoblastoma

Histogenesis

Not established; an origin from Schwann cells or mesenchymal cells has been proposed.

Diagnostic Features

- Often circumscribed and demarcated solid mass with little stroma; is composed of large epitheloid round to oval cells with large pale nuclei and abundant eosinophilic granular cytoplasm and smaller more fusiform cells with small dark uniform nuclei
- Small to medium-sized cytoplasmic granules are positive with PAS stain, are diastase-resistant, and are considered to be various stages of lysosomes present in almost every cell
- Expansive growth causes compression and atrophy of adjacent tissue
- Some tumors show a more infiltrative pattern of growth. These tumors tend to have a more fusiform morphology. In the absence of metastases or invasion into other organs these tumors are considered to be benign

Differential Diagnosis

Leiomyoma
: Lacks PAS-positive granules or epitheloid cells

Adenocarcinoma
: Formation of glands, epithelium often amphophilic or basophilic, lacks abundant PAS-positive granules

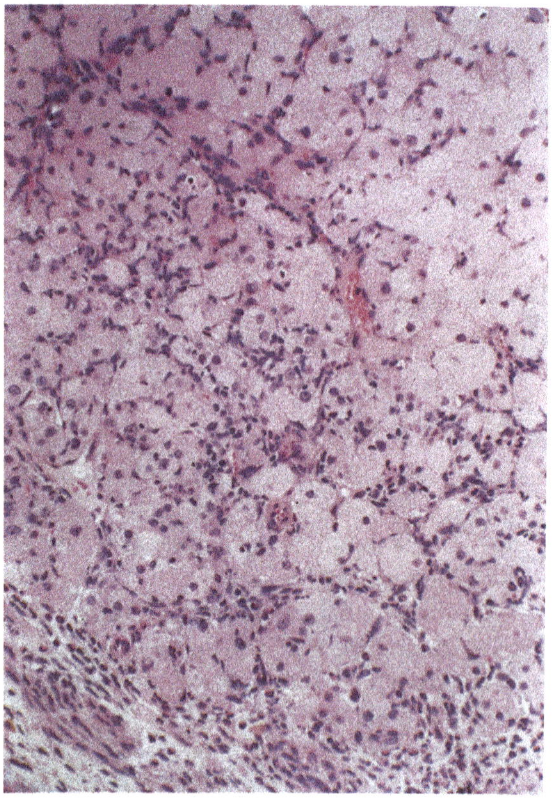

Fig. 38. Benign granular cell tumor of the vagina. H&E

Comment

Granular cell tumors generally occur as a single pink to pale yellow-gray mass. These tumors are frequently seen in the male and female genital tract, but may occur in other organs. Granular cell tumors react positively in immunocytochemical stains for NSE, S-100 protein and peripheral myelin proteins. A malignant counterpart of the benign granular cell tumor has not been described but cannot be excluded. In this classification evidence of distant metastasis or marked invasion beyond the normal organ structures should be present before such a diagnosis is warranted. The term hyperplasia does not apply even to very small accumulations/nodules since equivalent normal structures are unknown.

References

(See 6, 19, 25, 27)

Decidual Reaction *(H)*

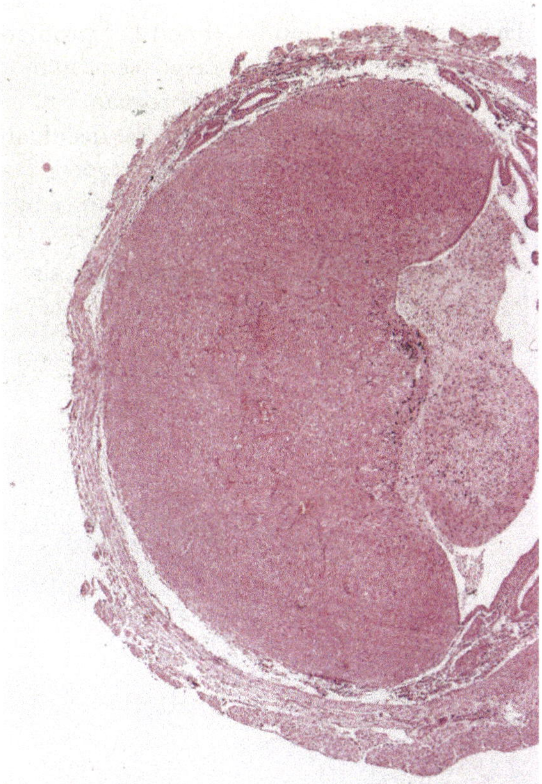

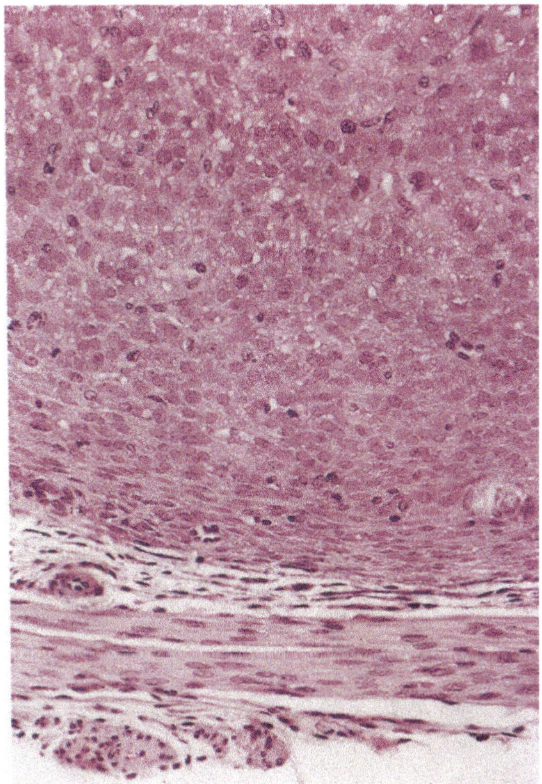

Synonym(s). Deciduoma

Histogenesis

Decidual tissue.

Diagnostic Features

- Consists of proliferating decidual tissue with high levels of structural organization and regional variation
- Nodular lesion with two distinct areas, antimesometrial region containing closely packed cells with small capillary channels and mesometrial region containing two cell types: spiny mesometrial cells with long cytoplasmic processes and abundant glycogen which are commonly binucleated and granulated metrial gland cells with extensive cytoplasm containing PAS-positive cytoplasmic granules

Differential Diagnosis

Sarcoma, Endometrial Stromal
No PAS-positive cytoplasmic granules, predominantly spindle-shaped cells

Fig. 39 *(above).* Decidual reaction, uterus

Fig. 40 *(below).* Decidual reaction, uterus

Comment

The decidual reaction is defined as a proliferation of decidual tissue in response to non-specific stimuli in the pseudopregnant rat. The normal life span of nodular rat decidual reaction is from days 4–16 of pseudopregnancy. The lesion can be induced in mice by sutures and surgery.
Focal/multifocal decidual reactions can also be found within other uterine lesions, such as cystic hyperplasia, polyps and other tumors. These lesions are characterized by the presence of PAS-positive eosinophilic granules in the myometrial cells, as well as swollen myometrial cells often more central showing abundant glycogen accumulations.

References

(See 12)

Hyperplasia, Squamous Cell *(H)*

Keratinizing • Non-keratinizing

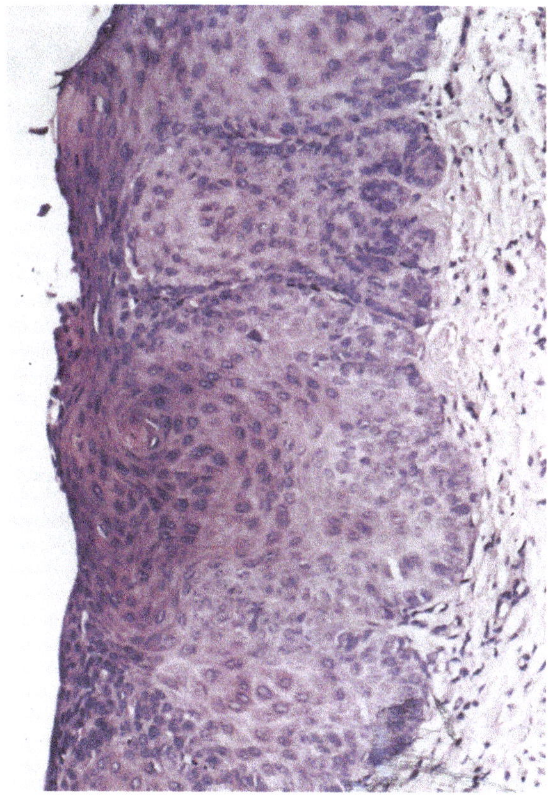

Fig. 41. Squamous cell hyperplasia of the vagina. H&E

Histogenesis

Vaginal epithelium.

Diagnostic Features

- Marked generally diffuse squamous hyperplasia of the vaginal epithelium
- Prominent rete are generally present, which are well-differentiated and do not invade the adjacent tissues

Differential Diagnosis

Papilloma, Squamous Cell or Carcinoma, Squamous Cell

Squamous hyperplasia is differentiated from squamous cell papilloma and carcinoma in that the hyperplasia is well-differentiated and restricted to the normal mucosal surface and does not protrude into the lumen or infiltrate the underlying tissue

Comment

Frequently seen after chronic estrogen stimulation.

References

(See 5, 7, 9, 11, 27)

Papilloma, Squamous Cell *(B)*

Keratinizing • Non-keratinizing

Histogenesis

Surface epithelium.

Diagnostic Features

- Marked focal papillary proliferation of the mucosa, generally with mild hyperkeratinization
- No invasion of the underlying stroma or cellular atypia present
- Almost always associated with chronic suppurative inflammation and some ulceration

Differential Diagnosis

Hyperplasia, Squamous Cell
Squamous hyperplasia is well-differentiated and is restricted to the normal mucosal surface and does not protrude papilliform into the lumen

Carcinoma, Squamous Cell
Squamous cell carcinoma is differentiated from the papilloma squamous cell in that the carcinoma shows marked atypia, infiltrative growth into the adjacent tissues, and a high mitotic rate

Comment

Squamous cell papillomas and carcinomas arising from the surface epithelium of the vagina are similar to those of the skin, forestomach, and oral cavity.

References

(See 3, 4, 5, 7, 9, 11, 27)

Carcinoma, Squamous Cell *(M)*

Keratinizing • Non-keratinizing

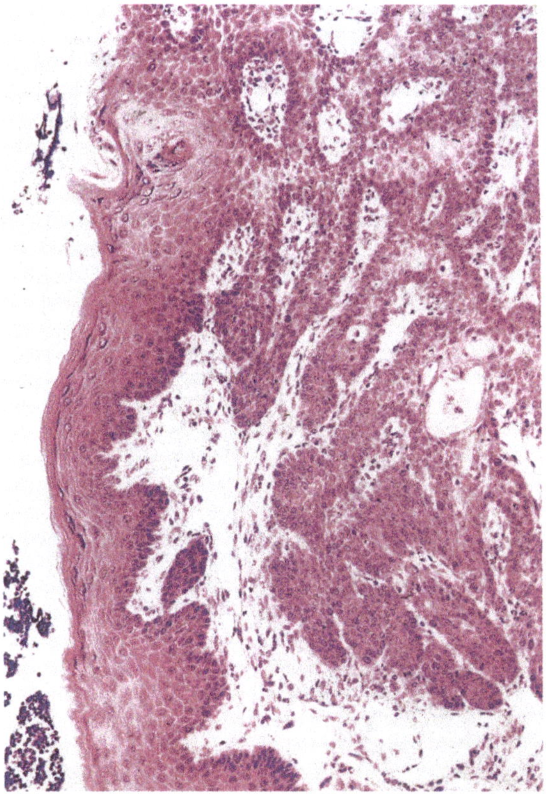

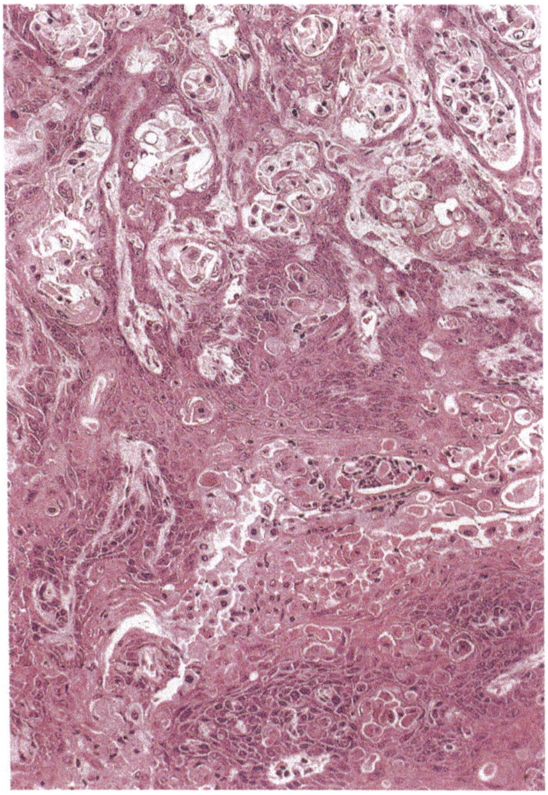

Synonym(s). Carcinoma, epidermoid

Histogenesis

Surface epithelium.

Diagnostic Features

- Usually well-differentiated and heavily infiltrated with leukocytes
- Tumor cells large and polygonal with prominent vesicular nuclei containing one or more nucleoli
- Cells arranged in cords and nests and invading deep into the adjacent tissues
- Epithelium on the luminal surface of the uterus markedly thickened, dysplastic and keratinized. Keratin pearls may be present also in the deeper tissue
- Stroma may be scant or abundant giving a scirrhous structure

Differential Diagnosis

Papilloma, Squamous Cell

Squamous cell carcinomas are differentiated from squamous cell papillomas by the infiltrative growth, high mitotic index, atypia, and/or metastasis

Fig. 42 *(above)*. Squamous cell carcinoma of the vagina. H&E

Fig. 43 *(below)*. Squamous cell carcinoma of the vagina. Higher magnification of anastomosing islands of neoplastic epithelial cells forming and surrounding keratin pearls. B6C3F1 mouse, age 105 weeks (courtesy of Dr. Shinji Yamamoto). H&E

Comment

Squamous cell papillomas and carcinomas arising from the surface epithelium of the vagina are similar to those of the skin, forestomach, and oral cavity. Rare spontaneous tumor but can be induced by estrogens.

References

(See 3, 4, 5, 7, 11, 27)

Keratoacanthoma *(B)*

Keratinizing • Non-keratinizing

Synonym(s). Epithelioma, intracutaneous cornifying

Histogenesis

Surface epithelium.

Diagnostic Features

- Similar to keratoacanthomas of the skin
- Well encapsulated tumors composed of the single or multiple cavities lined by stratified cornifying squamous cell epithelium
- Stratum granulosa is often lacking
- Papillary projection into the lumen or cavities may be found
- The cavities are mostly filled with a laminated concentrically arranged or homogenous horny material, which may be mixed with cholesterol crystals but also may contain proteinaceous fluid
- The nuclear to cytoplasmic ratio of the proliferating epithelium is low
- The epithelium is generally well-differentiated and does not invade the adjacent tissues

Differential Diagnosis

Carcinoma, Squamous Cell

The differentiation between keratoacanthoma and squamous cell carcinoma is made on the differentiation of the tissue in the keratoacanthoma, its low mitotic rate, and absence of invasion of the adjacent tissue

References

(See 3, 5, 7, 9, 11, 27)

Hyperplasia, Acinar Cell *(H)*

Histogenesis

Acinar epithelium.

Diagnostic Features

- Focal or multifocal increase of eosinophilic cells surrounded by dark flat basal cells
- Foci show lobulation similar to normal glands
- Morphology of hyperplastic acinar cells is similar to normal glandular tissue
- Central round large nucleus with one or rarely two large round nuclei
- Condensation of heterochromatin on the nuclear envelope
- Cytoplasm with lipid-filled vesicles and vacuoles depending on age, containing typical pink granules

Differential Diagnosis

Adenoma, Acinar Cell
Adenoma of acinar cells has compression and loss of normal structure. Growth pattern is solid and an ill-defined capsule may be present

Comment

This lesion occurs rarely spontaneously mainly in older animals (2 years old), but can be induced by several carcinogens.

References

(See 25, 26)

Adenoma, Acinar Cell *(B)*

Histogenesis

Acinar epithelium.

Diagnostic Features

- Growth is usually solid with irregular nodular formations that extend peripherally, mimicking acini of the normal gland; excretory ducts are infrequent and an ill-defined connective tissue capsule may be present
- Neoplastic acini are composed centrally of pale, foamy sebaceous type cells, and peripherally, of basophilic basal cells
- Cytoplasm may contain characteristic eosinophilic granules
- Squamous metaplasia may be present

Differential Diagnosis

Hyperplasia, Acinar Cell
Smaller and no tissue capsule with regular acini and uniformity of cells. Excretory ducts commonly present

Papilloma, Squamous Cell
Lacks glandular components

Adenocarcinoma, Acinar Cell
Invasion of surrounding connective tissue and variably sized acini. Tumor cells show cellular atypia and resemble immature acinar cells

Comment

Spontaneous adenomas are extremely rare, but they can be induced by several compounds. Cystic and papillary growth variants as seen in the rat have not been described in the mouse.

References

(See 26)

Adenocarcinoma, Acinar Cell *(M)*

Histogenesis

Acinar epithelium.

Diagnostic Features

- Variably sized nests and nodules differ distinctly from regular acinar structures
- Small nodules and epithelial cords, separated by dense connective tissue stroma may be present, resulting in a scirrhous appearance
- Squamous differentiation may be present
- Cystic areas, necrosis and inflammation may be present
- Acinar structures exhibit centrally pale, foamy sebaceous type cells, and peripherally, small, basophilic basal cells
- Mitotic figures are present
- Invasion of adjacent structures is present

Differential Diagnosis

Hyperplasia, Acinar Cell or Adenoma, Acinar Cell
No cellular atypia and invasion of adjacent tissues present

Carcinoma, Squamous Cell
Lack neoplastic sebaceous cells, composed of non-keratinizing or keratinizing squamous cells

Adenocarcinoma of the Mammary Gland
Extremely rare tumors in males, glandular structures are present

Comment

Adenocarcinoma is considered the preferred term since the clitoral gland is a skin appendage composed of modified sebaceous glands.

References

(See 26)

Hyperplasia, Squamous Cell *(H)*

Histogenesis

Squamous epithelial cells of central and excretory ducts – origin, squamous cell epithelium.

Diagnostic Features

- May occur focal, multifocal or generalized
- Squamous epithelium contains kerato-hyaline granules and is keratinized

Differential Diagnosis

Papilloma, Squamous Cell
Compression of surrounding duct structures, papillary growth into the duct lumina

Comment

This lesion is seen in older mice, occurs spontaneously in mice with dilatation of ducts and/or with chronic inflammation. They can be in both glands or only in one. It is not considered to be a pre-neoplastic change and there is no evidence for its progression.

References

(See 25)

Metaplasia, Squamous Cell (*M*)

Keratinizing • Non-keratinizing

Histogenesis

Basal cells of central and excretory ducts.

Diagnostic Features

- Focal or multifocal

Keratinizing
- Squamous epithelium contains kerato-hyaline granules
- Is highly keratinized

Non-keratinizing
- The squamous cells may be non-keratinized

Differential Diagnosis

Hyperplasia, Squamous Cell
Mostly papillary

Comment

Very often in older animals; glands should be cut in longitudinal direction to identify the main excretory duct and several side branches. Clitoral glands are very hard to recognize. They are situated on both sides of the clitoris and have a yellowish-white color. They are very small and weigh only 9 mg in contrast to the preputial glands which have a weight of 56 mg. The clitoral gland has less acinar glands than the preputial glands and their ducts are very often more cystic in appearance.

References

(See 25)

Papilloma, Squamous Cell *(B)*

Histogenesis

Squamous cell epithelium of ducts and basal cells.

Diagnostic Features

- Papillary structures with central connective tissue stalk lined by squamous epithelium with or without keratinization
- Mitotic figures are rare
- Central and excretory duct may also show squamous cell hyperplasia
- Papillomas may be single or multiple and can obstruct the ducts

Differential Diagnosis

Carcinoma, Squamous Cell
Destruction of basal membrane and invasion into surrounding acinar tissue. Cellular atypia and disorientation and frequent mitoses

Hyperplasia, Squamous Cell
Lacks central connective tissue stalk

Adenoma, Acinar Cell
Shows sebaceous differentiation

References

(See 25, 26)

Carcinoma, Squamous Cell *(M)*

Synonym(s). Carcinoma, epidermoid

Histogenesis

Duct epithelial cells. Acinar cells undergoing squamous cell differentiation.

Diagnostic Features

- Consists of irregular papillary fronds or nodules of pleomorphic poorly differentiated squamous cells with frank invasion of adjacent tissues
- Keratinization varies from keratohyoline granules, keratin pearls to highly keratinized material
- Cells may be spindle-shaped, show distinct intercellular bridges, have eosinophilic or amphophilic cytoplasm, or sebaceous cell differentiation
- Mitotic figures, necrosis and inflammation are common

Differential Diagnosis

Adenocarcinoma, Acinar Cell
Extensive sebaceous cell differentiation with smaller squamous component, lacks prominent intercellular bridges; poorly differentiated types show scirrhous interstitium

Papilloma, Squamous Cell
Lacks invasion of surrounding structures

Carcinoma, Squamous Cell of the Skin
Differentiation may be impossible if invading preputial gland; no clitoral gland duct origin

Adenoacanthoma, Malignant of the Mammary Gland
Extremely rare in male mice; molluscoid radiating growth pattern, presence of mammary gland acini

Comment

Non-tumorous acini may be atrophic or cystic. Distinction between skin, mammary and clitoral glands may be very difficult if characteristic features are not present.

References

(See 26)

References

1. Alison RH, Morgan KT (1987) Teratoma, ovary, mouse. In: Jones TC, Mohr U, Hunt RD (eds) Monographs on pathology of laboratory animals. Genital system. Springer, Berlin Heidelberg New York Tokyo, pp 46–52
2. Alison RH, Morgan KT, Haseman JK, Boorman GA (1987) Morphology and classification of ovarian neoplasms in F344 rats and (C57BL/6 x C3H)F1 mice. J Natl Cancer Inst 78: 1229–1243
3. Ashley DJB (1990) Evan's histological appearances of tumors. Churchill Livingstone, Edinburgh, pp 773–861
4. Davis BJ, Dixon D, Herbert RA (1999) Ovary, oviduct, uterus, cervix, and vagina. In: Maronpot RR, Boorman GA, Gaul BW (eds) Pathology of the mouse. Reference and atlas. Cache River Press, Vienna, pp 409–443
5. Faccini JM, Abbott DP, Paulus GJJ (1990a) Mouse histopathology. A glossary for use in toxicity and carcinogenicity studies. Elsevier, Amsterdam
6. Faccini JM, Abbott DP, Paulus GJJ (1990b) Mouse histopathology. A glossary for use in toxicity and carcinogenicity studies. Elsevier, Amsterdam, pp 147–168
7. Frith CH, Ward JM (1988) Color atlas of neoplastic and non-neoplastic lesions in aging mice. Elsevier, Amsterdam
8. Goodman DG, Hildebrandt PK (1987) Stromal polyp, endometrium, rat. In: Jones TC, Mohr U, Hunt RD (eds) Monographs on pathology of laboratory animals. Genital system. Springer, Berlin Heidelberg New York Tokyo, pp 146–148
9. Gopinath C, Prentice DE, Lewis DJ (1987) Atlas of experimental toxicological pathology. MTP Press, Lancaster, pp 94–97
10. Greaves P, Faccini JM (1984) Rat histopathology. A glossary for use in toxicity and carcinogenicity studies. Elsevier, Amsterdam, pp 173–175
11. Johnson LD (1978) Lesions of the female genital system caused by diethylstilbestrol in humans, subhuman primates, mice. In: Jones TC, Mohr U, Hunt RD (eds) Monographs on pathology of laboratory animals. Genital system. Springer, Berlin Heidelberg New York Tokyo, pp 84–109
12. Karbe E, Hartmann E, George C, Wadsworth P, Harleman J, Geiss V (1998) Similarities between the uterine decidual reaction and the "mesenchymal lesion" of the urinary bladder in aging mice. Exp Toxicol Pathol 50: 330–340
13. Leininger JR, Jokinen MP (1990) Oviduct, uterus and vagina. In: Boorman GA, Eustis SL, Elwell MR, Montgomery CA, Jr, MacKenzie WF (eds) Pathology of the Fischer rat. Reference and atlas. Academic Press, San Diego, pp 443–459
14. Lemon PG, Gubareva AV (1979) Tumours of the ovary. In: Turusov VS (ed) Pathology of tumours in laboratory animals, vol. II. Tumours of the mouse. IARC Scientific Publications No. 23, Lyon, pp 385–409
15. Madarame H, Sakurai H, Konno S (1989) Choriocarcinoma in a DDD mouse: a case report with immunohistochemical and ultrastructural studies. Lab Anim Sci 39: 255–258
16. Maita K, Hirano M, Harada T, Mitsumori K, Yoshida A, Takahashi K, Nakashima N, Kitazawa T, Enomoto A, Inui K, Shirasu Y (1988) Mortality, major cause of moribundity, and spontaneous tumors in CD-1 mice. Toxicol Pathol 16: 340–349
17. Majeed SK, Alison RH, Boorman GA, Gopinath C (1986) Ovarian yolk sac carcinoma in mice. Vet Pathol 23: 776–778
18. Morgan KT, Alison RH (1987a) Cystadenoma, ovary, mouse. In: Jones TC, Mohr U, Hunt RD (eds) Monographs on pathology of laboratory animals. Genital system. Springer, Berlin Heidelberg New York Tokyo, pp 42–52
19. Morgan KT, Alison RH (1987b) Tubular adenoma, ovary, mouse. In: Jones TC, Mohr U, Hunt RD (eds) Monographs on pathology of laboratory animals. Genital system. Springer, Berlin Heidelberg New York Tokyo, pp 36–41
20. Nielsen SW, Misdorp W, McEntee K (1976) Tumours of the ovary. Bull World Health Organ 53: 203–215
21. Rehm S, Dierksen D, Deerberg F (1984) Spontaneous ovarian tumors in Han:NMRI mice: histologic classification, incidence, and influence of food restriction. J Natl Cancer Inst 72: 1383–1395
22. Sass B, Rehm S (1994) Tumours of the ovary. In: Turusov VS, Mohr U (eds) Pathology of tumours in laboratory animals, vol 2. Tumours of the mouse, 2nd edn. IARC Scientific Publications No. 111, Lyon, pp 493–526
23. Serov SF, Scully RE, Sobin LH (1973) International histological classification of tumours No. 9. Histological typing of ovarian tumours. WHO, Geneva, pp 1–56
24. Squire RA, Goodman DG, Valerio MG, Fredrickson T, Strandberg JD, Levitt MH, Lingeman CH, Harshbarger JC, Dawe CJ (1978) Tumors. Female reproductive system. In: Benirschke K, Garner FM, Jones TC (eds) Pathology of laboratory animals, vol II. Springer, Berlin Heidelberg New York Tokyo, pp 1172–1194
25. Stolte M (1993) Histomorphological age changes and ultrastructural characteristics of the preputial and clitoral glands of mice. J Exp Anim Sci 35: 166–176

26. Toth B, Nagel D, Patil K (1980) Tumorigenic action of N-n-butyl-N-formylhydrazine in mice. Carcinogenesis 1: 589–593
27. Turusov VS, Munoz N, Dunn TB (1994) Tumours of the vagina and uterus. In: Turusov VS, Mohr U (eds) Pathology of tumours in laboratory animals, vol 2. Tumours of the mouse, 2nd edn. IARC Scientific Publications No. 111, Lyon, pp 451–491

Endocrine System

C.C. Capen[1] · E. Karbe[2,3] · U. Deschl[3] · C. George · P.-G. Germann[3]
C. Gopinath · J.F. Hardisty · J. Kanno · W. Kaufmann[3] · G. Krinke
K. Küttler · B. Kulwich · C. Landes · B. Lenz · L. Longeart[3] · I. Paulson
E. Sander · K. Tuch[3]

[1]Chair; [2]Co-chair; [3]Initial draft.

Hyperplasia, Pars Intermedia *(H)*

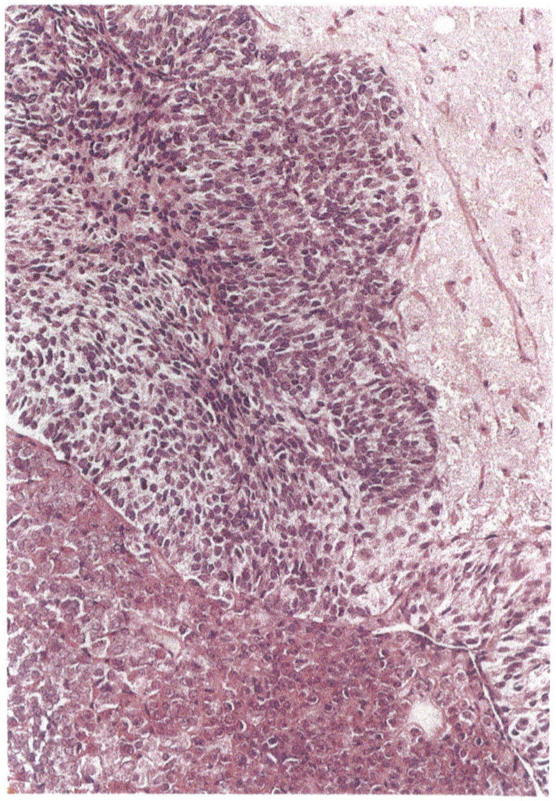

Fig. 1. Pars intermedia hyperplasia of the pituitary gland. H&E

Histogenesis

Glandular cells of the pars intermedia of the pituitary gland.

Diagnostic Features

Focal

- Focal increase in the number of glandular cells is present
- Compression is not present
- Lobular architecture of pars intermedia is maintained
- Cells are uniform
- Cells show some tinctorial differences from normal cells

Diffuse

- Diffuse increase in the number of glandular cells is present
- Cells may appear to infiltrate pars nervosa or pars distalis

Differential Diagnosis

Adenoma, Pars Intermedia

Lobular architecture is not maintained or Compression is present

Comment

Focal hyperplasia, probably a precursor of the adenoma, is rare in mice.

References

(See 2, 6, 8, 37)

Adenoma, Pars Intermedia *(B)*

Histogenesis

Glandular cells of the pars intermedia of the pituitary gland.

Diagnostic Features

- Compression may be present
- Lobular architecture is not maintained, lobular (or pseudofollicular) pattern is frequently expressed
- Cells are uniform or pleomorphic
- Cells are pale, eosinophilic, or granulated similar to those in regular tissue, but may have marked tinctorial differences
- Often extends into adjacent pars nervosa or pars distalis
- May dent or cave in adjacent brain
- Does not infiltrate organs outside the pituitary

Differential Diagnosis

Hyperplasia, Pars Intermedia (Focal)
Lobular architecture is maintained and
Compression is not present

Hyperplasia, Pars Intermedia (Diffuse)
Not focal, affects the whole pars intermedia

Carcinoma, Pars Intermedia
Distinct infiltration of the brain is present

Adenoma, Pars Distalis
Lobular pattern is rarely present and
Cells do not exhibit similarity to pars intermedia cells (with respect to the staining of cytoplasm)

Comment

Adenoma of pars intermedia in the mouse is much less common than in the pars distalis. Cells resemble those in the normal pars intermedia.
In tumors of the pars intermedia ACTH or precursor molecules can usually be demonstrated by immunostaining.

References

(See 6, 8, 37, 44, 51)

Figs. 2-4 see p. 272

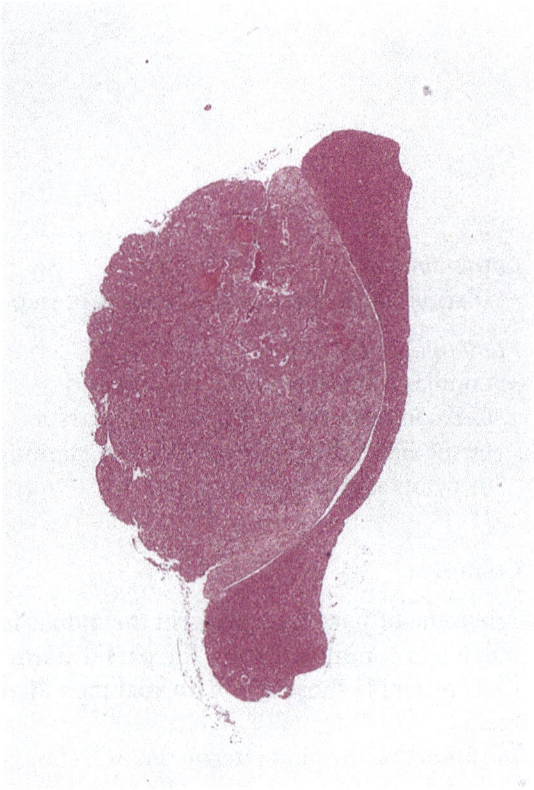

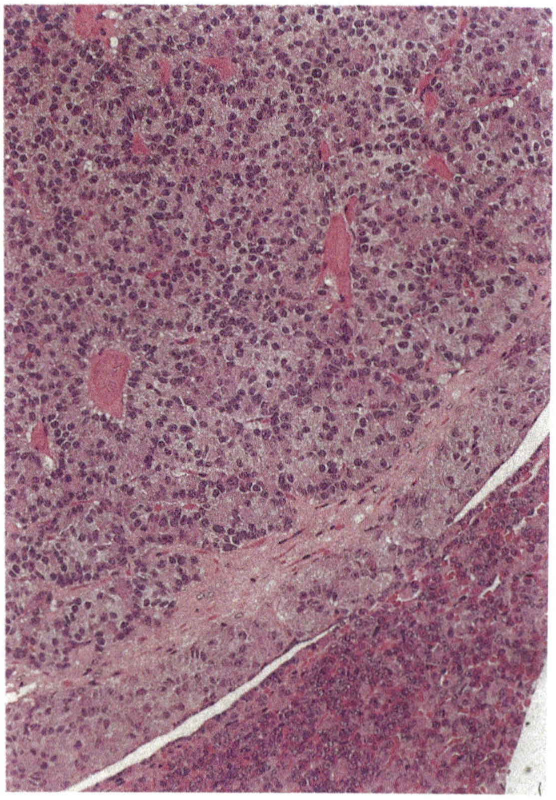

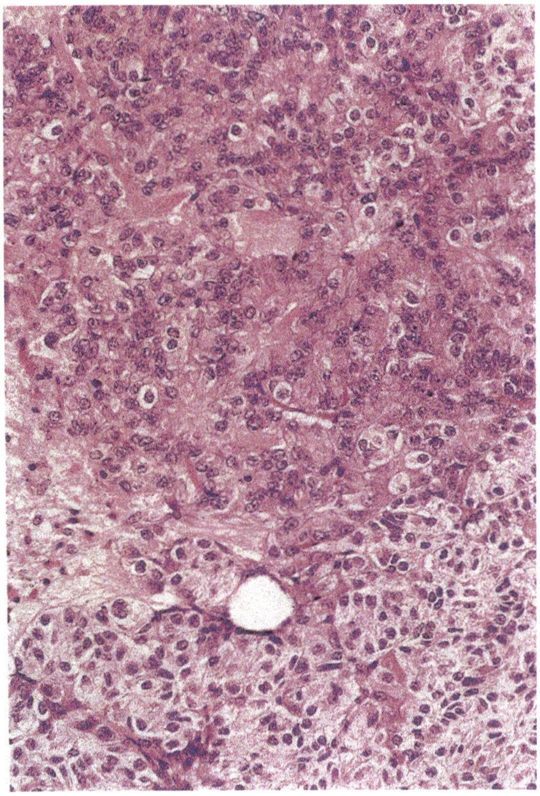

Fig. 2 *(upper left)*. Pars intermedia adenoma of the pituitary gland. B6C3F1 mouse, male, age 93 weeks. H&E (courtesy of Dr. Shinji Yamamoto)

Fig. 3 *(upper right)*. Pars intermedia adenoma of the pituitary gland. B6C3F1 mouse, male, age 93 weeks. H&E (courtesy of Dr. Shinji Yamamoto)

Fig. 4 *(below)*. Pars intermedia adenoma of the pituitary gland. H&E

Carcinoma, Pars Intermedia *(M)*

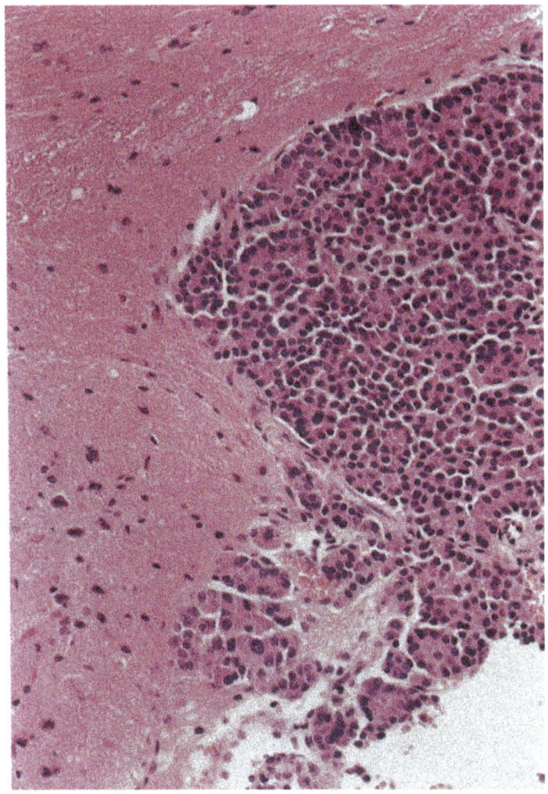

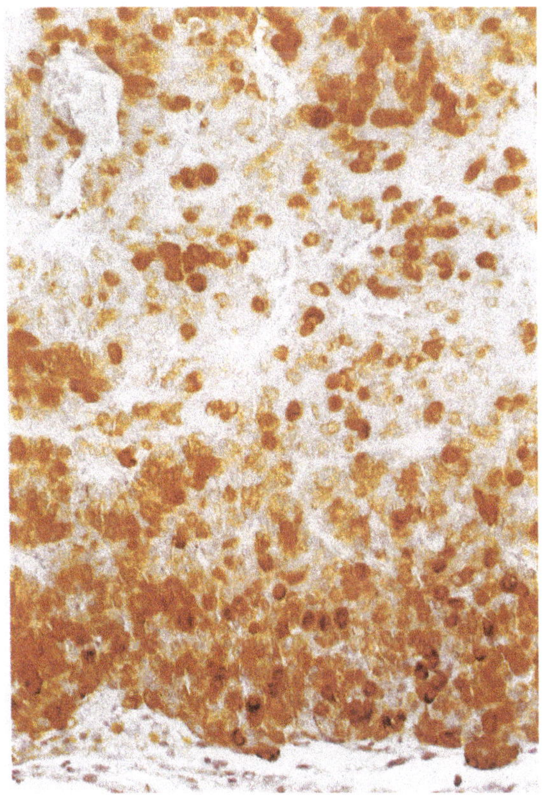

Histogenesis

Endocrine cells of the pars intermedia of the pituitary gland.

Diagnostic Features

- Compression may be present
- Lobular architecture is not maintained
- Cells are uniform or pleomorphic
- Cells are pale, eosinophilic, or granulated similar to those in the normal pituitary gland, but may have marked tinctorial differences
- Infiltrates organs outside the pituitary (brain). Surrounding of tumor cells by reactive meninges or compression of the adjacent brain is not a sign of malignancy

Differential Diagnosis

Adenoma, Pars Intermedia
Distinct infiltration of the brain is not present but infiltrative growth within gland including pars nervosa may be noted

Fig. 5 *(above)*. Pars intermedia carcinoma. Tumor cells invade adjacent brain. H&E (courtesy of Dr. Volker Geiss)

Fig. 6 *(below)*. Pars intermedia carcinoma shown in Fig. 5. Cytoplasm of tumor cells immunoreactive for ACTH. Monoclonal anti-ACTH, ABC method (staining by courtesy of Dr. Elke Hartmann)

Comment

Reports of carcinomas arising from the pars intermedia of the mouse pituitary gland are rare and equivocal. Diagnosis by Liebelt (1994) was made due to projection into a large cystic space and abutting against brain or bone or due to infiltration of the meninges, respectively. Such tumors do not qualify for carcinoma according to these criteria. However, one unpublished case was recently communicated to the compilers. This tumor showed immunoreactivity for ACTH and distinct infiltration of the brain.

References

(See 6, 37)

Hyperplasia, Pars Distalis *(H)*

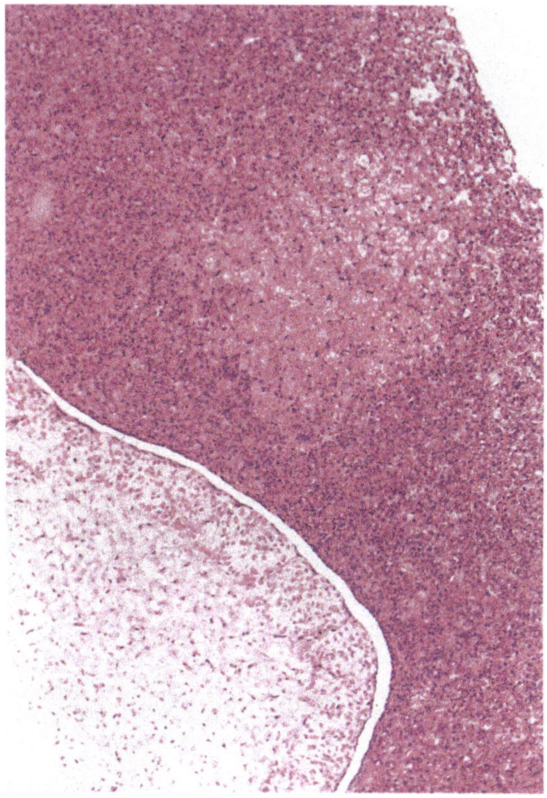

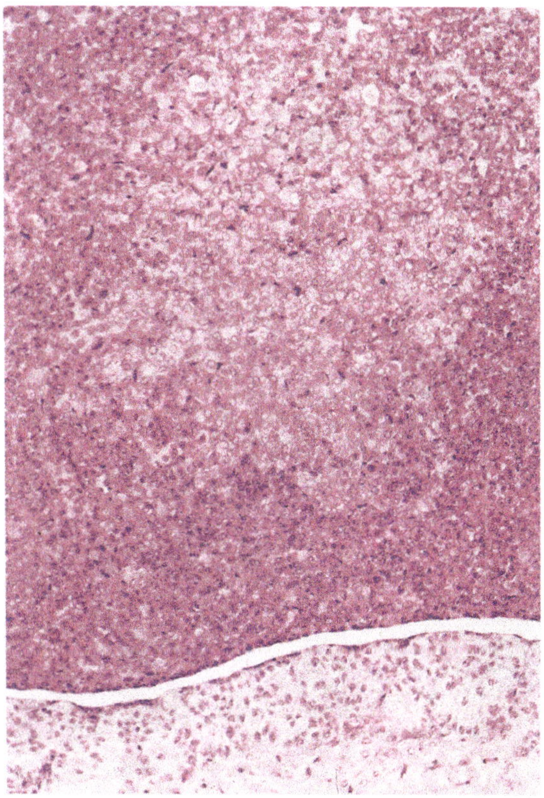

Histogenesis

Glandular cells of the pars distalis of the pituitary gland.

Diagnostic Features

Focal

- Focal increase in the number of glandular cells is present
- Is not well demarcated
- No compression or only minor compression affecting not more than one quadrant is present
- Sinusoids within the affected area may be dilated without affecting growth pattern
- Cells are enlarged
- Cells are uniform in shape
- Cellular pleomorphism or atypia is not present
- Diameter is not larger than 50% of the pituitary height

Diffuse

- Diffuse increase in the number of glandular cells is present, cellular crowding helps to identify the change
- Sinusoids within the affected area may be dilated without affecting growth pattern
- Affects all or most of inner portions of the lobes

Fig. 7 *(above)*. Focal pars distalis hyperplasia. H&E

Fig. 8 *(below)*. Focal pars distalis hyperplasia. H&E

Differential Diagnosis

Adenoma, Pars Distalis

Compression in more than one quadrant may be present or
Cells exhibit pleomorphism and atypia or
Diameter is larger than 50% of the pituitary height

Comment

The pituitary height is the maximal vertical distance seen on a transversal section. A round lesion with a diameter of 50% of the pituitary height will fill about one fourth of the lobe's cut surface.

In hyperplastic lesions, the cell type of origin cannot be determined with H&E staining. Specific identification needs immunocytochemical staining. Reticulin stains may be useful in differentiating hyperplasias from adenomas. The regular network of reticulin fibers separating cell nests in normal tissue is maintained in hyperplastic tissue but becomes abnormal in adenomas, however, in the mouse reticulin fibers are so scant that uneven distribution is often not discernible.

References

(See 2, 6, 8, 15, 27, 37)

Adenoma, Pars Distalis *(B)*

Histogenesis

Glandular cells of the pars distalis of the pituitary gland.

Diagnostic Features

- Compression in more than one quadrant is present in some cases
- May be highly vascular, angiomatous, cystic, hemorrhagic, trabecular, pseudofollicular or solid
- Cells and nuclei are enlarged
- Cellular pleomorphism and atypia may be present or diameter is larger than 50% of the pituitary height
- Does not infiltrate other organs, but may show infiltrative growth within gland including pars nervosa

Differential Diagnosis

Hyperplasia, Pars Distalis (Focal)
No compression or minimal compression in not more than one quadrant is present and
Cells do not exhibit pleomorphism and atypia and
Diameter is not larger than 50% of the pituitary height

Carcinoma, Pars Distalis
Distinct infiltration into other organs (brain) is present

Comment

Evidence suggests that focal hyperplasia is a precursor of the adenoma. The classic differentiation of adenomas into chromophobic, basophilic and acidophilic has lost its importance. This is due to methods allowing the identification of hormones produced by tumor cells. Immunohistochemical techniques permit the specific recognition of cell types in the majority of cases. However, these spontaneous neoplasms of mice have been generally less well characterized by immunocytochemistry than those found in rat or man. Compression is often absent even in large adenomas and even after staining for reticulin fibers (the pituitary of the mouse contains less reticulin fibers than that of the rat).

References

(See 6, 15, 27, 37, 51)

Figs. 9-11 see p. 278

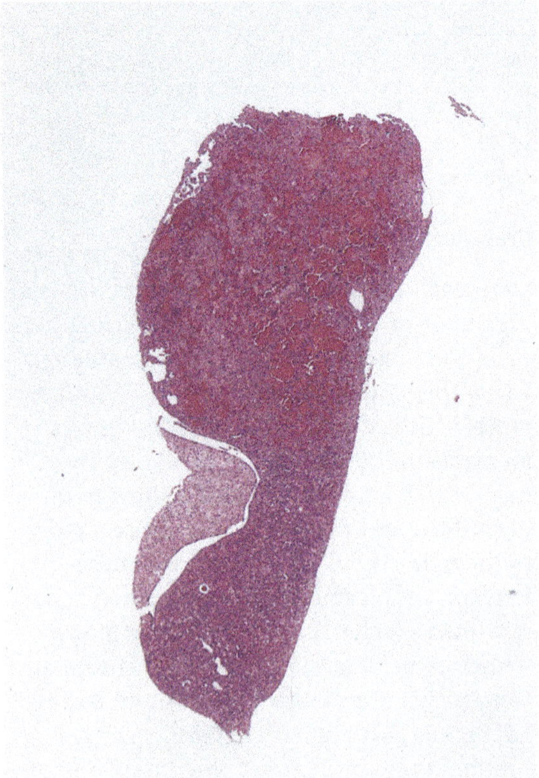

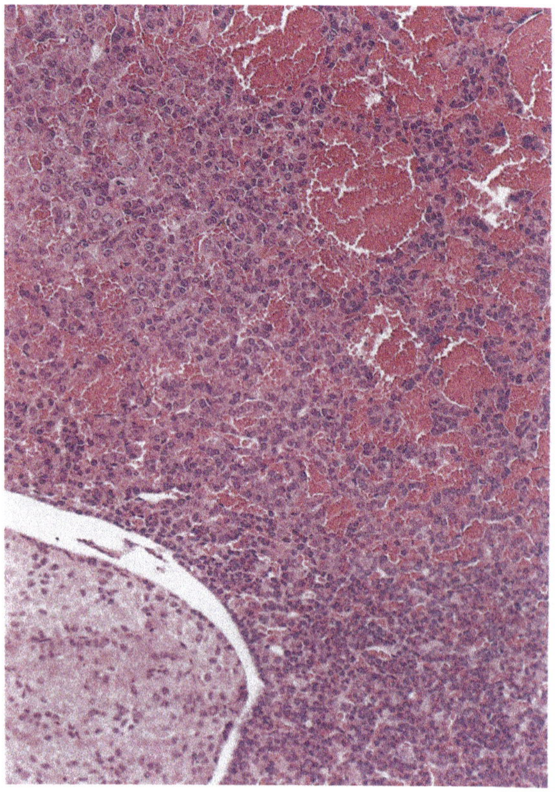

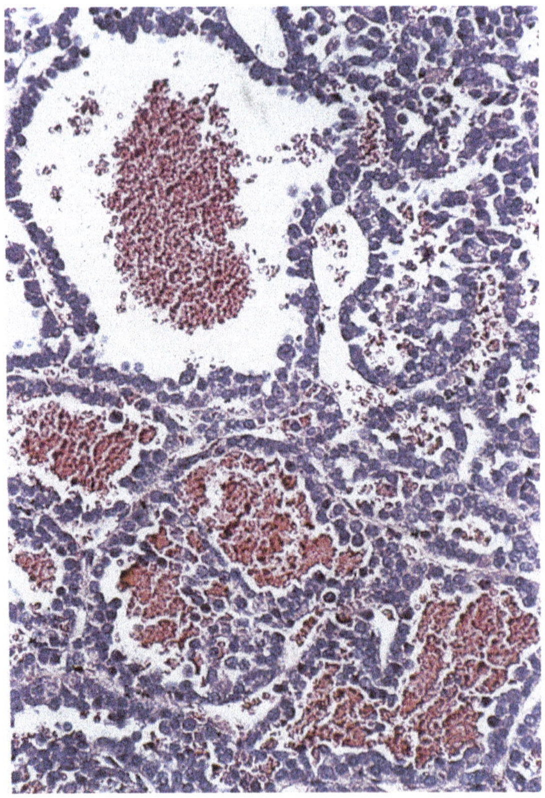

Fig. 9 *(upper left)*. Pars distalis adenoma of the pituitary gland. B6C3F1 mouse, female, age 105 weeks. H&E (courtesy of Dr. Shinji Yamamoto)

Fig. 10 *(upper right)*. Pars distalis adenoma of the pituitary gland. B6C3F1 mouse, female, age 105 weeks. H&E (courtesy of Dr. Shinji Yamamoto)

Fig. 11 *(below)*. Pars distalis adenoma of the pituitary gland. Pseudofollicular type. H&E

Carcinoma, Pars Distalis *(M)*

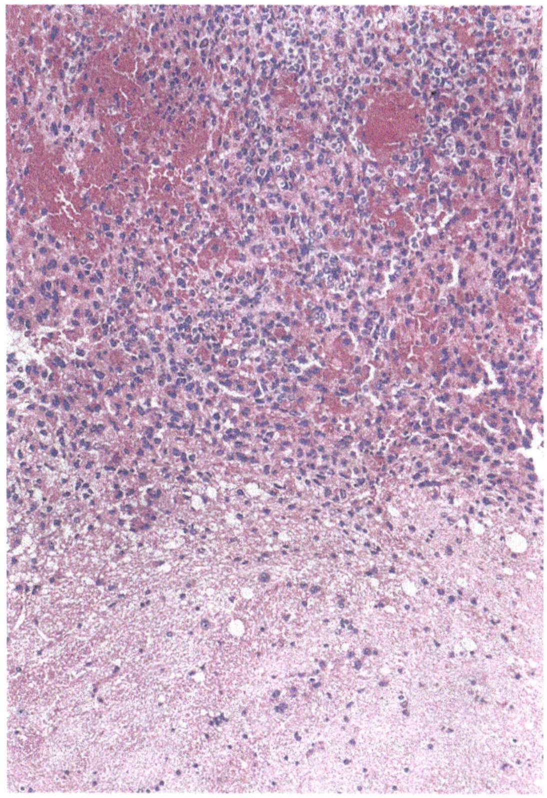

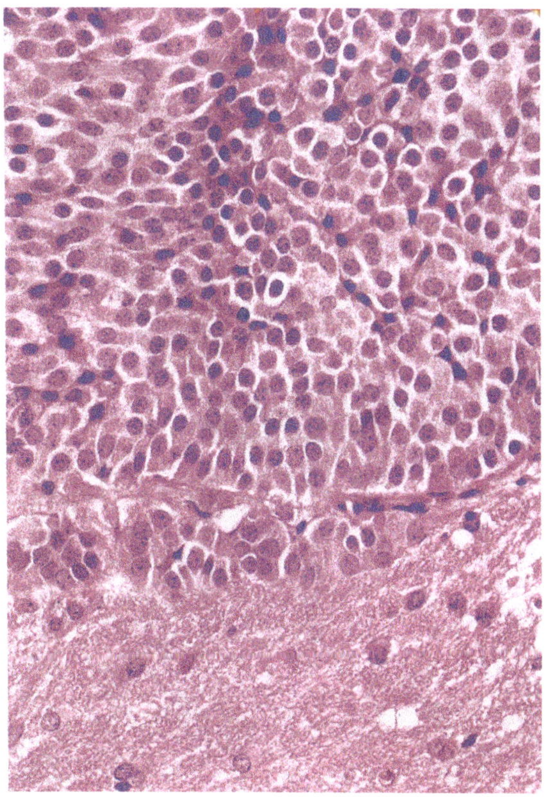

Histogenesis

Glandular cells of the pars distalis of the pituitary gland.

Diagnostic Features

- May be highly vascular, angiomatous, cystic, hemorrhagic, trabecular, pseudofollicular or solid
- Cells and nuclei are enlarged
- Cells may exhibit pleomorphism and atypia
- Infiltrates other organs such as brain or adjacent sphenoid bone, does not just extend onto brain following lines of least resistance (abutting) or is not just surrounded by reactive meninges

Differential Diagnosis

Adenoma, Pars Distalis
Distinct infiltration into other organs is not present, but infiltrative growth within gland including pars nervosa may be noted

Comment

This tumor is rare. The chance of diagnosing carcinoma is reduced if adjacent brain is not evaluated histopathologically. Metastasis of spontaneous pituitary carcinomas does not appear to occur in distant organs in mice.

References

(See 6, 15, 27, 37, 51)

Fig. 12 *(above).* Pars distalis carcinoma of the pituitary gland. p53 heterozygous mouse. H&E

Fig. 13 *(below).* Pars distalis carcinoma of the pituitary gland. H&E

Craniopharyngioma, Benign *(B)*

Histogenesis

Remnants of oro-pharyngeal epithelium of craniopharyngeal duct (Rathke's pouch).

Diagnostic Features

- Develops within or in close proximity to the pituitary gland
- May compress pituitary gland or brain stem
- Neoplastic cells form a keratinizing squamous epithelium
- Exhibits well-differentiated squamous epithelium with papillary and cystic formations, or cord-like proliferation associated with marked hyperkeratosis and parakeratosis
- Distinct infiltration of the brain or other organs outside the pituitary gland is absent
- Scattered peripheral tumor cells and fibrocytic proliferation may mimic infiltration

Differential Diagnosis

Craniopharyngeal Cyst(s)
Cyst(s) lined by cuboidal to pseudostratified columnar (ciliated) epithelium containing proteinaceous material

Craniopharyngioma, Malignant
Infiltration of organs (brain) outside the pituitary gland

Aberrant Craniopharyngeal Tissue
Not yet published in mice, but in rat and Syrian hamster

Metastatic Tumors
Metastasis of squamous cell carcinoma with a primary site in a different organ

Comment

Only one craniopharyngioma in mice has been published, which was diagnosed to be malignant. Thus, the benign counterpart is expected to occur in the mouse as it is known in the rat.

References

(See 6, 29, 41)

Craniopharyngioma, Malignant *(M)*

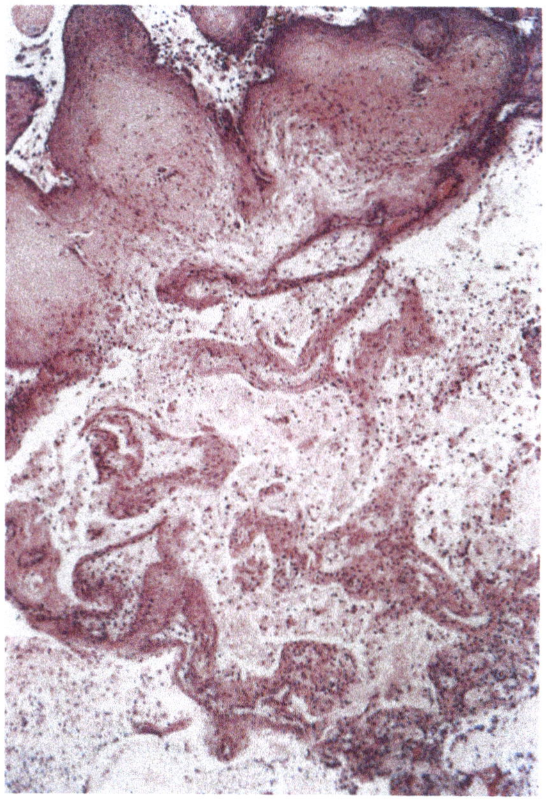

Fig. 14. Malignant craniopharyngioma of the pituitary gland showing stratified squamous component. H&E (section by courtesy of Dr. Katharina Heider)

Histogenesis

Remnants of oro-pharyngeal epithelium of craniopharyngeal duct (Rathke's pouch).

Diagnostic Features

- Develops within or in close proximity to the pituitary gland
- May compress pituitary gland or brain stem
- Neoplastic cells form a keratinizing squamous epithelium
- Exhibits well-differentiated squamous epithelium with papillary and cystic formations, or cord-like proliferation associated with marked hyperkeratosis and parakeratosis
- Distinct infiltration of the brain is present
- Scattered peripheral tumor cells and fibrocytic proliferation may mimic infiltration

Differential Diagnosis

Craniopharyngeal Cyst(s)
Cyst(s) lined by cuboidal to pseudostratified columnar (ciliated) epithelium containing proteinaceous material

Craniopharyngioma, Benign
No infiltration of organs (brain) outside the pituitary gland

Aberrant Craniopharyngeal Tissue
Not yet published in mice, but in rat and Syrian hamster

Metastatic Tumors
Metastasis of squamous cell carcinoma with a primary site in a different organ

Comment

Extremely rare tumor in mice with only one case published. In this tumor there were no cytological features of malignancy. However, it showed an invasive growth into cerebral structures and thus was classified as malignant.

References

(See 6, 29, 41)

Pituicytoma *(B)*

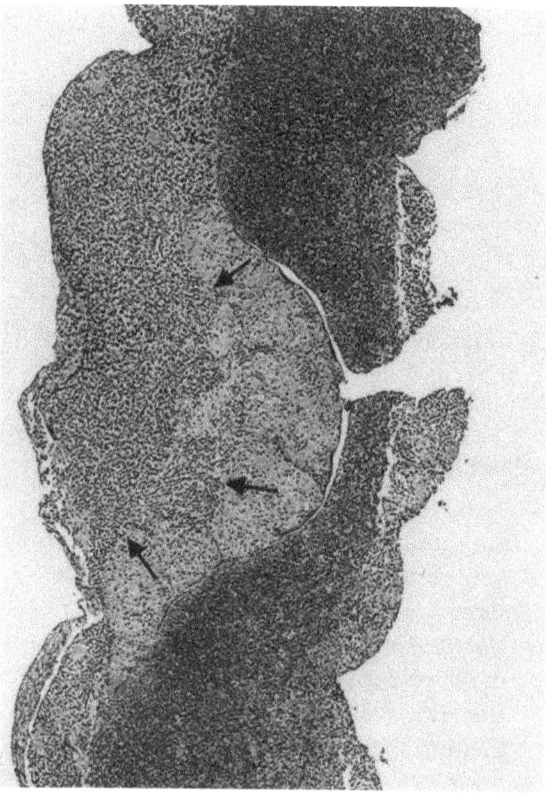

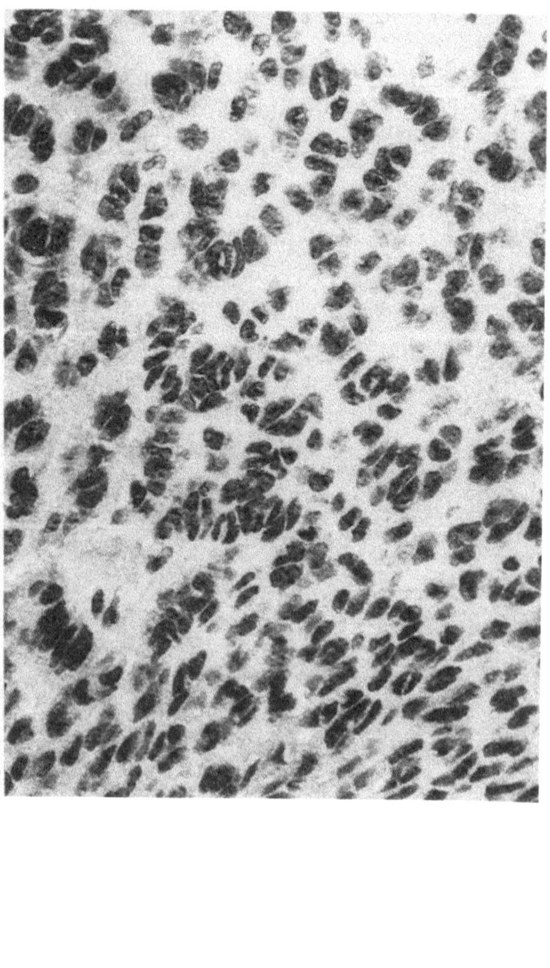

Histogenesis

Pituicytes (glial cells of neurohypophysis).

Diagnostic Features

- Originates in neurohypophysis
- Compression of adjacent pituitary or brain may be present
- Sheets of small spindle cells are closely packed
- Cell borders are inapparent
- No evidence of cellular pleomorphism is present
- Cytoplasm is eosinophilic, foamy to vacuolated
- Cytoplasmic granules positive in phosphotungstic acid hematoxylin stain may be present
- Nuclei are elongated and irregular
- Nuclear palisading is occasionally present
- Infiltration within the confines of the pituitary gland may be present

Fig. 15 *(above)*. Pituicytoma of the pars nervosa. Tumor cells invade adjacent pars intermedia. H&E (photomicrograph by courtesy of Dr. Daniel Morton and Toxicologic Pathology)

Fig. 16 *(below)*. Higher magnification of pituicytoma of the pars nervosa, shown in Fig. 15. H&E (photomicrograph by courtesy of Dr. Daniel Morton and Toxicologic Pathology)

Differential Diagnosis

Meningioma, Benign
Different location: meningeal thickening on the surface of brain and spinal cord or in the ventricular cavities

Astrocytoma, Benign or Oligodendroglioma, Benign or Glioma, Mixed, Benign
Different location: originating in the brain
Confined to one area of the brain
No evidence of invasion of the pituitary gland

Comment

The glial origin of pituicytoma can be demonstrated by positive GFAP reaction. The pituicytoma in the mouse is extremely rare. Only one case has been reported.

References

(See 6, 54, 55)

Pinealoma, Benign *(B)*

Histogenesis

Origin from differentiated cells of the pineal parenchyma.

Diagnostic Features

- Midline location at the dorsal brain surface
- Cell clusters are divided by a delicate fibrovascular stroma
- Lobulated pattern is present
- Pseudorosettes may occur
- Cells have a vascular orientation
- Low nuclear:cytoplasmic ratio is present
- Tumor cells have dark-staining nuclei with coarse chromatin
- Some mitotic activity is present

Differential Diagnosis

Pinealoma, Malignant
General features of malignancy are present, especially invasive growth

Ectopic Pineal Tissue
Occasional groups of pineal cells attached to meninges are not neoplastic

Teratoma, Benign
Composed of tissues representing the three germ layers, but foreign to the site of tumor origin

Comment

Only one pineal gland tumor in mice has been mentioned in literature, but no detailed description was provided. Therefore, this description is based on the rat manuscript.

References

(See 7, 16, 33, 49, 56)

Pinealoma, Malignant *(M)*

Histogenesis

Origin from differentiated cells or precursor cells of the pineal parenchyma.

Diagnostic Features

- Midline location at the dorsal brain surface
- Tumor cells are densely and diffusely arranged
- Lobulated pattern only occasionally discernible
- Pseudorosettes may occur
- High nuclear:cytoplasmic ratio is present
- Tumor cells are characterized by pleomorphism, irregular hyperchromatic nuclei and scanty cytoplasm
- High mitotic activity is present
- Necrotic areas may be a common feature
- Deeply infiltrative growth in adjacent brain parenchyma or ventricles is often seen

Differential Diagnosis

Pinealoma, Benign
General features of malignancy are absent.

Medulloblastoma
The site of origin are the folia of the cerebellar cortex

Comment

Pineal gland tumors are extremely rare in mice. Only one pineal gland tumor in mice has been mentioned in literature, but no detailed description was provided.
Recently, midline brain tumors in MSV-SV 40-transgenic mice were shown to originate from the pineal gland.
Metastasis of pineal tumors have not been reported in rodents, but have been recognized in human patients.

References

(See 7, 16, 25, 33, 34, 35, 49, 55, 56)

Hyperplasia, C-Cell *(H)*

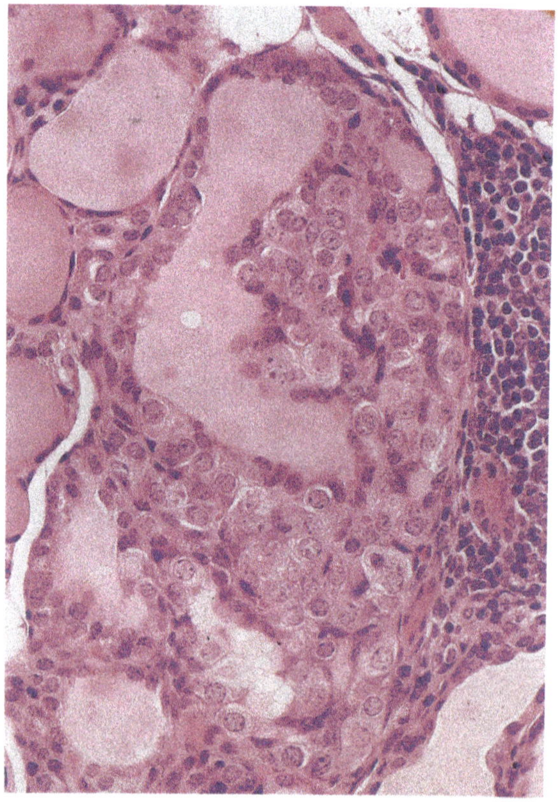

Fig. 17. C-cell hyperplasia of the thyroid gland. H&E

Histogenesis

Calcitonin producing C-cells of the thyroid gland.

Diagnostic Features

Focal

- Focal increase in parafollicularly located C-cells is present
- No evidence of compression of the adjacent tissue is present
- A capsule is not present
- Cells are arranged in solid nests
- Stromal septa are not present
- Cells are polygonal and pale
- Cell borders are indistinct
- Nuclei are round or oval
- C-cell cluster is not larger than the area of five average thyroid follicles

Diffuse

- Diffuse increase in parafollicularly located C-cells is present
- No evidence of compression of the adjacent tissue is present
- Cells are polygonal and pale.
- Cell borders are indistinct
- Nuclei are round or oval

Differential Diagnosis

Adenoma, C-Cell

Compression is present and
A thin capsule may be present and
Stromal septa may be present or
Larger than the area of five average thyroid follicles

Adenoma, Follicular Cell (Solid)

Small colloid follicles are present. Immunohistochemical proof can be provided for the exclusion of calcitonin production

References

(See 20, 21, 31)

Adenoma, C-Cell *(B)*

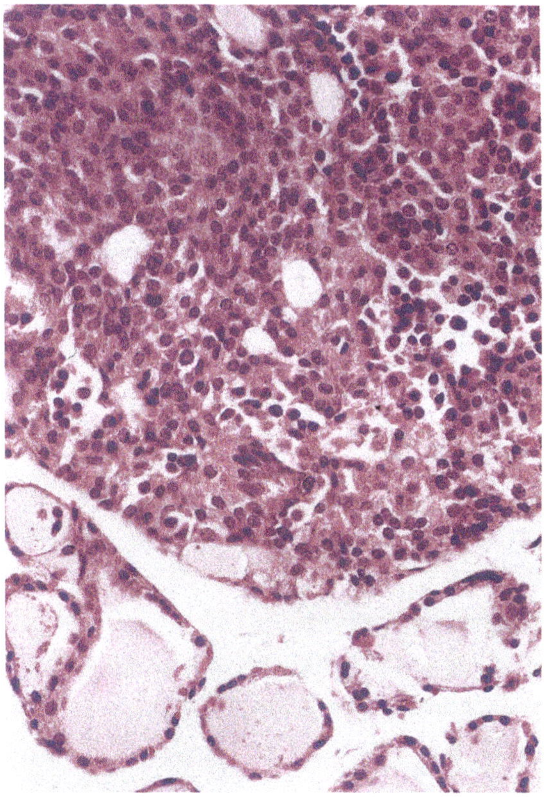

Fig. 18. C-cell adenoma of the thyroid gland. H&E

Synonym(s). Adenoma, parafollicular cell; tumor, C-cell, benign

Histogenesis

Calcitonin producing C-cells of the thyroid gland.

Diagnostic Features

- Increase is circumscribed and confined to a distinct area
- Compression of the adjacent tissue may be present
- A thin capsule may be present
- Cells are arranged in solid nests
- Stromal septa may be present
- Cells are polygonal and pale
- Cell borders are indistinct
- Nuclei are uniform round to oval
- Mitotic figures are rare
- C-cell cluster is larger than the area of five average thyroid follicles
- Central necrosis is not present
- Expansion tendency is low
- Fibroplasia is not present in the (surrounding) invaded tissue

Differential Diagnosis

Hyperplasia, C-Cell
Evidence of compression is not present and
A thin capsule is not present and
Stromal septa are not present and
Not larger than the area of five average thyroid follicles

Carcinoma, C-Cell
Invasion of extrathyroidal tissue is present and
Fibroplasia may be present in the (surrounding) invaded tissue

Adenoma, Follicular Cell (Solid)
Small colloid-containing follicles are present. Immunohistochemical proof can be provided for the exclusion of calcitonin production

Comment

C-cell adenomas in mice are rare. The differentiation between adenoma and carcinoma is not clearly defined. Invasion in the extrathyroidal tissue is therefore the only criterion of malignancy.

References

(See 31, 50, 57, 59)

Carcinoma, C-Cell *(M)*

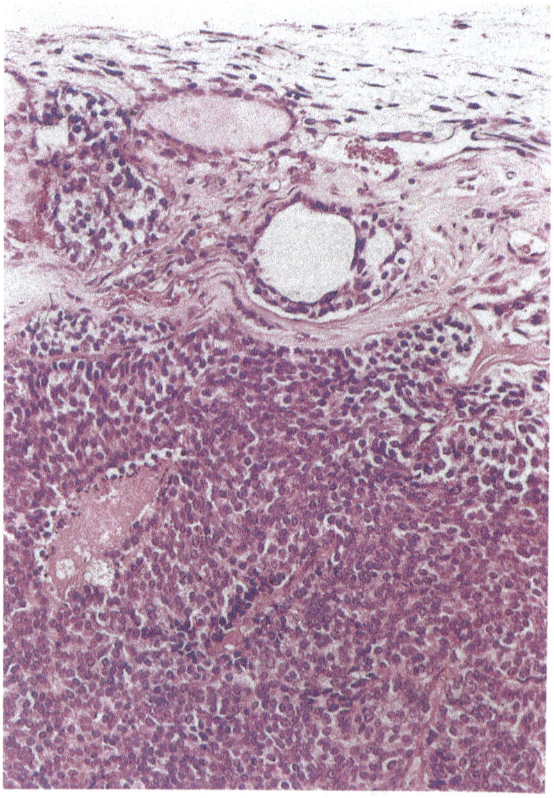

Synonym(s). Adenocarcinoma, parafollicular cell; adenocarcinoma, C-cell; tumor, C-cell, malignant

Histogenesis

Calcitonin producing C-cells of the thyroid gland.

Diagnostic Features

- Capsule may be present, but invaded
- Cells are arranged in solid nests and sheets
- Nests and sheets may be separated by fibrous stroma.
- Cells are polygonal and pale
- Cell borders are indistinct
- Nuclei are uniform round to oval
- Mitotic figures are numerous
- Central necrosis may be found
- Invades the adjacent extrathyroidal tissue
- Fibroplasia is present in the (surrounding) invaded tissue

Differential Diagnosis

Adenoma, C-Cell

Invasion of extrathyroidal tissue is not present and

Fibroplasia in the surrounding tissue is not present

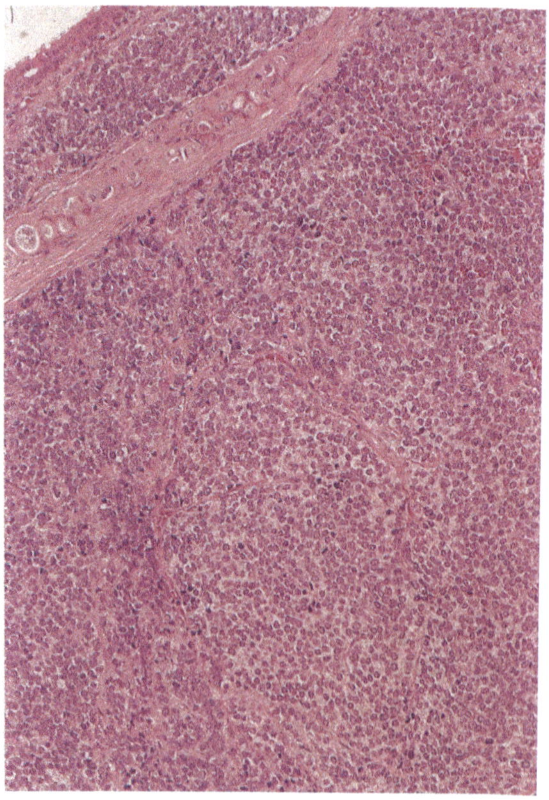

Fig. 19 *(above).* C-cell carcinoma of the thyroid gland. H&E

Fig. 20 *(below).* C-cell carcinoma of the thyroid gland. B6C3F1 mouse, male, age 87 weeks. H&E (courtesy of Dr. Shinji Yamamoto)

Comment

Generally the cellular morphology is the same in focal hyperplastic changes as in C-cell tumors. There seems to be a progression from benign to the malignant form. The transition is difficult to distinguish; malignancy is diagnosed when extrathyroidal invasion is present.

References

(See 31, 50, 57, 59)

Hyperplasia, Follicular Cell *(H)*

Cystic

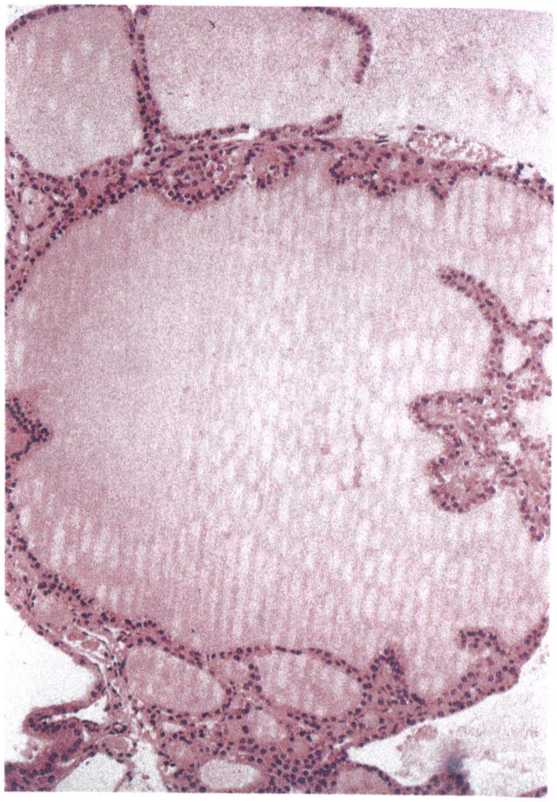

Histogenesis

Follicular cells of the thyroid follicles.

Diagnostic Features

- Is poorly demarcated
- Compression of the adjacent thyroidal tissue is not present
- A capsule is not present
- Epithelium is single-layered
- Growth pattern is not atypical, but papillary infoldings may be present
- Cells are cuboidal or low columnar
- Cytoplasm is more eosinophilic

Cystic

- Follicles are enlarged and appear cystic
- Follicular cells form papillary projections
- Large colloid-filled follicles may cause compression

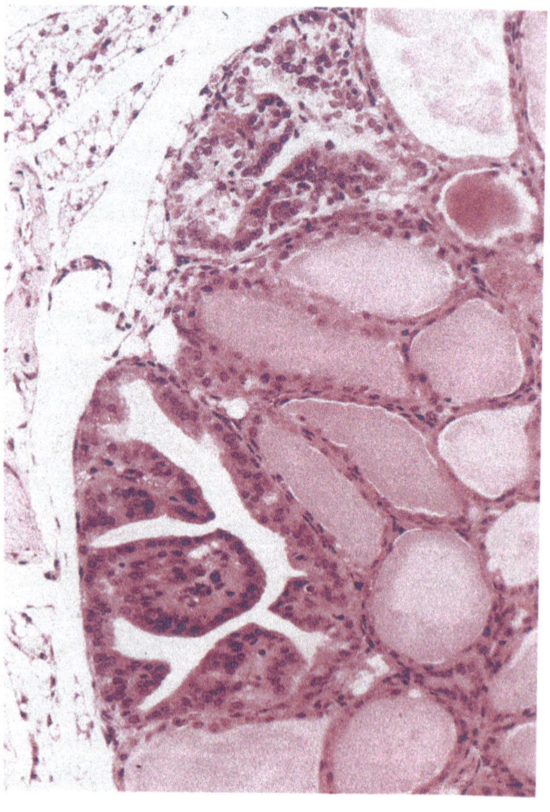

Differential Diagnosis

Large Cystic Follicles

Large dilated colloid-filled follicles, sometimes originating from several confluent follicles and
Epithelium is flattened and
Projections are not present

Fig. 21 *(above)*. Follicular cell hyperplasia, cystic, of the thyroid gland. H&E

Fig. 22 *(below)*. Follicular cell hyperplasia of the thyroid gland. H&E

Adenoma, Follicular Cell
Compression is present and
Growth pattern is atypical and
A capsule may be present and
Is well demarcated and
Slight cellular atypia and increased basophilia is present

Comment

The large cystic follicles may be ruptured and the broken septa may be misinterpreted as hyperplastic papillary feature, but cystic follicles are lined by flattened epithelium. Distinction between hyperplasia and adenoma is not always clear, but the adenoma shows compression and an atypical cellular growth pattern simultaneously.

References

(See 20, 21, 31, 43, 47, 59)

Adenoma, Follicular Cell *(B)*

Follicular • Papillary • Solid

Histogenesis

Follicular cells of the thyroid follicles.

Diagnostic Features

- Is well demarcated
- Compression of the adjacent thyroidal tissue is present
- A capsule may be present
- Epithelium is single-layered
- Growth pattern is atypical
- Cells are differentiated
- Slight cellular atypia is present
- Cells are cuboidal or low columnar
- Cytoplasm is eosinophilic or basophilic and reduced to a small rim
- Some cells may have abundant clear vacuolated cytoplasm
- Nuclei are enlarged oval or round with distinct nucleoli
- Mitotic figures are rare
- Size is not a criterion
- Growth pattern are papillary, follicular, and solid. These forms can be present simultaneously in the same animal and even in the same neoplasm

Follicular

- Small colloid-containing follicles are mixed with single regular sized follicles

Papillary

- Cells are organized to papillary projections in enlarged follicles. Fibrous stroma is branching
- Microfollicles or solid foci may be distributed among the papillary pattern
- Inflammatory cells and debris may be present in lumina

Solid

- Cells are arranged in solid sheets and densely packed nodules
- Some small follicles with less colloid are present
- Cells are polygonal to elongated with eosinophilic to clear cytoplasm
- Resemble C-cell tumors

Differential Diagnosis

Hyperplasia, Follicular Cell (Cystic)
Dilated colloid-filled follicles and
Epithelium is not flattened and
Small infoldings (projections without branching) may be present

Hyperplasia, Follicular Cell
Compression is not present and
Growth pattern is not atypical, but small papillary infoldings (projections) may be present and
A capsule is not present and
Is not well demarcated

Carcinoma, Follicular Cell
Is not well demarcated or
Epithelium is multiple-layered or
Distinct invasion is present and
Fibroplasia without capsule formation may be present in the periphery

Fig. 23 *(upper left).* Follicular cell adenoma of the thyroid gland. H&E ▶

Fig. 24 *(upper right).* Follicular cell adenoma of the thyroid gland. H&E

Fig. 25 *(lower left).* Follicular cell adenoma of the thyroid gland. H&E

Fig. 26 *(lower right).* Follicular cell adenoma of the thyroid gland. H&E

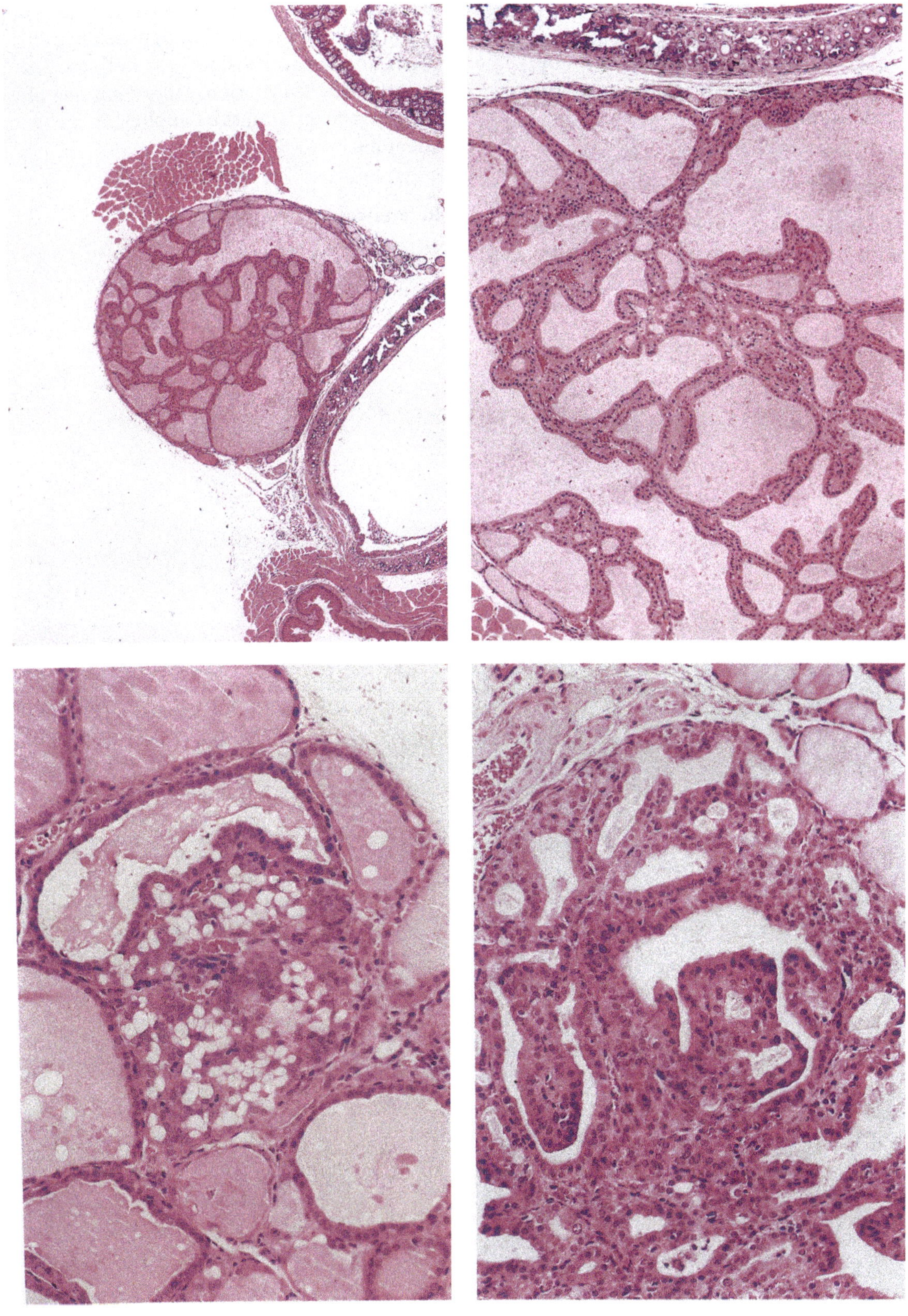

Adenoma, C-Cell

Small colloid-filled follicles are not present and
The C-cell origin could be shown by immunohistochemical demonstration of calcitonin

Comment

Spontaneous thyroid adenomas are rare in mice. There seems to be a morphologic continuum from the focal hyperplasia to carcinoma, as can be shown in induced thyroid tumors. The differentiation between hyperplasia and adenoma is not well defined. Size is not a reliable criterion. More than one of the given criteria must be applicable for an adenoma.

References

(See 20, 21, 31, 43, 47, 59)

Carcinoma, Follicular Cell *(M)*

Follicular • Papillary • Solid

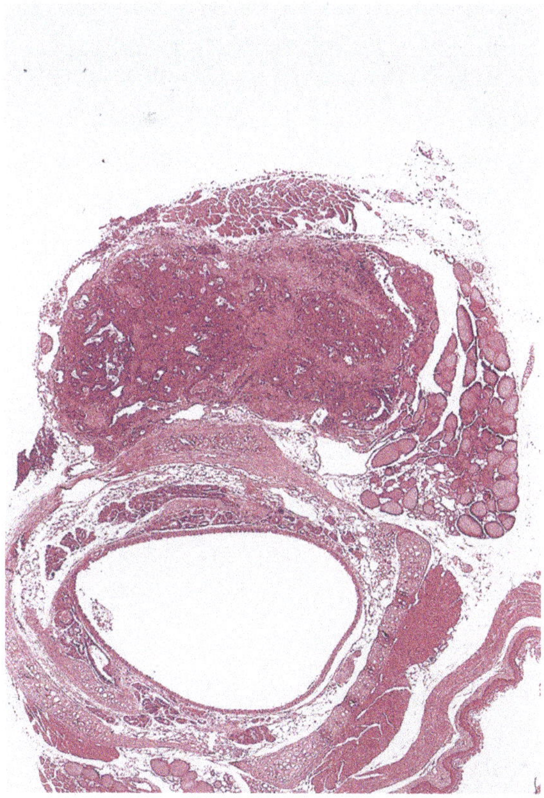

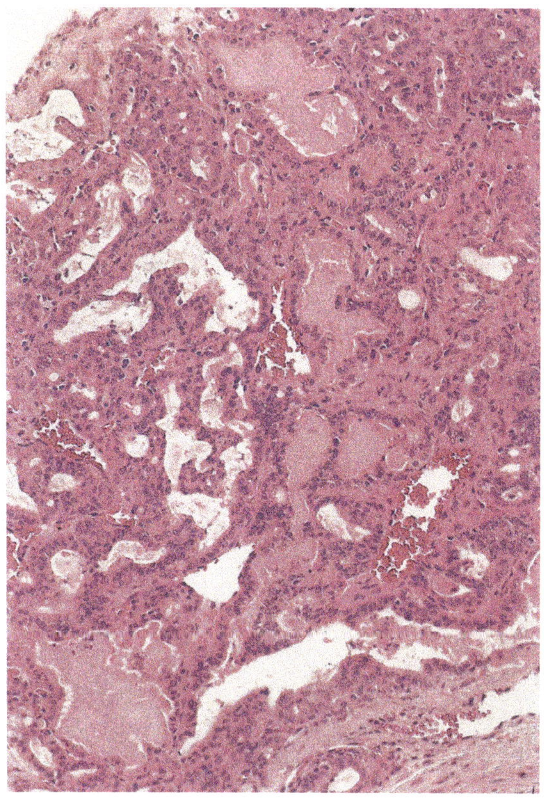

Histogenesis

Follicular cells of the thyroid follicles.

Diagnostic Features

- Borders are not well defined
- A capsule may be present
- Epithelium is multilayered, organized to solid clusters and sheets
- Predominant pattern is solid growth
- Fibroplasia is present in the surrounding tissue and capsule
- Tumor cells have atypia and pleomorphism indicated by enlarged or reduced size
- Nuclei are often large, round or elongated, sometimes clear
- Mitotic figures may be numerous
- Invades the adjacent tissue, distant metastases may be found
- Growth pattern are papillary, follicular, and solid, or a mixture of these patterns, often with solid growth predominating

Follicular

- Irregular variable-sized follicles are often lined with multiple layers of disorganized epithelium
- Sometimes solid clusters are present

Fig. 27 *(above)*. Follicular cell carcinoma of the thyroid gland. H&E (courtesy of Dr. Shinji Yamamoto)

Fig. 28 *(below)*. Follicular cell carcinoma of the thyroid gland. H&E (courtesy of Dr. Shinji Yamamoto)

Papillary

- Irregular papillary structures with one or multiple epithelial layers
- Dense packing may mimic tubular features

Solid

- Cells are arranged in sheets and nodules
- Cells are polygonal and atypical
- Mitotic figures are rare
- Resembles C-cell tumors

Differential Diagnosis

Adenoma, Follicular Cell

Is well demarcated and
Epithelium is single-layered and
Evidence of distinct invasion is not present and
Evidence of fibroplasia without capsule formation is not present in the periphery

Carcinoma, C-Cell

Small colloid filled follicles are not present and
The C-cell origin can be shown by immunohistochemical demonstration of calcitonin

References

(See 20, 21, 31, 43, 47, 53, 59)

Hyperplasia *(H)*

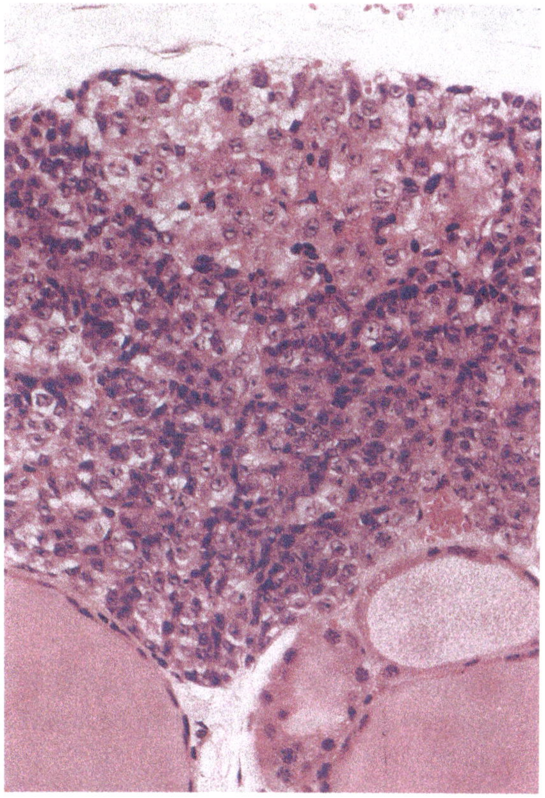

Fig. 29. Hyperplasia of the parathyroid gland. H&E

Histogenesis

Chief cells of the parathyroid gland.

Diagnostic Features

Focal

- Confined to single or multifocal areas within one or multiple glands
- Poor demarcation blending with surrounding parenchyma or no evidence of demarcation is present
- No evidence of compression of parathyroid tissue is present
- A capsule is not present
- Architecture is maintained
- Slight difference in growth pattern is present

Diffuse

- All parathyroids have homogeneous uniform enlargement
- Compressed rim of normal parathyroid parenchyma
- Thyroid gland may be compressed
- Both glands have similar growth pattern
- Cells appear homogeneous
- Nuclei are fusiform

Differential Diagnosis

Hypertrophy

The chief cells are pale and show an enlarged cytoplasm. A minimal compression of the surrounding tissue due to the increase in cell size is sometimes present

Adenoma

Architecture is not maintained or
Distinct demarcation is present or
Capsule is present or
Distinct compression is present or
Distinct difference in growth pattern is present

Comment

Diffuse hyperplasia may return to normal if the endocrine imbalance normalizes or, in the case of an induced lesion, the stimulus is withdrawn. Mitotic activity is not a helpful parameter to distinguish a hyperplastic lesion from a tumor. Diffuse hyperplasia is frequently associated with the "osteo-renal syndrome" (renal secondary hyperparathyroidism). Detection of fibrous osteodystrophy in bone is helpful to confirm the diagnosis of diffuse parathyroid hyperplasia. Dietary and renal diseases that alter calcium metabolism are etiological factors. Focal hyperplasia in laboratory animals is non-functional.

References

(See 4, 14, 36)

Adenoma *(B)*

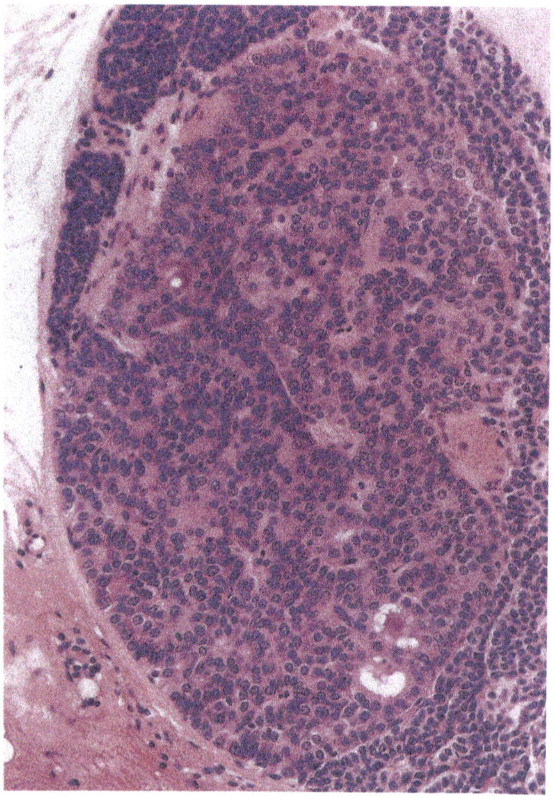

Fig. 30. Adenoma of the parathyroid gland. H&E

Histogenesis

Chief cells of the parathyroid gland.

Diagnostic Features

- Usually confined to a solitary nodule in one parathyroid gland
- Distinct demarcation is present
- Distinct compression is present
- A capsule may be present
- Architecture is not maintained
- The growth pattern is solid, but papillary, acinar or cystic pattern can also be found
- Cell borders are distinct
- Uniform cells exhibit pale eosinophilic cytoplasm
- Nuclei are slightly pleomorphic
- Diagnosis of adenoma is independent of the lesion size
- Evidence of invasion is not present

Differential Diagnosis

Hypertrophy
The cells are pale and show an enlarged cytoplasm. A minimal compression of the surrounding tissue due to the increase in cell size is sometimes present

Hyperplasia (Focal)
Architecture is maintained and
Demarcation is poor or not present and
Capsule is not present and
Compression of parathyroid tissue is not present and
Slight difference in growth pattern is present

Carcinoma
Invades parathyroid capsule, lymphatic vessels, blood vessels or connective tissue

Comment

Parathyroid tumors are rare in mice. There are only three reports available in the literature.

References

(See 4, 14, 36)

Carcinoma *(M)*

Histogenesis

Chief cells of the parathyroid glands.

Diagnostic Features

- Architecture is not maintained
- Cells are organized in interlacing sheets or nodules with central necrosis
- Cells may be spindle-shaped
- Nuclei are large and vesicular
- Invades parathyroid capsule, lymphatic vessels, blood vessels or connective tissue

Differential Diagnosis

Adenoma
 Invasion is not present

Comment

Parathyroid tumors are rare in mice. There are only three reports available in the literature.

References

(See 4, 14, 36)

Hyperplasia, Subcapsular Cell *(H)*

Type A • Type B • Mixed type

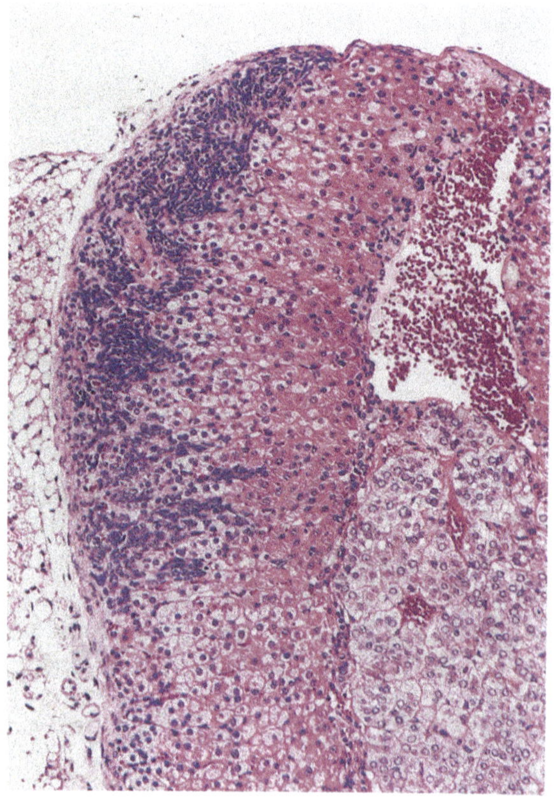

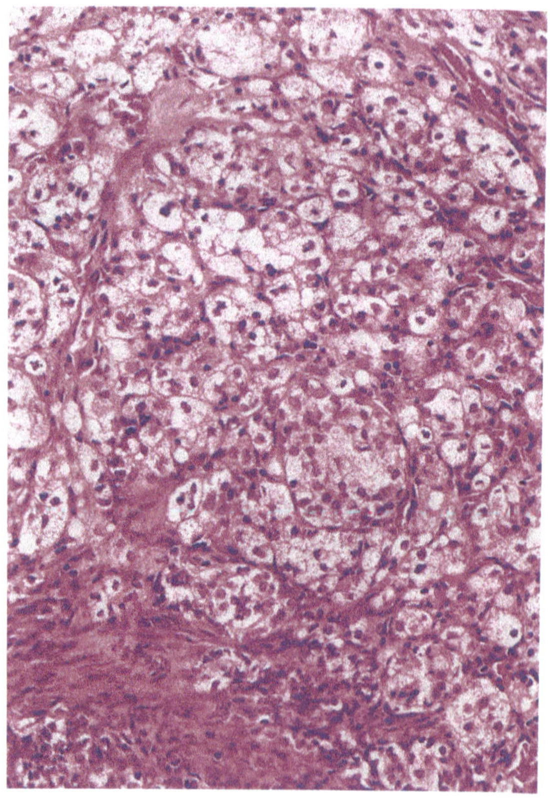

Histogenesis

Probably derived from a subcapsular reserve cell of the adrenal cortex.

Diagnostic Features

- Proliferation is surface-oriented
- Often extends towards the medulla
- Focal, multifocal to circumferential
- Focal hyperplasia may slightly bulge the surface of the adrenal gland
- May exhibit minimal compression of the underlying cortex
- Does not grow within/beyond the capsule
- Usually displays a packeted architecture
- Two cell types are intermingled: Type A (spindle) cells are dark, small, oval to spindle-shaped, cytoplasm is scant and free of lipid. Type B (lipid-laden) cells are large and polygonal, cytoplasm is clear and lipid-laden
- Mitotic figures are rare
- Not larger than the regular width of the cortex in a young mouse.
- May contain numerous mast cells
- Modifiers (type A [spindle], type B [lipid-laden] or mixed) may be used according to the predominant cell type (>70%). Mixed lesions are lesions in which no predominant cell type is present, which occurs rather often

Fig. 31 *(above)*. Focal subcapsular cell hyperplasia, type A, of the adrenal gland. H&E

Fig. 32 *(below)*. Focal subcapsular cell hyperplasia, mixed type, of the adrenal gland. H&E

Differential Diagnosis

Adenoma, Subcapsular Cell
Circumscribed lesion larger than the regular width of the cortex in a young mouse

Hyperplasia, Cortical
Composed of non-polygonal, non-spindle cells, usually eosinophilic or amphophilic. Arises within the zona fasciculata (not a surface-oriented lesion)

Comment

Very common lesion in aging mice. Gonadectomy and housing conditions (crowding and isolation) may increase the incidence. Due to cortical atrophy frequently observed in old mice, the size criterion should be handled cautiously. Differentiation between focal subcapsular hyperplasia and subcapsular adenoma is arbitrary and difficult.

References

(See 13, 15, 26, 28)

Adenoma, Subcapsular Cell *(B)*

Type A • Type B • Mixed type

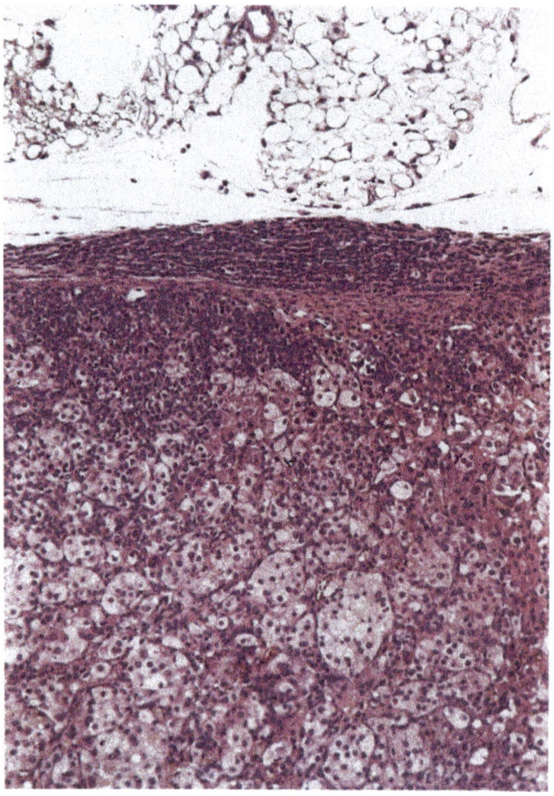

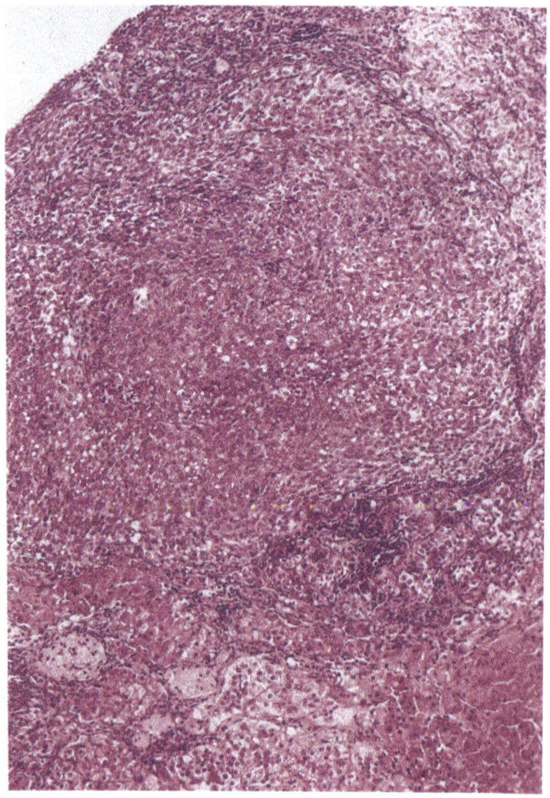

Histogenesis

Probably derived from a subcapsular reserve cell of the adrenal cortex.

Diagnostic Features

- Proliferation is surface-oriented
- Distinctly bulges the surface of the adrenal gland
- Exhibits some compression of the underlying cortex
- May dissect the capsule
- Usually displays a packeted architecture
- Two cell types are intermingled: Type A (spindle) cells are dark, small, oval to spindle-shaped, cytoplasm is scant and free of lipid. Type B (lipid-laden) cells are large and polygonal, cytoplasm is clear and lipid-laden
- Mitotic figures are rare
- Larger than the regular width of the cortex in a young mouse
- May contain numerous mast cells
- Modifiers (type A [spindle], type B [lipid-laden] or mixed) may be used according to the predominant cell type (>70%). Mixed lesions are lesions in which no predominant cell type is present, which occurs rather often

Fig. 33 *(above)*. Subcapsular cell adenoma, mixed type, of the adrenal gland. The proliferation is composed of a mixture of spindle-shaped and polygonal cells in the same proportion. "Pseudo-invasive growth": presence of cells apparently beyond the adrenal capsule is not regarded as evidence of malignancy. H&E

Fig. 34 *(below)*. Subcapsular cell adenoma, mixed type, of the adrenal gland. H&E

Differential Diagnosis

Hyperplasia, Subcapsular Cell (Focal)
Not larger than the regular width of the cortex in a young mouse

Carcinoma, Subcapsular Cell
Presence of metastases

Adenoma, Cortical
Composed of non-polygonal, non-spindle cells, usually eosinophilic or amphophilic. Arises within the zona fasciculata (not a surface-oriented lesion)

Comment

Due to cortical atrophy frequently observed in old mice, the size criterion should be handled cautiously. Growth beyond or dissection of the capsule is not considered a criterion for malignancy. Differentiation between focal subcapsular hyperplasia and subcapsular adenoma is arbitrary and difficult.

References

(See 13, 15, 17, 19, 24, 26, 28, 46)

Carcinoma, Subcapsular Cell *(B)*

Type A • Type B • Mixed type

Histogenesis

Probably derived from a subcapsular reserve cell of the adrenal cortex.

Diagnostic Features

- Proliferation is surface-oriented
- Distinctly bulges the surface of the adrenal gland
- Grows beyond the capsule
- Usually displays a packeted architecture
- Two cell types are intermingled: Type A (spindle) cells are dark, small, oval to spindle-shaped, cytoplasm is scant and free of lipid. Type B (lipid-laden) cells are large and polygonal, cytoplasm is clear and lipid-laden
- Mitotic figures are rare
- May contain numerous mast cells
- Distant metastases are present
- Modifiers (type A [spindle], type B [lipid-laden] or mixed) may be used according to the predominant cell type (>70%). Mixed lesions are lesions in which no predominant cell type is present, which occurs rather often

Differential Diagnosis

Adenoma, Subcapsular Cell
Absence of metastases

Carcinoma, Cortical
Composed of non-polygonal, non-spindle cells, usually eosinophilic or amphophilic

Comment

Metastases have not been reported to occur spontaneously for subcapsular cell proliferation in intact animals but apparently occur in gonadectomized animals. Metastases reported in the literature to occur spontaneously for so-called type B subcapsular tumors in our view are derived from adrenal cortical carcinomas.

References

(See 15, 19)

Hyperplasia, Cortical *(H)*

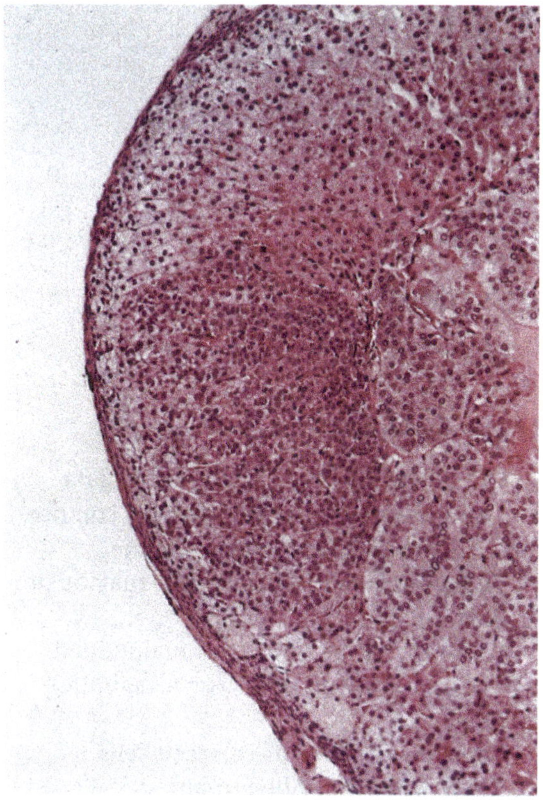

Fig. 35. Cortical hyperplasia of the adrenal gland. Sharply demarcated cortical nodule, composed exclusively of large eosinophilic cells. No intermingled spindle cells. Little compression. Fascicularis architecture is maintained. Size is smaller than normal cortical width. H&E

Histogenesis

Adrenocortical cell.

Diagnostic Features

- Usually located in the zona fasciculata
- No compression or only slight compression of the adjacent tissue
- Cortical architecture is maintained
- Never intermingled with type A (spindle) cells
- Usually composed of enlarged cells
- Karyomegaly may be present
- Cytoplasm may be eosinophilic or amphophilic
- Vacuolation may be present
- Cellular atypia is not present
- Mitotic figures usually are not present

Differential Diagnosis

Adenoma, Cortical
Distinct compression of the adjacent tissue
or
Architecture is not maintained

Hyperplasia, Subcapsular Cell (Type B)
Composed of large polygonal cells with clear cytoplasm (type B), admixed with fusiform, basophilic cells (type A)
Surface-oriented lesion

Hyperplasia, Medullary
Lipid droplets are not present
Cytoplasmic basophilia
Medullary cells react positively with antibodies against tyrosine hydroxylase, chromogranin A, synaptophysin, and neuron-specific enolase

Comment

Common in certain strains of mice (CD-1), particularly in males. One brief report by Faccini, Abbott and Paulus (1990) exists in the literature. Cortical hyperplasia is most likely lumped under the term "subcapsular cell hyperplasia, type B".

References

(See 13, 15, 24)

Adenoma, Cortical *(B)*

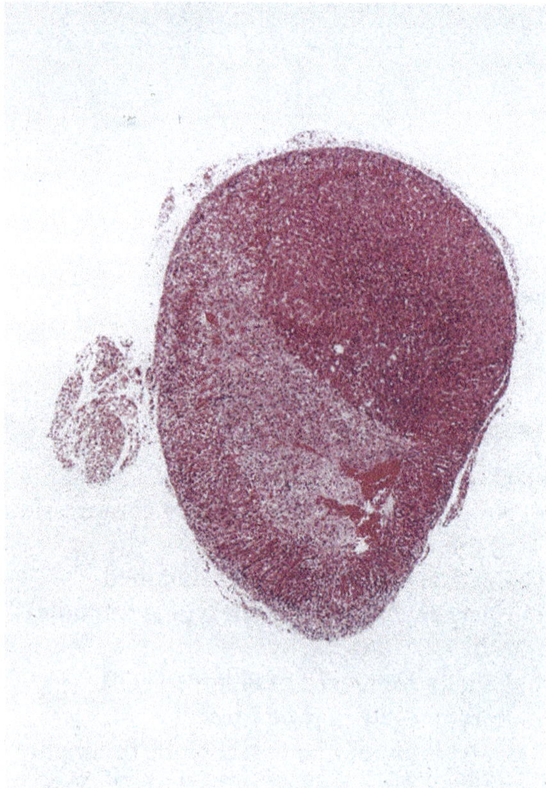

Fig. 36. Cortical adenoma of the adrenal gland. The proliferation is composed exclusively of large eosinophilic cells. No intermingled spindle cells. Marked compression, in particular towards the medulla. Fascicularis architecture is not maintained. H&E

Histogenesis

Adrenocortical cell.

Diagnostic Features

- Usually located in the zona fasciculata
- Well-circumscribed with distinct compression of the adjacent cortical tissue
- Thin connective tissue capsule may be present
- Cortical architecture is not maintained
- Not intermingled with type A (spindle) cells
- Usually composed of enlarged cells
- Karyomegaly may be present
- Cytoplasm may be eosinophilic or amphophilic
- Vacuolation may be present
- Cellular atypia may be present
- Mitotic figures may be present
- No evidence of invasion into or beyond the capsule nor distant metastasis is present

Differential Diagnosis

Hyperplasia, Cortical
No compression or only slight compression of the adjacent tissue and
Architecture is maintained

Carcinoma, Cortical
Invasive growth into and beyond the capsule or distant metastasis are present

Adenoma, Subcapsular Cell (Type B)
Composed of large polygonal cells with clear cytoplasm (type B), admixed with some fusiform, basophilic cells (type A)

Tumor, Medullary, Benign (Pheochromocytoma Type)
Lipid droplets are not present
Prominent blood vessels
Cytoplasmic basophilia
Tumor cells react positively with antibodies against tyrosine hydroxylase, chromogranin A, synaptophysin, and neuron-specific enolase

Comment

Some confusion has arisen from the literature. This lesion, which does not fit the classical morphological description of B cell proliferative changes (Dunn 1970; Russfield 1967), has been nevertheless lumped with it in newer classifications (Frith 1983a; Frith and Ward 1988). There are a number of cytological and architectural differences between these two types of proliferative changes that warrant separating the two lesions.

References

(See 13, 15, 17, 24, 46)

Carcinoma, Cortical *(M)*

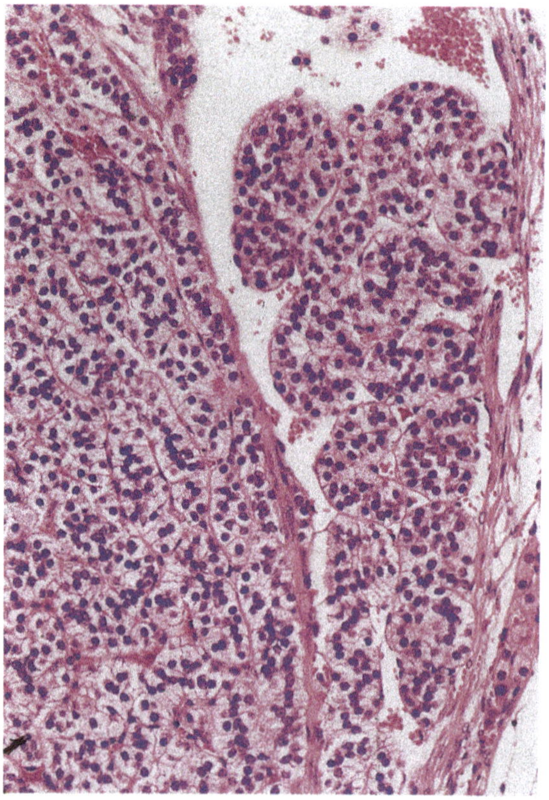

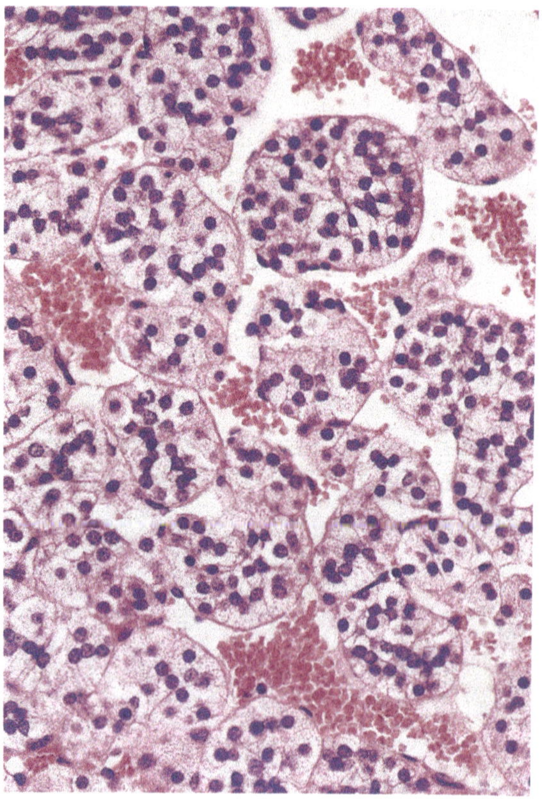

Histogenesis

Adrenocortical cell.

Diagnostic Features

- Cortical architecture is not maintained
- Not intermingled with type A (spindle) cells
- Usually composed of enlarged cells
- Karyomegaly may be present
- Cytoplasm is eosinophilic, rarely amphophilic
- Vacuolation may be marked
- Cellular atypia may be present
- Mitotic figures may be numerous
- Invasive growth into and beyond the capsule or distant metastasis is present

Differential Diagnosis

Adenoma, Cortical
Invasive growth into and beyond the capsule or metastasis are not present

Carcinoma, Subcapsular Cell (Type B)
Composed of large polygonal cells with clear cytoplasm (type B), admixed with fusiform, basophilic cells (type A)

Fig. 37 *(above)*. Cortical carcinoma of the adrenal gland. H&E

Fig. 38 *(below)*. Cortical carcinoma of the adrenal gland. H&E

Tumor, Medullary, Malignant (Pheochromocytoma Type)

Lipid droplets usually are not present
Prominent blood vessels
Cytoplasmic basophilia
Tumor cells react positively with antibodies against tyrosine hydroxylase, chromogranin A, synaptophysin, and neuron-specific enolase

Comment

Extracortical nodules of cortical cells must not be mistaken for evidence of invasive growth. Some confusion has arisen from the literature. This lesion, which does not fit the classical morphological description of B cell proliferative changes (Dunn 1970; Russfield 1967), has been nevertheless lumped with it in newer classifications (Frith 1983; Frith and Ward 1988). There are a number of cytological and architectural differences between these two types of proliferative changes that warrant separating the two lesions.

References

(See 13, 15, 52)

Hyperplasia, Medullary *(H)*

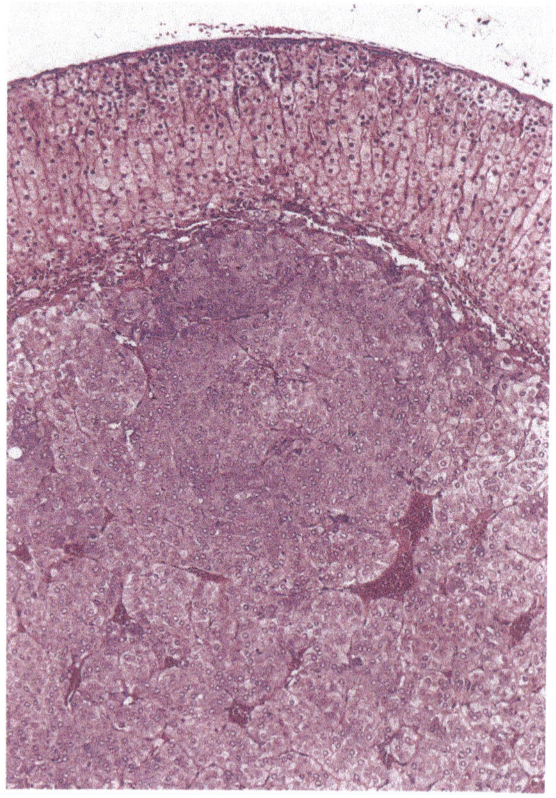

Fig. 39. Medullary hyperplasia of the adrenal gland. H&E

Histogenesis

Adrenal medullary (chromaffin) cell.

Diagnostic Features

Focal

- May impinge on the cortex
- Usually focal, infrequently multifocal
- No evidence of compression or minimal compression of medulla or cortex is present
- Architecture is maintained
- Cells may be larger or smaller or more basophilic than regular cells
- Nuclear:cytoplasmic ratio may be increased
- Cellular atypia is not present
- Mitotic figures are rare
- Section plane is smaller than 50% of the regular medulla

Diffuse

- Affects the whole medulla
- No recognizable zone of compression of the surrounding cortical tissue is present
- Architecture is maintained
- Nuclear:cytoplasmic ratio may be increased
- Cellular atypia is not present
- Mitotic figures are rare
- Frequently associated with cortical atrophy
- Widening of the hilus may occur

Differential Diagnosis

Tumor, Medullary, Benign (Pheochromocytoma Type)

Distinct compression at the periphery is present or
Architecture is not maintained or
Section plane is larger than 50% of the regular size of the medulla

Comment

Multiple focal medullary hyperplasias in the same medulla are not considered tumorous, even if their total section plane is larger than 50% of the normal medulla size.

References

(See 13, 15, 18, 19, 24, 39, 46)

Tumor, Medullary, Benign (*B*)

Pheochromocytoma type
Complex pheochromocytoma type
Ganglioneuroma type • NOS

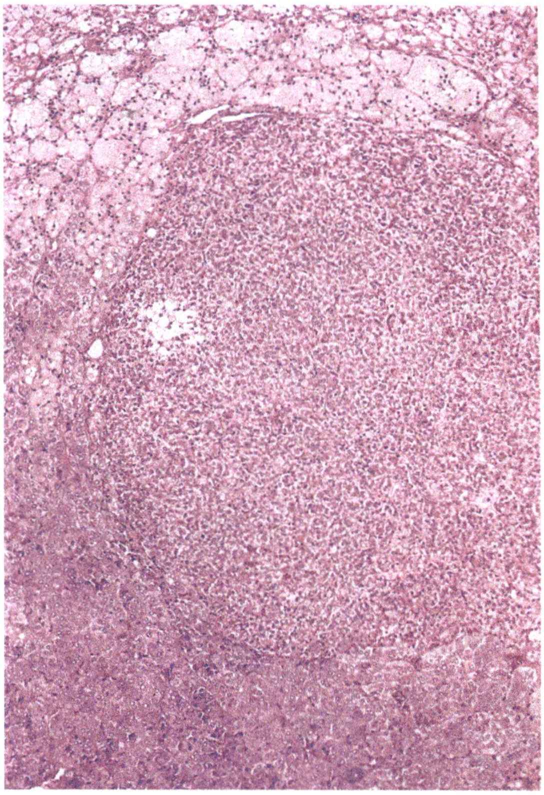

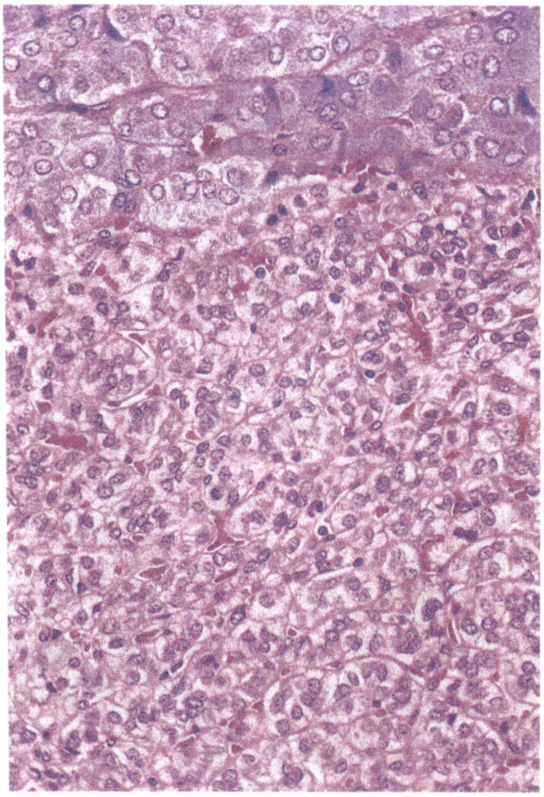

Histogenesis

Adrenal medullary cells with variable differentiation into predominantly secretory (chromaffin) or neural elements.

Diagnostic Features

- May replace the entire medulla
- Cellular atypia may be present
- Mitotic figures may be numerous
- Distinct compression of medulla or cortex is present or architecture is not maintained or section plane is larger than 50% of the normal medulla (pheochromocytoma type)
- Any size (all other types)
- Invasive growth (beyond the adrenal capsule) or distant metastasis is not present

Pheochromocytoma Type

- Composed of chromaffin cells organized in nests, rows and cords
- Cells arc large and rich in cytoplasm (so called "secretory" cells), or small, basophilic, with little cytoplasm
- Hemorrhage and necrosis may be present
- Numerous, sometimes distended blood vessels are present

Ganglioneuroma Type

- The tumor resembles nerve tissue containing large, well-differentiated ganglion cells and neurofibrils

Fig. 40 *(above)*. Benign medullary tumor of the adrenal gland. H&E

Fig. 41 *(below)*. Benign medullary tumor of the adrenal gland. II&E

Complex Pheochromocytoma Type

- Composed of areas resembling differentiated nerve tissue (ganglioneuroma type) and areas of chromaffin cells (pheochromocytoma type) intermingled or forming distinct areas

NOS (not otherwise specified)

- Cannot be classified due to the lack of characteristic features or for technical reasons

Differential Diagnosis

Hyperplasia, Medullary (Focal)

No evidence of, or only minimal, compression of the adjacent tissue is present and
Architecture is maintained and
Section plane is not larger than 50% of the regular medulla and
Cellular atypia is not present

Hyperplasia, Medullary (Diffuse)

No or slight compression of the surrounding tissue and
Architecture is maintained and
Cellular atypia is not present

Tumor, Medullary, Malignant

Invasive growth (beyond the capsule) or distant metastasis is present

Adenoma, Cortical

Lipid droplets are usually present
Negative immunohistochemical reaction for medullary cell markers

Comment

The adrenal medullary secretory cells have a positive chromaffin reaction, demonstrable with special stains such as tyrosine hydroxylase and chromogranin A, whereas cells with neural differentiation have a positive reaction with various antibodies against, for instance, synaptophysin and neuron-specific enolase.

References

(See 13, 15, 18, 19, 24, 39, 46, 58)

Tumor, Medullary, Malignant *(M)*

Pheochromocytoma type
Complex pheochromocytoma type
Ganglioneuroma type
Neuroblastoma type • NOS

Histogenesis

Adrenal medullary with variable differentiation into predominantly secretory (chromaffin) or neural elements.

Diagnostic Features

- Architecture is not maintained
- Cellular atypia may be present
- Mitotic figures may be numerous
- Invasive growth (beyond the adrenal capsule) or distant metastasis is present

Pheochromocytoma Type
- Composed of chromaffin cells organized in nests, rows and cords
- Cells are large and rich in cytoplasm (so called "secretory" cells), or small, basophilic, with little cytoplasm
- Hemorrhage and necrosis may be present
- Numerous, sometimes distended blood vessels are present

Ganglioneuroma Type
- The tumor resembles nerve tissue containing large, well-differentiated ganglion cells

Neuroblastoma Type
- Cells are small and hyperchromatic
- Cells are organized in dense sheets
- Cells occasionally form rosettes
- A fibrillary matrix may be present resembling tissue of the central nervous system
- Ganglion cells may be present

Complex Pheochromocytoma Type
- Composed of areas resembling differentiated nerve tissue (ganglioneuroma type) and areas of chromaffin cells (pheochromocytoma type) intermingled or forming distinct areas
- Small cells may metastasize to distant sites

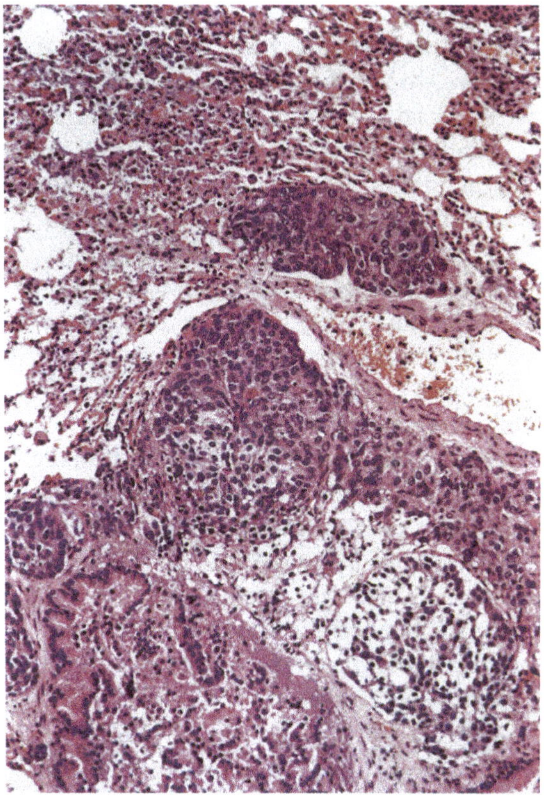

Fig. 42. Pulmonary metastasis of malignant medullary tumor of the adrenal gland. H&E

NOS (not otherwise specified)
- Cannot be classified due to the lack of characteristic features or for technical reasons

Differential Diagnosis

Tumor, Medullary, Benign
No evidence of invasive growth beyond the capsule or distant metastasis is present

Carcinoma, Cortical
Lipid droplets are usually present
Negative immunohistochemical reaction for medullary cell markers

Comment

The adrenal medullary secretory cells have a positive chromaffin reaction, demonstrable with special stains such as tyrosine hydroxylase and chromogranin A, whereas cells with neural differentiation have a positive reaction with various antibodies against, for instance, synaptophysin and neuron-specific enolase. The study of adrenal medullary cell lineages (Tischler and De Lellis 1988) shows that primitive sympathetic cells may develop either into neurons or chromaffin cells.
Only unequivocal evidence of invasion should be considered as indicative of malignancy. The occurrence of small groups of medullary cells within the cortex or in the "hilus area" of the adrenal gland is a regular anatomic feature. Only an invasion of blood vessels with a well-defined wall and active penetration of the vessel wall by tumor cells should be interpreted as angio-invasion. Subendothelial growth of tumor cells is a common microscopic feature in adrenal medullary tumor and should not be misinterpreted as vascular invasion.
The size of the primary mass is reported to correlate best with the capability for metastasis (Frith and Ward 1988).
Spontaneous neuroblastomas are rare (only one brief account by Maita et al. 1988). Aguzzi et al. (1990) reported that transgenic mice carrying a cDNA to the polyoma virus middle-T antigen linked to the thymidine kinase promoter developed neuroblastoma in multiple organs, including adrenal glands, by 2–3 months of age. These tumors expressed the N-myc oncogene and metastasized.

References

(See 1, 13, 15, 18, 19, 24, 39, 40, 46, 58)

Hyperplasia, Islet Cell *(H)*

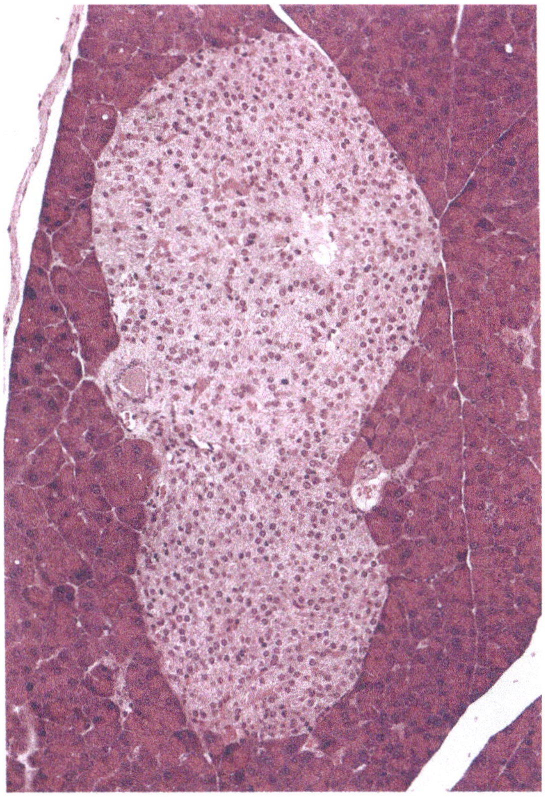

Histogenesis

Cells of islets of Langerhans.

Diagnostic Features

- Affecting multiple islets
- Islets are enlarged
- Compression of the adjacent acinar tissue is not present
- A capsule is not present
- Architecture of the islets is usually maintained
- Hyperplastic islets may be rounder than regular islets
- Central cystic cavities may be present in hyperplastic islets
- Cells may be slightly enlarged

Differential Diagnosis

Adenoma, Islet Cell

Compression or encapsulation is present or
Vascularity is distinctly increased or
Regular islet architecture is not maintained (abnormal growth in sheets, nests or ribbons) or
Usually solitary nodules

Comment

The specific cell type may be identified by immunohistochemistry (insulin, glucagon, somatostatin, pancreatic polypeptide).

References

(See 3, 5, 9, 10, 11, 12, 22, 23, 30, 32, 38, 42, 45, 48)

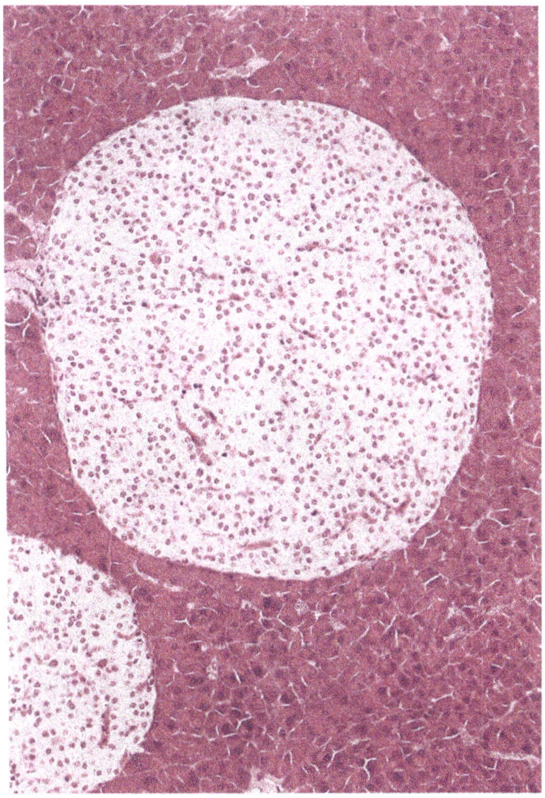

Fig. 43 *(above).* Islet cell hyperplasia of the pancreas. H&E

Fig. 44 *(below).* Islet cell hyperplasia of the pancreas. H&E

Adenoma, Islet Cell *(B)*

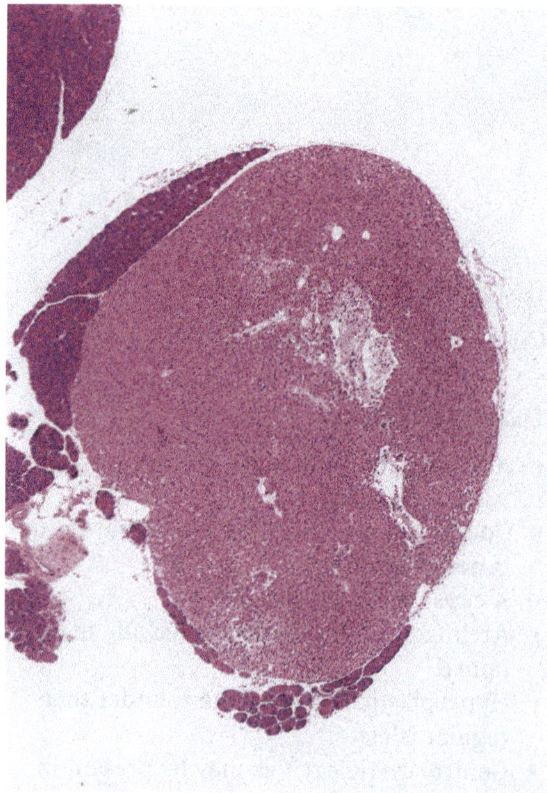

Synonym(s). Adenoma, endocrine pancreas; islet cell tumor, benign

Histogenesis

Cells of islets of Langerhans.

Diagnostic Features

- Usually affecting one islet changing it to a solitary, well-circumscribed nodule
- Compression of the adjacent acinar tissue may be present
- Partial, complete or no encapsulation may be present
- Architecture of the islets may not be maintained
- Growth pattern is ribbon- or cord-like along thin-walled vessels
- Vascularity may be increased
- Cells may be well-differentiated or various shapes (spindle-shaped, oval, fusiform) and sizes (larger or smaller) may be present
- Cytoplasm may be pale, eosinophilic or basophilic
- Karyomegaly may be present
- Mitotic figures are rare
- Growth into or beyond the tumor capsule or distant metastases are not present

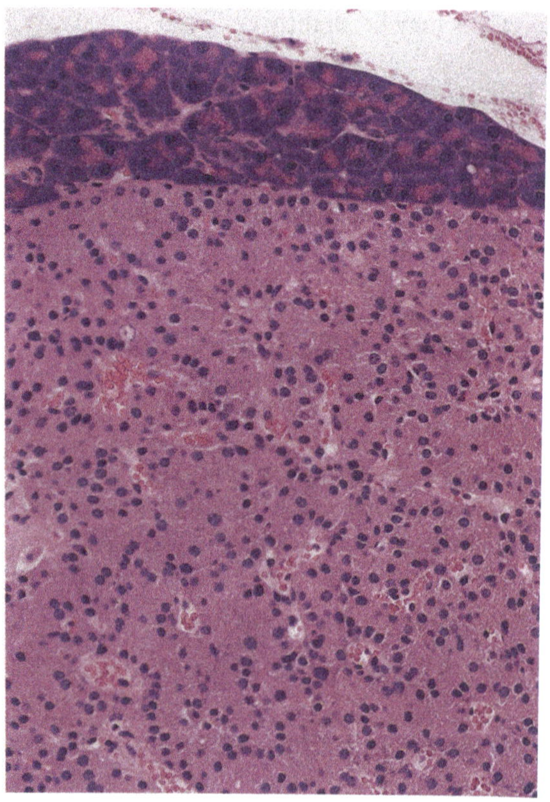

Fig. 45 *(above).* Islet cell adenoma of the pancreas. Well-circumscribed solitary nodule. H&E

Fig. 46 *(below).* Islet cell adenoma of the pancreas. H&E

Differential Diagnosis

Hyperplasia, Islet Cell
- Compression or encapsulation is not present and
- Vascularity is not increased and
- Regular islet architecture is maintained and
- Distribution is usually multifocal

Carcinoma, Islet Cell
- Local invasion is present, especially invasion of vessels at tumor periphery or
- Distant metastasis is present

Comment

Spontaneous islet cell adenomas are rare in mice, most of them produce insulin without inducing hypoglycemia.

References

(See 3, 5, 9, 10, 11, 12, 22, 23, 30, 32, 38, 42, 45, 48)

Carcinoma, Islet Cell *(M)*

Synonym(s). Adenocarcinoma, endocrine pancreas; islet cell tumor, malignant

Histogenesis

Cells of islets of Langerhans.

Diagnostic Features

- Generally large, even grossly visible tumors
- A fibrous capsule may be present
- May grow in sheets, nests or ribbons
- Cells vary from well-differentiated to extremely pleomorphic and anaplastic
- Cytoplasm is pale and eosinophilic
- Nuclei may be vesicular
- Nucleoli are prominent and may be multiple
- Mitotic figures may be numerous
- Growth into or beyond the tumor capsule may be present. Local invasion is present (affecting vessels, especially at the tumor periphery) or distant metastases (mainly into the liver)

Differential Diagnosis

Adenoma, Islet Cell
No evidence of invasion is present and
No evidence of metastasis is present

Comment

Diagnosis of some anaplastic carcinomas with islet cell origin can be supported by immunohistochemistry. Most of the islet cell tumors produce insulin without inducing hypoglycemia. Islet cell carcinomas are rare in all strains.

References

(See 3, 5, 9, 10, 11, 12, 22, 23, 30, 32, 38, 42, 45, 48)

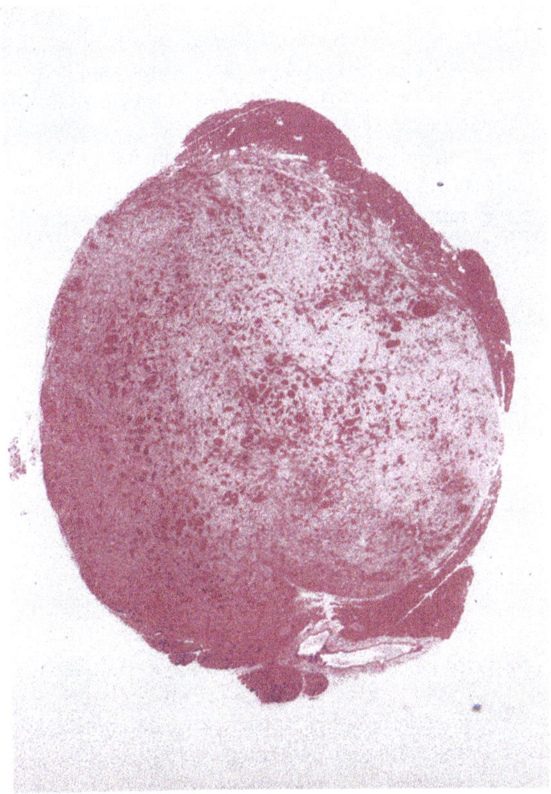

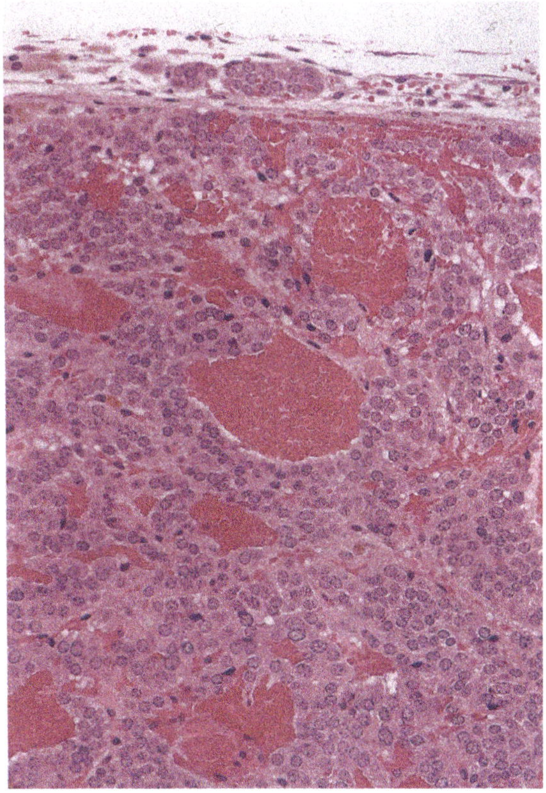

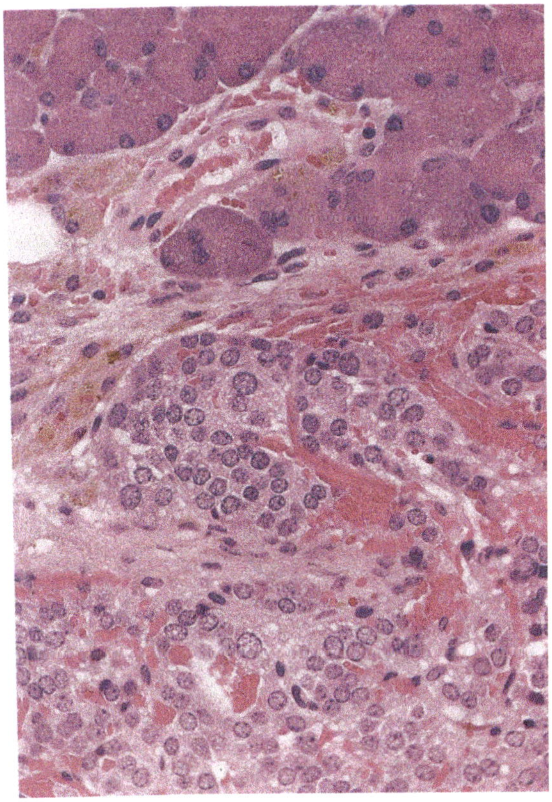

Fig. 47 *(upper left).* Islet cell carcinoma of the pancreas. Large and poorly circumscribed nodule. H&E

Fig. 48 *(upper right).* High power view of islet cell carcinoma. Note the high vascularity and the invasion of vessels at the tumor periphery. H&E

Fig. 49 *(below).* High power view of islet cell carcinoma with infiltration of the fibrous capsule. H&E

References

1. Aguzzi A, Wagner EF, Williams RL, Courtneidge SA (1990) Sympathetic hyperplasia and neuroblastomas in transgenic mice expressing polyoma middle T antigen. New Biol 2: 533–543
2. Anderson MP, Capen CC (1978) The endocrine system, pituitary. In: Benirschke K, Garner FM, Jones TC (eds) Pathology of laboratory animals, Vol. 1. Springer, Berlin Heidelberg New York Tokyo, pp 424–433
3. Belkin M (1942) The lack of influence of di-(hydroxymethyl)peroxide on the incidence and growth of transplanted, induced and spontaneous mouse tumors.
4. Bomhard E (1993) Frequency of spontaneous tumours in NMRI mice in 21-month studies. Exp Toxic Pathol 45: 269–289
5. Boorman GA, Sills RC (1999) Exocrine and endocrine pancreas. In: Maronpot RR, Boorman GA, Gaul BW (eds) Pathology of the mouse. Reference and atlas. Cache River Press, Vienna, pp 185–205
6. Botts S, Capen CC, DeLellis RA, Deschl U, Hartig F, Karbe E, Konishi Y, Krinke GJ, Landes C, Mettler F, Rebel W, Riley MGI, Tuch K, Urwyler H (1994) 6. Endocrine System. In: Mohr U, Capen CC, Dungworth DL, Griesemer RA, Ito N, Turusov VS (eds) International classification of rodent tumours, Part I: The rat. IARC Scientific Publications No. 122, Lyon, pp 1–11
7. Botts S, Capen CC, DeLellis RA, Deschl U, Hartig F, Karbe E, Konishi Y, Krinke GJ, Landes C, Mettler F, Rebel W, Riley MGI, Tuch K, Urwyler H (1994) 6. Endocrine System. In: Mohr U, Capen CC, Dungworth DL, Griesemer RA, Ito N, Turusov VS (eds) International classification of rodent tumours, Part I: The rat. IARC Scientific Publications No. 122, Lyon, pp 12–14
8. Capen CC (1983) Functional and pathologic interrelationships of the pituitary gland and the hypothalamus. In: Jones TC, Mohr U, Hunt RD (eds) Monographs on pathology of laboratory animals. Endocrine system. Springer, Berlin Heidelberg New York Tokyo, pp 101–120
9. Cardesa A, Bullon-Ramirez A, Levitt MH (1979) Tumours of the pancreas In: Turusov VS (ed) Pathology of tumours in laboratory animals, vol II IARC Scientific Publications No. 23, Lyon, pp. 235–241
10. Cloudman AM (1941) Spontaneous neoplasms in mice. In: Snell GD (ed) Biology of the laboratory mouse. Blakiston, Philadelphia, pp 168–233
11. Deringer MK (1951) Spontaneous and induced tumors in haired and hairless strain HR mice. J Natl Cancer Inst 12: 437–445
12. Deringer MK (1956) The effect of subcutaneous inoculation of 4-0-tolylazo-o-toluidine in strain HR mice. J Natl Cancer Inst 17: 533–539
13. Dunn TB (1970) Normal and pathologic anatomy of the adrenal gland of the mouse, including neoplasms. J Natl Cancer Inst 44: 1323–1389
14. Dunn TB (1979) Tumors of the parathyroid gland. In: Turusov VS (ed) Pathology of tumours in laboratory animals, vol II. Tumours of the mouse. IARC Scientific Publications No. 23, Lyon, pp 469–474
15. Faccini JM, Abbott DP, Paulus GJJ (1990) Endocrine glands. Mouse histopathology. A glossary for use in toxicity and carcinogenicity studies. Elsevier, Amsterdam, pp 169–186
16. Fraser H (1986) Brain tumours in mice, with particular reference to astrocytoma. Food Chem Toxicol 24: 105–111
17. Frith CH (1983a) Adenoma and carcinoma, adrenal cortex, mouse. In: Jones TC, Mohr U, Hunt RD (eds) Monographs on pathology of laboratory animals. Endocrine system. Springer, Berlin Heidelberg New York Tokyo, pp 49–56
18. Frith CH (1983b) Pheochromocytoma, adrenal medulla, mouse. In: Jones TC, Mohr U, Hunt RD (eds) Monographs on pathology of laboratory animals. Endocrine system. Springer, Berlin Heidelberg New York Tokyo, pp 27–30
19. Frith CH, Dunn TB (1994) Tumours of the adrenal gland. Pathology of tumours in laboratory animals, vol 2. Tumours of the mouse, 2nd edn. IARC Scientific Publications No. 111, Lyon, pp 595–609
20. Frith CH, Heath JE (1983) Adenoma, thyroid, mouse. In: Jones TC, Mohr U, Hunt RD (eds) Monographs on pathology of laboratory animals. Endocrine system. Springer, Berlin Heidelberg New York Tokyo, pp 184–187
21. Frith CH, Heath JE (1984) Morphological classification and incidence of thyroid tumors in untreated aged mice. J Gerontol 39: 7–10
22. Frith CH, Sheldon WD (1983) Hyperplasia, adenoma and carcinoma of pancreatic islets, mouse. In: Jones TC, Mohr U, Hunt RD (eds) Monographs on pathology of laboratory animals. Endocrine system. Springer, Berlin Heidelberg New York Tokyo, pp 297–303
23. Frith CH, Ward JM (1988a) Color atlas of neoplastic and non-neoplastic lesions in aging mice. Elsevier, Amsterdam
24. Frith CH, Ward JM (1988b) Color atlas of neoplastic and non-neoplastic lesions in aging mice. Endocrine system. Elsevier, Amsterdam, pp 33–37
25. Goetz W, Theuring F, Schachenmayr W, Korf HW (1992) Midline brain tumors in MSV-SV 40-

transgenic mice originate from the pineal organ. Acta Neuropathol (Berl) 83: 308–314
26. Goodman DG (1983) Subcapsular-cell hyperplasia, adrenal, mouse. In: Jones TC, Mohr U, Hunt RD (eds) Monographs on pathology of laboratory animals. Endocrine system. Springer, Berlin Heidelberg New York Tokyo, pp 66–68
27. Greaves P (1990a) Histopathology of preclinical toxicity studies: interpretation and relevance in drug safety evaluation. Elsevier, Amsterdam, pp 678–696
28. Greaves P (1990b) Histopathology of preclinical toxicity studies: interpretation and relevance in drug safety evaluation. Endocrine glands. Elsevier, Amsterdam, pp 677–755
29. Heider K (1986) Spontaneous craniopharyngioma in a mouse. Vet Pathol 23: 522–523
30. Hueper WC (1936) Islet-cell adenoma in the pancreas of a mouse. Arch Pathol 22: 220–221
31. Jokinen MP, Botts S (1994) Tumours of the thyroid gland. In: Turusov VS, Mohr U (eds) Pathology of tumours in laboratory animals, vol 2. Tumours of the mouse, 2nd edn. IARC Scientific Publications No. 111, Lyon, pp 565–594
32. Jones EE (1964) Spontaneous hyperplasia of the pancreatic islets associated with glucosuria in hybrid mice. In: Brolin SE (ed) The structure and metabolism of the pancreatic islets. Pergamon, New York, pp 189–191
33. Koestner A, Solleveld HA (1996) Tumors of the pineal gland, rat. In: Jones TC, Capen CC, Mohr U (eds) Monographs on pathology of laboratory animals. Endocrine system, 2nd edn. Springer, Berlin Heidelberg New York Tokyo, pp 205–213
34. Korf HW, Goetz W, Herken R, Theuring F, Gruss P, Schachenmayr W (1990) S-antigen and rod-opsin immunoreactions in midline brain neoplasms of transgenic mice: similarities to pineal cell tumors and certain medulloblastomas in man. J Neuropathol Exp Neurol 49: 424–437
35. Kuchelmeister K, von Borcke IM, Klein H, Bergmann M, Gullotta F (1994) Pleomorphic pineocytoma with extensive neuronal differentiation: report of two cases. Acta Neuropathol (Berl) 88: 448–453
36. Lewis DJ, Cherry CP (1982) Parathyroid adenoma in mice: a report of two cases. J Comp Pathol 92: 337–339
37. Liebelt AG (1994) Tumours of the pituitary gland. In: Turusov VS, Mohr U (eds) Pathology of tumours in laboratory animals, vol 2. Tumours of the mouse, 2nd edn. IARC Scientific Publications No. 111, Lyon, pp 527–563
38. Like AA, Steinke J, Jones EE, Cahill GF Jr (1965) Pancreatic studies in mice with spontaneous diabetes mellitus. Am J Pathol 46: 621–644
39. Longeart L (1996) Adrenal medullary tumors, mouse. In: Jones TC, Capen CC, Mohr U (eds) Monographs on pathology of laboratory animals. Endocrine system, 2nd edn. Springer, Berlin Heidelberg New York Tokyo, pp 421–427
40. Maita K, Hirano M, Harada T, Mitsumori K, Yoshida A, Takahashi K, Nakashima N, Kitazawa T, Enomoto A, Inui K, Shirasu Y (1988) Mortality, major cause of moribundity, and spontaneous tumors in CD-1 mice. Toxicol Pathol 16: 340–349
41. Morton D, Tekeli S (1997) "Have you seen this?" Pituitary cysts in a mouse. Toxicol Pathol 25: 333
42. Murphy ED (1966) Characteristic tumors In: Biology of the laboratory mouse, 2nd edn. McGraw-Hill, New York, pp 521–567
43. Neve P, Wollman SH (1972) Fine structure of a fifth type of epithelial cell in the thyroid gland of the C3H mouse. Anat Rec 172: 37–44
44. Rehm S, Rapp KG, Deerberg F (1985) Influence of food restriction and body fat on life span and tumour incidence in female outbred Han:NMRI mice and two sublines. Z Versuchstierkd 27: 240–283
45. Rowlatt UF (1967) Pancreatic neoplasms of rats and mice. In: Cotchin E, Roe FJC (eds) Pathology of laboratory rats and mice. Blackwell Scientific, Oxford, pp 85–103
46. Russfield AB (1967) Pathology of the endocrine glands, ovary and testis of rats and mice. In: Cotchin E, Roe FJC (eds) Pathology of laboratory rats and mice, Blackwell Scientific, Oxford, pp 391–467
47. Russfield AB (1982) Neoplasms of the endocrine system. In: Foster HL, Small JD, Fox JG (eds) American college of laboratory animal medicine series: The mouse in biomedical research, vol 4. Experimental biology and oncology. Academic Press, New York, pp 465–476
48. Sass B, Vernon ML, Peters RL, Kelloff GJ (1978) Mammary tumors, hepatocellular carcinomas, and pancreatic islet changes in C3H-Avy mice. J Natl Cancer Inst 60: 611–621
49. Schachner M, Huang SK, Ziegelmüller P, Bizzini B, Taugner R (1984) Glial cells in the pineal gland of mice and rats. Cell Tissue Res 237: 245–252
50. Sher SP (1982) Tumors in control hamsters, rats, and mice: literature tabulation. Crit Rev Toxicol 10: 49–79
51. Squire RA, Goodman DG, Valerio MG, Fredrickson T, Strandberg JD, Levitt MH, Lingeman CH, Harshberger JC, Dawe CJ (1978) Tumors, endocrine system. In: Benirschke K, Garner FM, Jones TC (eds) Pathology of laboratory animals, vol II. Springer, Berlin Heidelberg New York Tokyo, pp 1225–1228

52. Strandberg JD (1983) Adenocarcinoma, adrenal cortex, rat. In: Jones TC, Mohr U, Hunt RD (eds) Monographs on pathology of laboratory animals. Endocrine system. Springer, Berlin Heidelberg New York Tokyo, pp 46–49
53. Tamura S, Fujita H (1981) Fine structural aspects on the cold follicles in the aged mouse thyroid. Arch Histol Jpn 44: 177–188
54. Tekeli S, Morton D, Cusick PK (1997) Pituicytoma in a mouse. Toxicol Pathol 25: 516–517
55. Theuring F, Goetz W, Balling R, Korf HW, Schulze F, Herken R, Gruss P (1990) Tumorigenesis and eye abnormalities in transgenic mice expressing MSV-SV40 large T-antigen. Oncogene 5: 225–232
56. Walker VE, Morgan KT, Zimmerman HM, Innes JRM (1994) Tumours of the central and peripheral nervous system. In: Turusov VS, Mohr U (eds) Pathology of tumours in laboratory animals, vol 2. Tumours of the mouse, 2nd edn. IARC Scientific Publications No. 111, Lyon, pp 731–776
57. Ward JM, Goodman DG, Squire RA, Chu KC, Linhart MS (1979) Neoplastic and nonneoplastic lesions in aging (C57BL/6 N x C3H/HeN)F1 (B6C3F1) mice. J Natl Cancer Inst 63: 849–854
58. Wright JA, Wadsworth PF, Stewart MG (1990) Neuron-specific enolase immunoreactivity in rat and mouse phaeochromocytomas. J Comp Pathol 102: 475–478
59. Zwieten MJ Van, Frith CH, Nooteboom AL, Wolfe HJ, DeLellis RA (1983) Medullary thyroid carcinoma in female BALB/c mice. A report of 3 cases with ultrastructural, immunohistochemical, and transplantation data. Am J Pathol 110: 219–229

Central Nervous System

G. Krinke[1] · A. Fix[2] · W. Kaufmann[3] · L.J. Ackerman · R.H. Garman
C. George · B.S. Jortner · J.R. Leininger · K. Mitsumori · K.T. Morgan
I. Paulson · M. Robinson

[1]Chair; [2]Co-chair; [3]Initial draft.

Astrocytoma, Benign *(B)*

Synonym(s). Glioma, astrocytic, benign

Histogenesis

Astrocytes.

Diagnostic Features

- Lesion is confined to one major area of the central nervous system
- Borders are diffuse
- Lesion has a monomorphic appearance
- Cellularity is moderate to dense
- Perivascular cuffing may be present
- Necrosis and hemorrhages are not present
- Palisading of astrocytic cells around foci of necrosis is not present
- Cell borders are indistinct
- Cells may exhibit protoplasmic or fibrillary differentiation and prominent nuclei which are round or oval
- Cellular and nuclear atypia and pleomorphism are not features
- Neuronal cell bodies may persist within the neoplastic area
- Gemistocytes (reactive astrocytes) may be present

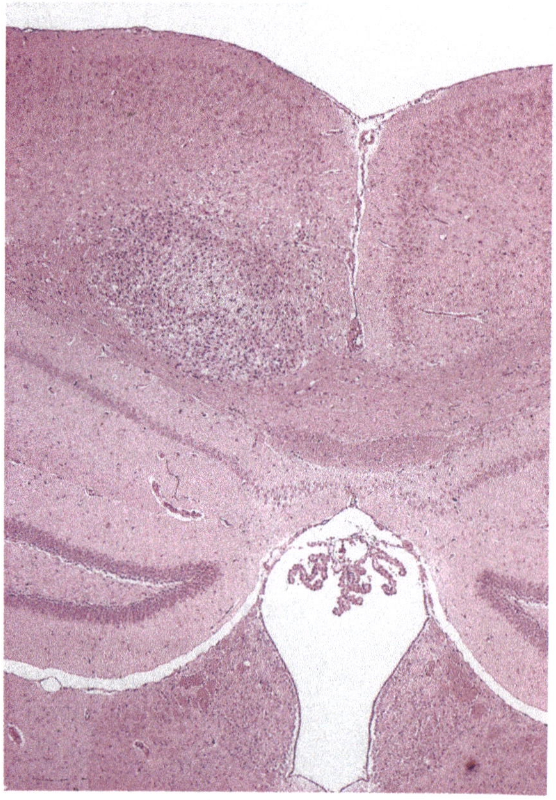

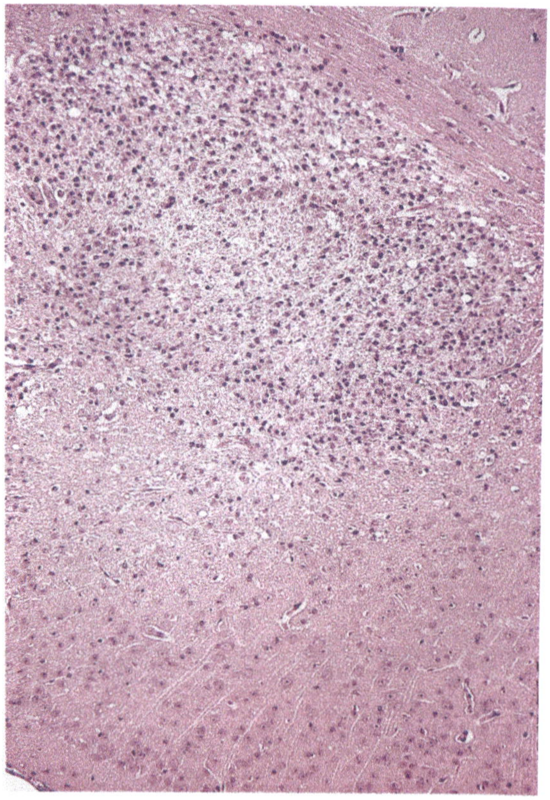

Fig. 1 *(above).* Benign astrocytoma in the retrosplenial cortex. C57BL mouse. H&E

Fig. 2 *(below).* Benign astrocytoma consists of a monomorphic population of astrocytic tumor cells, is confined to one major brain area and has indistinct borders. C57BL mouse. H&E

Differential Diagnosis

Astrocytoma, Malignant

Extends over multiple regions of the central nervous system and may be multicentric and

Invasive growth into perivascular spaces and meninges is characteristic and

Lesion has a polymorphic appearance due to a high degree of:

Cellular atypia and pleomorphism and

Necrosis and hemorrhage

Glioma, Mixed, Benign

Mixture of astrocytic and oligodendrocytic cells with at least 20% of either type

More then 80% of astrocytic or oligodendroglial neoplastic cells should be present for astrocytoma or oligodendroglioma, respectively

Reactive Gliosis

Differentiation of diffuse astrocytic glioma from a reactive (secondary) gliosis can be difficult, especially if the lesions are small in size, but identification of the cause of the reactive gliosis is helpful, as well as exclusive presence of reactive astrocytes

Comment

Astrocytomas in humans commonly give positive reaction for glial fibrillary acid protein (GFAP), but neoplastic astrocytes of mouse brain show mostly lack of reactivity, despite their obvious fibrillary differentiation.

Astrocytomas have been described in the fore-, mid- and hindbrain and the spinal cord. They are the predominant neoplasm of glial origin described in the VM and BRVR mouse strain.

The WHO international classification of human CNS tumors uses four grades for malignancy, grade 1 and 2 (low grade) for more "benign" tumors, and grade 3 and 4 (high grade) for malignant tumors. The term "benign" used for astrocytomas in this classification is used in the sense of a low grade malignancy tumor at the time point of diagnosis and it does not predict the biological behavior of an astrocytoma in mice.

References

(See 3, 7, 8, 9, 10, 13, 14, 15, 16, 17, 23, 25, 27, 31, 32, 33, 34, 35, 38, 39)

Astrocytoma, Malignant *(M)*

Synonym(s). Glioma, astrocytic, malignant

Histogenesis

Astrocytes.

Diagnostic Features

- Lesion extends over multiple areas of the central nervous system and may be multicentric
- Borders are diffuse
- Lesion has a polymorphic appearance
- Cellularity is high
- Perivascular invasion is characteristic
- Considerable necrosis and hemorrhage may occur
- Palisading of astrocytic cells around foci of necrosis may be present
- Cell borders are indistinct
- Cells may exhibit protoplasmic or fibrillary differentiation and prominent nuclei which are round or oval
- Cellular and nuclear atypia and pleomorphism is marked
- Neuronal cell bodies may persist within the neoplastic area
- Gemistocytes (reactive astrocytes) may be present
- Diffuse infiltrative growth is characteristic
- Invasion into meninges, ependyma, and ventricles may occur

Differential Diagnosis

Astrocytoma, Benign
Confined to a single region of the central nervous system and
Lesion has a monomorphic appearance due to a low degree of:
Cellular atypia and pleomorphism and
Necrosis and hemorrhage are absent

Glioma, Mixed, Malignant
Mixture of astrocytic and oligodendrocytic cells with at least 20% of either type
More than 80% of astrocytic or oligodendroglial neoplastic cells should be present for astrocytoma or oligodendroglioma, respectively

Comment

Astrocytomas have been described in the fore-, mid- and hindbrain and spinal cord. They are the predominant neoplasm of glial origin described in the VM and BRVR mouse strain.

References

(See 3, 7, 8, 9, 10, 13, 16, 17, 23, 25, 27, 31, 32, 33, 34, 35, 38, 39)

Fig. 3 *(upper left).* Malignant astrocytoma. H&E ▶

Fig. 4 *(upper right).* Malignant astrocytoma. H&E

Fig. 5 *(lower left).* Malignant astrocytoma. H&E

Fig. 6 *(lower right).* Malignant astrocytoma. H&E

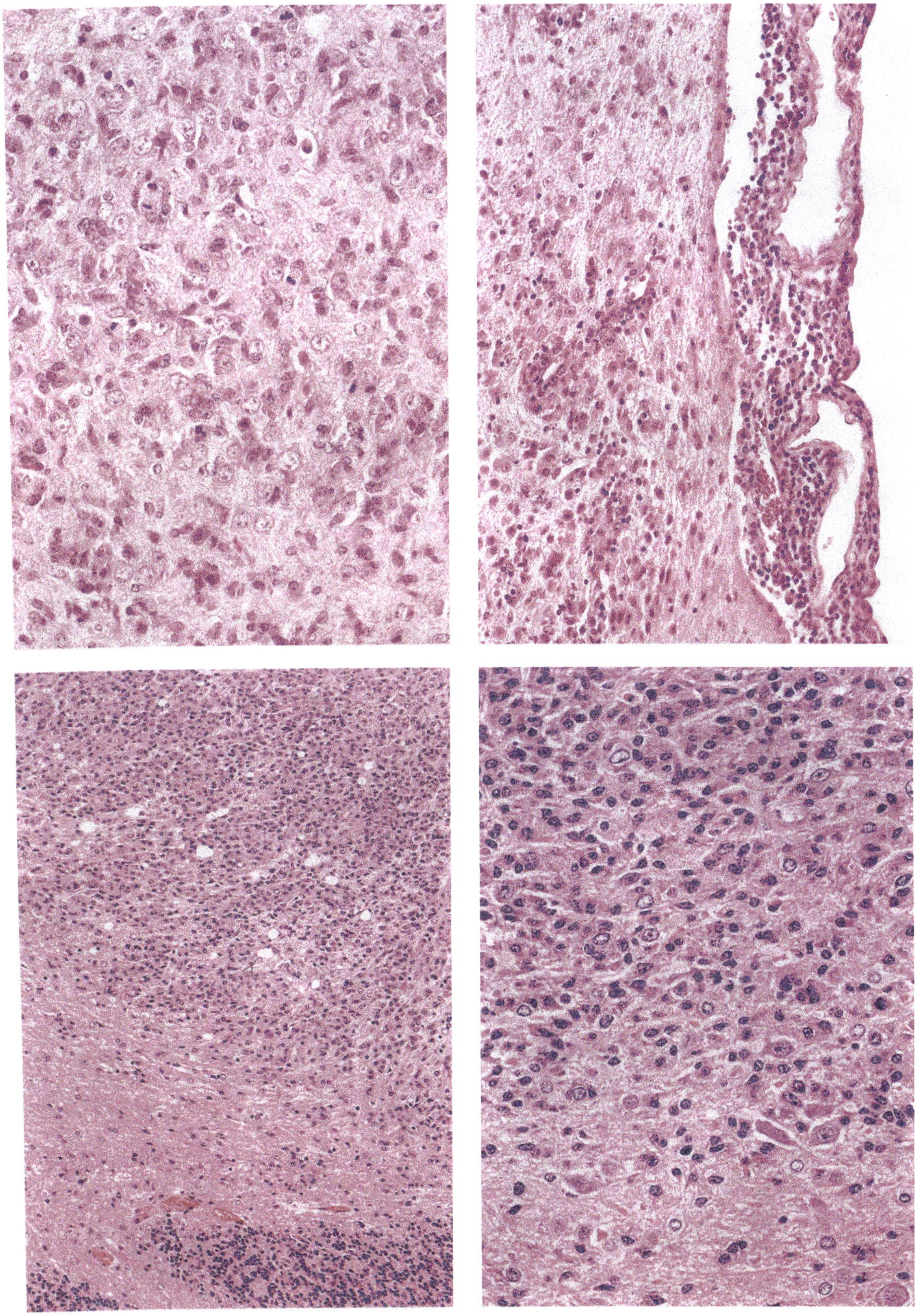

Oligodendroglioma, Benign *(B)*

Synonym(s). Glioma, oligodendrocytic, benign

Histogenesis

Oligodendrocytes.

Diagnostic Features

- Circumscribed lesion is confined to one major area of central nervous system
- Lesion has a monomorphic appearance
- Vascularized connective tissue septa often subdivide sheets into lobules or compartments
- Vascular endothelial hyperplasia is characteristic
- Necrosis with cystic changes and hemorrhages may occur
- Oligodendrocytic cells are monomorphic
- Cell borders are distinct
- Cytoplasm is scant or clear, especially in well-differentiated lesions, and cell membranes are distinct. This results in the appearance of a "perinuclear halo" and hence a honeycomb pattern

Differential Diagnosis

Oligodendroglioma, Malignant

Extends over multiple regions of the central nervous system and
Lesion has a polymorphic appearance due to:
Cellular atypia and pleomorphism and
Foci of necrosis and hemorrhage

Ependymoma, Benign

No evidence of significant numbers of oligodendroglial cells.
Cells are polygonal and arranged in rows and rosettes and
Confined to ventricles and periventricular regions of the brain

Glioma, Mixed, Benign

More then 80% of astrocytic or oligodendroglial neoplastic cells should be present for astrocytoma or oligodendroglioma, respectively. Lesions with 20% or more of both components are considered as mixed gliomas

Comment

Well-differentiated oligodendrogliomas are readily recognized by their typical honeycomb pattern formed by the well-defined cellular membranes and the clear cytoplasm. This clear appearance is usually regarded as post-mortem swelling due to autolysis or the fixation procedure.
Oligodendrogliomas were reported to occur in the forebrain/diencephalon area. They are the predominant chemically induced neoplasms of glial origin in many mouse strains, but they are rarely seen to occur naturally. They have been described to occur spontaneously in the BALB/c strain.
The WHO international classification of human CNS tumors uses grade 2 (low grade) for more "benign" oligodendrogliomas, and grade 3 (high grade) for malignant tumors. The term "benign" used for oligodendrogliomas in this classification is used in the sense of a low grade malignancy tumor at the time point of diagnosis and it does not predict the biological behavior of an oligodendroglioma in mice.

References

(See 3, 5, 6, 7, 10, 11, 13, 14, 16, 17, 21, 23, 25, 27, 28, 31, 33, 35, 36, 37, 38, 39)

Oligodendroglioma, Malignant *(M)*

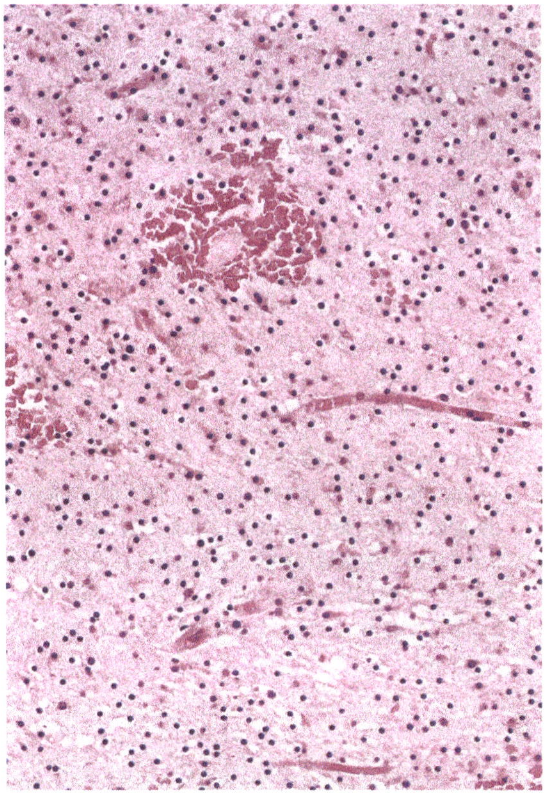

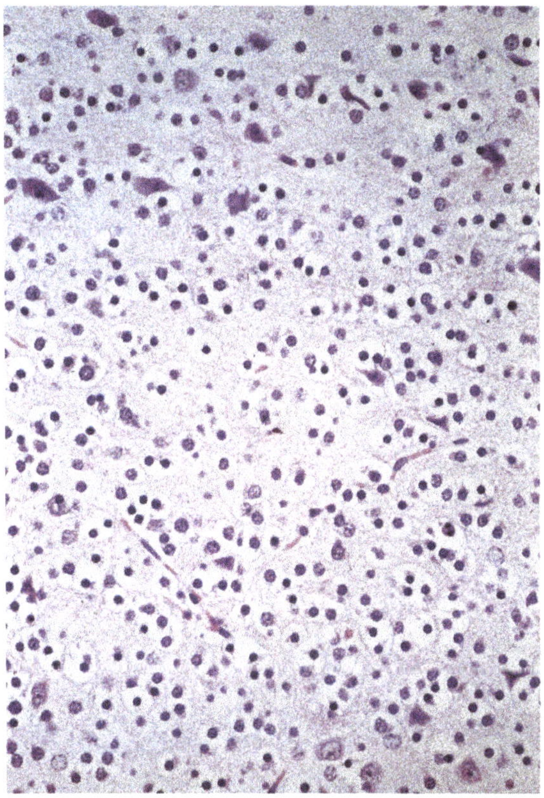

Synonym(s). Glioma, oligodendrocytic, malignant

Histogenesis

Oligodendrocytes.

Diagnostic Features

- Lesion extends over multiple areas of the central nervous system
- Lesion has a polymorphic appearance
- Oligodendrocytic cells are arranged in sheets, but other pattern, i.e., growth in rows or groups may be present
- Atypical vascular endothelial hyperplasia, especially at the tumor periphery, is characteristic
- Necrosis with cystic changes and hemorrhages is frequent
- Cellular atypia and pleomorphism are features
- Cell borders are distinct
- Cytoplasm is clear
- Nuclei are round, centrally located and hyperchromatic

Fig. 7 *(above).* Malignant oligodendroglioma of the cerebrum. H&E

Fig. 8 *(below).* Malignant oligodendroglioma of the cerebrum consists generally of uniform tumor cells with clear cytoplasm, distinct cell borders and round, small, central nuclei. H&E

Differential Diagnosis

Oligodendroglioma, Benign
Confined to one region of the central nervous system and
Lesion has a monomorphic appearance with no cellular atypia or pleomorphism

Ependymoma, Malignant
No evidence of significant numbers of oligodendroglial cells, cells are polygonal and cell borders are indistinct and
Confined to ventricles and periventricular regions of the brain

Glioma Mixed, Malignant
More then 80% of astrocytic or oligodendroglial neoplastic cells should be present for astrocytoma or oligodendroglioma, respectively. Lesions with 20% or more of both components are considered as mixed gliomas

Comment

In contrast to oligodendrogliomas described in rats, oligodendrogliomas in mice have been described as microscopically rather poorly demarcated lesions. Nevertheless they appear less diffuse than astrocytomas. Additionally, an atypical capillary endothelial hyperplasia ("garlands"), which is a characteristic feature in oligodendrogliomas in rats, appears to be less prominent in mice. Oligodendrogliomas were reported to occur in the cerebrum/diencephalon area. They are the predominant chemically induced neoplasms of glial origin in many mouse strains, but they are rarely seen to occur naturally. They have been described to occur spontaneously in the BALB/c strain.

References

(See 3, 5, 6, 7, 10, 13, 14, 16, 21, 23, 25, 27, 28, 31, 33, 35, 38, 39)

Ependymoma, Benign *(B)*

Histogenesis

Ependymal cells lining the ventricles and aqueduct of the brain and central canal of the spinal cord.

Diagnostic Features

- Located near the ventricles or aqueduct of the brain or central canal of the spinal cord
- Cells are polygonal
- Cell borders are indistinct
- Cells are arranged in rows or rosettes
- Pseudorosettes (cells arranged around blood vessels) may be formed
- Cellular atypia and pleomorphism are not observed
- Invasive growth is not observed

Differential Diagnosis

Ependymoma, Malignant
Invasive growth is present
Cellular atypia and pleomorphism are present

Oligodendroglioma, Benign
Sheets of cells with condensed round nuclei and clear cytoplasm are present
Cell borders are distinct
Can occur in extraventricular locations

Papilloma, Choroid Plexus
There is a papillary fibrovascular stroma lined a by single layer of cuboidal to columnar, non-ciliated epithelial cells
Growth in rows or rosettes is not a feature

Medulloblastoma
Location in cerebellum: no association with brain ventricles or central spinal canal
Cell type is different: cells resemble the neurones of the granular layer of cerebellar cortex

Comment

Ependymomas described in mice are very rare and mostly experimentally induced. Human ependymomas are expected to form luminal rosettes and be composed of ciliated cells; this has not been observed in mice.

References

(See 3, 12, 13, 14, 16, 27, 30, 31, 32, 35, 39)

Ependymoma, Malignant *(M)*

Histogenesis

Ependymal cells lining the ventricles and aqueduct of the brain and central canal of the spinal cord.

Diagnostic Features

- Located near the ventricles or aqueduct of the brain or central canal of the spinal cord
- Cells are polygonal
- Cell borders are indistinct
- Cells are arranged in rows or rosettes
- Pseudorosettes (cells arranged around blood vessels) may be formed
- Cellular atypia and pleomorphism occur
- Foci of necrosis can occur
- Multinucleated giant cells can be present
- Invasive growth is a feature

Differential Diagnosis

Ependymoma, Benign
No evidence of invasive growth
Cellular atypia and pleomorphism are not present

Oligodendroglioma, Malignant
Sheets of cells with condensed round nuclei and clear cytoplasm are present
Cell borders are distinct
Can occur in extraventricular locations

Comment

Ependymomas described in mice are very rare and mostly experimentally induced.

References

(See 3, 12, 13, 16, 20, 21, 22, 30, 31, 32, 35, 39)

Glioma, Mixed, Benign *(B)*

Synonym(s). Oligoastroglioma, benign

Histogenesis

Astrocytes and oligodendrocytes.

Diagnostic Features

- Lesion is confined to one major area of the central nervous system
- Oligodendrocytic and astrocytic cells are either diffusely intermingled or confined to areas composed predominantly of a single cell type
- Either glial cell type provides at least 20% of the neoplasm
- Lesion has a (dual) monomorphic appearance
- There is no cellular atypia and pleomorphism
- Foci of necrosis are not encountered

Differential Diagnosis

Glioma, Mixed, Malignant
Extends over multiple areas of the central nervous system and may be multi-centric and
Extensive invasive growth is present and
Lesion has a polymorphic appearance due to:
Cellular atypia, pleomorphism, and anaplasia and
Necrosis and hemorrhage

Comment

The description is based on the criteria established for the rat. In the literature, there are very few oligoastrogliomas mentioned in mice – one spontaneously occurring, two ENU-induced.
There is no scientific ground for setting the criterion for mixed gliomas at 20% for either cell population. Some pathologists are using this rule, and it has been quoted in the literature. It appears to be of practical value for diagnostic purposes and it implies that lesions diagnosed as astrocytic or oligodendroglial gliomas should consist of more then 80% of such neoplastic cells.
The WHO international classification of human CNS tumors uses four grades for malignancy, grade 1 and 2 (low grade) for more "benign" tumors, and grade 3 and 4 (high grade) for malignant tumors. The term "benign" used for mixed gliomas in this classification is used in the sense of a low grade malignancy tumor at the time point of diagnosis and it does not predict the biological behavior of a mixed glioma in mice.

References

(See 3, 14, 16, 21, 25, 27, 29, 31, 33, 35, 38)

Glioma, Mixed, Malignant *(M)*

Synonym(s). Oligoastroglioma, malignant; glioma, anaplastic; glioblastoma; glioblastoma multiforme

Histogenesis

Astrocytes and oligodendrocytes.

Diagnostic Features

- Lesion extents over multiple areas of the central nervous system
- Oligodendrocytic and astrocytic cells may be either diffusely intermingled or confined to areas composed predominantly of a single cell type
- Either glial cell type provides at least 20% of the neoplasm
- Lesion has a polymorphic appearance
- Cellular atypia and pleomorphism is marked
- Multinucleated giant cells may be numerous
- Astrocytic or oligodendrocytic differentiation may not be obvious throughout the tumor
- Vascular proliferation is marked, blood vessels may be occluded by thrombi, and hemorrhage may be present
- Foci of necrosis may occur
- Palisading of spindle-shaped cells around areas of necrosis may occur
- Bizarre mitotic figures may be present
- Growth is diffusely infiltrative

Differential Diagnosis

Glioma, Mixed, Benign
Confined to one region of the central nervous system and
Invasive growth is not present and
Lesion has a (dual) monomorphic appearance with little cellular atypia, pleomorphism, or anaplasia and
Necrosis and hemorrhage are absent

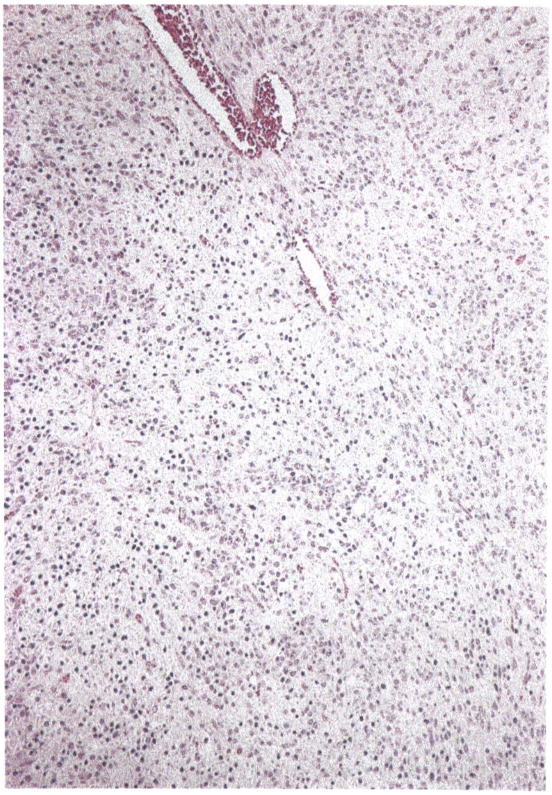

Fig. 9. Malignant mixed glioma shows varying proportions of astrocytic and oligodendrocytic tumor cells. B6C3F1 mouse. H&E

Comment

The "glioblastoma multiforme" is the most frequently reported induced neoplasm of the glioma group in mice. The histogenesis of such lesion, however, is disputed. According to some authors it represents a secondary progression of highly malignant astrocytoma.

References

(See 3, 4, 13, 14, 16, 18, 23, 25, 26, 31, 33, 35, 39)

Hamartoma, Lipomatous *(H)*

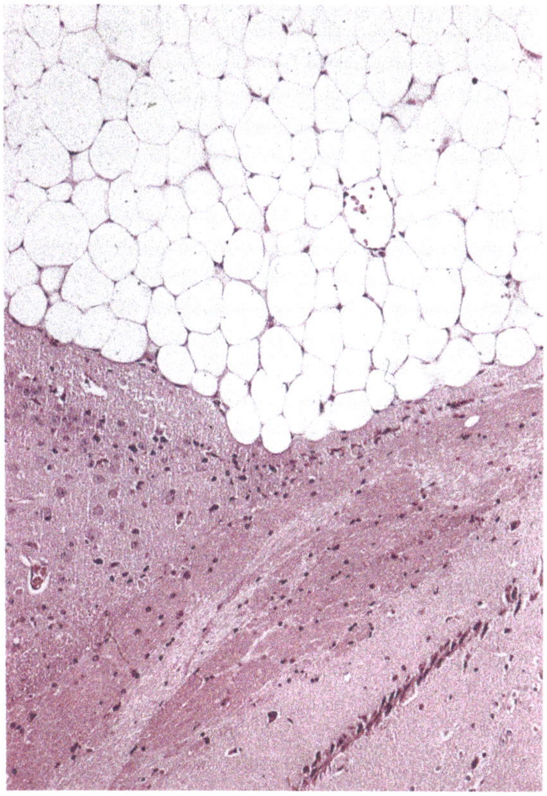

Fig. 10. Lipomatous hamartoma of the brain consists of mature adipocytes. C57BL mouse. H&E

Synonym(s). Lipoma, intracranial

Histogenesis

Adipocytes.

Diagnostic Features

- Predominantly located in the midline or ventricles of the brain
- Well-demarcated
- Contains single or multiple clusters of adipose cells
- May be associated with dysgenesis of corpus callosum and lateral displacement of adjacent blood vessels and brain tissue
- Cells are mature adipocytes containing one large fat droplet

Differential Diagnosis

Teratoma, Benign or Teratoma, Malignant
Not only adipose tissue, but tissues from all the three germ cell layers are present

Comment

These lesions result from a faulty development of the meninges or the choroid plexus. Because these lesions neither grow rapidly nor progress to malignant tumors and because they are related to abnormal development, the term lipomatous hamartoma rather than lipoma seems appropriate.

References

(See 1, 2, 16, 17, 24, 27, 31, 35)

Papilloma, Choroid Plexus *(B)*

Histogenesis

Epithelial cells of choroid plexus.

Diagnostic Features

- Located close to anatomic sites of the choroid plexus
- A fibrovascular stromal core is lined by a single layer of epithelium
- Organization of the epithelium to papillary projections in an arboriform pattern is well developed
- Piling up of epithelial cells is not present
- Cells are cuboidal to columnar and nonciliated
- Cytoplasm is abundant and eosinophilic
- Nuclei are round to oval

Differential Diagnosis

Carcinoma, Choroid Plexus
Invasive growth into the adjacent brain is present and
Cellular atypia and pleomorphism are present and
Piling up of epithelial cells is present

Ependymoma, Benign
Ependymomas form rosettes or grow in rows
A papillary fibrovascular stroma lined by a single layer of cuboidal to columnar, nonciliated epithelial cells is not a feature of ependymomas

Comment

Spontaneous choroid plexus tumors in mice are extremely rare. Such tumors are inducible in mice when a carcinogen is placed within the ventricles.

References

(See 3, 10, 16, 25, 27, 31, 33, 35, 38)

Carcinoma, Choroid Plexus *(M)*

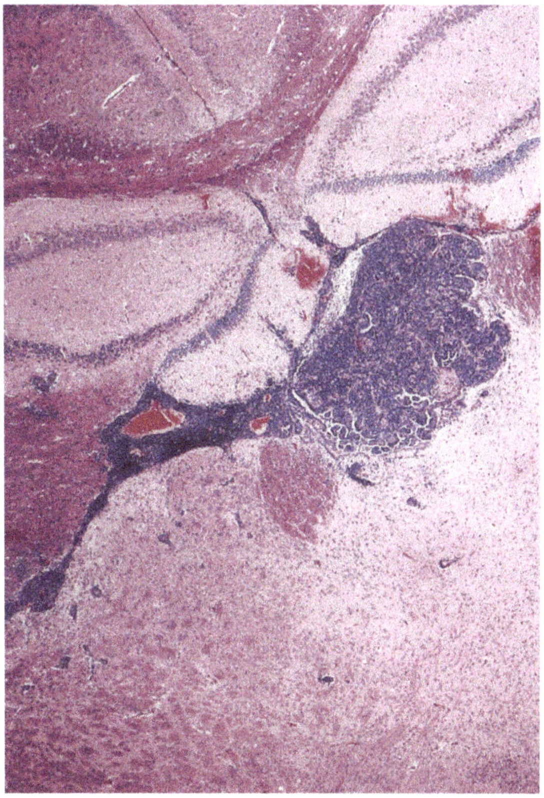

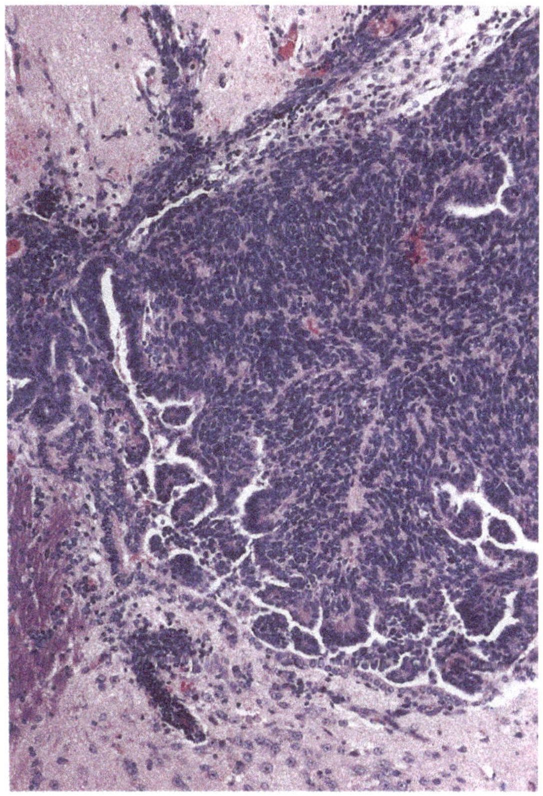

Histogenesis

Epithelial cells of choroid plexus.

Diagnostic Features

- Located close to anatomic sites of the choroid plexus
- A fibrovascular stromal core is lined by multiple rows (piling up) of epithelial cells
- Arrangement of the epithelium to papillary projections in an arboriform pattern is generally poor, although some well-developed areas may be present
- Piling up of epithelial cells is a feature
- Cellular atypia and pleomorphism occur
- Invasion into the adjacent brain is characteristic

Differential Diagnosis

Papilloma, Choroid Plexus
Invasive growth into the adjacent brain is absent and
There is no cellular atypia and pleomorphism and
Piling up of epithelial cells is not observed

Ependymoma, Malignant
A papillary fibrovascular stroma lined by a non-ciliated epithelial cells is not a feature of ependymomas

Fig. 11 *(above).* Choroid plexus carcinoma located in the foramen Monroe between the third and the lateral ventricle. H&E

Fig. 12 *(below).* Choroid plexus carcinoma with marginal papillary projections and central solid sheets of anaplastic cells. H&E

Comment

Spontaneous choroid plexus tumors in mice are extremely rare. Such tumors are inducible in mice when a carcinogen is placed within the ventricles.

References

(See 3, 10, 14, 16, 25, 31, 35, 39)

Medulloblastoma *(M)*

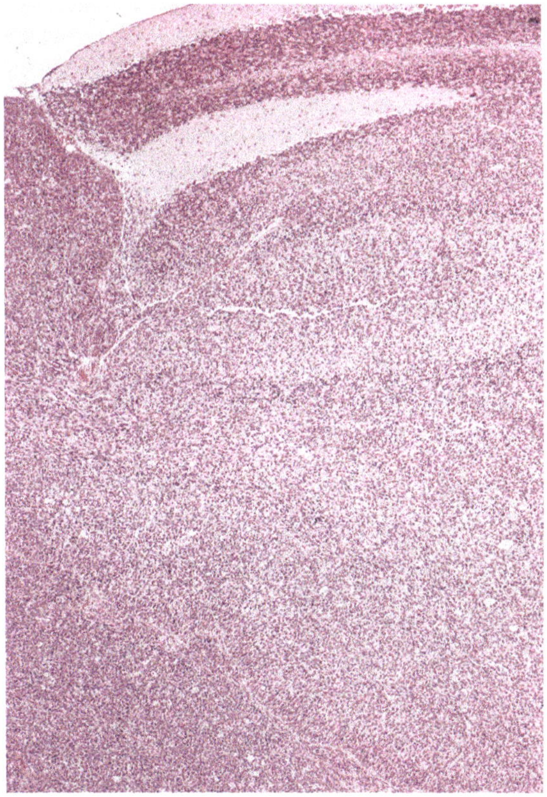

Fig. 13. Medulloblastoma showing transition of cerebellar cortical granular layer to tumor. B6C3F1 mouse. H&E

Histogenesis

Primitive neuroectodermal cells of cerebellum.

Diagnostic Features

- Cellularity is high
- Cells tend to form pseudorosettes around blood vessels
- Cells are quite uniform in appearance
- Cells are small, round to elongated, frequently resembling a "carrot-like" shape
- Nuclei are round and intensely stained similar to those of the granular layer of the cerebellar cortex
- Differentiation to neuron-like cells may occur
- Bizarre mitotic figures are frequent
- Growth pattern is invasive and the tumor replaces cerebellar folia

Differential Diagnosis

Ependymoma, Malignant

Located at a different site near the ventricles or aqueduct of the brain or central canal of the spinal cord
Cells are polygonal
Cells are arranged in rows and circles

Pinealoma, Malignant

Located at a different site in the midline of the dorsal brain surface

Comment

Medulloblastoma is an inducible lesion in mice, either by direct pellet implantation of carcinogenic compounds in the vermis and lateral lobes of the cerebellar cortex or via transplacental and neonatal induction with ethylnitrosurea. Spontaneous medulloblastomas are extremely rare in mice. Medulloblastoma is by definition located in the cerebellum. Similar tumors located elsewhere in the CNS are generally denominated "PNET" (primitive neuroectodermal tumors).

References

(See 3, 4, 6, 13, 14, 16, 17, 26, 27, 29, 31, 32, 33, 35, 38, 39)

Meningioma, Benign *(B)*

Fibrous type

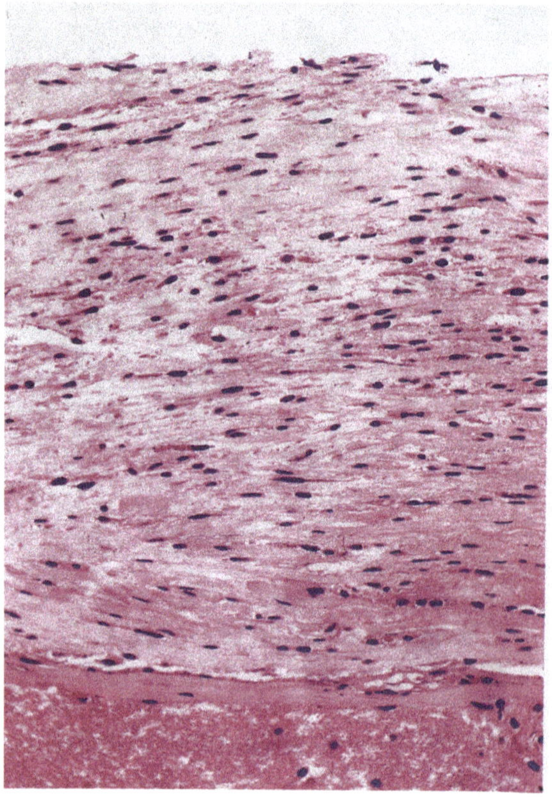

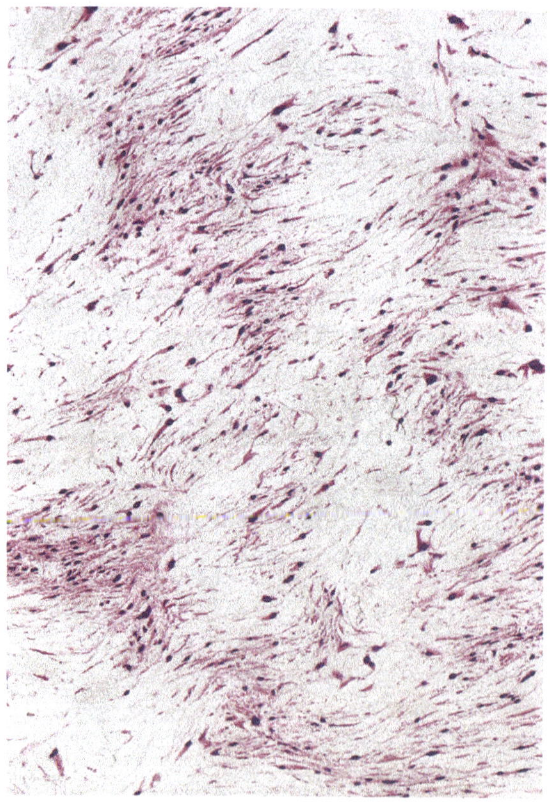

Histogenesis

Cellular elements of the meninges.

Diagnostic Features

- Located at the surface of brain, optic nerve, or spinal cord as meningeal thickening, or in the ventricular cavities
- Well-demarcated
- Mitotic figures are rare
- Invasion of the surrounding tissues is not a feature but expansive growth compressing the adjacent brain may occur

Fibrous type

- Exhibits a regular pattern of loosely interwoven bundles
- Irregular palisading may be present in some areas
- Some neoplasms may be distinctly myxomatous in appearance
- Cells are delicate and spindle-shaped
- Nuclei are small and hyperchromatic

Differential Diagnosis

Meningioma, Malignant

Exhibits invasive growth and
There are cellular atypia and pleomorphism and
Mitotic figures are frequent and
Multinucleated cells may be present

Fig. 14 *(above)*. A benign meningioma located at the surface of the brain as distinct meningeal thickening. B6C3F1 mouse. H&E

Fig. 15 *(below)*. The benign fibrous meningioma is composed of a regular pattern of loosely interwoven bundles of delicate spindle-shaped cells. B6C3F1 mouse. H&E

Comment

Although in man and other species there are numerous morphological variants of meningiomas, this is not known for mice, perhaps due to the relative paucity of described cases. In rats granular cells occur in some syncytial meningiomas. Neoplasms in which predominantly granular cells are observed are classified as "tumor, granular cell". Only benign granular cell tumors are described in rats. For this reason if similar cases occurred in mice, they should be classified similarly to rats.

References

(See 3, 7, 10, 14, 16, 25, 27, 31, 33, 35, 36, 37, 38)

Meningioma, Malignant *(M)*

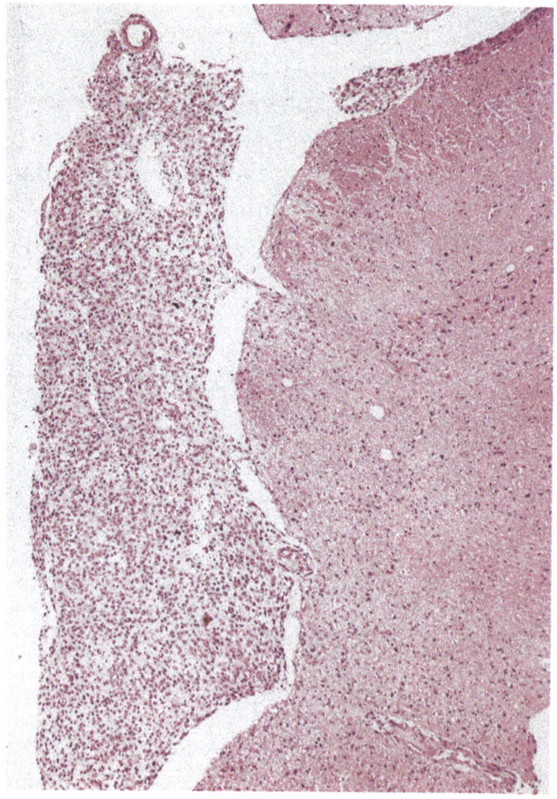

Synonym(s). Sarcoma, meningeal

Histogenesis

Cellular elements of the meninges.

Diagnostic Features

- Located at the surface of brain, optic nerve, or spinal cord as meningeal thickening with invasive growth, or in the ventricular cavities
- Demarcation is poor
- Cellularity is usually high
- Cellular atypia and pleomorphism are characteristic
- Multinucleated cells may be present
- Mitotic figures are numerous
- Invasion to both the neural parenchyma and the overlying dura, skull and scalp is characteristic and may be extensive
- There may be an extensive infiltration along the adventitia of small blood vessels

Differential Diagnosis

Meningioma, Benign
Expansive growth compressing the adjacent brain, but no invasion of the surrounding tissues and
Cellular morphology is uniform and
Mitotic figures are not frequent

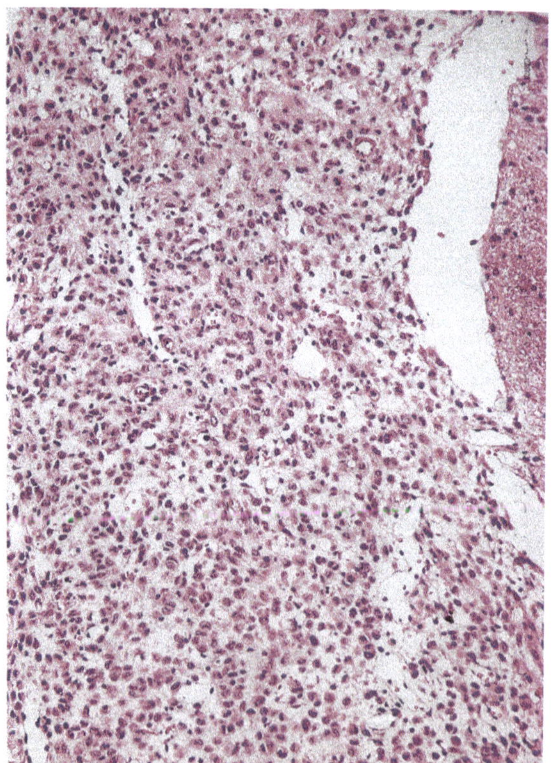

Fig. 16 *(above).* A malignant meningioma with high cellularity and invasive growth into the adjacent brain. H&E

Fig. 17 *(below).* Malignant meningioma. Small, round, dark, lymphocyte-like cells and a high mitotic activity may be present. H&E

Sarcoma, Histiocytic
Well-differentiated histiocytic and multinucleated giant cells with little fibroblastic differentiation are present
Pseudopalisading surrounding necrotic areas can occur

Comment

Malignant meningiomas are inducible by implantation of carcinogenic chemicals in the mouse brain (leptomeninges). Descriptions of spontaneous "malignant meningiomas" are very rare and based on the distinct invasiveness.

References

(See 3, 7, 10, 14, 16, 19, 25, 31, 33, 35, 38, 39)

References

1. Adkison DL, Sundberg JP (1991) "Lipomatous" hamartomas and choristomas in inbred laboratory mice. Vet Pathol 28: 305–312
2. Budka H (1974) Intracranial lipomatous hamartomas (intracranial "lipomas"). A study of 13 cases including combinations with medulloblastoma, colloid and epidermoid cysts, angiomatosis and other malformations. Acta Neuropathol (Berl) 28: 205–222
3. Cardesa A, Carlton WW, Dungworth DL, Enomoto M, Halm S, Koestner A, Krinke GJ, Render JA, Rittinghausen S, Ruben Z, Solleveld H, Turusov VS, Weisse I, Yoshitomi K (1994) 7. Central nervous system, eye, heart, and mesothelium. In: Mohr U, Capen CC, Dungworth DL, Griesemer RA, Ito N, Turusov VS (eds) International classification of rodent tumours, Part I: The rat. IARC Scientific Publications No. 122, Lyon
4. Cloudman AM (1941) Spontaneous neoplasms in mice. In: Snell GD (ed) Biology of the laboratory mouse. Blakiston, Philadelphia, pp 168–233
5. Cottier H, Luginbühl H (1961) Oligodendrogliom des Großhirns bei einer weißen Maus. Acta Neuropathol (Berl) 1: 198–200
6. Denlinger RH, Koestner A, Wechsler W (1974) Induction of neurogenic tumors in C3HeB-FeJ mice by nitrosourea derivatives: observations by light microscopy, tissue culture, and electron microscopy. Int J Cancer 13: 559–571
7. Faccini JM, Abbott DP, Paulus GJJ (1990) Mouse histopathology. XI. Nervous system and special sense organs. Elsevier, Amsterdam, pp 187–210
8. Fraser H (1971) Astrocytomas in an inbred mouse strain. J Pathol 103: 266–270
9. Fraser H (1986) Brain tumours in mice, with particular reference to astrocytoma. Food Chem Toxicol 24: 105–111
10. Frith CH, Ward JM (1988) Color atlas of neoplastic and non-neoplastic lesions in aging mice. Central nervous system. Elsevier, Amsterdam, pp 93–99
11. Haseman JK, Elwell MR, Hailey JR (1999) Neoplasm incidences in B6C3F1 mice: NTP historical data. In: Maronpot RR, Boorman GA, Gaul BW (eds) Pathology of the mouse. Reference and atlas. Cache River Press, Vienna, pp 679–689
12. Horn HA, Stewart HL (1953) A review of some spontaneous tumors in noninbred mice. J Natl Cancer Inst 13: 591–603
13. Jortner BS, Percy DH (1978) The nervous system. In: Benirschke K, Garner FM, Jones TC (eds) Pathology of laboratory animals, vol 1. Springer, Berlin Heidelberg New York Tokyo, pp 320–421
14. Kleihues P, Cavenee WK (1997) Pathology and genetics of tumours of the nervous system. IARC, Lyon
15. Krinke GJ (1997) Critical remarks on the international WHO classification of rodent central nervous system (CNS) tumours. Physiol Res 46: 89–91
16. Krinke GJ, Kaufmann W (1996) Neoplasms of the central nervous system. In: Mohr U, Dungworth DL, Capen CC, Carlton WW, Sundberg JP, Ward JM (eds) Pathobiology of the aging mouse, vol 2. ILSI Press, Washington, DC, pp 69–81
17. Krinke GJ, Kaufmann W, Mahrous AT, Schaetti P (2000) Morphologic characterization of spontaneous nervous system tumors in mice and rats. Toxicol Pathol 28: 178–192
18. Luginbühl H (1964) A comparative study of neoplasms of the central nervous system in animals. Acta Neurochir (Wien) (Suppl) 10: 30–42
19. Maita K, Hirano M, Harada T, Mitsumori K, Yoshida A, Takahashi K, Nakashima N, Kitazawa T, Enomoto A, Inui K, Shirasu Y (1988) Mortality, major cause of moribundity, and spontaneous tumors in CD-1 mice. Toxicol Pathol 16: 340–349
20. Mandybur TI, Alvira MM (1982) Ultrastructural findings in so-called ependymal rat tumors induced by transplacental administration of ethylnitrosourea (ENU). Acta Neuropathol (Berl) 57: 51–58
21. Mennel HD (1988) Geschwülste des zentralen und peripheren Nervensystems. III. Ätiologische Konzepte, Neuroonkologie und allgemeine Kanzerologie. 4. Experimentelle Hirntumoren. In: Doerr W, Seifert G, Uehlinger E (eds) Spezielle pathologische Anatomie. Mennel HD, Solcher H (eds) Pathologie des Nervensystems III, Entzündliche Erkrankungen und Geschwülste. Springer, Berlin Heidelberg New York London Paris Tokyo, pp 258–272
22. Mennel HD, Simon H (1985) Morphology of early stages of ENU-induced brain tumors in rats. Exp Pathol 28: 207–214
23. Morgan KT, Alison RH (1988) Gliomas, mouse. In: Jones TC, Mohr U, Hunt RD (eds) Monographs on pathology of laboratory animals. Nervous system. Springer, Berlin Heidelberg New York Tokyo, pp 123–130
24. Morgan KT, Sheldon WG (1988) Lipoma, brain, mouse. In: Jones TC, Mohr U, Hunt RD (eds) Monographs on pathology of laboratory animals. Nervous system. Springer, Berlin Heidelberg New York Tokyo, pp 130–134
25. Morgan KT, Frith CH, Swenberg JA, McGrath JT, Zulch KJ, Crowder DM (1984) A morphologic classification of brain tumors found in several strains of mice. J Natl Cancer Inst 72: 151–160

26. Peters RL, Rabstein LS, Spahn GJ, Madison RM, Huebner RJ (1972) Incidence of spontaneous neoplasms in breeding and retired breeder BALB-cCr mice throughout the natural life span. Int J Cancer 10: 273–282
27. Radovsky A, Mahler JF (1999) Nervous system. In: Maronpot RR, Boorman GA, Gaul BW (eds) Pathology of the mouse. Reference and atlas. Cache River Press, Vienna, pp 445–470
28. Rubinstein LJ (1972) Tumors of the central nervous system. In: Atlas of tumor pathology. Armed Forces Institute of Pathology, Washington DC, pp 86–87
29. Searle CE, Jones EL (1972) Tumours of the nervous system in mice treated neonatally with N-ethyl-N-nitrosourea. Nature 240: 559–560
30. Slye M, Holmes HF, Wells HG (1931) Intracranial neoplasms in lower animals. Studies in the incidence and inheritability of spontaneous tumors in mice. Am J Cancer 15: 1387–1400
31. Solleveld HA, Bigner DD, Averill DR Jr, Bigner SH, Boorman GA, Burger PC, Gillepsie Y, Hubbard GB, Laerum OD, McComb RD, McGrath JT, Morgan KT, Peters A, Rubinstein LJ, Schoenberg BS, Schold SC, Swenberg JA, Thompson MB, Vandevelde M, Vinores SA (1986) Brain tumors in man and animals: reports of a workshop. Environ Health Perspect 68: 155–173
32. Squire RA, Goodman DG, Valerio MG, Fredrickson T, Strandberg JD, Levitt MH, Lingeman CH, Harshberger JC, Dawe CJ (1978) Tumors. In: Benirschke K, Garner FM, Jones TC (eds) Pathology of laboratory animals, vol II. Springer, Berlin Heidelberg New York Tokyo, pp 1235–1242
33. Swenberg JA (1982) Neoplasms of the nervous system. In: Foster HL, Small JD, Fox JG (eds) The mouse in biomedical research, vol IV. Experimental biology and oncology. Academic Press, San Diego, pp 529–537
34. Vandenberg SR (1992) Current diagnostic concepts of astrocytic tumors. J Neuropathol Exp Neurol 51: 644–657
35. Walker VE, Morgan KT, Zimmerman HM, Innes JRM (1994) Tumours of the central and peripheral nervous system. In: Turusov VS, Mohr U (eds) Pathology of tumours in laboratory animals, vol 2. Tumours of the mouse, 2nd edn. IARC Scientific Publications No. 111, Lyon, pp 731–776
36. Ward JM, Rice JM (1982) Naturally occurring and chemically induced brain tumors of rats and mice in carcinogenesis bioassays. Ann N Y Acad Sci 381: 304–319
37. Ward JM, Goodman DG, Squire RA, Chu KC, Linhart MS (1979) Neoplastic and nonneoplastic lesions in aging (C57BL/6 N x C3H/HeN)F1 (B6C3F1) mice. J Natl Cancer Inst 63: 849–854
38. Wechsler W, Rice JM, Vesselinovitch SD (1979) Transplacental and neonatal induction of neurogenic tumors in mice: comparison with related species and with human pediatric neoplasms. Natl Cancer Inst Monogr 51: 219–226
39. Zimmerman HM, Innes JRM (1979) Tumours of the central and peripheral nervous system. In: Turusov VS (ed) Pathology of tumours in laboratory animals, vol II. Tumours of the mouse. IARC Scientific Publications No. 23, Lyon, pp 629–654

Eye and Harderian Gland

G. Krinke[1] · A. Fix[2] · M. Jacobs[3] · J. Render · I. Weisse (†)

[1]Chair; [2]Co-chair; [3]Initial draft.

Melanoma, Uveal, Malignant *(M)*

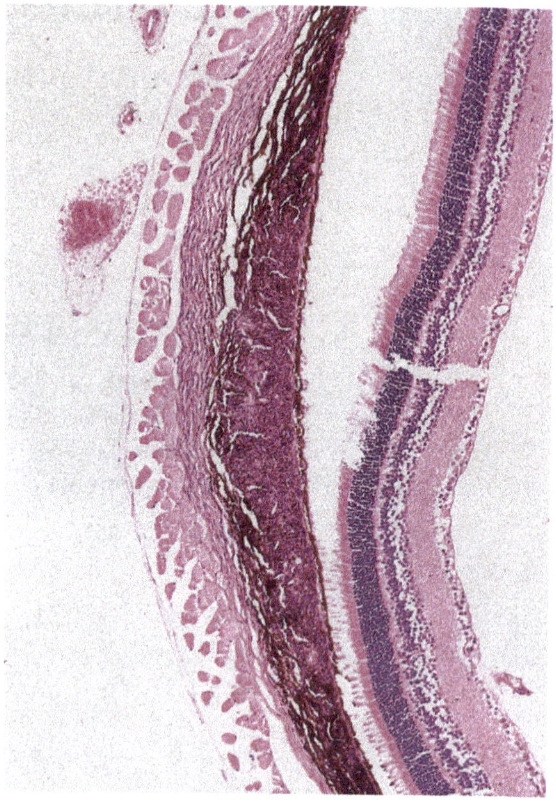

Synonym(s). Melanoma, intraocular, malignant; melanocytic tumor, malignant

Histogenesis

Neuroectodermal from uveal melanocytes.

Diagnostic Features

- Arising generally unilaterally in the choroid or iris
- Cells are spindle-shaped, epithelioid or mixed
- Cell borders are indistinct
- Melanin granules are present in the cytoplasm in varying numbers
- Spindle-shaped cells have slender elongated nuclei
- Epitheloid cells have large oval or pleomorphic nuclei
- Nucleoli may be distinct
- Mitotic figures may be present
- Invasion into adjacent tissue may occur

Differential Diagnosis

Uveal Leiomyoma or Schwannoma, Intraocular, Malignant or Melanoma, Benign (Nevoid Type)

These tumors have not yet been described in mice. Discrimination might be supported by electron microscopy or immunohistochemistry.

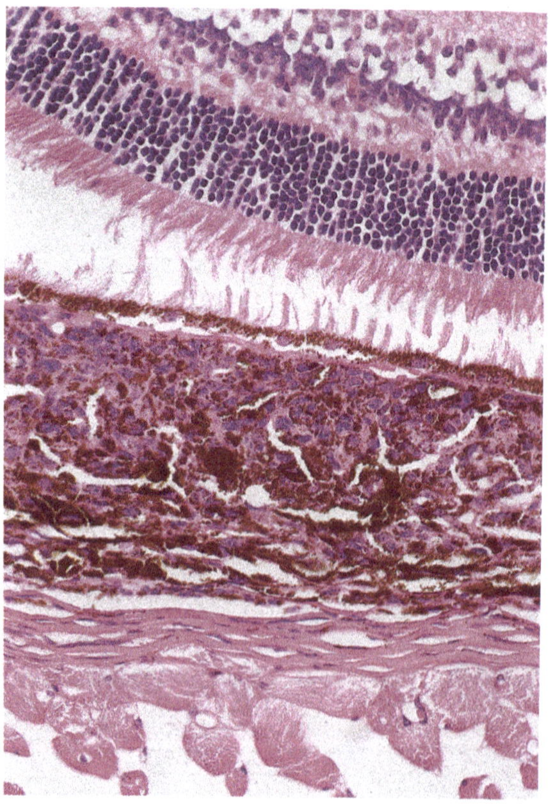

Fig. 1 *(above).* Malignant uveal melanoma of the choroid. B6C3F1 mouse, female. H&E

Fig. 2 *(below).* Malignant uveal melanoma of the choroid. B6C3F1 mouse, female. H&E

Comment

Spontaneous malignant intraocular melanomas are extremely rare in pigmented mice. As yet, malignant amelanotic melanomas have not been reported in albino mice. Benign melanomas have neither been reported in albino nor in pigmented mice.
In rats the diagnosis of uveal melanoma is aided by immunohistochemical demonstration of a positive reaction in the neoplastic cells for S-100 protein and vimentin intermediate filaments and a negative reaction for cytokeratin, desmin and glial fibrillary acid protein (GFAP). Ellipsoid premelanosomes (stage II melanosomes) could be demonstrated in electron micrographs of cells of the rat amelanotic uveal melanoma. Since the main differential diagnosis is malignant schwannoma, which can be positive for the S-100 protein as well, electron microscopy appears to be the best diagnostic method for differentiating these two neoplasms. Schwannomas, especially the "melanotic variant" may contain melanosomes: convoluted, moderately thin cytoplasmic processes lined by a continuous basal lamina are diagnostic for schwannomas.

References

(See 12, 17, 18, 26, 59)

Hyperplasia *(H)*

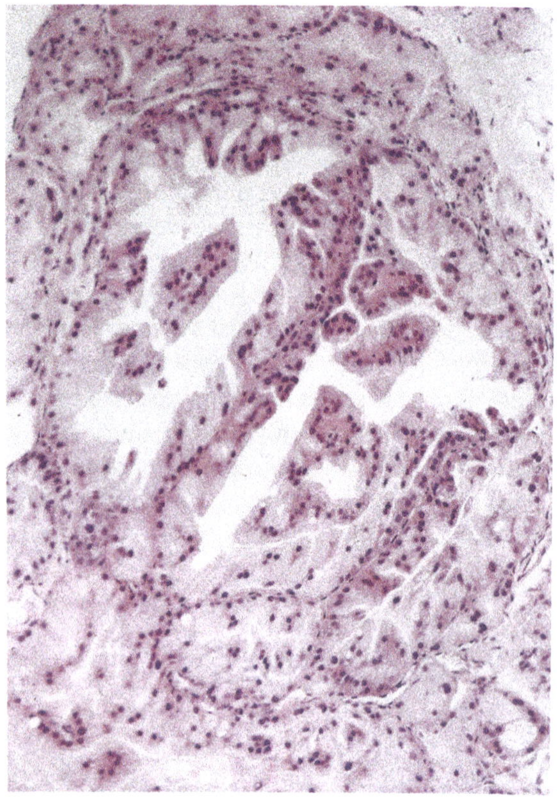

Fig. 3. Hyperplasia of the Harderian gland. H&E

Histogenesis

Acinar epithelium of the Harderian gland.

Diagnostic Features

- Unilateral or bilateral cellular proliferation confined to solitary or multiple areas
- There is no evidence of compression of adjacent parenchyma
- Architecture of the gland is maintained
- Alveolar pattern is retained, but there is an increased number of cells lining the acini
- Epithelium is one layer thick
- Epithelium may project into the acinar lumen, but evidence of piling-up of cells within the acinus is not present
- Cells closely resemble those of normal gland, they have no atypia
- Cellular hypertrophy of varying degree of severity can be associated with hyperplasia

Differential Diagnosis

Adenoma

Architecture of the gland is not maintained within the tumor or
Compression of the adjacent glandular tissue is present or
Cytoplasmic changes with increased nuclear/cytoplasmic ratio are present (acinar adenoma)

Comment

In mice, unilateral or bilateral focal hyperplasia of the Harderian gland can be observed occasionally.
In rats, hyperplasia of the Harderian gland may occur in association with adenitis, glandular atrophy and squamous metaplasia of ducts. Hyperplasia may be a regenerative response following damage to the gland caused by various factors, including viral sialodacryoadenitis and trauma of orbital bleeding. Primary hyperplasia of the Harderian gland may occur in rats and mice in the absence of adenitis.

References

(See 3, 4, 13, 27)

Adenoma (*B*)

Papillary • Cystic • Cystic-papillary • Acinar

Histogenesis

Acinar epithelium of the Harderian gland.

Diagnostic Features

- Generally confined to solitary areas
- Well-demarcated
- Compression of the surrounding gland is present
- Encapsulation may be present, especially in large tumors
- Architecture of the gland is not maintained within the tumor (except in acinar type)
- Epithelium is usually one layer thick
- The acini are lined by an increased number of cells
- Piling-up may be present and clusters of epithelial cells may project into the acinar lumina on the surface of the lining cells
- The cytoplasm retains its normal foamy appearance
- Nuclei are fairly uniform (except in acinar type)
- Mitotic figures are rare (except in acinar type)

Papillary
- There are multiple branching papillary projections with a fibrovascular stroma.
- Epithelium is one or occasionally more layers thick but still attached to the basal lamina

Cystic
- Multiple distended cystic spaces are lined by epithelium
- Papillary projections may be present, but are not prominent
- Mostly large tumors have a prominent capsule

Cystic-papillary
- Composed of both, papillary and cystic areas
- Cystic areas appear empty except for cellular debris and remnants of papillary projections
- The epithelial lining may be flattened

Acinar
- Growth pattern resembles acinar glandular architecture despite the marked proliferation of the epithelial cells
- Encapsulation and compression may be absent
- Papillary projections are not prominent
- Cells with single large cytoplasmic vacuoles may be present
- Cytoplasm may be eosinophilic and homogeneous without foamy appearance
- Nuclear to cytoplasmic ratio is increased in comparison to the other types
- Cells are more pleomorphic and mitotic figures are more numerous than in the other types

Differential Diagnosis

Hyperplasia
Architecture is maintained and
No evidence of compression of the adjacent glandular tissue is present

Adenocarcinoma
Well-differentiated:
Local invasion or
Prominent cellular atypia or
Metastasis
Poorly differentiated (anaplastic):
Growth pattern is solid or
Local invasion or metastasis

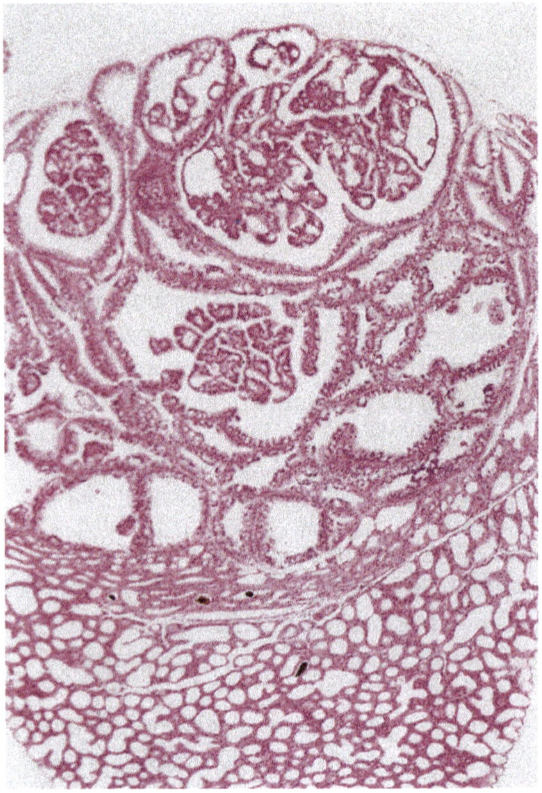

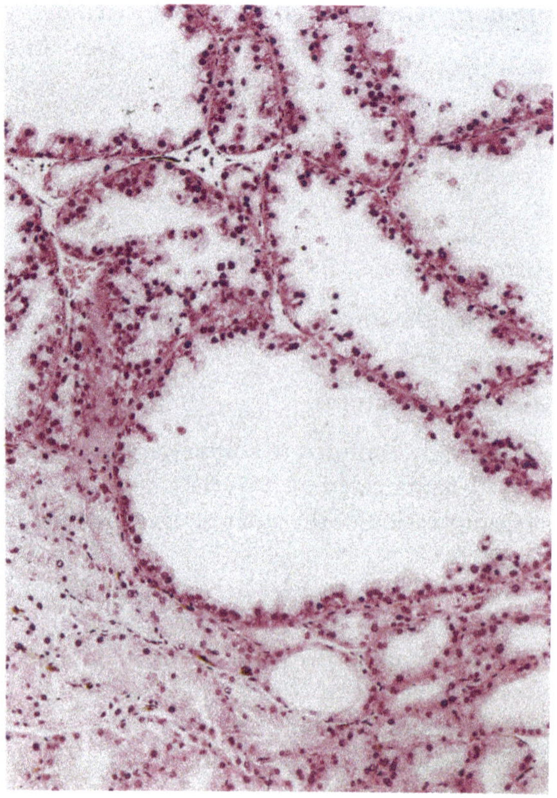

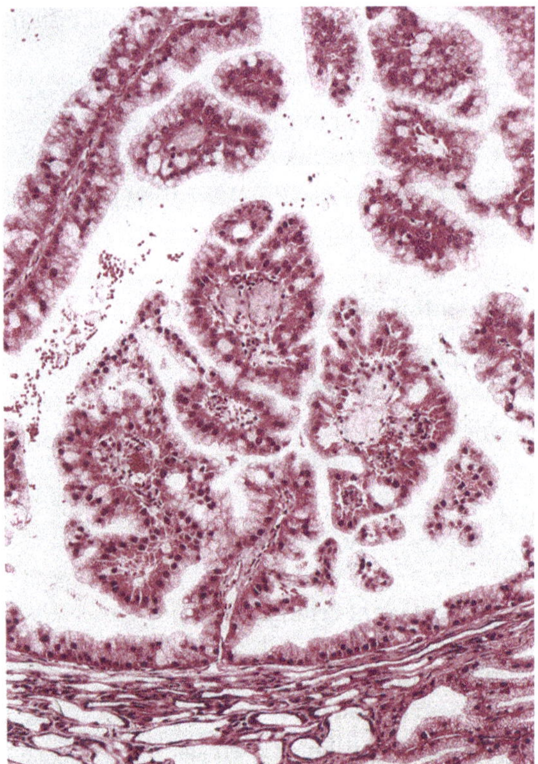

Fig. 4 *(upper left)*. Adenoma, cystic-papillary, of the Harderian gland. H&E

Fig. 5 *(upper right)*. Adenoma, cystic, of the Harderian gland. H&E

Fig. 6 *(below)*. Adenoma, papillary, of the Harderian gland. H&E

Comment

In contrast to these tumors in rats, spontaneous Harderian gland adenomas are frequently observed in old albino and pigmented mice. The incidence varies with strain and significant sex differences have been described. The incidence of spontaneous adenomas given in the literature varies from 0.5 to 14.9% (Table 1).

The recent report on spontaneous Harderian gland adenomas in B6C3F1 mice from 2-year feeding and inhalation studies at NTP (59) gives cumulative incidence for untreated males 4.7% and for "chamber controls" 5.4%, and maximum incidence for both 18%; cumulative incidence for untreated females 3.3%, for female "chamber controls" 3.1%, and maximum incidence for untreated females 10% and for female "chamber controls" 16%.

An increase of Harderian gland neoplasms in mice was described in RFM/Un mice treated with gamma radiation, in C57BL/6xC3HF1 mice treated with X-irradiation and in B6C3F1 mice treated with fission neutron irradiation. With neutron irradiation, the data indicate an induction of Harderian gland tumors in RF mice. Treatment with benzidine dihydrochloride, nitrogen mustard nitrous oxide, urethane, ethylnitrosourea, 1,3 butadiene, ethylenethiourea in combination with sodium nitrate and 1,2,3 trichloropropane also resulted in an increase in Harderian gland neoplasms. With benzidine dihydrochloride, induction of cystadenomas was seen in C57BL/6JxC3HeB/FeF1 (B6C3F1) mice dependent on the route of administration and study design. At the end of their life span, ddI mice showed an induction of Harderian gland tumors after subcutaneous injection of nitrogen mustard nitrous oxide at the age of 7 days for 4 weeks. The same was reported for urethane. Additionally, Harderian gland tumor-induction by urethane was seen in CTM, BRA/2eB, MRC and C57BLxC3HF1 mice with various study designs and modes of application. After one single i.p. injection of ethylnitrosourea, B6C3F1 and C3AF1 mice developed Harderian gland adenomas and carcinomas. The tumor rate was 27% in B6C3F1 and 12% in C3AF1 mice; however, the controls showed an incidence up to 5%. A 2-year-inhalation study with 1,3 butadiene showed a dose-related increase of Harderian gland adenomas. Ethylenethiourea in combination with sodium nitrate showed dose-dependent increase of Harderian gland adenomas after 10 weekly oral administrations and an 18-month follow-up period. Oral administration of trichloropropane for 2 years increased the incidence of Harderian gland adenomas in B6C3F1 mice. Induced tumors occur earlier, grow more rapidly and appear to be more frequently malignant than spontaneous tumors.

Table 1. Cumulative incidences of spontaneous Harderian gland adenomas for various strains

Strain	Incidence (%) up to Males	Females
BALB7c	9.2	14.3
C57BL/6	3.4	4.9
C3Hf/He	1.7	14.9
NMRI	6.0	2.7
CD-1	2.9	1.7
Athymic nude	1.8	1.8
SWR/J	4.4	—
SENCAR	—	5.1
B6C3F1	6.5	6.5

References

(See 1, 2, 3, 4, 5, 6, 7, 8, 9, 10, 11, 14, 15, 16, 19, 20, 21, 22, 23, 24, 25, 27, 28, 29, 30, 31, 32, 33, 34, 35, 36, 38, 39, 40, 41, 42, 43, 44, 45, 46, 47, 48, 49, 50, 51, 52, 53, 54, 55, 56, 57, 58)

Adenocarcinoma *(M)*

Histogenesis

Acinar epithelium of the Harderian gland.

Diagnostic Features

- Demarcation is poor
- Architecture of the gland is not maintained within the tumor
- Epithelium is more than one cell layer thick
- Detachment from the basal lamina and piling up of cells are prominent features
- Medullary (solid) growth pattern is common
- Cells are small and round
- Nuclear to cytoplasmic ratio is increased
- Cellular and nuclear pleomorphism is present
- Giant cells may be present
- Cells with or without foamy cytoplasm occur
- Necrotic foci may be present
- Some cells may contain large eosinophilic cytoplasmic inclusions
- Differentiation of cells varies from well-differentiated to anaplastic
- Mitotic figures may be numerous
- Can be locally invasive into the orbital and periorbital tissue
- Metastases may be formed

Differential Diagnosis

Adenoma

Well-demarcated and
Mitotic figures are rare and
Cytoplasm of the cells retains its normal foamy appearance but in acinar adenomas cytoplasmatic changes may be seen

Comment

In old albino and pigmented mice, spontaneous Harderian gland adenocarcinomas are less frequently observed than Harderian gland adenomas. The incidence varies with strain and significant sex differences have been described. The incidence of carcinomas in Charles River CD-1 mice is 0.2–0.3%, in NMRI mice 0–4%, in male C3Hf/He mice 3.6% and in male SWR/J mice 1.5%. Metastases have been observed in the lungs, regional lymph nodes, thymus or liver.
The recent report on spontaneous Harderian gland carcinomas in B6C3F1 mice from 2-year feeding and inhalation studies at NTP (59) gives cumulative incidence for untreated males 0.7% and for "chamber controls" 0.5%, and maximum incidence for

Fig. 7 *(upper left).* Adenocarcinoma of the Harderian gland still showing alveolar pattern with increased nucleus/cytoplasm ratio and multiple cell layers. NMRI mouse, female. H&E ▶

Fig. 8 *(upper right).* Adenocarcinoma of the Harderian gland still showing alveolar pattern with increased nucleus/cytoplasm ratio and numerous mitotic figures of cells arranged in multiple cell layers. NMRI mouse, male. H&E

Fig. 9 *(lower left).* Adenocarcinoma of the Harderian gland. H&E

Fig. 10 *(lower right).* Metastasis of an adenocarcinoma of the Harderian gland in the lung with alveolar pattern and foamy appearance of the cytoplasm of the metastatic cells. NMRI mouse, male. H&E

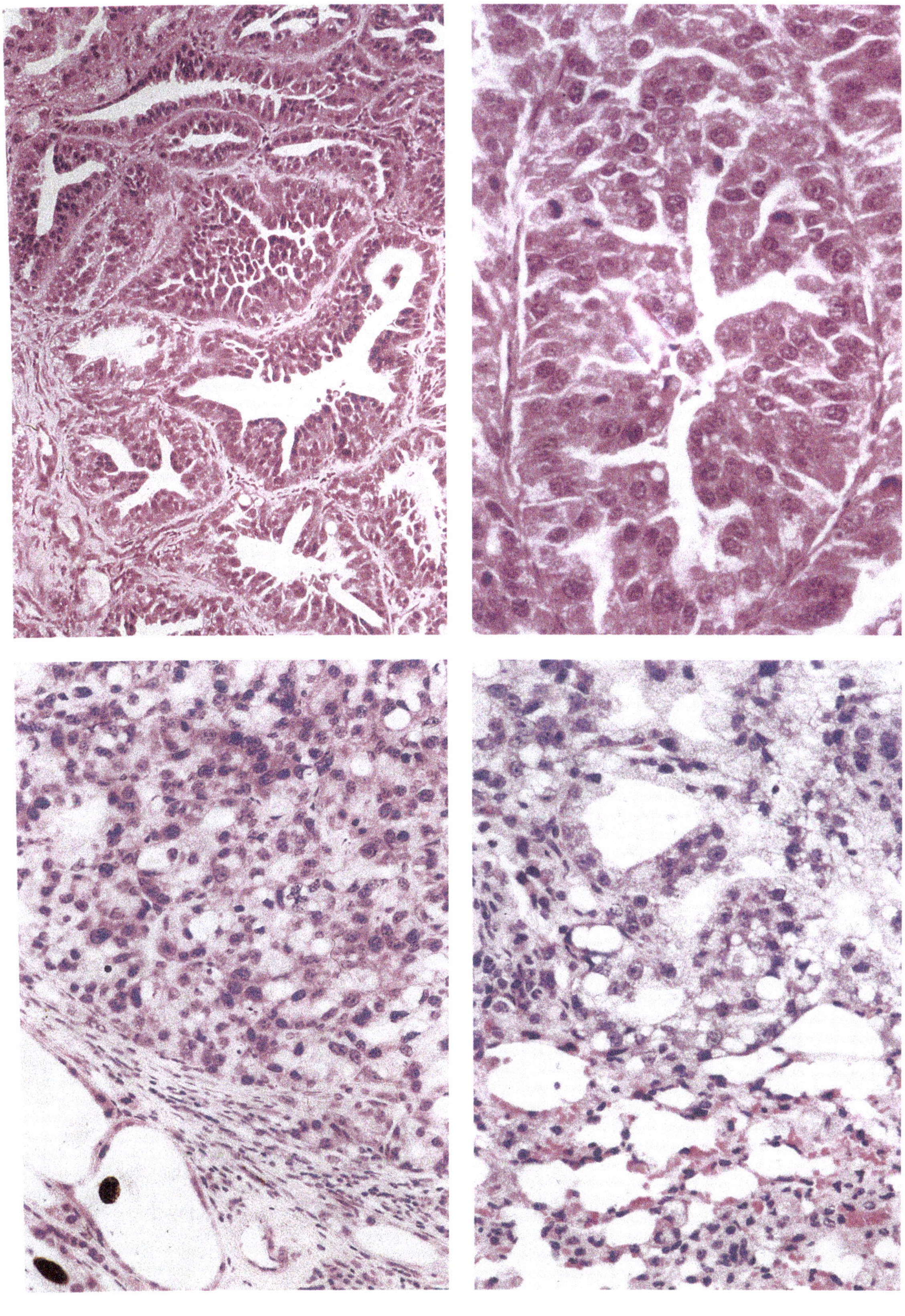

untreated males 4% and for male "chamber controls" 6%; cumulative incidence for untreated females 0.7%, for female "chamber controls" 0.8%, and maximum incidence for both untreated females and for female "chamber controls" 4%.

References

(See 1, 2, 3, 4, 5, 6, 7, 8, 9, 10, 11, 14, 15, 16, 20, 21, 22, 23, 25, 27, 29, 30, 35, 36, 37, 38, 39, 40, 41, 42, 43, 44, 45, 49, 50, 51, 52, 53, 57)

References

1. Bomhard E (1993) Frequency of spontaneous tumours in NMRI mice in 21-month studies. Exp Toxic Pathol 45: 269–289
2. Bomhard E, Mohr U (1989) Spontaneous tumors in NMRI mice from carcinogenicity studies. Exp Pathol 36: 129–145
3. Botts S, Jokinen M, Gaillard ET, Elwell MR, Mann PC (1999) Salivary, Harderian and lacrimal glands. In: Maronpot RR, Boorman GA, Gaul BW (eds) Pathology of the mouse. Reference and atlas. Cache River Press, Vienna, pp 49–79
4. Cardesa A, Carlton WW, Dungworth DL, Enomoto M, Halm S, Koestner A, Krinke GJ, Render JA, Rittinghausen S, Ruben Z, Solleveld H, Turusov VS, Weisse I, Yoshitomi K (1994) 7. Central nervous system, eye, heart, and mesothelium. In: Mohr U, Capen CC, Dungworth DL, Griesemer RA, Ito N, Turusov VS (eds) International classification of rodent tumours, Part I: The rat. IARC Scientific Publications No. 122, Lyon
5. Carlton WW, Render JA (1991) Adenoma and adenocarcinoma, Harderian gland, mouse, rat, hamster. In: Jones TC, Mohr U, Hunt RD (eds) Monographs on pathology of laboratory animals. Eye and ear. Springer, Berlin Heidelberg New York Tokyo, pp 133–137
6. Chandra M, Frith CH (1992a) Spontaneous neoplasms in aged CD-1 mice. Toxicol Lett 61: 67–74
7. Chandra M, Frith CH (1992b) Spontaneous neoplasms in B6C3F1 mice. Toxicol Lett 60: 91–98
8. Della Porta G, Capitano J, Montipo W, Parmi L (1963) Studio sull'azione cancerogena dell'uretano nel topo. Tumori 49: 413–428
9. Deringer MK (1959) Occurrence of tumors, particularly mammary tumors, in agent-free strain C3HeB mice. J Natl Cancer Inst 22: 995–1002
10. Deringer MK (1962) Development of tumors, especially mammary tumors, in agent-free strain DBA/2eB mice. J Natl Cancer Inst 28: 203–210
11. Deringer MK (1965) Response of strain DBA-2eBDe mice to treatment with urethan. J Natl Cancer Inst 34: 841–847
12. Ernst H, Rittinghausen S, Mohr U (1991) Melanoma of the eye, mouse. In: Jones TC, Mohr U, Hunt RD (eds) Monographs on pathology of laboratory animals. Eye and ear. Springer, Berlin Heidelberg New York Tokyo, pp 44–47
13. Faccini JM, Abbott DP, Paulus GJJ (1990) Harderian gland, ear. Mouse histopathology. A glossary for use in toxicity and carcinogenicity studies. Elsevier, Amsterdam, pp 213–219
14. Frith CH, Ward JM (1988) Color atlas of neoplastic and non-neoplastic lesions in aging mice. Elsevier, Amsterdam
15. Frith CH, Littlefield NA, Umholtz R (1981) Incidence of pulmonary metastases for various neoplasms in BALB/cStCrlfC3H/Nctr female mice fed N-2-fluorenylacetamide. J Natl Cancer Inst 66: 703–712
16. Frith CH, Highman B, Burger G, Sheldon WD (1983) Spontaneous lesions in virgin and retired breeder BALB/c and C57BL/6 mice. Lab Anim Sci 33: 273–286
17. Geiss V, Yoshitomi K (1999) Eyes. In: Maronpot RR, Boorman GA, Gaul BW (eds) Pathology of the mouse. Reference and atlas. Cache River Press, Vienna, pp 471–489
18. Ghadially FN (1985) Diagnostic electron microscopy of tumours, Butterworth, London
19. Grahn D, Lombard LS, Carnes BA (1992) The comparative tumorigenic effects of fission neutrons and cobalt-60 gamma rays in the B6C3F1 mouse. Radiat Res 129: 19–36
20. Haseman JK, Elwell MR, Hailey JR (1999) Neoplasm incidences in B6C3F1 mice: NTP historical data. In: Maronpot RR, Boorman GA, Gaul BW (eds) Pathology of the mouse. Reference and atlas. Cache River Press, Vienna, pp 679–689
21. Holland JM, Fry RJM (1982) Neoplasms of the integumentary system and Harderian gland. In: Foster HL, Small JD, Fox JG (eds) The mouse in biomedical research, vol IV. Academic Press, New York, pp 513–528
22. Holland JM, Mitchell TJ, Gipson LC, Whitaker MS (1978) Survival and cause of death in aging germfree athymic nude and normal inbred C3Hf/He mice. J Natl Cancer Inst 61: 1357–1361
23. Ihara M, Tajima M, Yamate J, Shibuya K (1994) Morphology of spontaneous Harderian gland tumors in aged B6C3F1 mice. J Vet Med Sci 56: 775–778
24. Irwin RD, Haseman JK, Eustis SL (1995) 1,2,3-Trichloropropane: a multisite carcinogen in rats and mice. Fundam Appl Toxicol 25: 241–252
25. Kircher CH (1978) Comparative pathology of ocular tumors. In: Benirschke K, Garner FM, Jones TC (eds) Pathology of laboratory animals, vol I. Springer, Berlin, pp 647–662
26. Kleihues P, Cavenee WK (1997) Pathology and genetics of tumours of the nervous system. IARC, Lyon
27. Krinke GJ, Schaetti PR, Krinke AL (1996) Nonneoplastic and neoplastic changes in the Harderian and lacrimal glands. In: Mohr U, Dungworth DL, Capen CC, Carlton WW, Sundberg JP, Ward JM (eds) Pathobiology of the aging mouse, vol 2. ILSI Press, Washington, DC, pp 139–152
28. Lohrke H, Hesse B, Goerttler K (1984) Spontaneous tumors and lifespan of female NMRI mice of the outbred stock Sut:NMRT during a

lifetime study. J Cancer Res Clin Oncol 108: 192–196
29. Majeed SK, Gopinath C (1990) Harderian gland tumours in mice. Arzneimittelforschung 40: 1265–1267
30. Matsuyama M, Suzuki H, Nakamura T (1969) Carcinogenesis in dd-I mice injected during suckling period with urethane, nitrogen mustard N-oxide, and nitroso-urethane. Br J Cancer 23: 167–171
31. Matthews HB, Eustis SL, Haseman J (1993) Toxicity and carcinogenicity of chronic exposure to tris(2-chloroethyl)phosphate. Fundam Appl Toxicol 20: 477–485
32. Melnick RL, Huff J, Chou BJ, Miller RA (1990) Carcinogenicity of 1,3-butadiene in C57BL/6 x C3H F1 mice at low exposure concentrations. Cancer Res 50: 6592–6599
33. Melnick RL, Sills RC, Roycroft JH, Chou BJ, Ragan HA, Miller RA (1994) Isoprene, an endogenous hydrocarbon and industrial chemical, induces multiple organ neoplasia in rodents after 26 weeks of inhalation exposure. Cancer Res 54: 5333–5339
34. Percy DH, Jonas AM (1971) Incidence of spontaneous tumors in CD (R)-1 HaM-ICR mice. J Natl Cancer Inst 46: 1045–1065
35. Rabstein LS, Peters RL, Spahn GJ (1973) Spontaneous tumors and pathologic lesions in SWR-J mice. J Natl Cancer Inst 50: 751–758
36. Sheldon W (1994) Tumours of the Harderian gland. In: Turusov VS, Mohr U (eds) Pathology of tumours in laboratory animals, vol 2. Tumours of the mouse, 2nd edn. IARC Scientific Publications No. 111, Lyon, pp 101–113
37. Sheldon WG, Greenman DL (1980) Spontaneous lesions in control BALB/C female mice. J Environ Pathol Toxicol 3: 155–167
38. Sheldon WG, Curtis M, Kodell RL, Weed L (1983) Primary harderian gland neoplasms in mice. J Natl Cancer Inst 71: 61–68
39. Sher SP (1982) Tumors in control hamsters, rats, and mice: literature tabulation. Crit Rev Toxicol 10: 49–79
40. Squire RA, Goodman DG, Valerio MG, Fredrickson T, Strandberg JD, Levitt MH, Lingeman CH, Harshbarger JC, Dawe CJ (1978) Tumors. In: Benirschke K, Garner FM, Jones TC (eds) Pathology of laboratory animals, vol II. Springer, Berlin Heidelberg New York Tokyo, pp 1242–1245
41. Tannenbaum A, Silverstone H (1958) Urethan (ethyl carbamate) as a multipotential carcinogen. Cancer Res 18: 1225–1231
42. Tucker MJ (1979) Tumours of the Harderian gland. In: Turusov VS (ed) Pathology of tumours in laboratory animals, vol II. Tumours of the mouse. IARC Scientific Publications No. 23, Lyon, pp 135–146
43. Ullrich RL, Storer JB (1979a) Influence of gamma irradiation on the development of neoplastic disease in mice. II. Solid tumors. Radiat Res 80: 317–324
44. Ullrich RL, Storer JB (1979b) Influence of gamma irradiation on the development of neoplastic disease in mice. III. Dose-rate effects. Radiat Res 80: 325–342
45. Upton AC, Furth J, Christenberry KW (1954) Late effects of thermal neutron irradiation in mice. Cancer Res 14: 682–690
46. Vesselinovitch SD, Mihailovich N (1967) The neonatal and infant age periods as biologic factors which modify multicarcinogenesis by urethan. Cancer Res 27: 1422–1429
47. Vesselinovitch SD, Mihailovich N (1968) The development of neurogenic neoplasms, embryonal kidney tumors, harderian gland adenomas, anitschkow cell sarcomas of the heart, and other neoplasms in urethan-treated newborn rats. Cancer Res 28: 888–897
48. Vesselinovitch SD, Mihailovich N, Rao KVN, Itze L (1971a) Perinatal carcinogenesis by urethan. Cancer Res 31: 2143–2147
49. Vesselinovitch SD, Simmons EL, Mihailovich N, Rao KVN, Lombard LS (1971b) The effect of age, fractionation, and dose on radiation carcinogenesis in various tissues of mice. Cancer Res 31: 2133–2142
50. Vesselinovitch SD, Rao KVN, Mihailovich N, Rice JM, Lombard LS (1974) Development of broad spectrum of tumors by ethylnitrosourea in mice and the modifying role of age, sex, and strain. Cancer Res 34: 2530–2538
51. Vesselinovitch SD, Rao KVN, Mihailovich N (1975) Factors modulating benzidine carcinogenicity bioassay. Cancer Res 35: 2814–2819
52. Vesselinovitch SD, Itze L, Mihailovich N, Rao KVN (1980) Modifying role of partial hepatectomy and gonadectomy in ethylnitrosourea-induced hepatocarcinogenesis. Cancer Res 40: 1538–1542
53. Ward JM (1983) Background data and variations in tumor rates of control rats and mice. Prog Exp Tumor Res 26: 241–258
54. Ward JM, Goodman DG, Squire RA, Chu KC, Linhart MS (1979) Neoplastic and nonneoplastic lesions in aging (C57BL/6 N x C3H/HeN)F1 (B6C3F1) mice. J Natl Cancer Inst 63: 849–854
55. Ward JM, Quander R, Devor D, Wenk ML, Spangler EF (1986) Pathology of aging female SENCAR mice used as controls in skin two-stage carcinogenesis studies. Environ Health Perspect 68: 81–89

56. Watanabe H, Okamoto T, Yamada K, Ando Y, Ito A, Hoshi M, Sawada S (1993) Effects of dose rate and energy level on fission neutron (252Cf) tumorigenesis in B6C3F1 mice. J Radiat Res (Tokyo) 34: 235–239
57. Weisse I, Koellmer H, Tilov T, Stoetzer H (1975) Spontantumoren bei einem NMRI/Mäuse-Auszuchtstamm. Z Versuchstierkd 17: 91–98
58. Yoshida A, Harada T, Maita K (1993) Tumor induction by concurrent oral administration of ethylenethiourea and sodium nitrite in mice. Toxicol Pathol 21: 303–310
59. Yoshitomi K, Boorman GA (1991) Spontaneous amelanotic melanomas of the uveal tract in F344 rats. Vet Pathol 28: 403–409

Soft Tissue and Skeletal Muscle

H. Ernst[3] · W.W. Carlton[1] · C. Courtney · M. Rinke[3] · P. Greaves[2]
K.R. Isaacs · G. Krinke · Y. Konishi · G.M. Mesfin · G. Sandusky

[1]Chair; [2]Co-chair; [3]Initial draft.

Hyperplasia, Angiomatous *(H)*

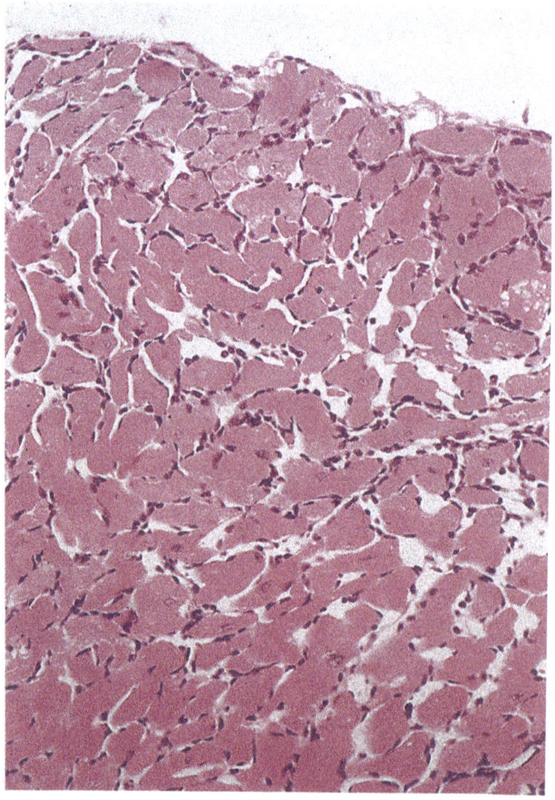

Fig. 1. Angiomatous hyperplasia, heart. The widened vascular spaces between the muscle fibers are lined by a single layer of prominent endothelial cells. B6C3F1 mouse, female. H&E (photograph by kind permission of H. Iwata and Y. Nomura)

Synonym(s). Hyperplasia, capillary; hyperplasia, endothelial

Histogenesis

Endothelial cells of blood or lymph vessels.

Diagnostic Features

- No or only minimal compression of the surrounding tissues; preservation of the normal organ architecture
- Proliferation and (cystic) ectasia of capillary-sized vessels
- Widened vascular space lined by a normal appearing endothelium or an endothelium with plump nuclei

Differential Diagnosis

Angiectasis
Vessels are not increased in number, have normal structure, and are lined by well-differentiated endothelial cells

Hemangioma
Usually causes compression of the surrounding tissues, has a greater size, and slight cytological abnormalities of the endothelial cells such as hypertrophy

Comment

Angiomatous hyperplasia has been described as a spontaneous lesion in B6C3F1 female mice with a low incidence. Other published cases include induction of endothelial cell hyperplasia in the heart of B6C3F1 mice chronically exposed by inhalation to 1,3 butadiene.

References

(See 6, 19, 36)

Hemangioma *(B)*

Capillary type • Cavernous type

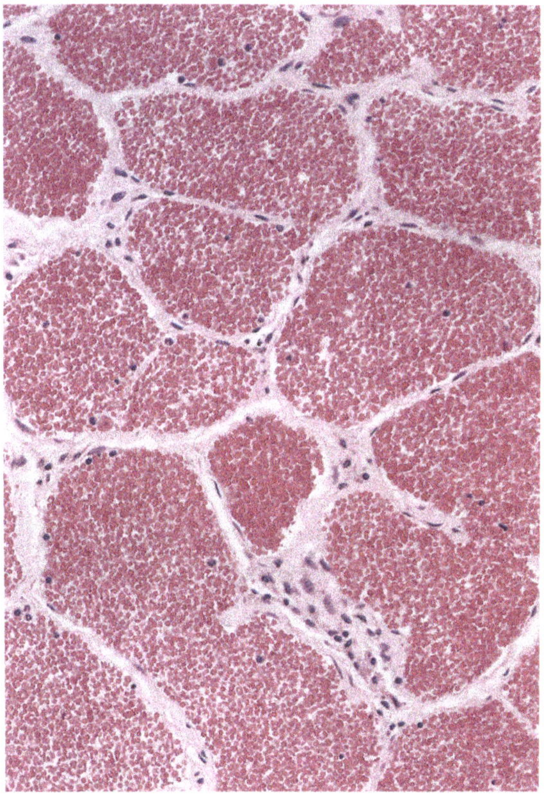

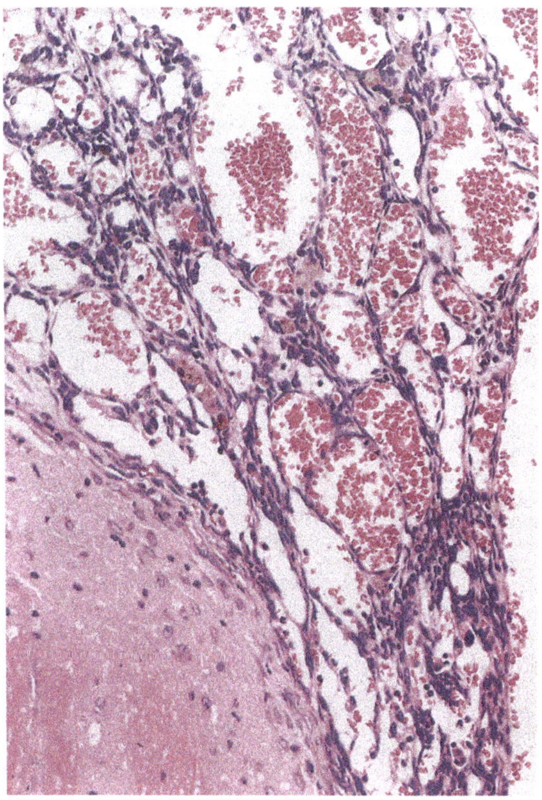

Synonym(s). Hemangioendothelioma, benign

Histogenesis

Endothelial cells of blood vessels (multicentric development).

Diagnostic Features

- Moderate compression of surrounding tissues is usually seen
- The tumor is rarely encapsulated
- The monolayer of lining endothelial cells is prominent with hyperchromatic nuclei
- Focal thrombosis can occur (ovary, uterus)
- Mitotic figures are rarely present
- No sharp separation between capillary and cavernous type

Capillary type

- Network of capillary-sized vessels
- Intervascular tissue can be acellular and collagenous
- Thickened reticular fibers at or in basement membranes

Cavernous type

- Large vascular channels
- Fibrous (collagenous) supporting stroma of variable thickness
- Hemosiderin-laden macrophages can be scattered through cavernous areas

Fig. 2 *(above).* Cavernous hemangioma of the skin consisting of large blood-filled vascular channels separated by fibrous trabeculae. CD-1 mouse, male, treated, age 39 weeks. H&E

Fig. 3 *(below).* Hemangioma, predominantly capillary type, uterus, showing large focal thrombosis. NMRI mouse, age 76 weeks. H&E

Differential Diagnosis

Angiectasis

Vessels are not increased in number, have normal structure and well-differentiated endothelial cells

Hyperplasia, Angiomatous

The hyperplastic vessels are increased in number and are lined by normal-appearing to plump endothelium and cause no or only minimal compression of the surrounding tissues

Hemangiosarcoma

Cytologic and histologic features of malignancy such as cellular pleomorphism, increased mitotic activity, multilayering, tissue invasion or metastases are present

Comment

Hemangiomas can occur within any tissue but are more commonly observed in the liver, uterus, skin, spleen and bone marrow. They can be multicentric. In the ovary, differentiation from angiectasis is not always clear particularly as focal thrombosis can occur in both.

Hemangiomas of the uterus appear to be rare in B6C3F1 mice as none were recorded in over 2500 mice of this strain. In CD-1 mice, reported overall incidences of hemangiomas were about 0.5–4% and the incidence of uterine hemangioma was about 1–2%. In some cases of hepatic hemangioma (cavernous type), it is difficult to distinguish the vascular neoplasm from a reactive angiectasis. Antibodies to collagen type IV and Factor VIII-related antigen can be useful immunomarkers for identifying well differentiated tumors.

References

(See 4, 10, 12, 13, 16, 21, 25, 30, 37, 46, 47)

Hemangiosarcoma *(M)*

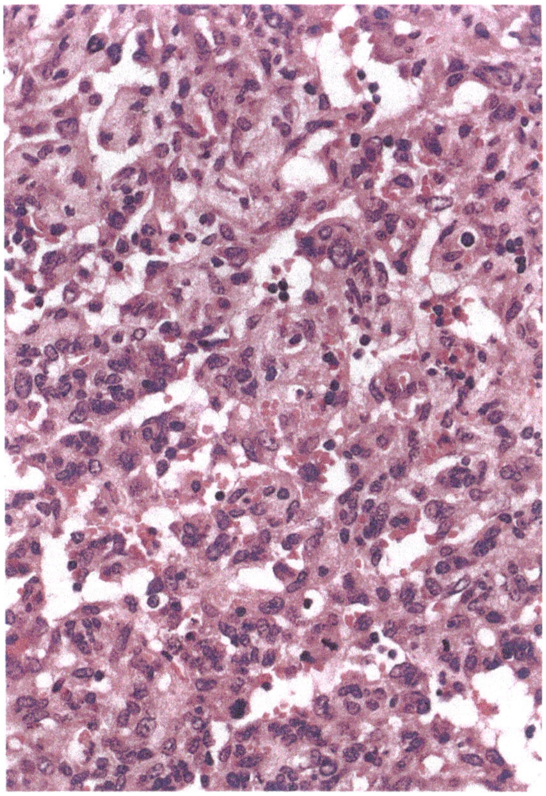

Fig. 4. Metastasizing hemangiosarcoma, spleen. Irregular vascular channels are lined by pleomorphic endothelial tumor cells. CBA/J mouse, male, age 72 weeks. H&E

Synonym(s). Hemangioendothelioma, malignant

Histogenesis

Pluripotential mesenchymal stem cells; endothelial cells of blood vessels.

Diagnostic Features

- Atypical endothelial cells form both vascular channels (capillary to cavernous) and solid cellular masses supported by variably developed fibrovascular stroma
- Tumor cells can be round, polygonal or extremely irregular in form; commonly they are spindle-shaped
- Tumor cells have single or multiple irregular shaped and sized nuclei that are often large and lobated and have large masses of chromatin
- Mitotic figures can be numerous and frequently are bizarre
- Local invasion and metastases are often present

Differential Diagnosis

Hemangioma
Usually no or only few mitotic figures. No invasiveness and no metastases. One layer of endothelium has nearly normal nuclei or nuclei have slight hyperchromasia and karyomegaly

Granulation Tissue
The newly-formed blood vessels are typically arranged perpendicular to fibroblasts, collagen bundles, and surfaces with no cytologic and histologic features of neoplasia

Fibrosarcoma
The tumor lacks a distinct vascular pattern and prominent endothelial cells, but consists of spindle-shaped cells with nuclear pleomorphism

Comment

Peripherally, a wide reactive hemorrhagic zone that usually is devoid of neoplastic cells can surround the tumor. Unless care is taken to include the center of the tumor mass in the material sectioned, the neoplastic nature of the lesion can be overlooked. Hemangiosarcomas are encountered most frequently in the skin, spleen, liver and bone marrow. In the liver, the tumor is composed of elongated, flattened, spindle-shaped or polyhedral endothelial cells that line vascular clefts and spaces and sometimes form solid sheets. The neoplastic cells can surround hepatocytes which can be larger than hepatocytes away from the neoplasm. The neoplasms are locally invasive and can metastasize to other organs, especially to the lungs. It is often difficult to determine whether hemangiosarcomas seen in multiple organs are multicentric or represent metastases. Hemangiosarcomas of the ovary often invade the ovarian capsule and bursa and occasionally metastasize to the lungs.

Antibodies to collagen type IV and Factor VIII-related antigen can be useful immunomarkers for identifying well-differentiated tumors.

References

(See 4, 10, 12, 13, 14, 16, 21, 25, 27, 30, 32, 34, 36, 37, 46, 47)

Lipoma (*B*)

Angiolipomatous type • Hibernomatous type

Histogenesis

Lipoblasts, lipocytes (normal fat cells); lipoblasts or lipocytes of brown fat.

Diagnostic Features

- Nodular, well-demarcated tumor, which can have red brown discoloration and central necrosis if pedunculated
- Surrounding tissues are sometimes compressed
- Tumor cells are arranged in lobules and separated by fibrous septa
- Typically, differentiated lipocytes have a single large prominent cytoplasmic vacuole with eccentrically located nucleus

Angiolipomatous Type

- Areas of vascular proliferation are prominent within the tumor

Hibernomatous Type

- Areas or lobules of round, oval or polygonal cells with variable degree of cytoplasmic vacuolation and central dense nuclei

Differential Diagnosis

Fat Necrosis or Granulomatous Steatitis

Because necrotic changes often occur in lipomas, distinguishing fat necrosis and lipogranulomatous inflammation from lipomas can be difficult. However, differentiation is based on the absence of unequivocal criteria of neoplasia.

Liposarcoma

Cytologic and histologic features of malignancy, such as cellular pleomorphism, increased mitotic activity, tissue invasion or metastases are present

Comment

Subcutaneous lipoma is a very rare tumor in mice. In contrast, lipomas of the brain are sometimes present at greater incidence in this species. In the brain, the lipomatous lesions can represent lipomatous hamartomas rather than neoplasms. Intracranial lipomas are usually present in the interstitium of the choroid plexus of the lateral ventricles.

References

(See 10, 12, 16, 26, 27, 30)

Figs. 5-7 see p. 368

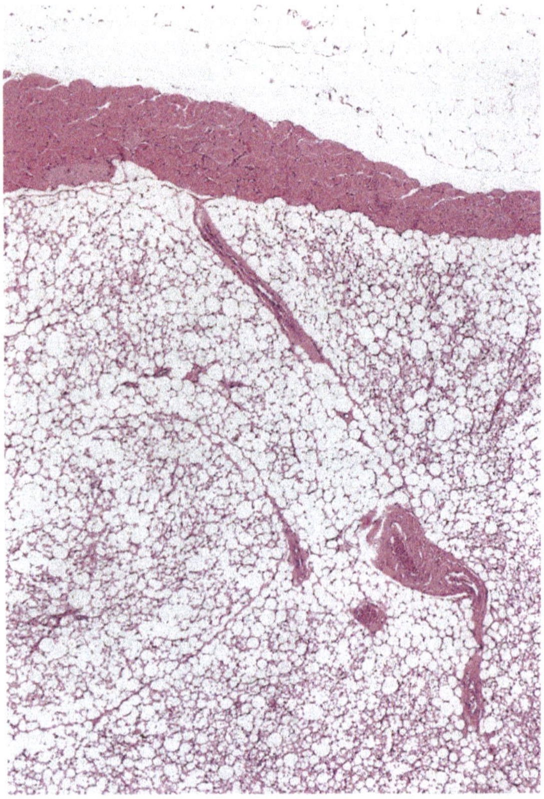

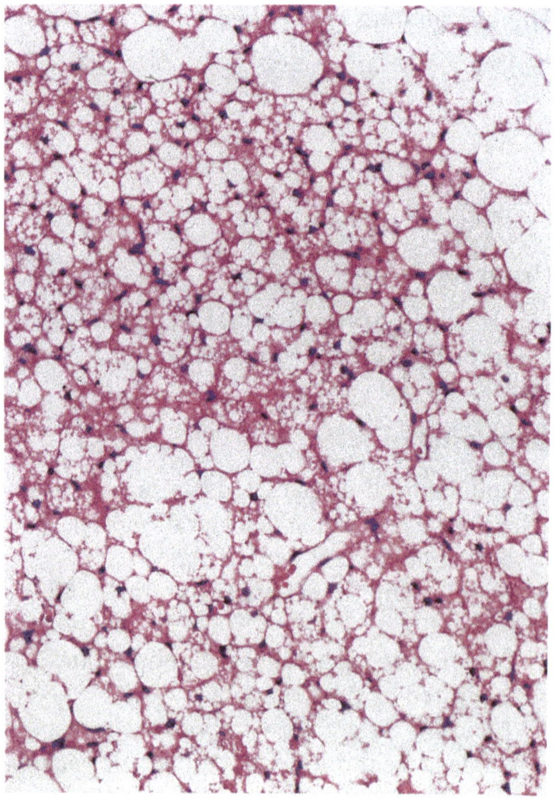

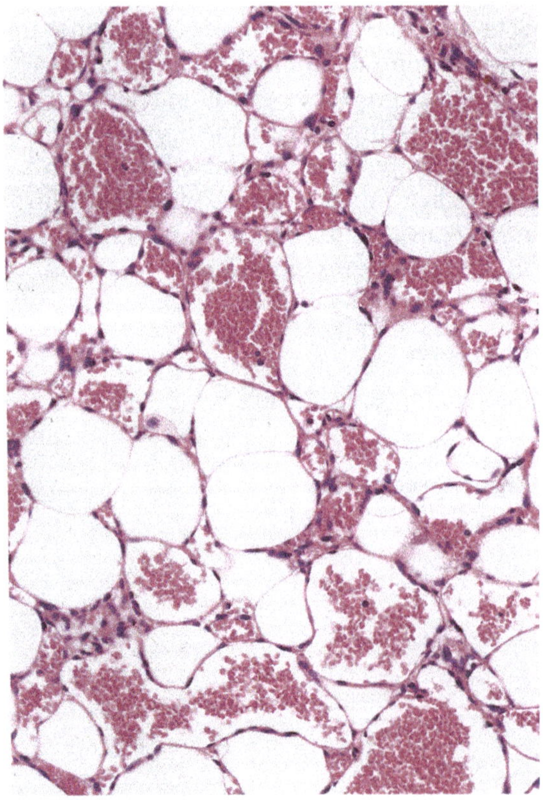

Fig. 5 *(upper left)*. Lipoma, skin. The overlying muscular layer is slightly distorted by the tumor mass. CD-1 mouse, male, age 85 weeks. H&E (section by kind permission of S.R. Frame)

Fig. 6 *(upper right)*. Lipoma, skin, consisting of well-differentiated cells with single or multiple fat vacuoles. CD-1 mouse, male, age 85 weeks. H&E (section by kind permission of S.R. Frame)

Fig. 7 *(below)*. Lipoma, angiolipomatous type, skin. Numerous blood-filled vessels are present within the lipomatous tumor. NMRI mouse, female, age 69 weeks. H&E

Liposarcoma *(M)*

Pleomorphic type • Hibernomatous type

Histogenesis

Pluripotential mesenchymal stem cells; lipoblasts, lipocytes.

Diagnostic Features

- Cells are round to oval with large central or eccentric nucleus and multivacuolated cytoplasm
- Lipid vacuoles of variable size are characteristic
- Mitotic activity can be high
- Various other cell types including "brown" fat cells, foam cells, giant cells, myxoid cells, fibroblast-like cells and stellate cells can also be present
- Infiltration of the surrounding tissues and distant metastases are common
- Myxoid change within the stroma can be prominent
- Rapidly growing liposarcomas frequently have foci of necrosis

Pleomorphic Type
- This variant is composed of lipoblasts in different developmental stages
- Fat staining methods are useful to demonstrate lipids in these undifferentiated tumors

Hibernomatous Type
- Areas or lobules of pleomorphic cells with single to multiple cytoplasmic vacuoles and nuclei varying in size and shape and eccentrically to centrally located

Differential Diagnosis

Lipoma
No cytologic (cellular pleomorphism, increased mitotic activity) and histologic features (tissue invasion or metastases) of malignancy are present

Fibrosarcoma (Myxomatous Type)
These neoplasms are negative for lipoblasts and lipocytes when stains for fat are used and are composed of pleomorphic spindle-shaped cells arranged in bundles or sheets

Histiocytoma, Fibrous, Malignant (Pleomorphic Type)
In reticulin-stained sections, reticulin fibers encircle clusters of tumor cells and not individual tumor cells as occurs in liposarcoma

Steatitis, Granulomatous
Lipid-laden macrophages without histologic features of neoplasia are usually accompanied by infiltrates of other inflammatory cells

Comment

Liposarcomas are rare neoplasms in mice. These neoplasms have been described in the subcutis and mesentery.

References

(See 10, 16, 30, 37)

Figs. 8-10 see p. 370

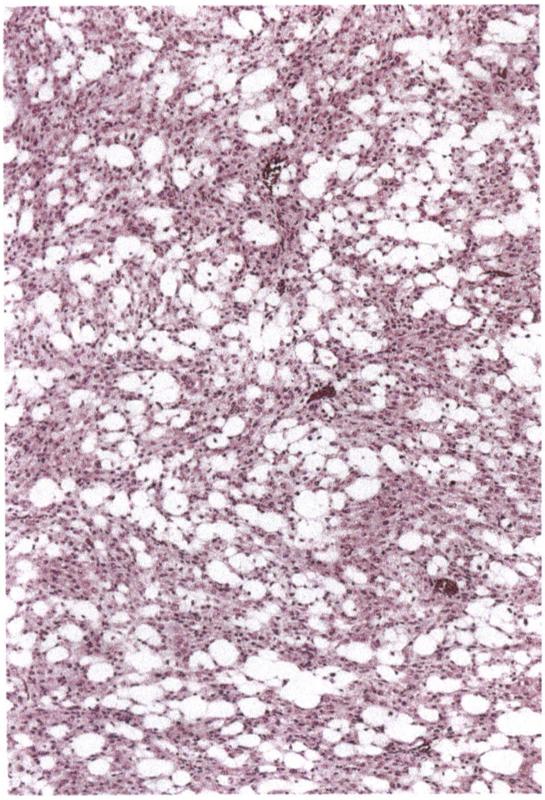

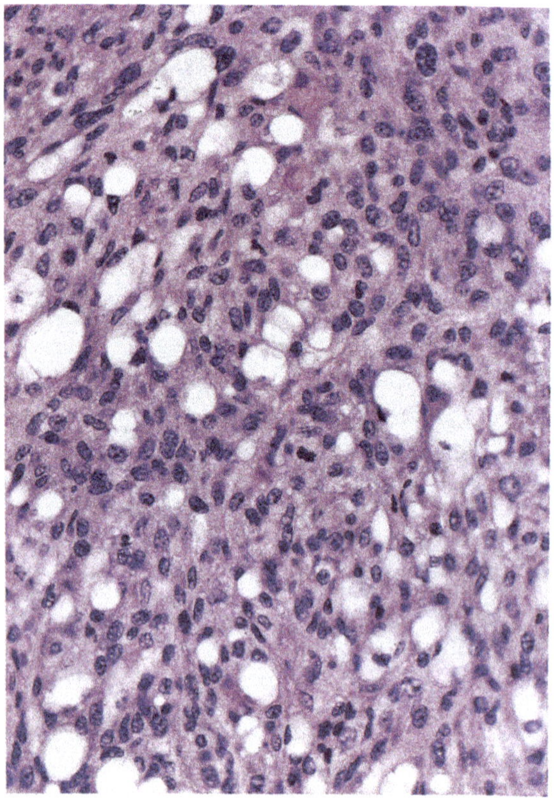

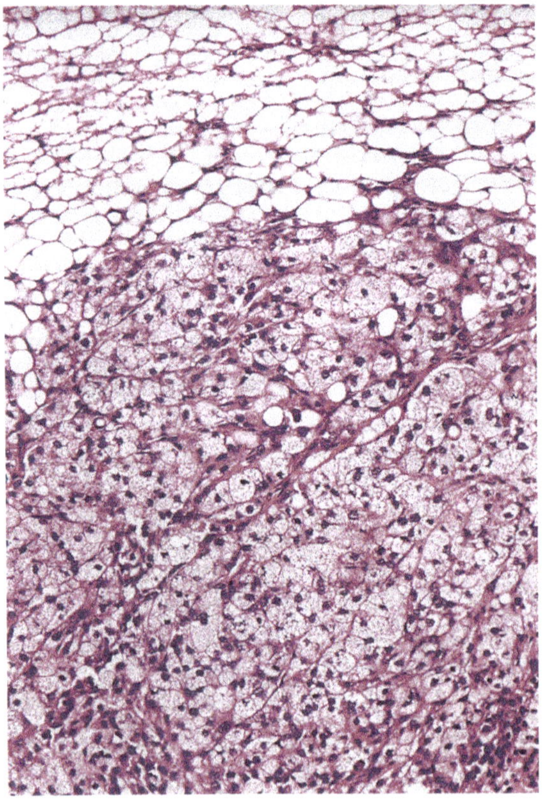

Fig. 8 *(upper left).* Liposarcoma, skin, consisting of lipoblasts in different developmental stages. B6C3F1 mouse, female. Masson trichrome

Fig. 9 *(upper right).* Liposarcoma, skin. The neoplastic lipoblasts are pleomorphic and show a variable degree of cytoplasmic vacuolation. NMRI mouse, female, age 97 weeks. H&E

Fig. 10 *(below).* Liposarcoma, hibernomatous type, mesentery. Part of the tumor consists of 'brown' fat cells with multiple cytoplasmic vacuoles and central nuclei. NMRI mouse, female, age 103 weeks. H&E

Fibroma *(B)*

Myxomatous type

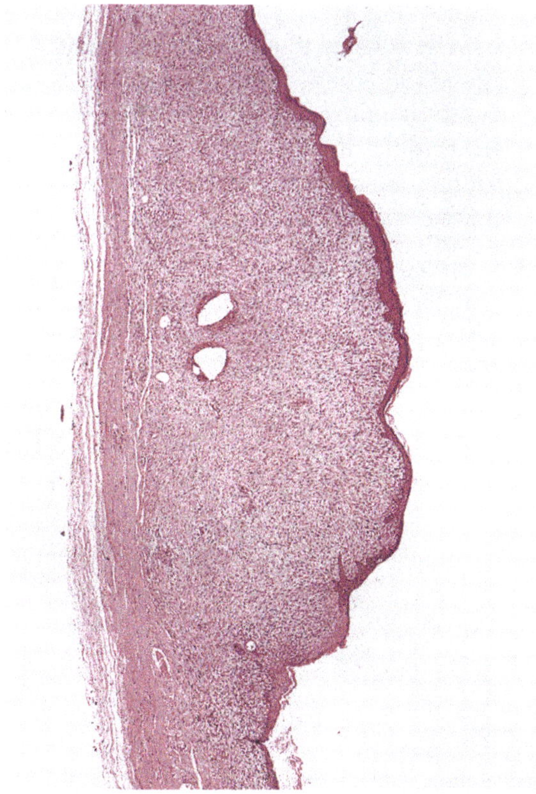

Histogenesis

Fibroblasts, fibrocytes.

Diagnostic Features

- Poorly to moderately cellular tumor that causes compression of adjacent tissue
- Interlacing bundles of well-differentiated fibroblasts are major component
- Cells are fusiform with elongated tapered hyperchromatic or vesicular nuclei and one or more nucleoli
- Interwoven bands of mature collagen are present
- Mitotic figures are rare

Myxomatous Type

- Tumor has myxomatous areas and stellate cells

Differential Diagnosis

Reactive Fibrosis or Scars

These lesions do not have characteristics of tumor and can be associated with wounds, ulcers, or inflammatory lesions

Histiocytoma, Fibrous, Benign

This tumor generally has a greater cellularity than the fibroma and typically has a storiform or cartwheel pattern of more rounded tumor cells

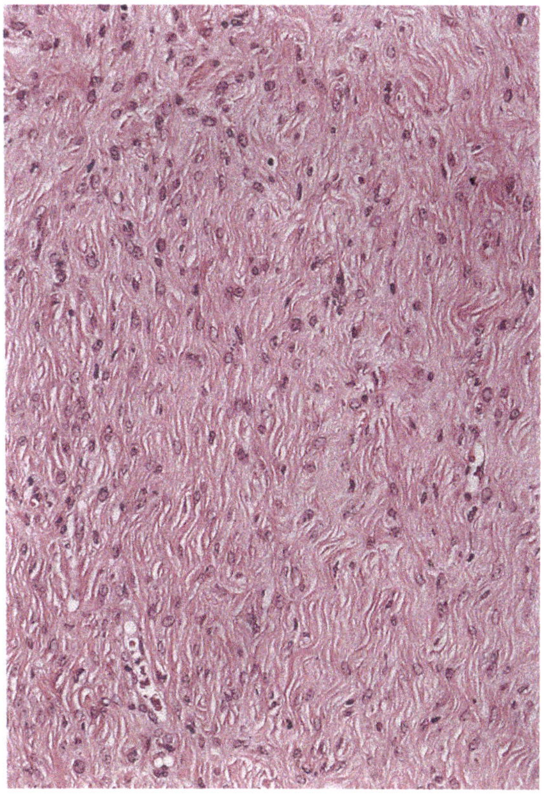

Fig. 11 *(above)*. Fibroma, skin, appearing as nodular growth within the dermis. B6C3F1 mouse, male, age 93 weeks. H&E

Fig. 12 *(below)*. Fibroma, skin. Moderately cellular tumor with dense bundles of collagen. B6C3F1 mouse, male, age 93 weeks. H&E

Leiomyoma
Tumor cells have cylindrical, blunt-ended or cigar-shaped nuclei. Special connective tissue stains allow differentiation of collagen-rich fibromas from collagen-poor leiomyomas that usually have positive immunoreactivity for desmin and smooth muscle actin

Schwannoma, Benign
Usually positive S-100 protein immunoreactivity. Variable proportions of Antoni type A tissue with Verocay bodies or Antoni type B tissue with cystic spaces can be present; the cells have typical wavy, buckled or comma-shaped nuclei

Fibrosarcoma
Cytologic and histologic features of malignancy are present, such as cellular pleomorphism, increased mitotic activity, tissue invasion or metastases

Comment

Fibromas are rare tumors in mice and occurrence has been reported in the dermis and subcutis, but they do also occur in other locations.

References

(See 2, 3, 16, 30)

Fibrosarcoma *(M)*

Myxomatous type

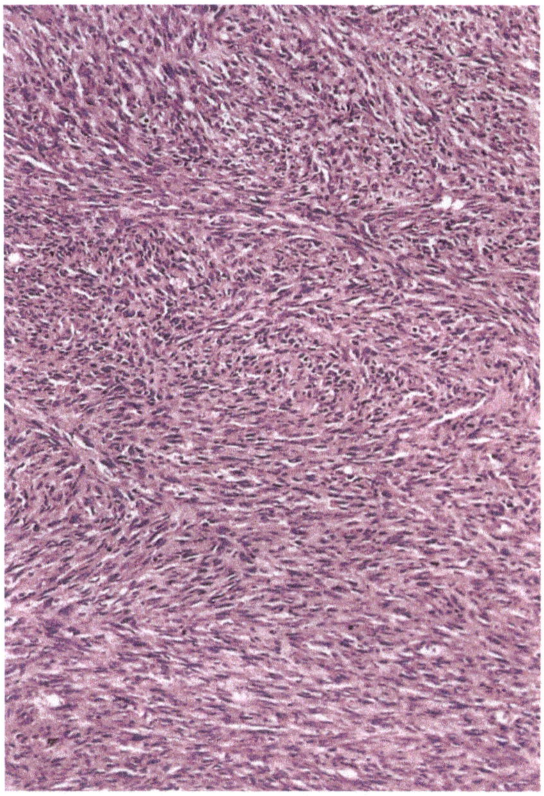

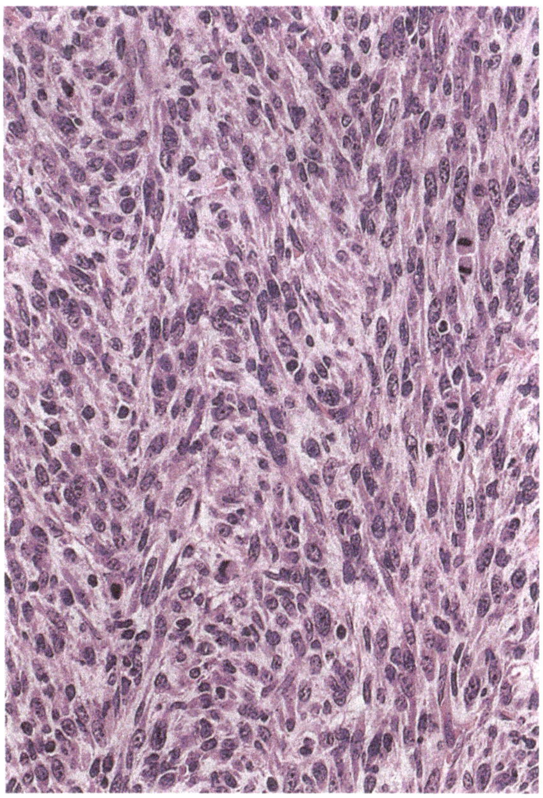

Histogenesis

Pluripotential mesenchymal stem cells; fibroblasts, fibrocytes.

Diagnostic Features

- Pleomorphic spindle-shaped cells often having a characteristic "herring-bone" cellular pattern or form interlacing bundles
- Various amounts of collagen bundles are interspersed between the closely applied bands of cells depending on the degree of differentiation
- Mitotic figures can be plentiful
- Areas of hemorrhage and necrosis can be present
- Local invasion can be extensive, but metastatic spread is relatively late and infrequent
- Metaplastic cartilage and bone can be observed

Myxomatous Type

- Tumor has large myxomatous areas with stellate-shaped cells

Fig. 13 *(above).* Fibrosarcoma, skin, consisting of interlacing bundles of spindle-shaped tumor cells. NMRI mouse, female, age 76 weeks. H&E

Fig. 14 *(below).* Fibrosarcoma, skin. Fascicles of fibroblast-like tumor cells are arranged in a "herring bone" pattern. CBA/J mouse, male, age 88 weeks. H&E

Differential Diagnosis

Fibroma
No cytologic and histologic features of malignancy, such as cellular pleomorphism, increased mitotic activity, tissue invasion or metastases, are present

Hemangiosarcoma
Abundant vascular channels lined by prominent and pleomorphic endothelial cells are characteristic

Histiocytoma, Fibrous, Malignant (MFH)
The MFH can have a characteristic storiform or cartwheel arrangement of the tumor cells. The cellular pleomorphism in MFH is even more pronounced than in poorly differentiated fibrosarcomas. Histiocytic markers like cathepsin B, alpha-1-antitrypsin (AT) and alpha-1-antichymotrypsin (ACT) are present in MFH

Leiomyosarcoma
The cells have a more distinct cytoplasm, which varies tinctorially from pink to deep red in H&E-stained sections. The nuclei are usually centrally located and blunt-ended or "cigar-shaped". A leiomyosarcoma is usually immunoreactive for desmin and smooth muscle actin

Schwannoma, Malignant
Usually positive S-100 protein immunoreactivity. Variable proportions of Antoni type A tissue with Verocay bodies or Antoni type B tissue with cystic spaces can be present; the cells have typical wavy, buckled or comma-shaped nuclei

Comment

Some authors extend the histologic features of fibrosarcoma to include tumors that have pleomorphic cells, multinucleated cells, giant cells and "strap cells". However, these pleomorphic tumors probably represent a more primitive phenotype which are widely regarded as malignant fibrous histiocytomas, particularly in human pathology. Subcutaneous implantation of microchips has been reported to induce fibrosarcomas in mice.

References

(See 1, 2, 3, 10, 16, 24, 30, 37, 44)

Histiocytoma, Fibrous, Benign *(B)*

Histogenesis

Pluripotential mesenchymal stem cells.

Diagnostic Features

- Characteristic storiform or cartwheel pattern of neoplastic cells
- Abundant collagen production of generally short thin bundles
- Composed mainly of well-differentiated fibroblast-like cells, but has population of histiocytic cells
- Little nuclear pleomorphism and mitotic activity
- Occasionally phagocytic activity can be observed
- Inflammatory cells can be scattered throughout the tumor

Differential Diagnosis

Fibroma
This tumor lacks the storiform or cartwheel arrangement of tumor cells, has less cellularity, and no histiocyte-like tumor cells.

Schwannoma, Benign
Variable proportions of Antoni type A tissue with Verocay bodies or Antoni type B tissue with cystic spaces; the cells have typical wavy, buckled or comma-shaped nuclei and positive S-100 protein immunoreactivity

Histiocytoma, Fibrous, Malignant (Fibrous Type)
Cytologic and histologic features of malignancy, such as cellular pleomorphism, increased mitotic activity, tissue invasion or metastases, are present

Comment

These are rare neoplasms in most strains of mice.

References

(See 9, 10, 30)

Histiocytoma, Fibrous, Malignant *(M)*

Fibrous type • Pleomorphic type

Histogenesis

Pluripotential mesenchymal stem cells.

Diagnostic Features

Fibrous type

- Composed predominantly of fibroblastic spindle cells arranged in a storiform or cartwheel pattern
- Abundant collagen production
- Well-defined foci of necrosis and inflammatory cells are present
- Infiltrative growth is present and metastases are uncommon
- Mitotic index is high and nuclei are pleomorphic

Pleomorphic type

- Predominantly rounded histiocyte-like cells are present
- Bizarre multinucleated tumor giant cells are numerous
- Pleomorphic fibroblast-like cells, xanthoma cells and undifferentiated cells are present
- Focal collagen production is present
- Foci of necrosis are present
- Infiltrative growth is present and metastases are present
- Mitotic activity is variable
- Myxomatous and hemangiopericytoma-like areas can also be present

Differential Diagnosis

Fibrous Type of MFH or Histiocytoma, Fibrous, Benign

No cytologic and histologic features of malignancy, such as cellular pleomorphism, increased mitotic activity, tissue invasion or metastases, are present

Pleomorphic Type of MFH or Fibrosarcoma (Poorly Differentiated)

The cellular pleomorphism is less pronounced and histiocytic markers like cathepsin B, α_1-antitrypsin (AT) and α_1-antichymotrypsin (ACT) are usually not present in these tumors

Sarcoma, Histiocytic

This tumor is composed of well-differentiated histiocytic and multinucleated giant cells with little fibroblastic differentiation; necrotic areas are surrounded by pseudopalisaded tumor cells

Rhabdomyosarcoma

This tumor typically is composed of pleomorphic strap-like or spindle-shaped cells with cross-striations, abundant eosinophilic cytoplasm, and positive desmin, actin, myoglobin, and myosin immunoreactivity

Comment

Malignant fibrous histiocytoma is a rare tumor in most strains of mice and the incidence in mice is lower than in rats. MFHs are mesenchymal tumors that have fibroblastic and histiocytic differentiation, as reflected by collagen production and phagocytosis. Both fibrous and pleomorphic regions can occur in the same neoplasm.

References

(See 9, 10, 16, 25, 30, 43)

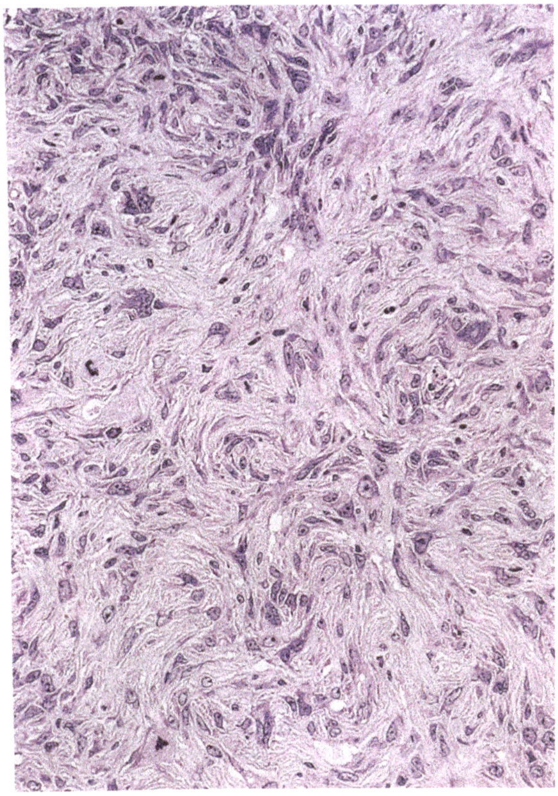

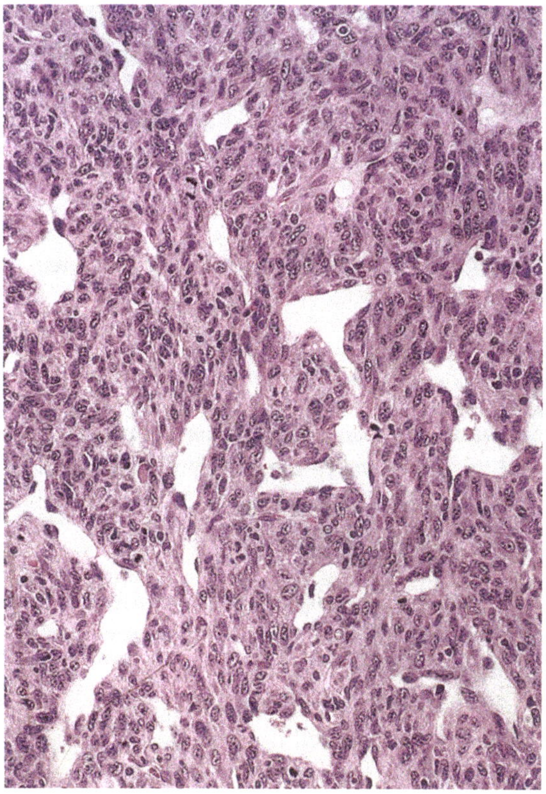

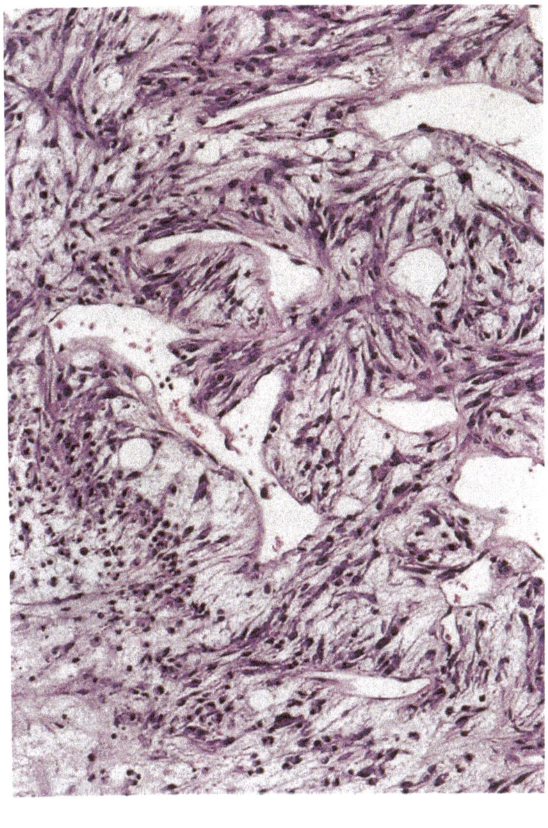

Fig. 15 *(upper left).* Malignant fibrous histiocytoma, fibrous/pleomorphic type, skin. The tumor consists of whorls of fibroblastic and pleomorphic histiocyte-like cells and has abundant collagenous stroma. CD-1 mouse, male, treated, age 76 weeks. H&E

Fig. 16 *(upper right).* Malignant fibrous histiocytoma, skin, showing an area with hemangiopericytoma-like differentiation. CBA/J mouse, female, age 83 weeks. H&E

Fig. 17 *(below).* Malignant fibrous histiocytoma, skin. The tumor contains areas of myxomatous change and degeneration. B6C3F1 mouse. H&E

Schwannoma, Benign *(B)*

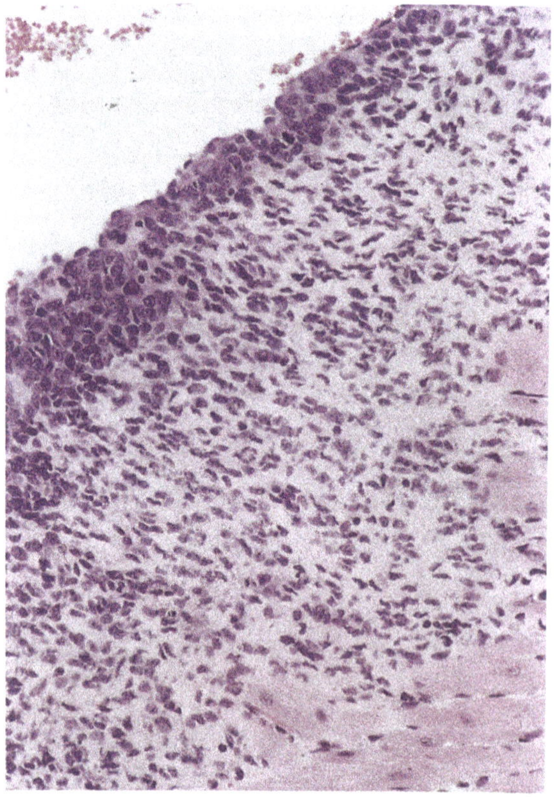

Fig. 18. Benign schwannoma, heart, growing along the endocardium of the left ventricle. CD-1 mouse, H&E

Synonym(s). Neurilemmoma, benign; neurinoma, benign

Histogenesis

Schwann cell, considered to be neuroectodermal with facultative mesenchymal features.

Diagnostic Features

- Expansive, compressing lesion and a capsule derived from the perineurium can be prominent
- Anaxonal Schwann cells are elongated, have indistinct cell borders, and form nuclear palisades (Antoni A pattern)
- Two adjacent palisades of nuclei and the intervening cytoplasm form a "Verocay body"
- Portions of the tumor can be sparsely cellular, with a clear matrix (Antoni B pattern)
- S-100 protein immunoreactivity and the presence in electron micrographs of basal lamina support the diagnosis of Schwannoma

Differential Diagnosis

Fibroma
: Benign tumor composed of interlacing bundles of collagen producing neoplastic fibroblasts without positive S-100 protein immunoreactivity

Leiomyoma
: Benign tumor composed of eosinophilic, spindle-shaped cells with blunt-ended nuclei and positive immunostaining for desmin and smooth muscle actin. Bundles of cells in leiomyoma are usually oriented perpendicular to each other

Schwannoma, Malignant
: Cellular atypia and invasive growth or distant metastases are present

Comment

Spontaneously-occurring Schwann cell neoplasms are rare in most strains of mice.

References

(See 41, 45)

Schwannoma, Malignant *(M)*

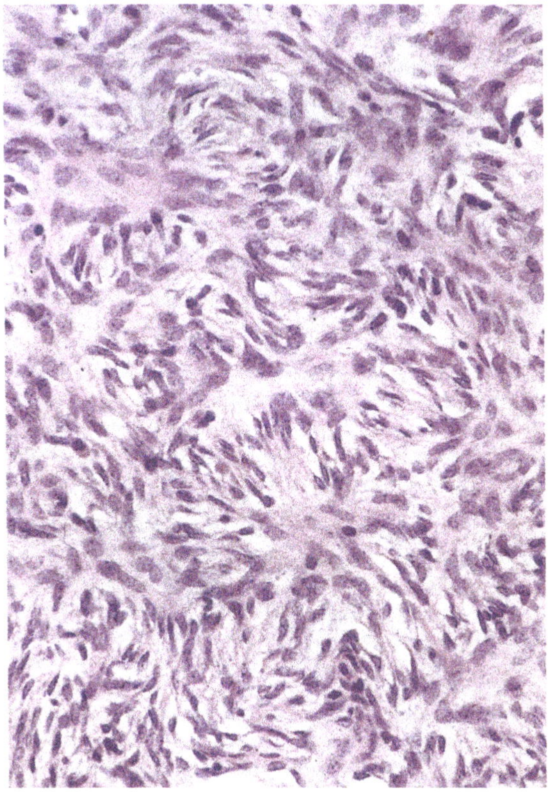

Synonym(s). Neurilemmoma, malignant; neurinoma, malignant

Histogenesis

Schwann cell, considered to be neuroectodermal with facultative mesenchymal features.

Diagnostic Features

- Neoplastic cells can be elongated and arranged in interlacing fascicles (Antoni A pattern), or cells can be more loosely arranged within a clear matrix (Antoni B pattern)
- Antoni A and B patterns are not always apparent
- Tumor can have cystic spaces containing proteinaceous fluid and blood cells, lined by epithelial-like, cuboidal cells
- S-100 protein immunoreactivity and the presence in electron micrographs of basal lamina support the diagnosis of Schwannoma
- Cellular or mitotic atypia, invasive growth, or distant metastases are features of malignancy

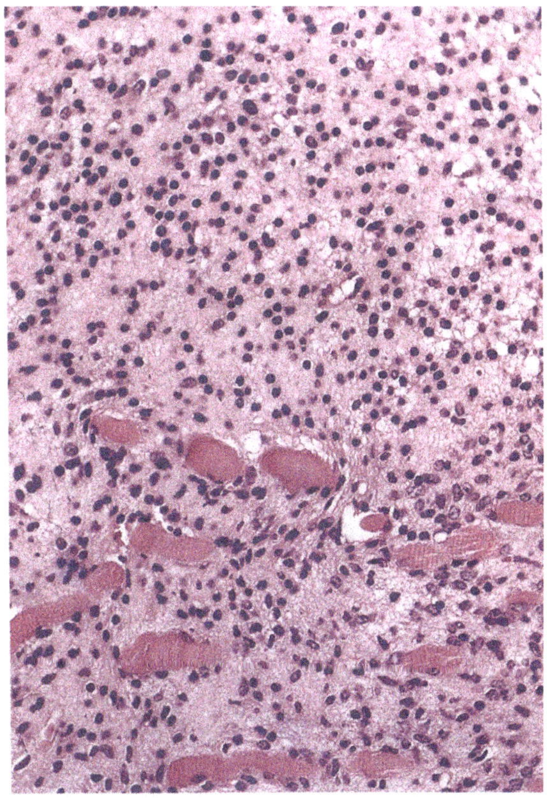

Fig. 19 *(above).* Malignant schwannoma, uterus. Elongated tumor cells are arranged in short bundles and tend to form nuclear palisades (Antoni type A pattern). B6C3F1 mouse. H&E

Fig. 20 *(below).* Malignant schwannoma, skin. Loosely arranged tumor cells with small nuclei in an edematous matrix (Antoni type B pattern). NMRI mouse, male. H&E

Differential Diagnosis

Fibrosarcoma
Malignant tumor is composed of collagen-producing neoplastic fibroblasts without positive S-100 protein immunoreactivity; no basal lamina present

Leiomyosarcoma
Malignant tumor is composed of eosinophilic, spindle-shaped cells that have blunt-ended nuclei and a positive immunostaining for desmin and smooth muscle actin and negative S-100 protein immunoreactivity

Schwannoma, Benign
No cellular atypia, invasive growth, or distant metastases are present

Comment

These are rare neoplasms in most strains of mice. An exception is the NHO strain in which an incidence of malignant schwannomas of 8.1% has been reported. In this strain, the neoplasms were found in the uterus, epididymis, and either spinal or cranial nerves.

References

(See 22, 30, 38, 41, 45)

Leiomyoma *(B)*

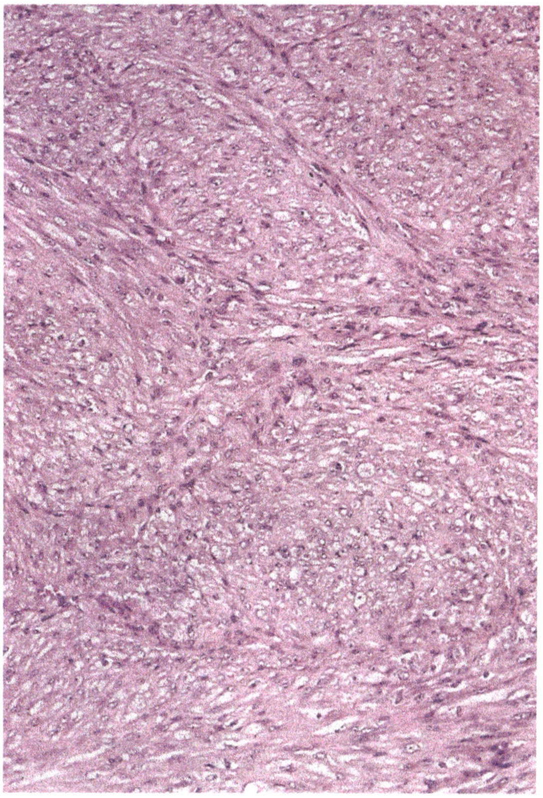

Fig. 21. Leiomyoma, uterus, composed of interlacing bundles of elongated tumor cells with eosinophilic cytoplasm and prominent perinuclear clear spaces. NMRI mouse, age 108 weeks. H&E

Histogenesis

Smooth muscle cells.

Diagnostic Features

- Well circumscribed, nodular tumor
- Interlacing bundles and whorls of uniform spindle cells resemble normal smooth muscle cells
- A characteristic criss-cross pattern of bundles is often observed
- Varying degrees of vascularity and amounts of fibrous tissue occur within the tumor
- Tumor cells have abundant eosinophilic cytoplasm, distinct cell boundaries, and cylindrical, blunt-ended or cigar-shaped nuclei
- Tumor cells frequently have a perinuclear clear space
- Longitudinal myofibrils can be demonstrated in the cytoplasm using the phosphotungstic acid-hematoxylin (PTAH) stain
- Nuclear pleomorphism and mitotic figures are minimal

Differential Diagnosis

Fibroma
The fibroma has significant amounts of collagenous tissue that can be demonstrated by special stains, such as van Gieson's or Masson-trichrome stain; cells have no longitudinal myofibrils

Schwannoma, Benign
Usually S-100 protein immunoreactivity is positive. Variable proportions of Antoni type A tissue with Verocay bodies or Antoni type B tissue with cystic spaces can be present; the cells have typical wavy, buckled, or comma-shaped nuclei

Leiomyosarcoma
Poorly circumscribed invasive tumor is composed of interwoven bundles of spindle-shaped cells with prominent nuclear pleomorphism and mitotic activity

Comment

As a further diagnostic aid, most leiomyomas have a positive immunohistochemical staining for desmin and smooth muscle actin. Leiomyomas are uncommon in mice and most of those described have occurred in the uterus.

References

(See 7, 10, 12, 16, 28, 30)

Leiomyosarcoma *(M)*

Histogenesis

Pluripotential mesenchymal stem cells; smooth muscle cells.

Diagnostic Features

- Ill-defined invasive tumor composed of interwoven fascicles of elongated spindle-shaped eosinophilic cells with blunt-ended oval nuclei
- High mitotic activity and prominent nuclear pleomorphism and hyperchromasia are present in some neoplasms
- Bizarre polygonal cells with double to multiple nuclei as well as vacuolated cells also can be observed
- Longitudinal myofibrils can be demonstrable in the cytoplasm using the phosphotungstic acid-hematoxylin (PTAH) stain
- Tumor cells frequently have a perinuclear clear space
- Varying degrees of vascularity are present
- Abundant reticulin fibers run parallel to the tumor cells
- Necrotic or cystic changes, mineralization, and distant metastases are common

Differential Diagnosis

Leiomyoma
Well-circumscribed, nodular appearance, little cellular pleomorphism, and low mitotic index are features of leiomyoma

Fibrosarcoma
Greater amounts of collagenous fibers are produced; negative immunostaining for desmin

Histiocytoma, Fibrous, Malignant
Storiform or cartwheel arrangement of the tumor cells; negative immunostaining for desmin and positive immunoreactivity for histiocytic markers such as cathepsin B, α_1-antitrypsin (AT) and α_1-antichymotrypsin (ACT)

Schwannoma, Malignant
Usually positive S-100 protein immunoreactivity. Variable proportions of Antoni type A tissue with Verocay bodies or Antoni type B tissue with cystic spaces can be present; the cells have typical wavy, buckled, or comma-shaped nuclei

Rhabdomyosarcoma
Cellular cross-striations are detectable; greater pleomorphism of the tumor cells including strap and racket cells

Comment

As a further diagnostic aid, most leiomyosarcomas have a positive immunohistochemical staining for desmin and smooth muscle actin. Leiomyosarcomas are rare in mice and usually have been found in the uterus, but occur in the urinary bladder and in the intestine as well.

References

(See 7, 8, 10, 12, 16, 20, 28, 30)

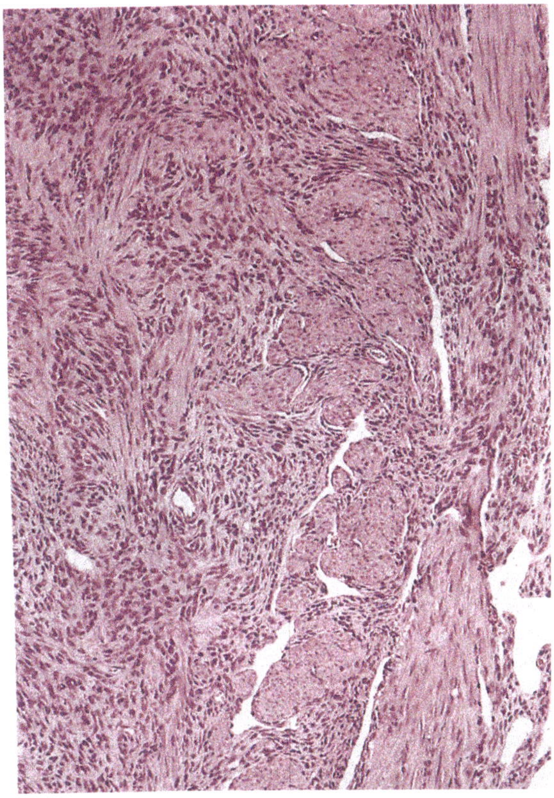

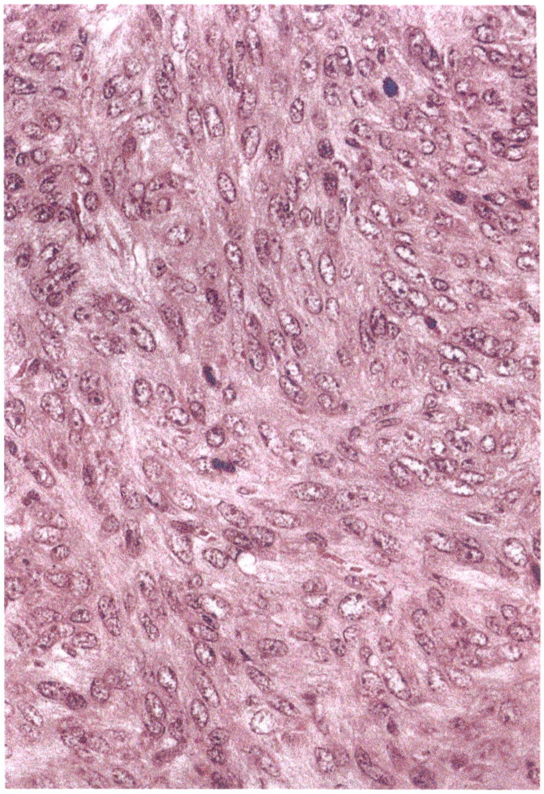

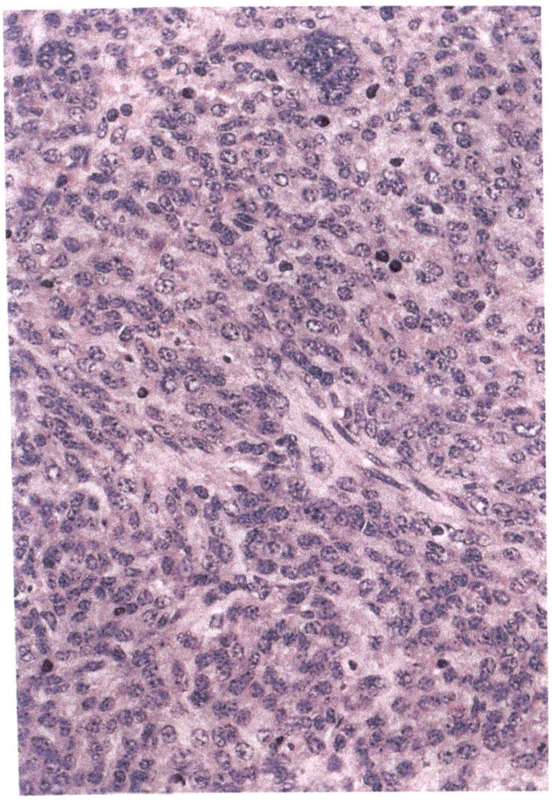

Fig. 22 *(upper left).* Leiomyosarcoma of the uterus infiltrating the myometrium. NMRI mouse, age 98 weeks. H&E

Fig. 23 *(upper right).* Leiomyosarcoma, uterus. Fascicles of elongated eosinophilic cells with large, blunt-ended nuclei. B6C3F1 mouse, age 79 weeks. H&E

Fig. 24 *(below).* Leiomyosarcoma, uterus. Less differentiated tumor with nuclear pleomorphism, high mitotic rate and occasional multinucleated cells. B6C3F1 mouse. H&E

Sarcoma, NOS *(M)*

Synonym(s). Sarcoma, undifferentiated

Histogenesis

Pluripotent mesenchymal stem cells.

Diagnostic Features

- Neoplasm composed of sheets of undifferentiated round to spindle-shaped pleomorphic or anaplastic cells
- Cytologic features by light microscopy do not allow for more definitive diagnosis into other categories of mesenchymal neoplasms
- Electron microscopic features and specific immunohistochemical markers could be sufficient to place some sarcomas, NOS under a specific diagnostic category
- Some sarcomas, NOS have no distinguishing ultrastructural or immunohistochemical features

Differential Diagnosis

Liposarcoma (Pleomorphic Type)
Pleomorphic cells with some having vacuolated cytoplasm positive for lipids with fat stains

Fibrosarcoma
Pleomorphic spindle-shaped cells forming interlacing bundles and producing varying amounts of collagen

Leiomyosarcoma
Composed of interwoven fascicles of spindle-shaped eosinophilic cells with blunt-ended to oval nuclei and positive immunoreactivity for desmin

Schwannoma, Malignant
Variable composition of interlacing fascicles of elongated cells or loosely arranged cells with clear matrix and cystic cavities; positive S-100 protein immunoreactivity

Rhabdomyosarcoma
Entangled to parallel bundles of pleomorphic spindle-shaped cells, giant cells and strap-like cells with cross striations in some cells and positive immunoreactivity for myoglobin

Comment

Unclassifiable sarcomas were found at a high incidence (10%) in p53$^{+/-}$ transgenic mice due to implantation of transponders, whereas in CD-1 and B6C3F1 mice this method of identification resulted only in non-neoplastic changes.

References

(See 1, 30)

Rhabdomyosarcoma *(M)*

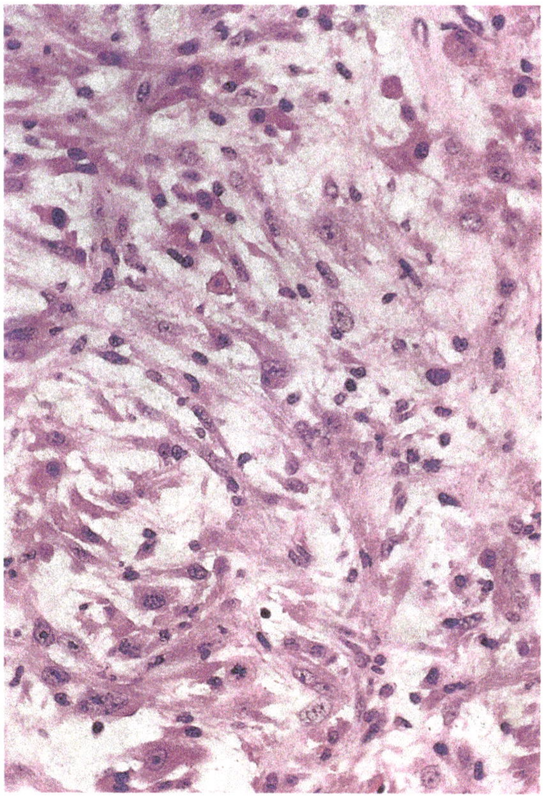

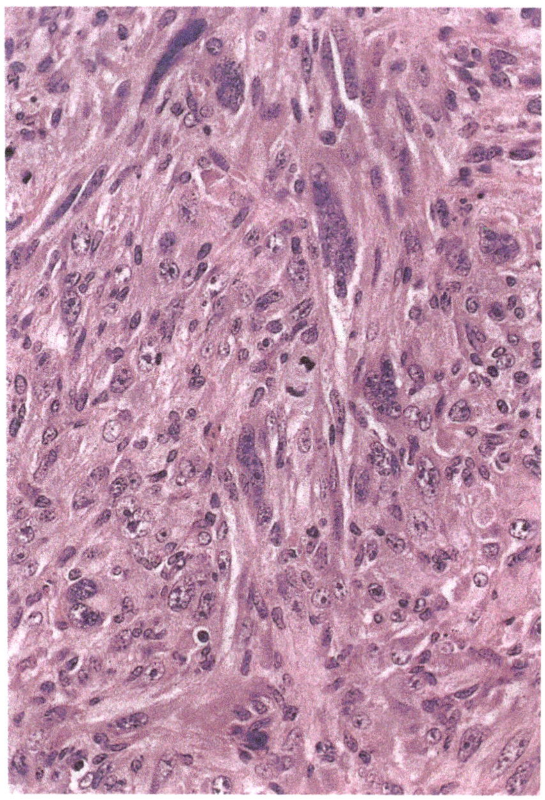

Histogenesis

Pluripotential mesenchymal stem cells; striated muscle cells.

Diagnostic Features

- Highly pleomorphic tumor composed of interlacing to entangled to parallel bundles of myoblast-like cells
- Various proportions of large pleomorphic strap-like cells, spindle-shaped cells, multinucleated giant cells, or round cells with one or two eccentrically located nuclei
- Tumor cells typically have moderate to abundant amounts of granular or filamentous eosinophilic cytoplasm with distinct cross-striations in PTAH-stained sections and PAS-positive glycogen inclusions
- High mitotic activity and numerous abnormal mitotic figures are present
- Areas of necrosis and hemorrhage are common
- Infiltrative growth and distant metastases are frequent

Fig. 25 *(above)*. Rhabdomyosarcoma, skeletal muscle, composed of myoblast-like cells and exhibiting myxoid appearance. NMRI mouse, male, age 77 weeks. H&E

Fig. 26 *(below)*. Rhabdomyosarcoma, skeletal muscle. Poorly differentiated tumor composed of highly pleomorphic cells having abundant granular eosinophilic cytoplasm. CBA/J mouse, male, age 70 weeks. H&E

Differential Diagnosis

Histiocytoma, Fibrous, Malignant
No cellular cross-striations are detectable; no positive myoglobin, myosin and desmin immunoreactivity

Leiomyosarcoma
No cellular cross-striations are detectable, less pleomorphism of elongated spindle-shaped tumor cells. Usually positive immunoreactivity for desmin and smooth muscle actin

Comment

This is a rare spontaneous tumor in mice and those of skeletal muscle occur more often than those of the heart. However, it can be induced experimentally by exposure to a variety of viruses, metals, and chemical carcinogens. The phosphotungstic acid-hematoxylin (PTAH) stain is suitable for identification of cellular cross-striations. The demonstration of *Z*-lines in electron micrographs can be a further aid to diagnosis. Myoglobin is a specific marker for rhabdomyosarcoma.

References

(See 5, 11, 12, 15, 16, 17, 18, 23, 24, 29, 31, 33, 35, 37, 39, 40, 42)

References

1. Blanchard KT, Barthel C, French JE, Holden HE, Moretz R, Pack FD, Tennant RW, Stoll RE (1999) Transponder-induced sarcoma in the heterozygous p53+/- mouse. Toxicol Pathol 27: 519–527
2. Bogovski P (1979) Tumours of the skin. In: Turusov VS (ed) Pathology of tumours in laboratory animals, vol II, Tumours of the mouse. IARC Scientific Publications No. 23, Lyon, pp 1–41
3. Boorman GA, Eustis SL, Elwell MR (1989) Fibrosarcoma, dermis and subcutis, mouse. In: Jones TC, Mohr U, Hunt RD (eds) Monographs on pathology of laboratory animals. Integument and mammary glands. Springer, Berlin Heidelberg New York Tokyo, pp 95–100
4. Booth CJ, Sundberg JP (1996) Hemangiomas and hemangiosarcomas. In: Mohr U, Dungworth DL, Capen CC, Carlton WW, Sundberg JP, Ward JM (eds) Pathobiology of the aging mouse, vol 1. Cardiovascular system. ILSI Press, Washington DC, pp 393–401
5. Carlton WW, Engelhardt JA (1991) Rhabdomyosarcoma, mouse and hamster. In: Jones TC, Mohr U, Hunt RD (eds) Monographs on pathology of laboratory animals. Cardiovascular and musculoskeletal systems. Springer, Berlin Heidelberg New York Tokyo, pp 135–140
6. Carlton WW, Ernst H, Faccini JM, Greaves P, Krinke GJ, Long PH, Maekawa A, Newsholme SJ, Weisse G (1992) 2. Soft tissue and musculoskeletal system. In: Mohr U, Capen CC, Dungworth DL, Griesemer RA, Ito N, Turusov VS (eds) International classification of rodent tumours, Part I: The rat. IARC Scientific Publications No. 122, Lyon
7. Chandra M, Frith CH (1991) Spontaneously occurring leiomyosarcomas of the mouse urinary bladder. Toxicol Pathol 19: 164–167
8. Dawson PJ, Brooks RE, Fieldsteel AH (1974) Unusual occurrence of endometrial sarcomas in hybrid mice. J Natl Cancer Inst 52: 207–214
9. Enzinger FM, Weiss SW (1988) Soft tissue tumors, 2nd edn. Mosby, St Louis, pp 223–300
10. Faccini JM, Abbott DP, Paulus GJJ (1990) Mouse histopathology. A glossary for use in toxicity and carcinogenicity studies. Elsevier, Amsterdam
11. Freeman AI, Johnson WW (1968) A comparative study of childhood rhabdomyosarcoma and virus-induced rhabdomyosarcoma in mice. Cancer Res 28: 1490–1500
12. Frith CH, Ward JM (1988) Color atlas of neoplastic and non-neoplastic lesions in aging mice. Elsevier, Amsterdam
13. Frith CH, Wiley L (1982) Spontaneous hepatocellular neoplasms and hepatic hemangiosarcomas in several strains of mice. Lab Anim Sci 32: 157–162
14. Giddens WE, Renne RA (1985) Haemangiosarcoma, nasal cavity, mouse. In: Jones TC, Mohr U, Hunt RD (eds) Monographs on pathology of laboratory animals. Respiratory systems. Springer, Berlin Heidelberg New York Tokyo, pp 72–74
15. Gregson RL (1984) A metastasizing cardiac rhabdomyosarcoma in a CD 1 strain mouse. J Comp Pathol 94: 477–480
16. Heider K, Eustis SL (1994) Tumours of the soft tissues. In: Turusov VS, Mohr U (eds) Pathology of tumours in laboratory animals, vol 2. Tumours of the mouse, 2nd edn. IARC Scientific Publications No. 111, Lyon, pp 611–649
17. Hoch-Ligeti C, Stewart HL (1984) Cardiac tumors of mice. J Natl Cancer Inst 72: 1449–1456
18. Hoch-Ligeti C, Restrepo C, Stewart HL (1986) Comparative pathology of cardiac neoplasms in humans and in laboratory rodents: a review. J Natl Cancer Inst 76: 127–142
19. Iwata H, Nomura Y, Enomoto M (1994) Spontaneous hemangioendothelial cell hyperplasia of the heart in B6C3F1 female mice. Toxicol Pathol 22: 423–429
20. Jacobs JB, Cohen SM, Arai M, Friedell GH, Bulay O, Urman HK (1976) Chemically induced smooth muscle tumors of the mouse urinary bladder. Cancer Res 36: 2396–2398
21. Jones G, Butler WH (1975) Morphology and spontaneous neoplasia. In: Butler WM, Newberne PM (eds) Mouse hepatic neoplasia. Elsevier, Amsterdam, pp 9, 21–59
22. Krinke G (1996) Nonneoplastic and neoplastic changes in the peripheral nervous system. In: Mohr U, Dungworth DL, Capen CC, Carlton WW, Sundberg JP, Ward JM (eds) Pathobiology of the aging mouse, vol 2. Nervous system. ILSI Press, Washington DC, pp 83-93
23. Leininger JR (1999) Skeletal muscle. In: Maronpot RR, Boorman GA, Gaul BW (eds) Pathology of the mouse. Reference and atlas. Cache River Press, Vienna, pp 637–643
24. Lombard LS (1982) Neoplasms of musculoskeletal system. In: Foster HL, Small JD, Fox JG (eds) The mouse in biomedical research, vol IV. Experimental biology and oncology. Academic Press, San Diego, pp 501–511
25. Maita K, Hirano M, Harada T, Mitsumori K, Yoshida A, Takahashi K, Nakashima N, Kitazawa T, Enomoto A, Inui K, Shirasu Y (1988) Mortality, major cause of moribundity, and spontaneous tumors in CD-1 mice. Toxicol Pathol 16: 340–349

26. Morgan KT, Sheldon WG (1988) Lipoma, brain, mouse. In: Jones TC, Mohr U, Hunt RD (eds) Monographs on pathology of laboratory animals. Nervous system. Springer, Berlin Heidelberg New York Tokyo, pp 130–134
27. Morgan KT, Frith CH, Swenberg JA, McGrath JT, Zulch KJ, Crowder DM (1984) A morphologic classification of brain tumors found in several strains of mice. J Natl Cancer Inst 72: 151–160
28. Munoz N, Dunn TB, Turusov VS (1979) Tumours of the vagina and uterus. In: Turusov VS (ed) Pathology of tumours in laboratory animals, vol II. Tumours of the mouse. IARC Scientific Publications No. 23, Lyon, pp 359–383
29. Nettleship A (1943) Study of a spontaneous mouse rhabdomyosarcoma. J Natl Cancer Inst 3: 563–568
30. Peckham JC, Heider K (1999) Skin and subcutis. In: Maronpot RR, Boorman GA, Gaul BW (eds) Pathology of the mouse. Reference and atlas. Cache River Press, Vienna, pp 555–612
31. Perk K, Moloney JB (1966) Pathogenesis of a virus-induced rhabdomyosarcoma in mice. J Natl Cancer Inst 37: 581–599
32. Pozharisski KM, Turusov VS (1991) Angiosarcoma of the renal capsule, mouse. In: Jones TC, Mohr U, Hunt RD (eds) Monographs on pathology of laboratory animals. Cardiovascular and musculoskeletal systems. Springer, Berlin Heidelberg New York Tokyo, pp 91–97
33. Reznik M, Nameroff MA, Hansen JL (1970) Ultrastructure of a transplantable murine rhabdomyosarcoma. Cancer Res 30: 601–610
34. Sakamoto M, Takayama S, Hosoda Y (1989) Hemangioendothelial sarcoma in brown adipose tissue of mouse induced by carcinogenic heterocyclic amine, Glu-P-1. Toxicol Pathol 17: 754–758
35. Saxen EA (1953) On the factor of age in the production of subcutaneous sarcomas in mice by 20-methylcholanthrene. J Natl Cancer Inst 14: 547–569
36. Solleveld HA, Miller RA, Banas DA, Boorman GA (1988) Primary cardiac hemangiosarcomas induced by 1,3-butadiene in B6C3F1 hybrid mice. Toxicol Pathol 16: 46–52
37. Stewart HL (1979) Tumours of the soft tissues. In: Turusov VS (ed) Pathology of tumours in laboratory animals, vol II. Tumours of the mouse. IARC Scientific Publications No. 23, Lyon, pp 487–526
38. Stewart HL, Deringer MK, Dunn TB, Snell KC (1974) Malignant schwannomas of nerve roots, uterus, and epididymis in mice. J Natl Cancer Inst 53: 1749–1758
39. Sundberg JP, Adkison DL, Bedigian HG (1991) Skeletal muscle rhabdomyosarcomas in inbred laboratory mice. Vet Pathol 28: 200–206
40. Sundberg JP, Prattis S, Dudley M (1996) Skeletal muscle rhabdomyosarcomas in inbred laboratory mice. In: Mohr U, Dungworth DL, Capen CC, Carlton WW, Sundberg JP, Ward JM (eds) Pathobiology of the aging mouse, vol 2. Musculoskeletal system. ILSI Press, Washington DC, pp 415–423
41. Swenberg JA (1982) Neoplasms of the nervous system. In: Foster HL, Small JD, Fox JG (eds) The mouse in biomedical research, vol IV. Experimental biology and oncology. Academic Press, San Diego, pp 529–537
42. Szepsenwol J, Boschetti NV (1975) Primary and secondary heart tumors in mice maintained on various diets. Oncology 32: 58–72
43. Takahashi K, Maita K, Shirasu Y (1990) Eosinophilic globule cells in mouse MFH-like sarcomas. Light and electron microscopic studies. Virchows Arch B 59: 367–376
44. Tillmann T, Kamino K, Dasenbrock C, Ernst H, Kohler M, Morawietz G, Campo E, Cardesa A, Tomatis L, Mohr U (1997) Subcutaneous soft tissue tumours at the site of implanted microchips in mice. Exp Toxicol Pathol 49: 197–200
45. Walker VE, Morgan KT, Zimmerman HM, Innes JRM (1994) Tumours of the central and peripheral nervous system. In: Turusov VS, Mohr U (eds) Pathology of tumours in laboratory animals, vol 2. Tumours of the mouse, 2nd edn. IARC Scientific Publications No. 111, Lyon, pp 731–776
46. Ward JM, Goodman DG, Squire RA, Chu KC, Linhart MS (1979) Neoplastic and nonneoplastic lesions in aging (C57BL/6 N x C3H/HeN)F1 (B6C3F1) mice. J Natl Cancer Inst 63: 849–854
47. Yamate J, Tajima M, Ihara M, Shibuya K, Kudow S (1988) Spontaneous vascular endothelial cell tumors in aged B6C3F1 mice. Jpn J Vet Sci 50: 453–461

Skeletal System and Teeth

H. Ernst[3] · P.H. Long[1] · P.F. Wadsworth · J.R. Leininger[2] · S. Reiland
Y. Konishi

[1]Chair; [2]Co-chair; [3]Initial draft.

Osteoma *(B)*

Synonym(s). Cancellous osteoma; compact osteoma; juxta cortical (parosteal) osteoma; medullary osteoma (enostosis); spongious osteoma; trabecular osteoma

Histogenesis

Osteoblasts, osteocytes.

Diagnostic Features

- Preferentially arise from the periosteal surface of cortical bone
- Expansile growth pattern with sharp demarcation from surrounding tissue
- Composed of very dense bone with a predominantly woven structure
- Many lacunae lack cells
- Lacks spindle cell stroma described for osteofibroma
- Smooth outline with inactive or active osteoblasts
- Fatty or hematopoietic marrow may be present

Differential Diagnosis

Bony Exostosis or Reactive Bone
Distinguishable only by the essential absence of any tumor characteristics

Osteofibroma
Spindle cells are present between the bony trabeculae

Osteosarcoma
Low degree of tissue differentiation (immature or atypical bone) and presence of cytological criteria of malignancy

Comment

Spontaneous osteomas are rare in most strains of mice. They most often occur in the skull and seem to be more common in females than in males. The influence of female sexual hormones has been demonstrated. They can be induced in mice by several viruses.

References

(See 2, 7, 14, 17, 19, 24, 25, 40)

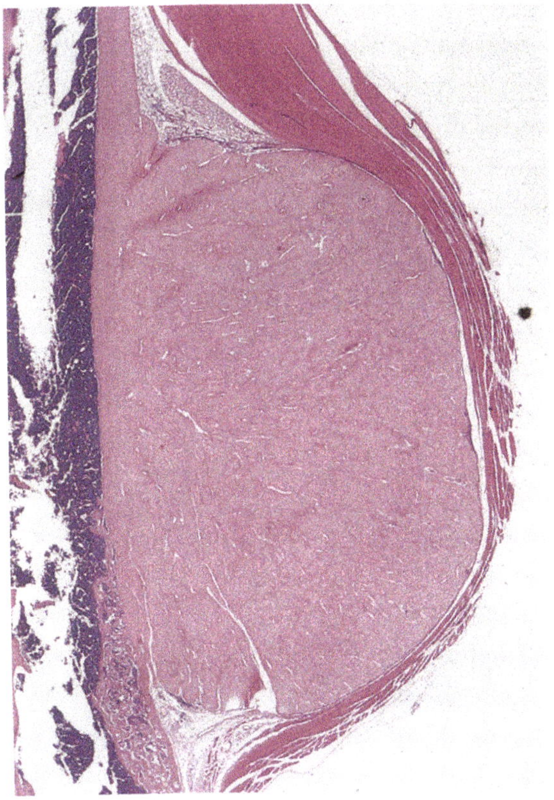

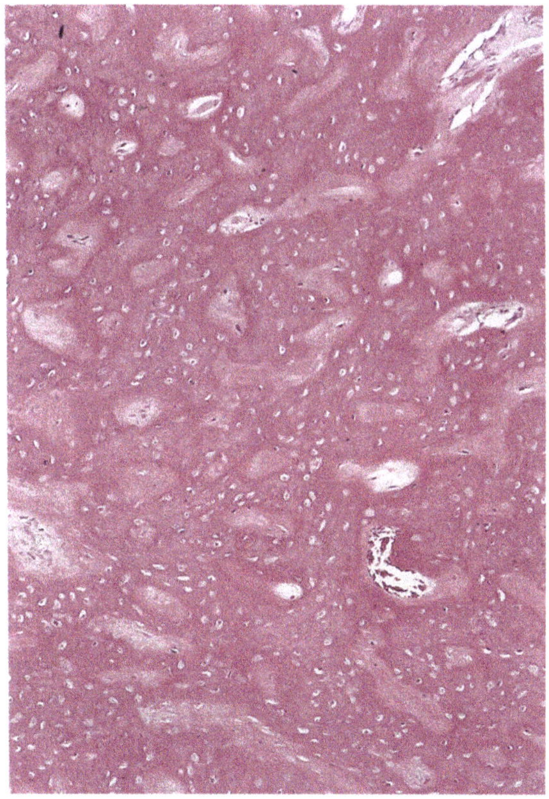

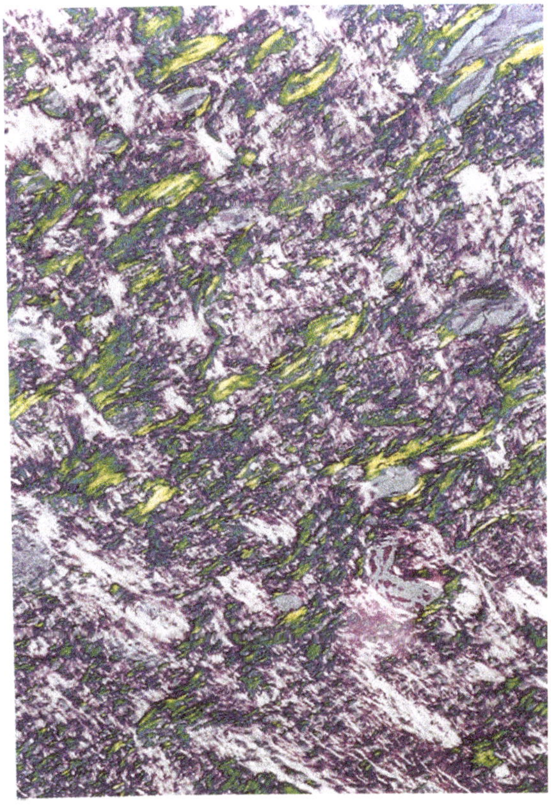

Fig. 1 *(upper left)*. Osteoma, skull. Expansively growing compact bone mass with smooth contours. B6C3F1 mouse. H&E

Fig. 2 *(upper right)*. Osteoma, skull. Dense and poorly cellular woven bone. B6C3F1 mouse. H&E

Fig. 3 *(below)*. Osteoma, skull, showing the typical irregular criss-cross pattern of woven bone. B6C3F1 mouse. Polarized light microscopy. Masson trichrome

Osteosarcoma *(M)*

Synonym(s). Osteogenic sarcoma

Histogenesis

Pluripotential mesenchymal stem cells, osteoblasts, osteocytes.

Diagnostic Features

- Highly invasive and destructive tumor arising in the skeleton
- Metastases are common, predominantly in liver, spleen and lungs
- The formation of tumor osteoid is an essential feature of osteosarcoma
- Bone formed is typically immature woven bone
- Progressive maturation from newer to older portions of the tumor may be observed
- Marrow cavities with hematopoietic tissue and osteoclasts may be present
- Lysis of normal bone may occur simultaneously with neoplastic osteogenesis
- Foci of collagen, cartilage and mucoid ground substance may occur in different areas of the same tumor
- Proliferating cells may be pleomorphic, spindle-shaped or multinucleate
- Foci of necrosis and hemorrhage may be extensive
- Mitotic figures are variable
- For subclassification, one or more of the following optional qualifiers may be used depending upon the morphologic characteristics of the tumor:
  - Eburnating (osteoma-like, but with pleomorphic and infiltrating cells at the tumor periphery)
  - Osteoblastic (highly differentiated with varying amounts of osteoid and bone)
  - Fibroblastic (large, basophilic spindle-shaped cells and variable amount of osteoid)
  - Osteoclastic (predominantly osteoclast-like giant cells)
  - Chondroblastic (associated areas of immature cartilage and bone)
  - Vascular (large blood-filled sinuses present)
  - Anaplastic (very low degree of tissue differentiation)
  - Mixed (combination of two or more histologic types)

Figs. 8-10 see p. 394

Fig. 4 *(upper left).* Osteosarcoma, eburnating type, skull. Tumor is mainly composed of tumor osteoid/bone. C57BL mouse. H&E ▶

Fig. 5 *(upper right).* Osteosarcoma, osteoblastic type, vertebra. Tumor is composed mainly of neoplastic osteoblasts. C57BL mouse. H&E

Fig. 6 *(lower left).* Osteosarcoma, osteoblastic type, metastasis, liver. C57BL mouse. H&E

Fig. 7 *(lower right).* Osteosarcoma, fibroblastic type, femur. Small clumps of osteoid are surrounded by elongated tumor cells simulating the pattern of fibrosarcoma. C57BL mouse. H&E

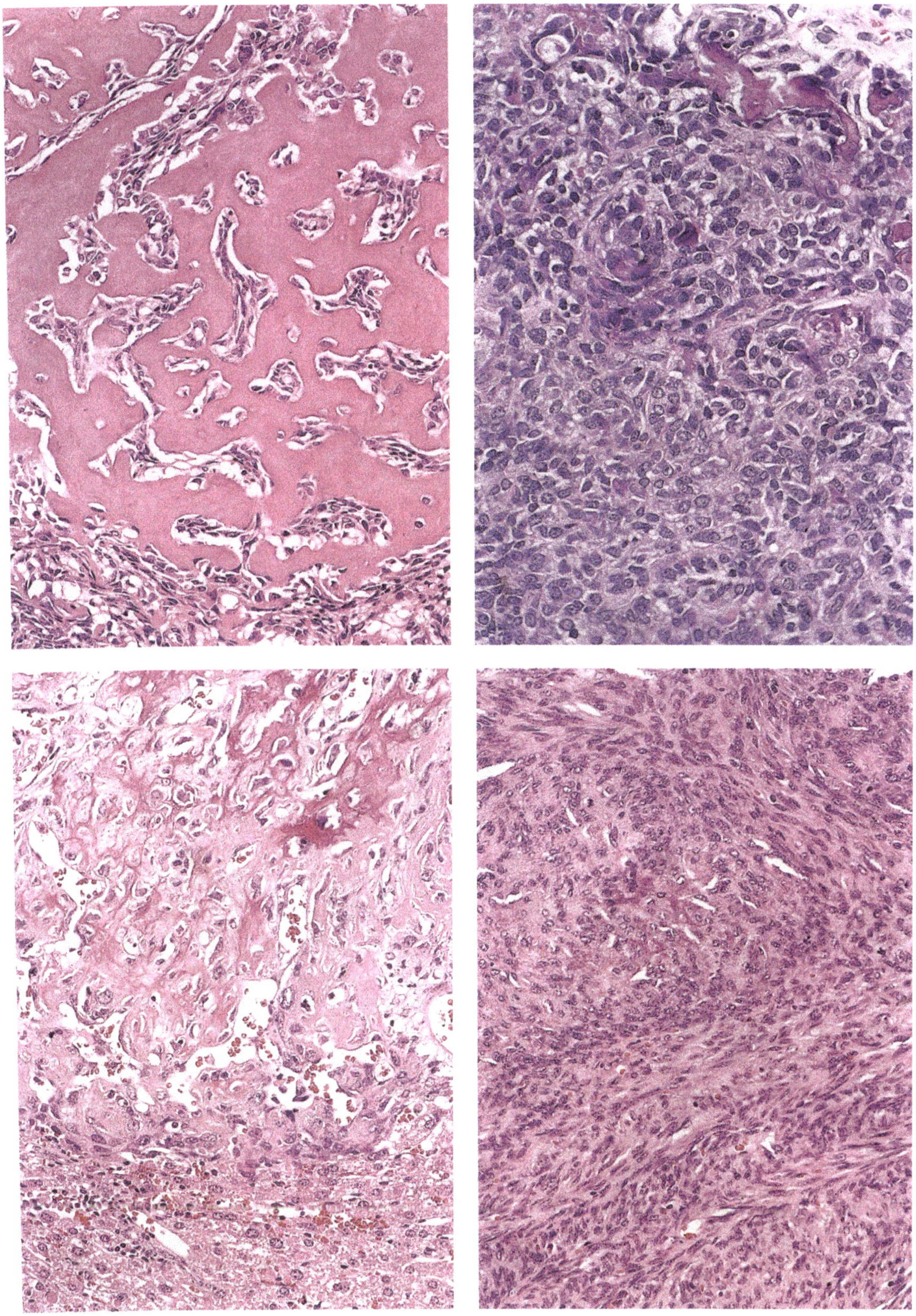

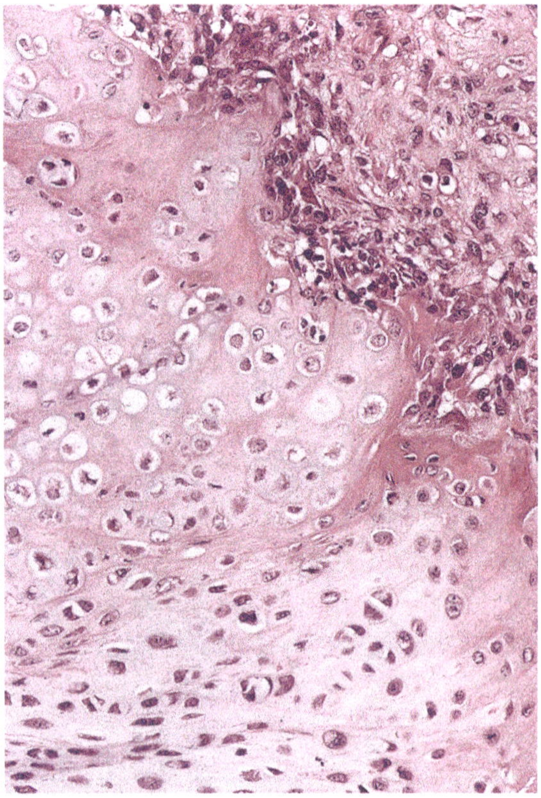

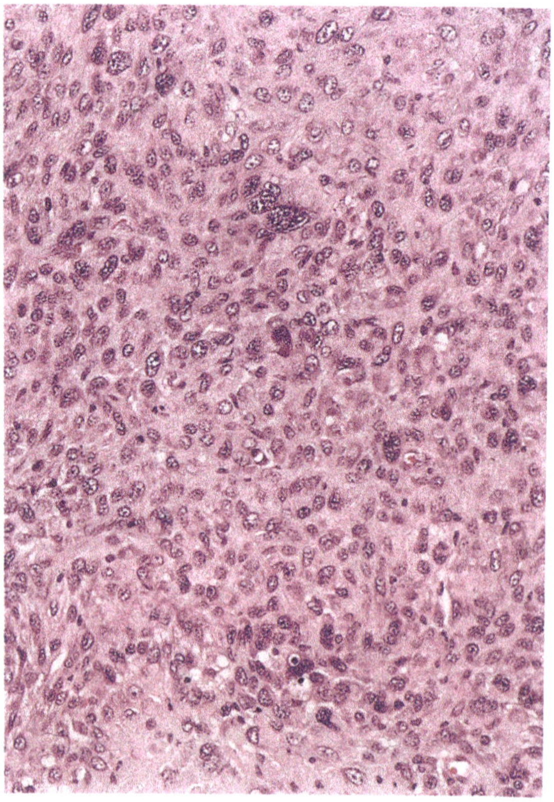

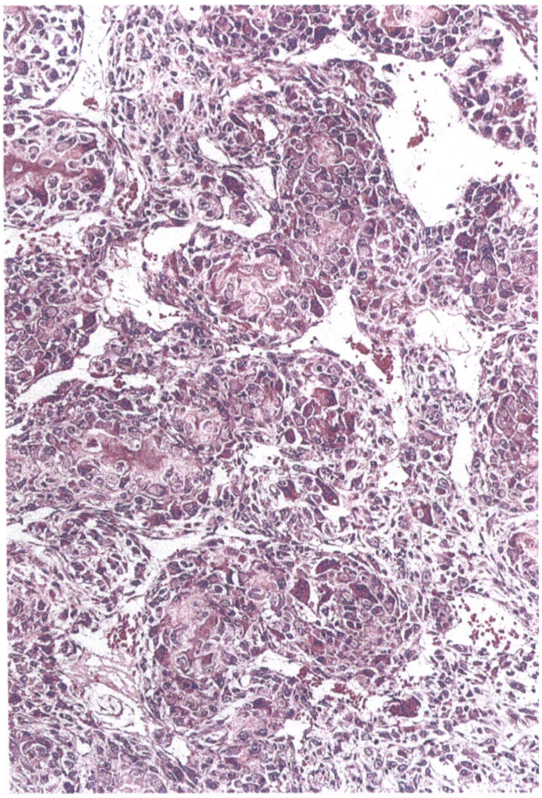

Fig. 8 *(upper left).* Osteosarcoma, chondroblastic type, vertebra. Composed of both cartilagenous and osteoid tumor tissue. C57BL mouse. H&E

Fig. 9 *(upper right).* Osteosarcoma, anaplastic type, vertebra. Tumor consists mainly of pleomorphic cells with only little osteoid formation. C57BL mouse. H&E

Fig. 10 *(below).* Osteosarcoma, vascular type, vertebra. Tumor tissue contains large, blood-filled and branching vascular channels. p53 heterozygous mouse, male, age 32 weeks. H&E

Differential Diagnosis

Osteoma
Well-defined compact osseous tumor consisting of dense woven bone and lacking pleomorphic, infiltrating tumor cells indicative of malignancy

Osteofibroma
Well-differentiated spindle cells between bony trabeculae and lacking cytological criteria of malignancy

Chondrosarcoma or Hemangiosarcoma or Fibrosarcoma
Tumor cells do not form osteoid

Comment

Most osteosarcomas originate from the metaphysis. There is no obvious correlation between histologic differentiation or tumor size and the probability of metastasis. Tumors of the caudal vertebrae have the highest frequency of metastasis.
Subclassification of osteosarcoma is possible but not required for risk assessment purposes.

References

(See 1, 2, 3, 6, 7, 9, 10, 11, 19, 23, 25, 28, 38, 39)

Osteofibroma *(B)*

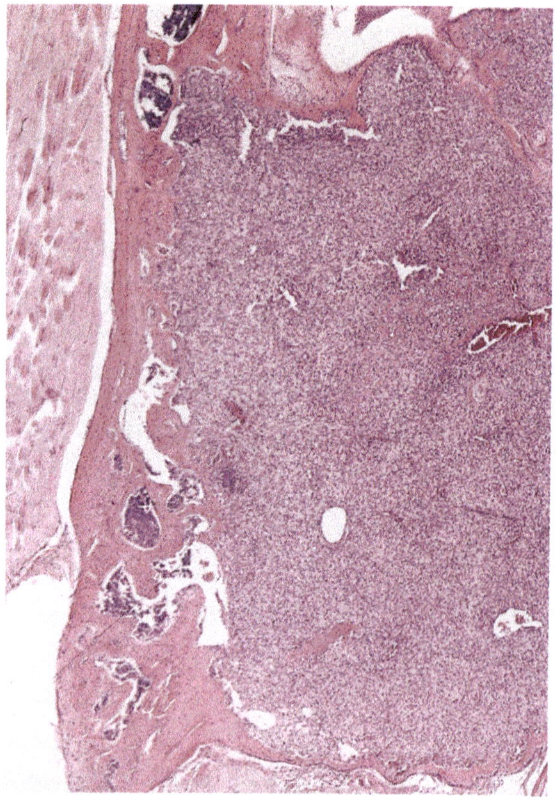

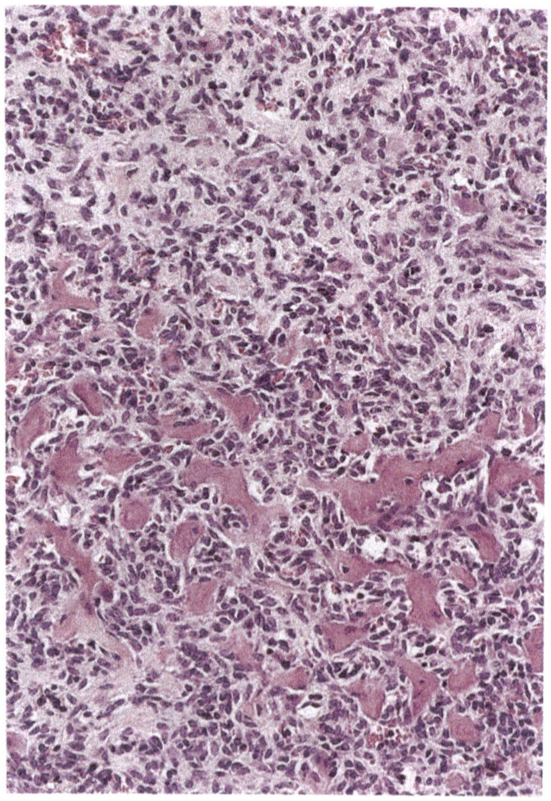

Synonym(s). Osteoblastoma

Histogenesis

Pluripotential mesenchymal stem cells, osteoblasts.

Diagnostic Features

- Expansively growing mass arising preferentially in the vertebral column
- May appear as osteosarcoma radiographically
- Appear to arise from medullary cavity (as opposed to periosteal surface)
- Composed of mature bone trabeculae separated by stroma of spindle cells
- Scattered highly cellular areas with osteoid spiculae are also typical
- Both active osteoblasts and osteoclasts may be present
- Blood-forming marrow may occur
- Stromal cells are not highly pleomorphic and do not invade surrounding tissue

Fig. 11 *(above).* Osteofibroma, vertebra. NMRI mouse, female, treated, age 101 weeks. H&E (section by kind permission of A. Luz)

Fig. 12 *(below).* Osteofibroma, vertebra. Tumor consists of spindle-shaped cells surrounding thin osteoid spiculae. NMRI mouse, female, treated, age 101 weeks. H&E (section by kind permission of A. Luz)

Differential Diagnosis

Bony Exostosis or Reactive Bone
Distinguishable by the absence of any tumor characteristics

Fibro-Osseous Lesion
Characterized by proliferation of fibrovascular stroma, osteoblasts, and osteoclasts. Bone turnover is accelerated with spreading resorption encroaching on the center of the medullary cavity followed by bone formation

Osteoma
Lacks spindle cell stroma

Ossifying Fibroma
The spindle cell component is more prominent and the bone trabeculae show letter-like profiles

Osteosarcoma
Low degree of tissue differentiation (immature or atypical bone), infiltrative growth and presence of cytological criteria of malignancy

Comment

Osteofibromas have only been reported in mice treated with radionuclides. They may progress to osteosarcomas. Due to their preferential location in the vertebral column, they may lead to paralysis

References

(See 6, 13, 19, 22, 25)

Ossifying Fibroma *(B)*

Synonym(s). Cementifying fibroma

Histogenesis

Pluripotential mesenchymal stem cells, osteoblasts (cementoblasts).

Diagnostic Features

- Well-demarcated tumor occurring exclusively in the jaw
- Appearance is suggestive of a fibroma in which bone/cementum forms by osseous metaplasia of the fibrous connective tissue component
- The proliferating component consists of spindle cells resembling fibroblasts that undergo transformation to cuboidal osteoblasts/cementoblasts and form multiple rounded cementicle-like bodies or cementum-like trabeculae with blue-staining borders
- Cell processes run perpendicular to the hard-tissue surface
- Bone spicules are composed almost entirely of woven bone
- Bone spicules often appear as letter-like "C" and "Y" profiles
- Usually separated from surrounding tissue by a thin shell of newly-formed cortical bone
- Low rate of mitotic figures

Differential Diagnosis

Fibro-Osseous Lesion
Distinguishable by the heterogeneous cell population and absence of any tumor criteria

Osteoma
Lacks spindle cell stroma

Osteofibroma
Tumor has much less spindle cells and fibrous tissue and lacks letter-like shape of bony trabeculae

Osteosarcoma
Low degree of tissue differentiation (immature or atypical bone), infiltrative growth and presence of cytological criteria of malignancy

Comment

Spontaneous ossifying fibromas are rare in mice and occur in the jaws. In this location they often surround an incisor tooth, contain cementicles and are considered to originate from the periodontal membrane.

References

(See 7, 20, 21, 25)

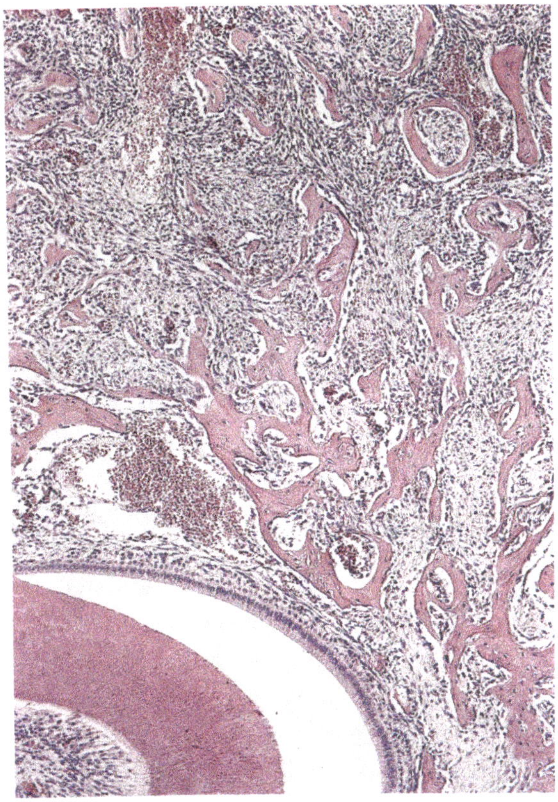

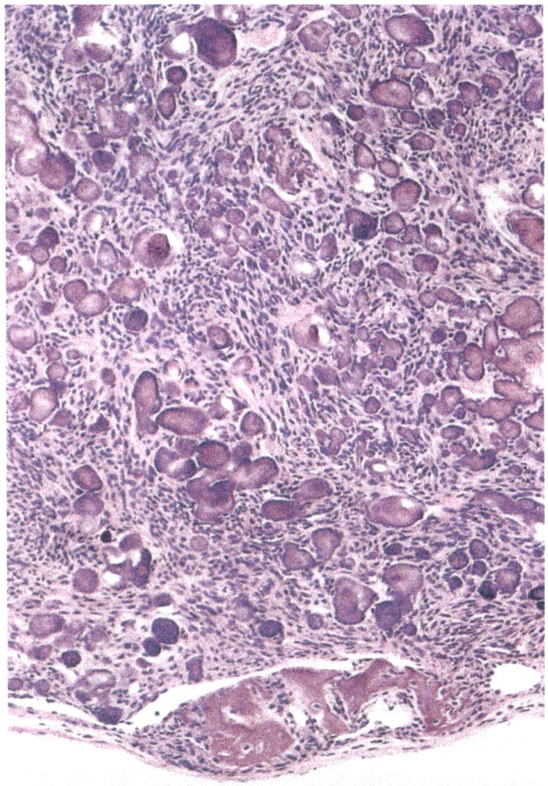

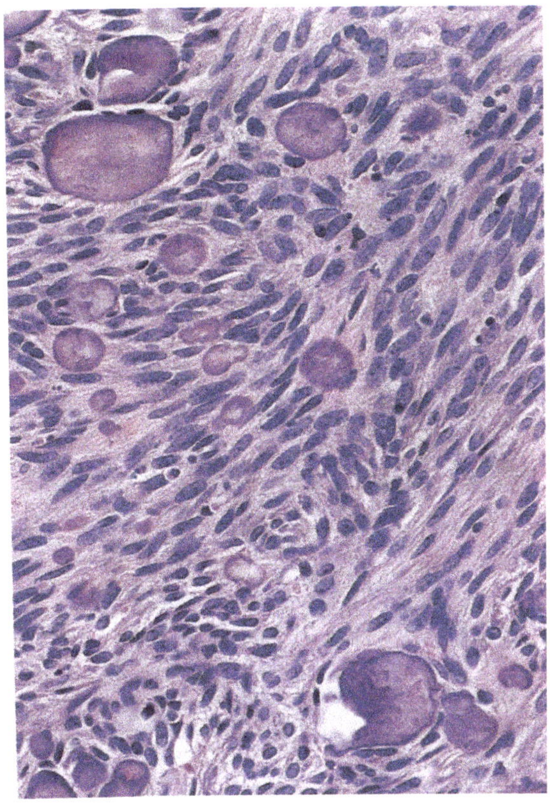

Fig. 13 *(upper left).* Ossifying fibroma, lower jaw, consisting of letter-like bone spicules and spindle cells. NMRI mouse, female, treated, age 101 weeks. H&E (section by kind permission of A. Luz)

Fig. 14 *(upper right).* Cementifying fibroma, upper jaw, eroding the maxillary bone. NMRI mouse, female. H&E

Fig. 15 *(below).* Cementifying fibroma, upper jaw. The tumor is composed of fibroblast-like spindle cells and cementicles which form in the intercellular matrix. NMRI mouse, female. H&E

Chondroma *(B)*

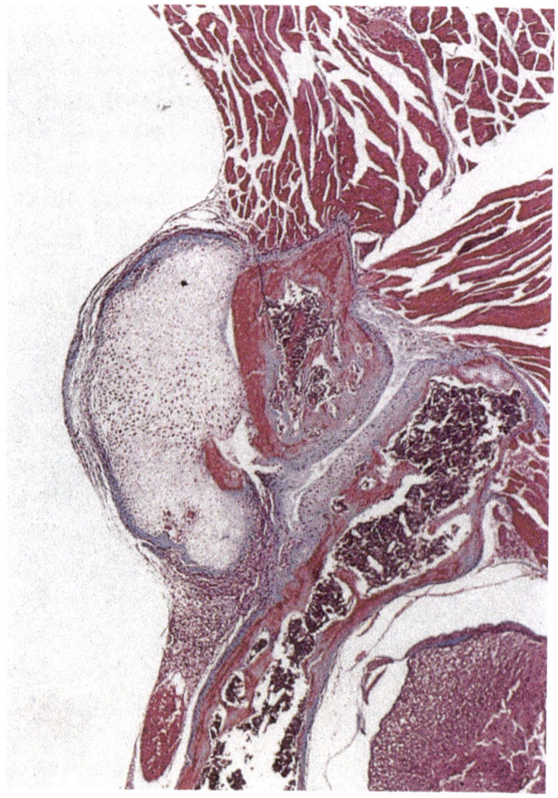

Fig. 16. Chondroma, vertebra. Expansile tumor composed of mature hyaline cartilage. B6C3F1 mouse. Masson trichrome

Histogenesis

Chondroblasts, chondrocytes.

Diagnostic Features

- Well-circumscribed expansile tumor consisting of lobules of hyaline cartilage
- Well-differentiated chondrocytes lying individually in lacunae
- No cellular and nuclear pleomorphism
- Matrix may have areas of osseous metaplasia

Differential Diagnosis

Chondrosarcoma

Pleomorphic, binucleated or multinucleated cartilage cells may be seen in addition to infiltrative growth or metastases

Comment

Chondromas are extremely rare in mice.

References

(See 19, 25, 35, 39)

Chondrosarcoma *(M)*

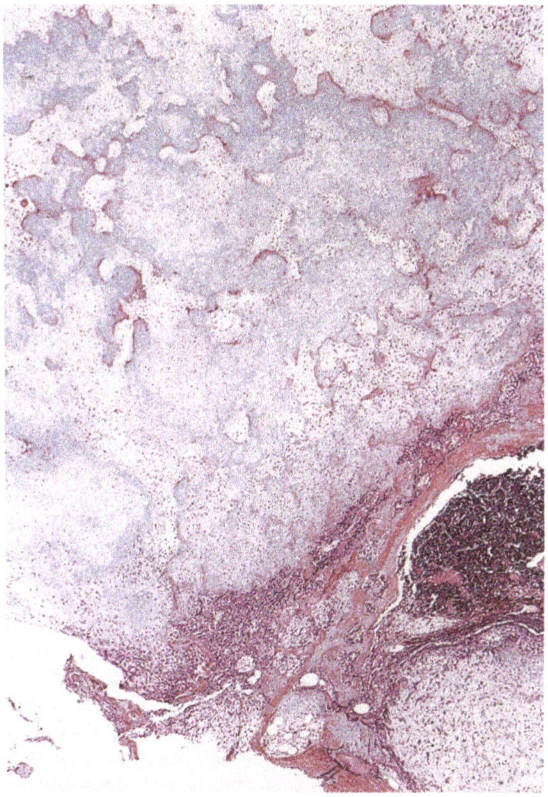

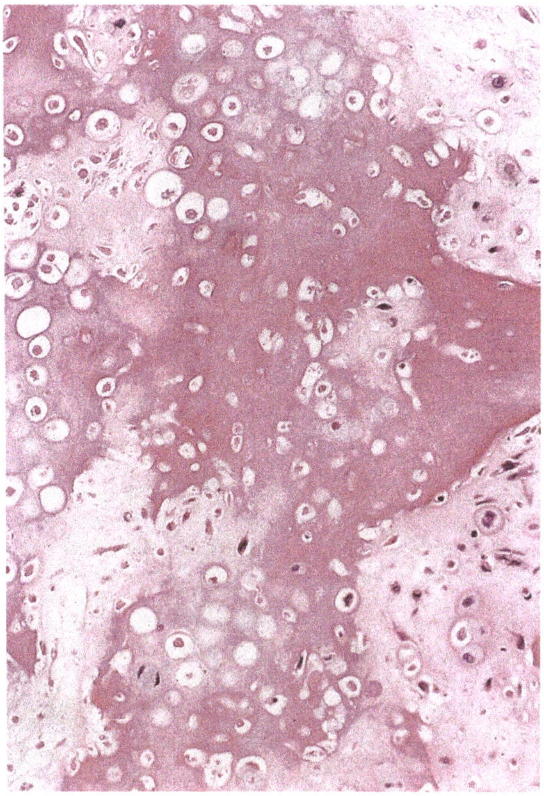

Histogenesis

Pluripotential mesenchymal stem cells, chondroblasts, chondrocytes.

Diagnostic Features

- Lobulated, highly cellular tumor with minimal tendency to invade surrounding tissues
- Well-differentiated tumors have large basophilic chondrocytes with large nuclei and prominent nucleoli lying individually in lacunae and surrounded by a hyaline ground substance
- Occasional pleomorphic, binucleated or multinucleated giant cartilage cells
- Mitotic figures are usually rare
- Poorly differentiated chondrosarcomas have spindle-shaped cells reminiscent of fetal chondroblasts
- No formation of bone or tumor osteoid but sometimes abundant foci of calcium, extensive necrosis and cyst formation may occur

Fig. 17 *(above)*. Chondrosarcoma, vertebra. The cartilagenous matrix shows irregular areas of mineralization. B6C3F1 mouse. Masson trichrome

Fig. 18 *(below)*. Chondrosarcoma, vertebra. Partially necrotic tumor tissue containing variably sized lacunae with pleomorphic tumor cells. B6C3F1 mouse. H&E

Differential Diagnosis

Osteosarcoma
Although chondrosarcomatous areas may occur, the production of tumor osteoid or bone is the *sine qua non* of osteosarcoma

Chondroma
Well-differentiated tumor without cytologic and histologic features of malignancy

Comment

Chondrosarcomas are extremely rare in mice. They may also arise from extraskeletal cartilage in the third eyelid, the external ear and the larynx.

References

(See 10, 19, 25, 35, 39)

Sarcoma, Synovial *(M)*

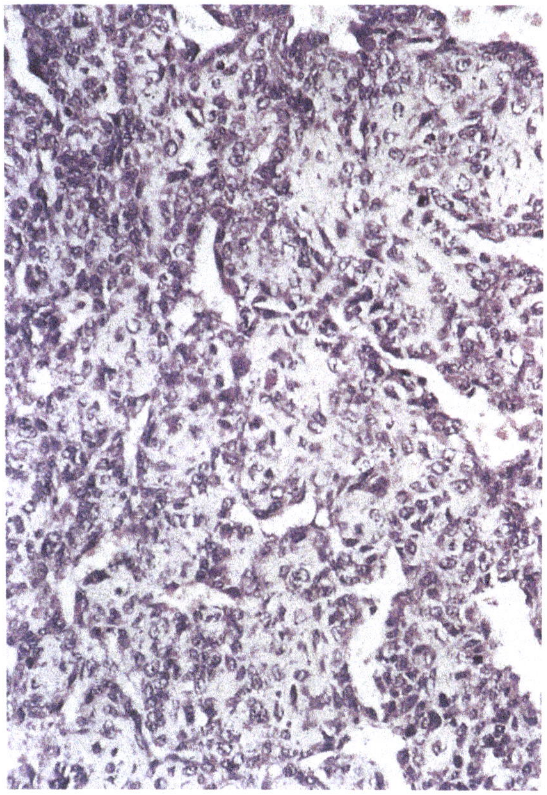

Fig. 19. Synovial sarcoma mainly composed of branching sheets of plump ovoid or polyhedral (epitheloid) cells. H&E

Histogenesis

Pluripotential mesenchymal stem cells.

Diagnostic Features

- Close association with tendon sheaths, bursae, and joint capsules
- Destroys and invades joint structures and adjacent bones
- Small uniform cells with indistinct cytoplasmic borders forming loosely organized fronds or finger-like projections
- Biphasic pattern with a fibrosarcomatous and an epithelioid component, which may form pseudoglandular spaces with PAS-positive material may be observed
- Monophasic pattern with spindle cells resembling fibrosarcoma may also occur
- Multinucleated giant cells are occasionally observed

Differential Diagnosis

Fibrosarcoma or Histiocytoma, Fibrous, Malignant
Usually not intimately associated with joints and have no biphasic pattern

Osteosarcoma
The production of tumor osteoid or bone is the *sine qua non* of osteosarcoma

Comment

Synovial sarcomas are extremely rare in mice. However, they have been induced in mice by subcutaneous injections of a cannabinoid.

References

(See 19, 25, 29, 34, 36)

Ameloblastoma *(B)*

Acanthomatous

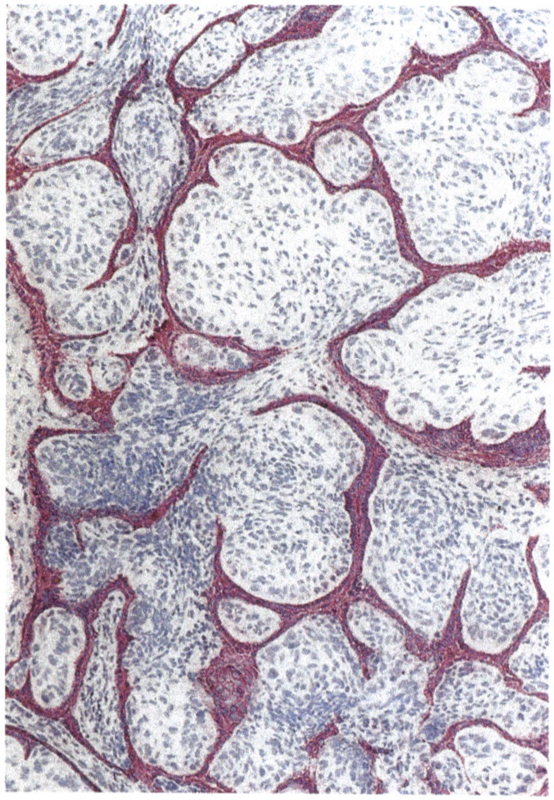

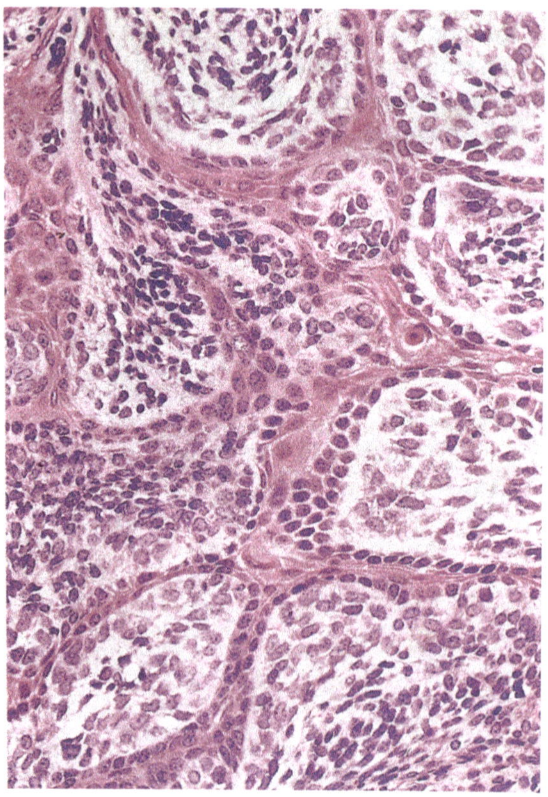

Synonym(s). Adamantinoma

Histogenesis

Epithelium of dentigerous cysts; remnants of the dental lamina and of the enamel organ; basal cell layer of the oral mucosa.

Diagnostic Features

- Usually large and concentrically growing tumor which is locally invasive and destructive but which does not metastasize
- Composed of islands (follicular type), nests or anastomosing strands (plexiform type) of epithelial cells embedded in a collagenous stroma
- Epithelium consists of a peripheral layer of tall columnar cells resembling the inner enamel epithelium and loosely arranged central cells similar to stellate reticulum
- The epithelial islands may be solid (solid type) or may show degenerative changes in the stellate cells resulting in cyst formation (cystic type)
- The stroma may be focally hyalinized, but dental hard tissues (enamel, dentin, or cementum) are not produced

Fig. 20 *(above).* Ameloblastoma, upper jaw. v-Ha-ras mouse, male, age 38 weeks. Anastomosing cords of neoplastic epithelium embedded in mesenchymal stroma (plexiform type). Positive immunostaining for pan-cytokeratin

Fig. 21 *(below).* Ameloblastoma, upper jaw. Between the epithelial strands are stellate reticulum-like cells which partly show squamous metaplasia. v-Ha-ras mouse, male, age 38 weeks. H&E

Acanthomatous
- Epithelial cells of the stellate reticulum may undergo squamous metaplasia

Differential Diagnosis

Fibro-Odontoma, Ameloblastic
Small amounts of non-aggressive odontogenic epithelium; formation of dental hard tissues occurs

Odontoma, Ameloblastic
In addition to proliferation of ameloblastic (ameloblastoma-like) epithelium, formation of dental hard tissues occurs

Odontoma (Complex or Compound)
Consists almost entirely of dental hard tissues with no or only small amounts of ameloblastic epithelium associated with the hard tissues

Fibroma, Odontogenic
Composed of pulp-like mesenchyme with indistinct rests of non-neoplastic odontogenic epithelium

Comment

Spontaneous ameloblastomas are extremely rare in mice. Ameloblastoma-like tumors (ameloblastic carcinoma, adamantoblastoma), however, have been induced in mice by polyoma virus and by local application of 3-methylcholanthrene.

References

(See 4, 14, 15, 16, 20, 32, 33, 37, 42)

Odontoma, Ameloblastic *(B)*

Synonym(s). Odontoameloblastoma

Histogenesis

Epithelial cells (ameloblasts) and ectomesenchymal cells (odontoblasts, cementoblasts) of the tooth germ.

Diagnostic Features

- Usually well-circumscribed, concentrically growing tumor
- Different degrees of morphologic differentiation ranging from primitive tooth buds to mature tooth structures
- All dental tissues are found including ameloblastic epithelium, stellate reticulum-like cells, enamel matrix, enamel, dentin, odontoblasts, cementum, cementoblasts and pulpal tissue
- The proliferating ameloblastic (ameloblastoma-like) epithelium is located in the tumor periphery and contains no or minimal amounts of dental hard tissues
- Formation of the dental hard tissues occurs within the central areas of the tumor
- Degenerating keratinized ("ghost cells") and calcified odontogenic epithelium may also be observed
- Formation of reactive multinucleated giant cells within the tumor tissue in response to accumulation of keratin debris may occur
- Ameloblastic odontomas usually behave locally aggressive (invasive and destructive) but do not metastasize

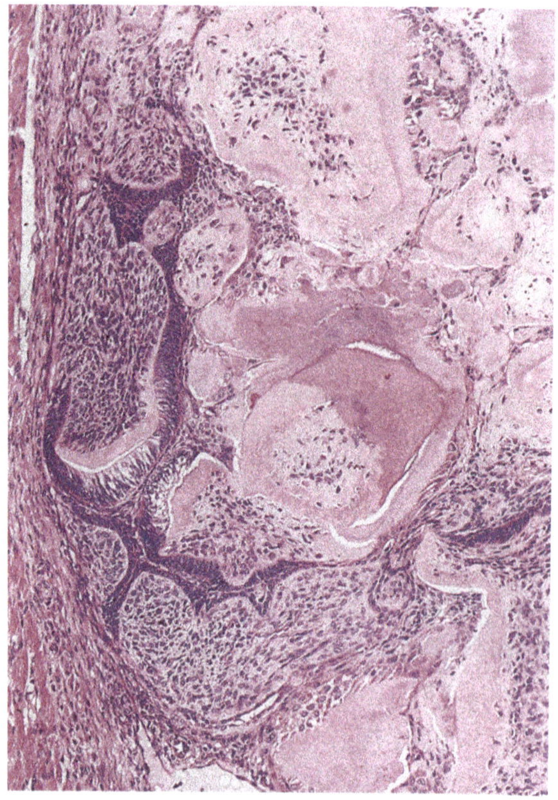

Fig. 22. Ameloblastic odontoma, upper jaw. The ameloblastoma-like epithelium in the periphery of the hard tissue components of the tumor is very proliferative and tends to behave aggressively. v-Ha-ras mouse, female, age 64 weeks. H&E

Differential Diagnosis

Ameloblastoma
Ameloblastomas do not produce dental hard tissues

Fibro-Odontoma, Ameloblastic
Mainly composed of mesodermal tissue resembling that of the dental papilla

Odontoma (Complex or Compound)
The bulk of tumor tissue consists of differentiated mineralized dental hard tissues

Fibroma, Odontogenic
Composed of pulp-like mesenchyme, no formation of enamel and dentin

Dental Dysplasia
Usually involves incisor teeth, is associated with trauma and/or inflammation, and consists mainly of dentin-like material but may contain other odontogenic tissues

Comment

Spontaneous ameloblastic odontomas are rare in mice. Similar "odontogenic tumors" have been observed in genetically altered mice carrying albumin-myc and albumin-ras transgenes.

References

(See 12, 20, 41)

Fibro-Odontoma, Ameloblastic *(B)*

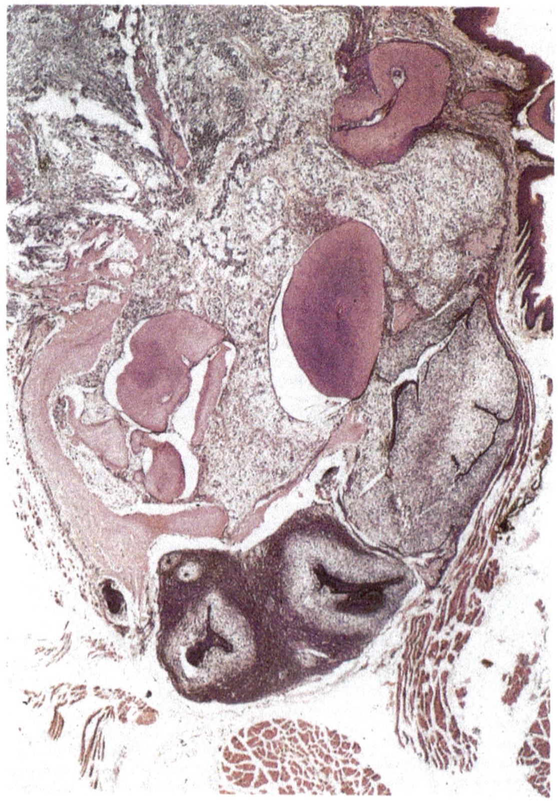

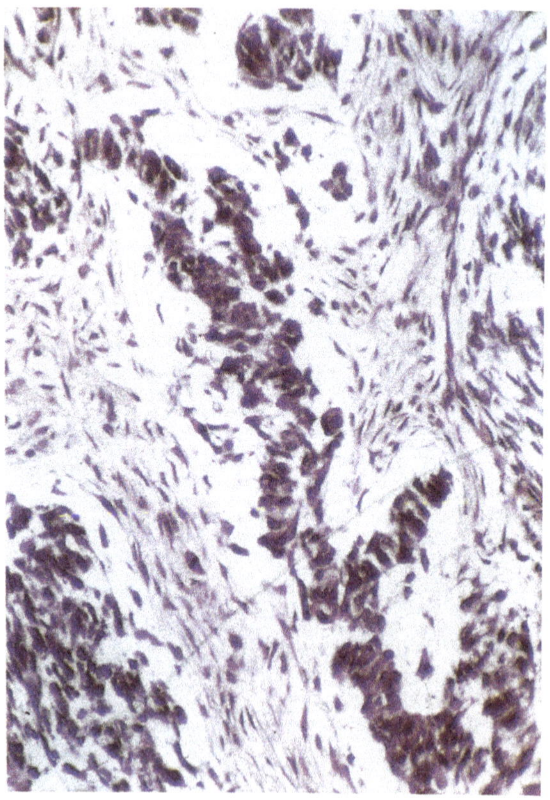

Histogenesis

Epithelial cells (ameloblasts) and ectomesenchymal cells (odontoblasts, cementoblasts) of the tooth germ.

Diagnostic Features

- Well-circumscribed, concentrically growing tumor
- Composed mainly of sheets and bundles of spindle-shaped, angular or polygonal cells resembling those seen in the dental papilla
- Cords of odontogenic epithelium are embedded within the highly cellular mass and are arranged predominantly in two layers with no stellate cells between them
- Well-defined areas of calcified tissue resembling (dysplastic) dentin and enamel are present

Differential Diagnosis

Ameloblastoma

Ameloblastomas do not produce dental hard tissues

Odontoma, Ameloblastic

Has proliferating ameloblastic (ameloblastoma-like) epithelium in the tumor periph-

Fig. 23 *(above)*. Ameloblastic fibro-odontoma, lower jaw. Tumor is composed of areas of dental hard tissues and cords of odontogenic epithelium. CD-1 mouse, female, age 73 weeks. H&E (photograph by kind permission of A. Nyska)

Fig. 24 *(below)*. Ameloblastic fibro-odontoma, lower jaw. The epithelial cords consist of two cell layers with no stellate cells between them. CD-1 mouse, female, age 73 weeks. H&E (photograph by kind permission of A. Nyska)

ery which often shows aggressive behavior and contains significant amounts of dental hard tissues

Odontoma (Complex or Compound)
The bulk of tumor tissue consists of differentiated mineralized dental hard tissues

Fibroma, Odontogenic
Composed of pulp-like mesenchyme; no formation of enamel and dentin

Dental Dysplasia
Usually involves incisor teeth, is associated with trauma and/or inflammation, and consists mainly of dentin-like material but may contain other odontogenic tissues

References

(See 20, 27)

Odontoma *(B)*

Complex • Compound

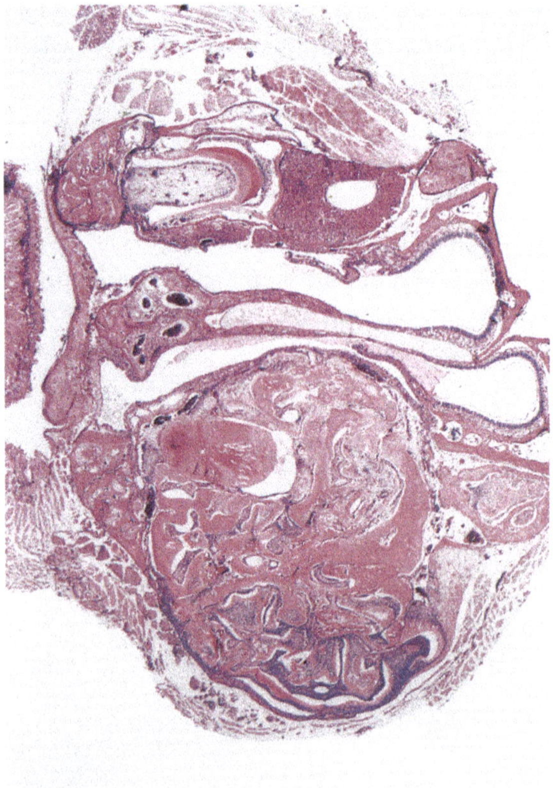

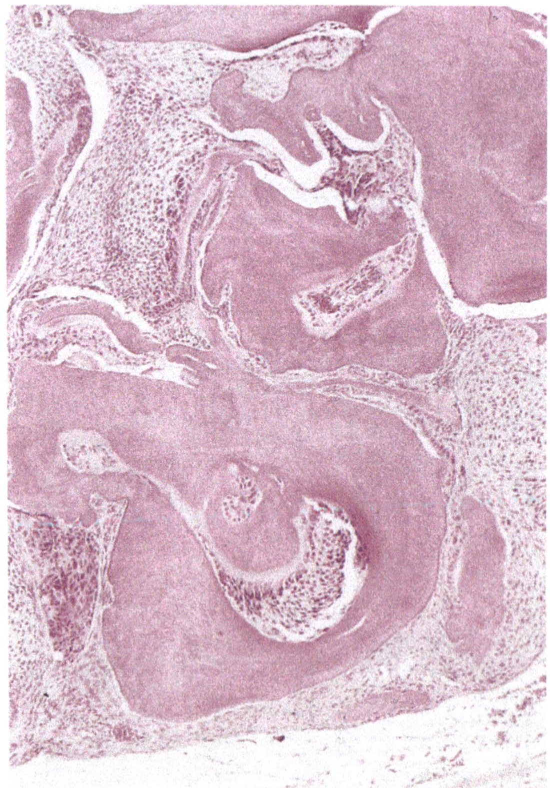

Histogenesis

Epithelial cells (ameloblasts) and ectomesenchymal cells (odontoblasts, cementoblasts) of the tooth germ.

Diagnostic Features

Complex

- Dental tissues show a characteristic poor morphodifferentiation with little resemblance to normal teeth
- All dental hard tissues such as enamel, dentin and cementum as well as odontoblasts, cementoblasts and dental pulp mesenchymal cells are present
- Absence of distinct areas of ameloblastic (ameloblastoma-like) epithelium unassociated with dental hard tissues
- Not uncommonly associated with an unerupted tooth

Compound

- Dental tissues exhibit a high degree of morphodifferentiation and histodifferentiation similar to normal teeth
- Usually small and well-circumscribed lesion

Fig. 25 *(above)*. Odontoma, complex type, upper jaw. Nodular tumor consisting mainly of irregularly arranged dental hard tissues. NMRI mouse, female, age 78 weeks. H&E (section by kind permission of A. Luz)

Fig. 26 *(below)*. Odontoma, complex type, upper jaw. Dentin structures with little resemblance to normal teeth embedded in pulp-like mesenchyme. CD-1 mouse, male, age 103 weeks. H&E

- All dental hard tissues such as enamel, dentin and cementum as well as odontoblasts, cementoblasts and dental pulp mesenchymal cells are present
- Absence of distinct areas of ameloblastic (ameloblastoma-like) epithelium unassociated with dental hard tissues
- Not uncommonly associated with an unerupted tooth

Differential Diagnosis

Ameloblastoma
Ameloblastomas do not produce dental hard tissues. Radiographs (if available) reveal a radiolucent lesion

Odontoma, Ameloblastic
Has proliferating ameloblastic (ameloblastoma-like) epithelium in the tumor periphery which is unassociated with dental hard tissues and often shows aggressive behavior

Fibro-Odontoma, Ameloblastic
Mainly composed of mesodermal tissue resembling that of the dental papilla

Fibroma, Odontogenic
Composed of dental follicle-like mesenchyme; no formation of enamel and dentin

Dental Dysplasia
Usually involves incisor teeth, is associated with trauma and/or inflammation, and consists mainly of dentin-like material but may contain other odontogenic tissues

Comment

Complex as well as compound odontomas are considered to be malformations (hamartomas) rather than true neoplasms. They represent that stage of tooth development where the degeneration of the ameloblastic epithelium occurs soon after hard tissue formation. The odontogenic tumors described by Nozue and Kayano (1978) and by Goessner and Luz (1994) are considered to represent odontomas rather than dentinomas.
Odontomas of the jaws may be diagnosed as "cementomas", when cementum-like bony substance is present. If the cementum-like structures are surrounded by fibrous tissue with cementoblasts, such tumors may be diagnosed as "cementoblastomas".

References

(See 5, 8, 12, 14, 18, 20, 26, 30, 31)

Fibroma, Odontogenic *(B)*

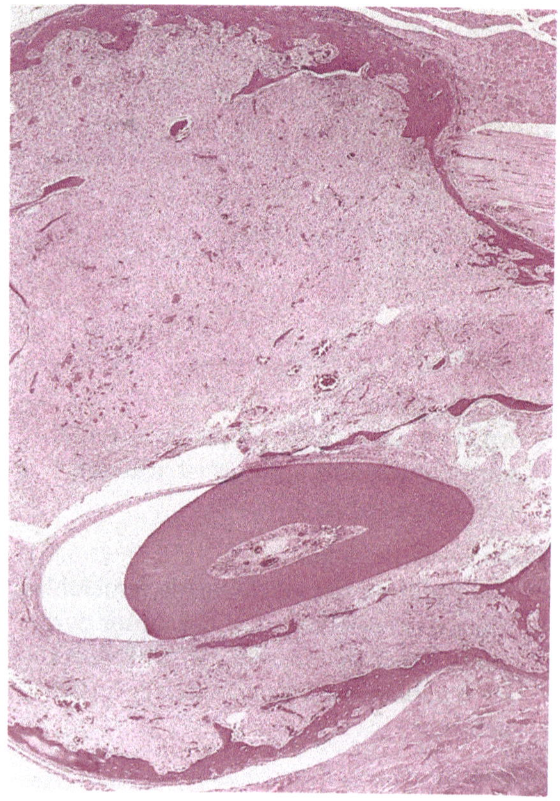

Histogenesis

Mesenchymal cells of the periodontal ligament or of the developing dental apparatus, the dental papilla or dental sac.

Diagnostic Features

- Well-circumscribed expansile tumor, usually associated with the pulp of a continuously erupting incisor tooth
- Mainly composed of cellular whorls of primitive-appearing dental follicle-like mesenchyme separated by distinct areas of collagen formation
- Strands and islands of non-neoplastic, small and mostly undifferentiated epithelial cells, representing remains of the dental epithelium, are scattered throughout the tumor
- Formation of round or irregular foci of cementum-like material can also be observed within the mesenchymal tissue

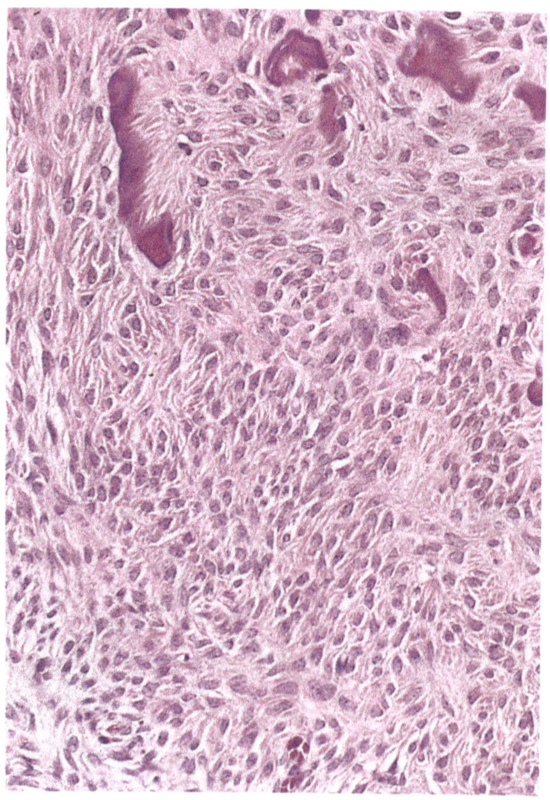

Fig. 27 *(above)*. Fibroma, odontogenic, lower jaw. Expansile tumor associated with an incisor tooth and composed of dental pulp-like mesenchyme. CD-1 mouse, female, age 88 weeks. H&E

Fig. 28 *(below)*. Fibroma, odontogenic, lower jaw. Focal formation of cementum-like material within collagenized mesenchyme. CD-1 mouse, female, age 88 weeks. H&E

Differential Diagnosis

Ameloblastoma
The odontogenic epithelium is neoplastic and proliferating, cuboidal to columnar in shape and encloses a stellate reticulum

Odontoma, Ameloblastic
In addition to proliferation of ameloblastic (ameloblastoma-like) epithelium, formation of dental hard tissues also occurs

Fibro-Odontoma, Ameloblastic
Formation of dentin and enamel occurs

Odontoma (Complex or Compound)
Both lesions contain differentiated and mineralized tooth-like dental hard tissues

References

(See 14, 20)

References

1. Albassam MA, Courtney CL (1996) Nonneoplastic and neoplastic lesions of the bone. In: Mohr U, Dungworth DL, Capen CC, Carlton WW, Sundberg JP, Ward JM (eds) Pathobiology of the aging mouse, vol 2. Musculoskeletal system. ILSI Press, Washington DC, pp 425–437
2. Charles RT, Turusov VS (1974) Bone tumours in CF-1 mice. Lab Anim 8: 137–144
3. Cockmann-Thomas RA, Dunn DG, Innskeep W II, Mondy WL, Swearengen JR (1994) Spontaneous osteosarcoma in a C57BL/6 J mouse. Lab Anim Sci 44: 531–533
4. Dawe CJ, Law LW, Dunn TB (1959) Studies of parotid-tumor agent in cultures of leukemic tissues of mice. J Natl Cancer Inst 23: 717–798
5. Dayan D, Waner T, Harmelin A, Nyska A (1994) Bilateral complex odontoma in a Swiss (CD-1) male mouse. Lab Anim 28: 90–92
6. EULEP Committee on Pathology (1991) Bone tumors in mice and rats. European Late Effects Project Group, Gesellschaft fuer Strahlen- und Umweltforschung, Neuherberg (EULEP pathology atlas)
7. Faccini JM, Abbott DP, Paulus GJJ (1990) Mouse histopathology. A glossary for use in toxicity and carcinogenicity studies. Elsevier, Amsterdam
8. Finkel MP, Lombard LS, Staffeldt EF, Duffy PH (1979) Odontomas in *Peromyscus leucopus*. J Natl Cancer Inst 63: 407–411
9. Franks LM, Rowlatt C, Chesterman FC (1973) Naturally occurring bone tumors in C57BL-Lcrf mice. J Natl Cancer Inst 50: 431–438
10. Frith CH, Ward JM (1988) Color atlas of neoplastic and non-neoplastic lesions in aging mice. Elsevier, Amsterdam
11. Frith CH, Johnson BP, Highman B (1982) Osteosarcomas in BALB/c female mice. Lab Anim Sci 32: 60–63
12. Gibson CW, Lally E, Herold RC, Decker S, Brinster RL, Sandgren EP (1992) Odontogenic tumors in mice carrying albumin-myc and albumin-ras transgenes. Calcif Tissue Int 51: 162–167
13. Goessner W (1986) Pathology of radiation-induced bone tumors. Leuk Res 10: 897–904
14. Goessner W, Luz A (1994) Tumours of the jaws. In: Turusov VS, Mohr U (eds) Pathology of tumours in laboratory animals, vol 2. Tumours of the mouse, 2nd edn. IARC Scientific Publications No. 111, Lyon, pp 141–165
15. Gollard RP, Slavkin HC, Snead ML (1992) Polyoma virus-induced murine odontogenic tumors. Oral Surg Oral Med Oral Pathol 74: 761–767
16. Greene GW, Collins DA, Bernier JL (1960) Response of embryonal odontogenic epithelium in the lower incisor of the mouse to 3-methylcholanthrene. Arch Oral Biol 1: 325–332
17. Hoger H, Gialamas J, Jelinek F (1994) Multiple osteomas in mice. Vet Pathol 31: 429–434
18. Humphreys ER, Robins MW, Stones VA (1985) Age-related and 224Ra-induced abnormalities in the teeth of male mice. Arch Oral Biol 30: 55–64
19. Long PH, Leininger JR (1999a) Bones, joints, and synovia. In: Maronpot RR, Boorman GA, Gaul BW (eds) Pathology of the mouse. Reference and atlas. Cache River Press, Vienna, pp 645–678
20. Long PH, Leininger JR (1999b) Teeth. In: Maronpot RR, Boorman GA, Gaul BW (eds) Pathology of the mouse. Reference and atlas. Cache River Press, Vienna, pp 13–28
21. Luz A, Goessner W, Murray AB (1991a) Ossifying fibroma, mouse. In: Jones TC, Mohr U, Hunt RD (eds) Monographs on pathology of laboratory animals. Cardiovascular and musculoskeletal systems. Springer, Berlin Heidelberg New York Tokyo, pp 228–232
22. Luz A, Goessner W, Murray AB (1991b) Osteofibroma, mouse. In: Jones TC, Mohr U, Hunt RD (eds) Monographs on pathology of laboratory animals. Cardiovascular and musculoskeletal systems. Springer, Berlin Heidelberg New York Tokyo, pp 224–228
23. Luz A, Goessner W, Murray AB (1991c) Osteosarcoma, spontaneous and radiation-induced, mouse. In: Jones TC, Mohr U, Hunt RD (eds) Monographs on pathology of laboratory animals. Cardiovascular and musculoskeletal systems. Springer, Berlin Heidelberg New York Tokyo, pp 202–213
24. Luz A, Murray AB, Schmidt J (1991d) Osteoma, spontaneous and virus-induced, mouse. In: Jones TC, Mohr U, Hunt RD (eds) Monographs on pathology of laboratory animals. Cardiovascular and musculoskeletal systems. Springer, Berlin Heidelberg New York Tokyo, pp 182–190
25. Nilsson A, Stanton MF (1994) Tumours of the bone. In: Turusov VS, Mohr U (eds) Pathology of tumours in laboratory animals, vol 2. Tumours of the mouse, 2nd edn. IARC Scientific Publications No. 111, Lyon, pp 681–729
26. Nozue T, Kayano T (1978) Effects of mitomycin C in postnatal tooth development in mice with special reference to neural crest cells. Acta Anat 100: 85–94
27. Nyska A, Waner T, Tal H, Dayan D (1991) Spontaneous ameloblastic fibro-odontoma in a female mouse. J Oral Pathol Med 20: 250–252
28. Pybus FC, Miller EW (1938) Spontaneous bone tumours of mice. Am J Cancer 33: 98–111

29. Rabstein LS, Peters RL (1973) Tumors of the kidneys, synovia, exocrine pancreas and nasal cavity in BALB-cf-Cd mice. J Natl Cancer Inst 51: 999–1006
30. Robins MW, Rowlatt C (1971) Dental abnormalities in aged mice. Gerontologia 17: 261–272
31. Sokoloff L, Zipkin I (1967) Odontogenic hamartomas in an inbred strain of mouse (STR-1 N). Proc Soc Exp Biol Med 124: 147–149
32. Stanley HR, Dawe CJ, Law LW (1964) Oral tumors induced by polyoma virus in mice. Oral Surg 17: 547–558
33. Stanley HR, Baer PN, Kilham L (1965) Oral tissue alterations in mice inoculated with the Rowe substrain of polyoma virus. Periodontics 3: 178–183
34. Stanton MF (1979) Tumours of the bone. In: Turusov VS (ed) Pathology of tumours in laboratory animals, vol II. Tumours of the mouse. IARC Scientific Publications No. 23, Lyon, pp 577–609
35. Stewart HL (1979) Tumours of the soft tissues. In: Turusov VS (ed) Pathology of tumours in laboratory animals, vol II. Tumours of the mouse. IARC Scientific Publications No. 23, Lyon, pp 487–526
36. Szepsenwol J, Fletcher J, Casales EA, Murison GL (1985) Experimentally produced synovial sarcoma in mice. Oncology 42: 61–67
37. Van Rijssel TG, Mühlbock O (1955) Intramandibular tumors in mice. J Natl Cancer Inst 16: 659–689
38. Wadsworth PF (1989) Tumours of the bone in C57BL/10J mice. Lab Anim 23: 324–327
39. Wadsworth PF (1996) Classification of bone neoplasms and presentation of data on the incidences and types in C57BL/10 J mice. In: Mohr U, Dungworth DL, Capen CC, Carlton WW, Sundberg JP, Ward JM (eds) Pathobiology of the aging mouse, vol 2. Musculoskeletal system. ILSI Press, Washington DC, pp 439–443
40. Wilson JT, Hauser RE, Ryffel B (1985) Osteomas in OF-1 mice: no alteration in biologic behavior during long-term treatment with cyclosporine. J Natl Cancer Inst 75: 897–903
41. Wright JT, Hansen L, Mahler J, Szczesniak C, Spalding JW (1995) Odontogenic tumours in the v-Ha-ras (TG.AC) transgenic mouse. Arch Oral Biol 40: 631–638
42. Zegarelli EV (1944) Adamantoblastomas in the Slye stock of mice. Am J Pathol 20: 23–87

Hematopoietic System

C.H. Frith[2] · J.M. Ward[2] · J.H. Harleman[3] · P.C. Stromberg[1] · S. Halm
T. Inoue · J.A. Wright

[1]Chair; [2]Co-chair; [3]Initial draft.

Lymphoma, Malignant *(M)*

Small lymphocyte • Lymphoblastic
Pleomorphic/follicular • Plasmacytic
Immunoblastic • Marginal zone • NOS

Synonym(s). Lymphosarcoma; malignant lymphoma; lymphoma

Histogenesis

Lymphocytes or their precursors.

Small lymphocyte: B- or T-cell origin.

Lymphoblastic: B- or T-cell origin, usually of T-cell origin in the thymus.

Pleomorphic: B-cell origin but may be admixed with T-helper CD4+-cells.

Plasmacytic: Origin from immunoblasts.

Diagnostic Features

Small lymphocyte

- Architecture of involved organs is not maintained
- Cells are uniform and non-cohesive
- No evidence of tingible body macrophages
- Cells are small to medium-sized well-differentiated lymphocytes
- Cells differ little, if any at all, from the regular circulating small lymphocytes
- Cytoplasm is present in a narrow rim
- Nuclear chromatin in the small cells is densely clumped
- Some slightly larger cells with somewhat larger irregular nuclei may be present, corresponding to small cleaved follicular center cells
- Mitotic figures are rare

Lymphoblastic

- Cells are non-cohesive but form homogeneous sheets
- Often starry-sky appearance is present with tingible body macrophages and apoptosis
- Cells are medium-sized lymphoblasts
- Cytoplasm is moderate in amount, basophilic and may be vacuolated
- Nuclear to cytoplasmic ratio is high
- Nucleus is round
- A central nucleolus is prominent
- Mitotic figures are numerous
- May be leukemic and invade the central nervous system
- Aggressive with diffuse infiltration of liver, kidneys, ovaries and along the vascular tree in the lungs, similar to immunoblastic lymphoma, but pleomorphic lymphoma is less invasive

Pleomorphic/Follicular

- Origin is single or multicentric
- The growth pattern is diffuse or follicular. Especially in the spleen, one or more follicles may be involved resulting in a nodular appearance
- Cells are medium-sized to large cells
- Lesions express a varying degree of pleomorphism due to a cell population consisting of several cell types including small lymphocytes, small and large follicular center cells (centrocytes and centroblasts), macrophages and immunoblasts
- The proportion of each cell type varies in each lymphoma and even sometimes in each anatomic site
- Nucleoli are prominent
- Multinucleated cells may be present

Plasmacytic

- Cells are mature plasma cells or cells with an obvious plasmacytoid differentiation
- Nuclei are round and show a cartwheel appearance
- Cytoplasm is basophilic, and pyroninophilic, and a small perinuclear halo (Golgi apparatus) may be present

- Mitotic figures are rare
- Immunocytochemically positive for cytoplasmic immunoglobulins in paraffin sections

Immunoblastic
- Cells are non-cohesive and monotypic
- Cells are large
- Cytoplasm is conspicuously amphophilic and immunoglobulin staining
- Nuclei are large and vesicular with one large central nucleolus
- Plasmacytoid cells and plasma cells may be present
- Mitotic figures are numerous
- Rare in most strains
- Pattern of organ involvement is aggressive with diffuse infiltration of liver, kidney, ovaries and along the vascular tree in the lungs similar to that of pleomorphic lymphoma, but immunoblastic lymphoma is more invasive

Marginal zone
- An early lesion of lymphoma in some strains of mice restricted to the spleen
- Widening of the marginal zone with diffuse invasion of the red pulp by a very homogeneous population of medium sized cells
- May progress to larger cells with vesicular nuclei as marginal zones increase in width and start to coalesce
- In advanced cases, the PALS are compressed consisting only of small lymphocytes and the red pulp is obliterated

NOS (not otherwise specified)
- NOS may be used if further classification is precluded by the presence of autolysis, poor fixation, etc.

Differential Diagnosis

Lymphoid Hyperplasia

Reactive lesion which follows and remains within the regular organization of the lymphatic tissue and

A cause for the reactive lesion is present, such as inflammation etc.

Sarcoma, Histiocytic

Cytoplasm is eosinophilic and abundant and

Structure may be fibrous and sometimes large multinucleated foreign body type cells are present and

Anatomic distribution is different, often has a primary site retroperitoneally, in the mesenteric lymph nodes, uterus or in the liver and

Distribution rarely follows a primarily lymphoid pattern

Comment

Malignant lymphomas are often systemic tumors and in most cases more than one site is involved. However, they may originate in a single site before spread occurs. In some instances, single-site lesions are encountered. They are most common in the mesenteric lymph nodes and the spleen with pleomorphic lymphoma. The diagnosis of single-site lesion can be made if a marked infiltration leading to structural changes is present. An inflammatory cause should not be present. The term pleomorphic describes the cytology on H&E best, other terminology used for this type of tumor is follicular center cell or follicular lymphoma.

The qualifier "leukemic" may optionally be used when there is an obvious leukemia present judged from the peripheral blood smears, but this is a phase of a solid lymphoma and not a primary leukemia.

Tumors of the hematopoietic tissues have a relatively high incidence in most common mice strains. The follicular center lymphoma is the most common lymphoma type in most mouse strains. Plasmacytic lymphomas are rare tumors in most mouse strains (more common in NZB mice and inducible in BALB/c mice).

Malignant lymphomas are also inducible with various chemicals, viruses, and radiation. These treatments can induce specific subtypes of lymphoma such as lymphoblastic lymphoma with a primary site in the thymus. There is a high incidence of spontaneous hematopoietic neoplasms in aged mice for which the etiology is unknown. Many of these neoplasms may be due to endogenous murine leukemia viruses.

In some strains of mice, an early lesion of lymphoma can be distinguished in the spleen as widening of the marginal zone with diffuse invasion of the red pulp. In these cases, the initially very homogeneous population of medium-sized cells may progress to larger cells with vesicular nuclei as marginal zones increase in width and start to coalesce. In advanced cases, the PALS are compressed consisting only of small lymphocytes and the red pulp is obliterated. This lesion is in some classifications is referred to as marginal zone lymphoma.

In routine studies with formalin-fixed tissues and H&E stained sections, the characterization and classification of these tumors in mice is based solely on morphologic criteria, as methods with specific immunologic and cytochemical markers are not generally suitable for the formalin-based fixatives used in routine studies, however, some do work with formalin-fixed tissues (e.g., B220, a B-cell marker). It is possible to further immunotype malignant lymphomas in mice. The above morphologic classification can be further expanded by immunotyping, using T/B cell markers. A pan-T-cell marker (CD3) can be used on Bouins, B-5, Zenkers or paraformaldehyde-fixed tissues.

References

(See 1, 4, 6, 8, 9, 10, 11,16, 18, 19, 21, 24, 25, 28, 36, 37, 40)

Fig. 1 *(upper left).* Lymphocytic (small lymphocyte) lymphoma of a lymph node. H&E ▶

Fig. 2 *(upper right).* Lymphoblastic lymphoma of the thymus. H&E

Fig. 3 *(lower left).* Lymphoblastic lymphoma. *Top* H&E *bottom* CD3. Bouins fixed

Fig. 4 *(lower right).* Lymphoblastic lymphoma – blood smear. Wright-Giemsa

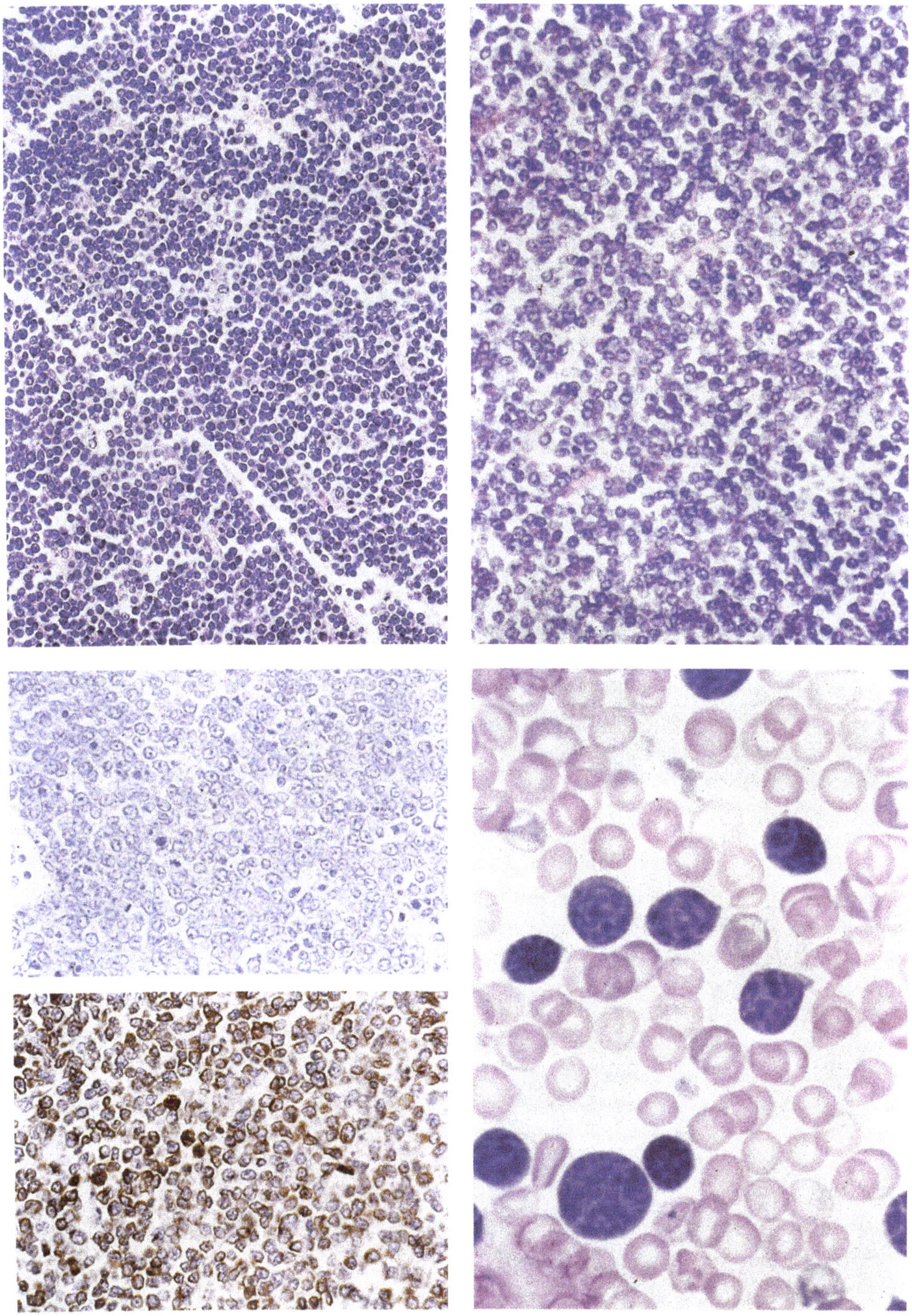

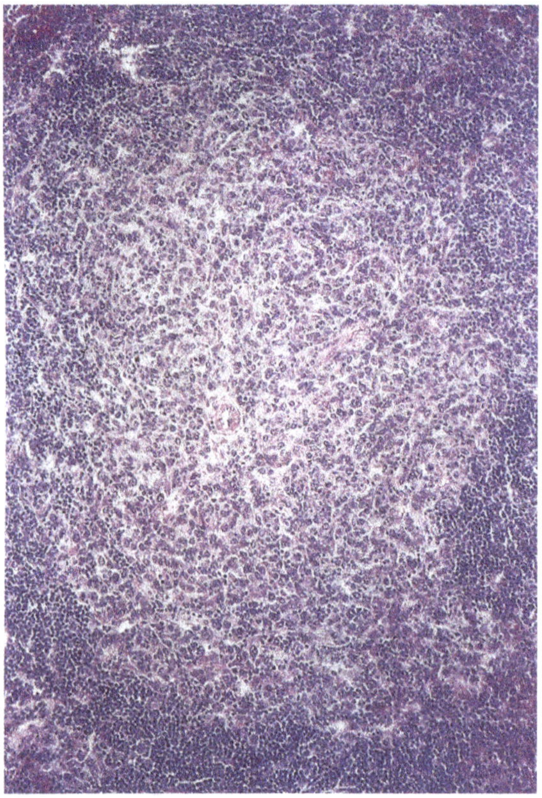

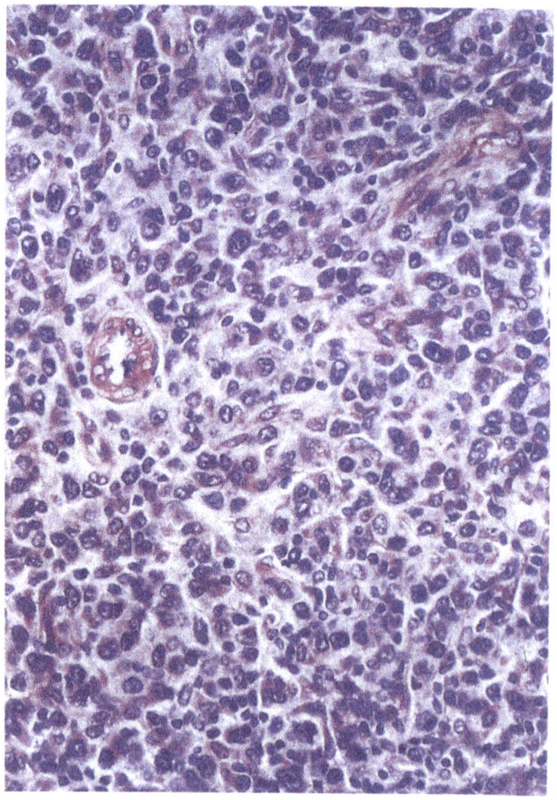

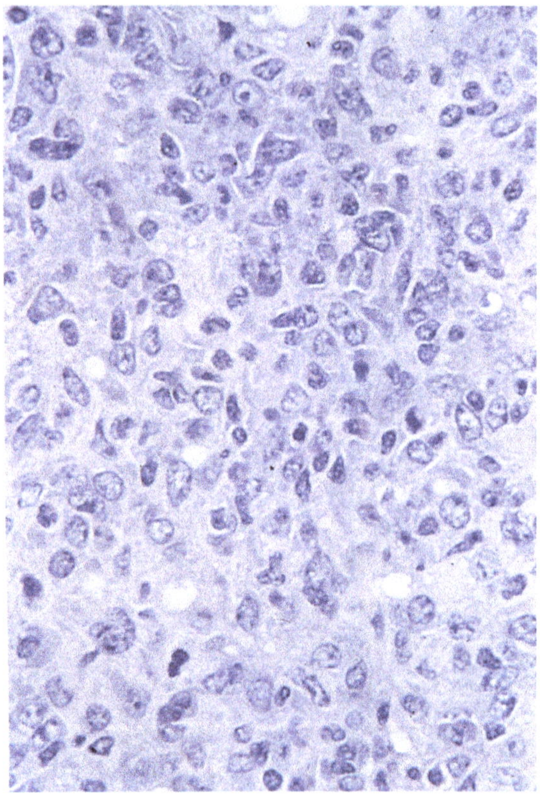

Fig. 5 *(upper left).* Early pleomorphic (follicular center cell) lymphoma of the spleen. H&E

Fig. 6 *(upper right).* Early pleomorphic (follicular center cell) lymphoma of the spleen. H&E

Fig. 7 *(below).* Pleomorphic (follicular center cell) lymphoma of the spleen. H&E

Fig. 8 *(upper left).* Pleomorphic (follicular center cell) lymphoma of the spleen. IgG ▶

Fig. 9 *(upper right).* Plasmacytic lymphoma of a lymph node. H&E

Fig. 10 *(lower left).* Immunoblastic lymphoma of a lymph node. H&E

Fig. 11 *(lower right).* Pleomorphic lymphoma of the liver. p53 wild-type mouse, male, age 84 weeks. H&E

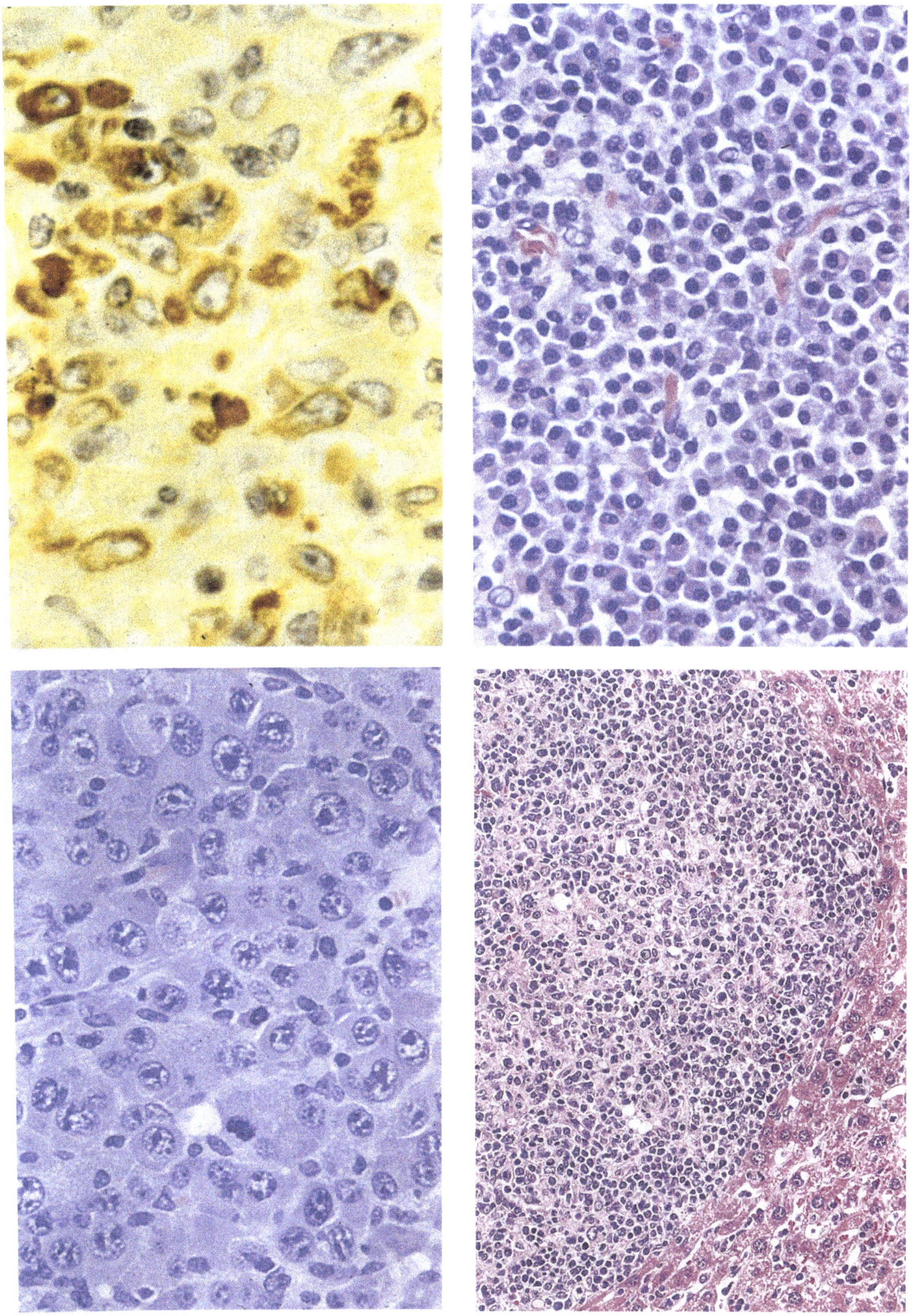

Leukemia, Granulocytic *(M)*

Basophilic • Eosinophilic
Neutrophilic • NOS

Synonym(s). Chloroleukemia; granulocytic leukemia; leukemia, myeloid; myeloid leukemia; myeloblastic leukemia

Histogenesis

Develops from the myeloid cell lineage in the spleen or bone marrow.

Diagnostic Features

- Arises in the red pulp of the spleen spreading preferentially to the liver
- Primarily involves the spleen with secondary involvement in liver, both of which may become very large and have a characteristic dull red color
- Leukemia is frequently present but tissue masses may also be produced
- Main sites of hepatic infiltration are the periportal areas often with fairly well-differentiated cells
- Various degrees of cellular maturation and differentiation may be present
- Differentiation may vary from immature (poorly differentiated) to mature cells showing a proportion from blastic to segmented neutrophils
- Azurophilic granules are frequently present in myeloblasts, but they are not visible in H&E sections
- Nuclei have prominent nucleoli
- Cells are positive for lysosome markers, peroxidase, chloroacetate and granulocytic antigen markers
- Macroscopically, affected organs exhibit a characteristic dull red color (with greenish tinge)

Basophilic, Eosinophilic, Neutrophilic

- Modifiers indicating the lineage of the cells are based upon the tinctorial characteristics of the secondary granules

NOS (not otherwise specified)

- Used when a differentiation is not made or is unfeasible

Differential Diagnosis

Extramedullary Hematopoiesis

Neutrophils are differentiated and
Increased megakaryocytosis and erythropoiesis usually accompany splenic reaction. Involvement of lymph nodes is rare and
Follows anatomical boundaries and does not show aggression and
Foci of erythroid cells are often present
Part of an inflammatory response (e.g., pyogenic bacterial infections, chronic ulcerated mammary tumor, repeated blood loss, necrotic tumors, and increased red cell destruction)
Areas of extramedullary hematopoiesis may occur in a variety of organs including the medullary cords of lymph nodes

Comment

The differential diagnosis between extramedullary hematopoiesis and granulocytic leukemia can be very difficult.
By far the most common form is granulocytic leukemia, neutrophilic. This tumor has a low spontaneous incidence in most mouse strains of aged mice, but may be induced by radiation, chemicals and viruses. Eosinophilic and basophilic types have not been reported in the mouse.

References

(See 1, 4, 16, 18, 20, 31, 35, 39)

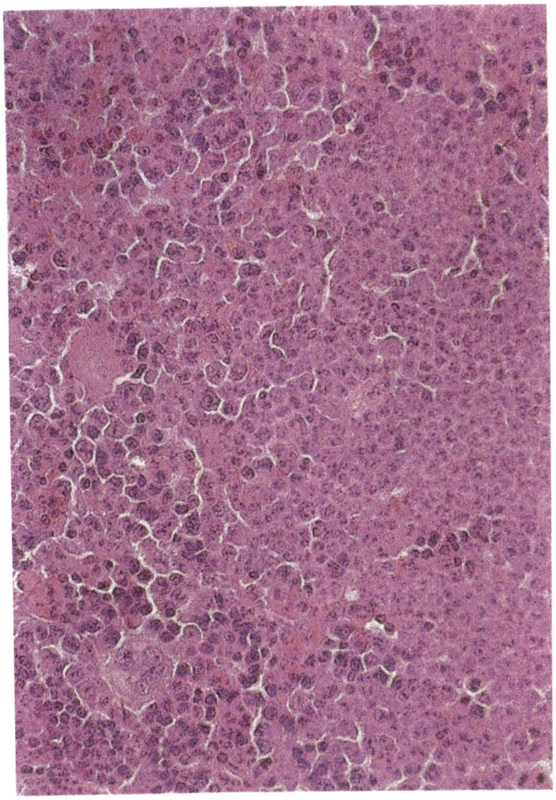

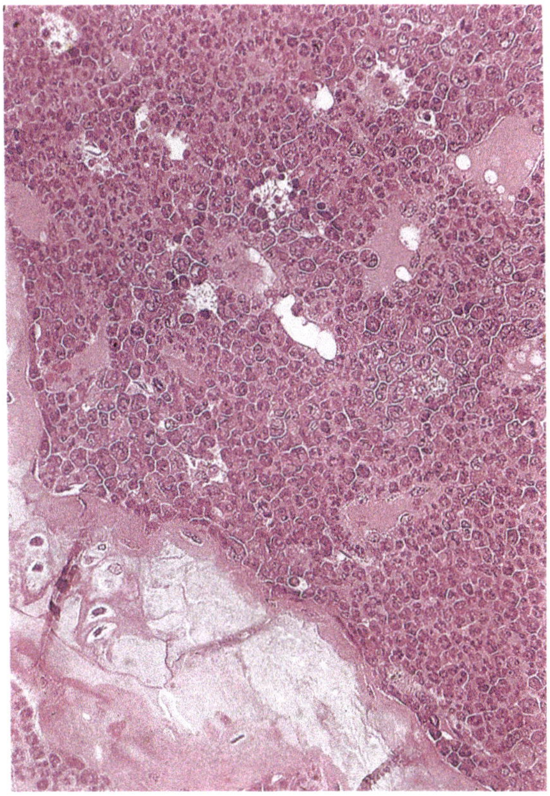

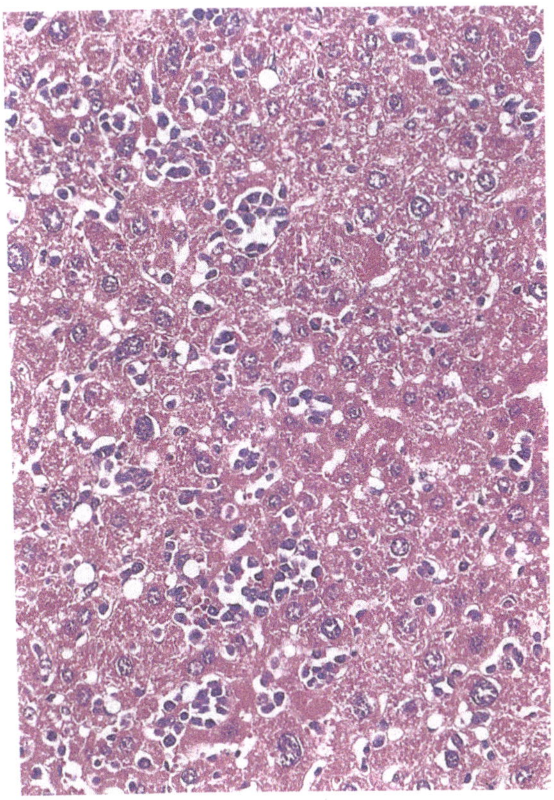

Fig. 12 *(upper left)*. Granulocytic leukemia, spleen. B6C3F1 mouse, male, age 95 weeks. H&E (courtesy of Dr. Shinji Yamamoto)

Fig.13 *(upper right)*. Granulocytic leukemia, bone marrow. B6C3F1 mouse, male, age 95 weeks. H&E (courtesy of Dr. Shinji Yamamoto)

Fig. 14 *(below)*. Granulocytic leukemia, liver. H&E

Leukemia, Erythroid *(M)*

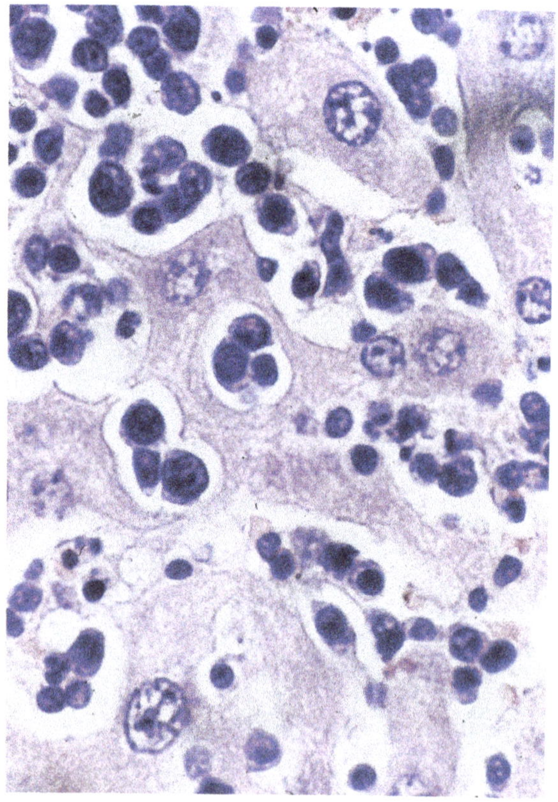

Fig. 15. Erythroleukemia of the liver. H&E

Synonym(s). Erythroid leukemia; erythroleukemia; erythroblastic leukemia

Histogenesis

Develops from the erythroid cell lineage in the spleen.

Diagnostic Features

- Arises in red pulp of the spleen and spreads preferentially to the liver sinusoids
- White pulp of the spleen is compressed and atrophic
- Lymph nodes are not involved in contrast to splenomegaly and hepatomegaly which may be severe
- Leukemia with erythroblasts and normoblasts
- Proliferation of erythroblastic cells is excessive
- Enlarged spleens may have numerous subcapsular hematomas

Differential Diagnosis

Lymphoma, Malignant

No evidence of erythroid differentiation and
A non-sinusoidal pattern of infiltration is present in the liver and
A mixture of erythroblasts and easily distinguishable normoblasts is present and
Undifferentiated lesions can only be distinguished by immunohistochemistry

Extramedullary Hematopoiesis

A stimulus for the response is present such as inflammation and
Mixture of cell lineages is present

Comment

Severe extramedullary hematopoiesis (EMH) involving both the liver and spleen may be difficult to distinguish from erythroid leukemia.
If differentiation of the erythroid leukemia is arrested at a very primitive stage of erythroid differentiation, a diagnosis may be difficult. In such cases a definitive diagnosis is generally only possible with the use of specific antibodies recognizing antigens expressed during normal erythropoiesis, e.g., carbonic anhydrase, hemoglobin.
To identify erythrocytic lineage, it is recommended to use a specific antibody to erythrocytic glycoprotein. To stain immature erythroid lineage, use benzidine (p-diaminobiphenyl). Erythroid leukemia has been described as a rare spontaneous lesion in the mouse, and is inducible with MuLVs. One of the MuLVs is the Friend virus which induces Friend leukemia. Erythroid leukemia is also induced by whole body radiation in C3H/He mice as well as RF mice.

References

(See 1, 5, 17, 20, 22, 24, 30, 39)

Leukemia, Megakaryocytic *(M)*

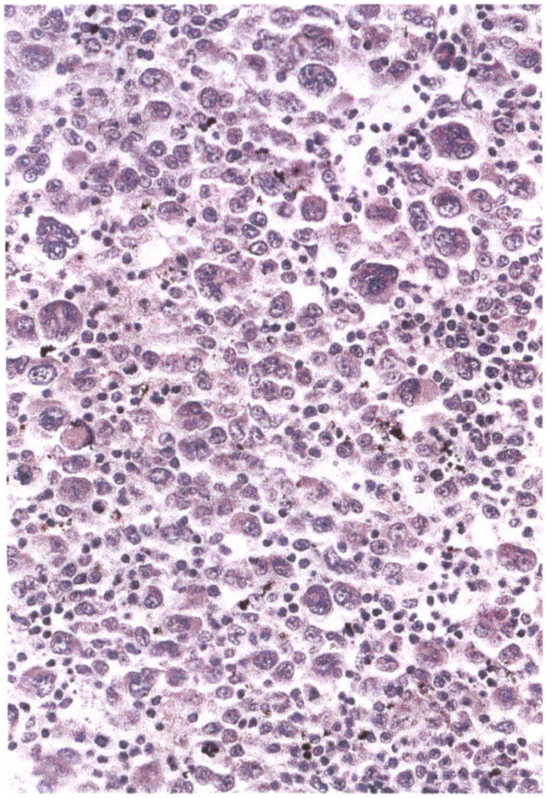

Fig. 16. Megakaryocytic leukemia of the spleen. H&E

Synonym(s). Megakaryoblastic leukemia; megakaryocytic myelosis

Histogenesis

Develops from megakaryocytic cell lineage in the spleen and bone marrow.

Diagnostic Features

- Arises in red pulp of the spleen and is present in hematopoietic tissue including bone marrow, spleen, liver and kidneys
- Lymph nodes are not always directly involved, but megakaryocytes may be present in medullary sinuses
- Increase of large mononuclear cells is marked including scattered megakaryocytic cells

Differential Diagnosis

Extramedullary Hematopoiesis

Inflammatory responses in the red pulp usually associated with an increase in mature megakaryocytes but no evidence of blast cells and

Antibody to platelet glycoprotein Iib/IIIa can be used to stain fresh tissue. Also acetylcholine esterase is positive in most cases

Comment

Only reported as an induced lesion by a recombinant retrovirus (MuLV).

References

(See 6, 7)

Tumor, Mast Cell, Malignant *(M)*

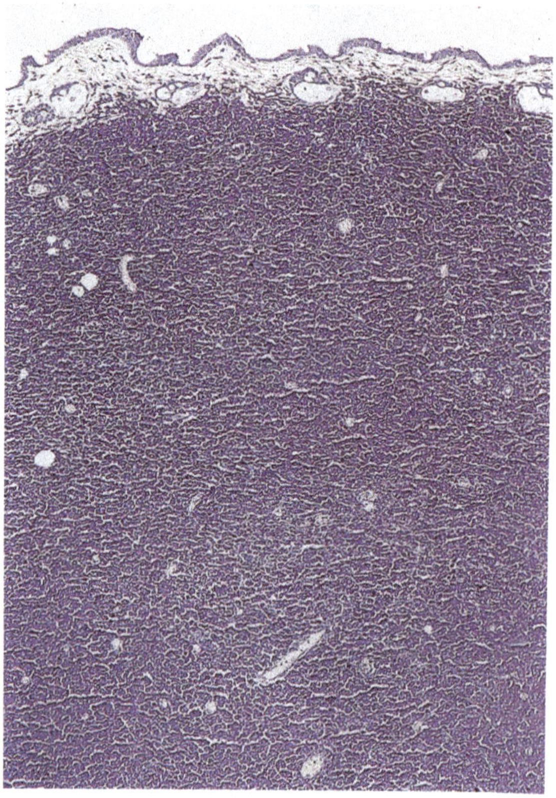

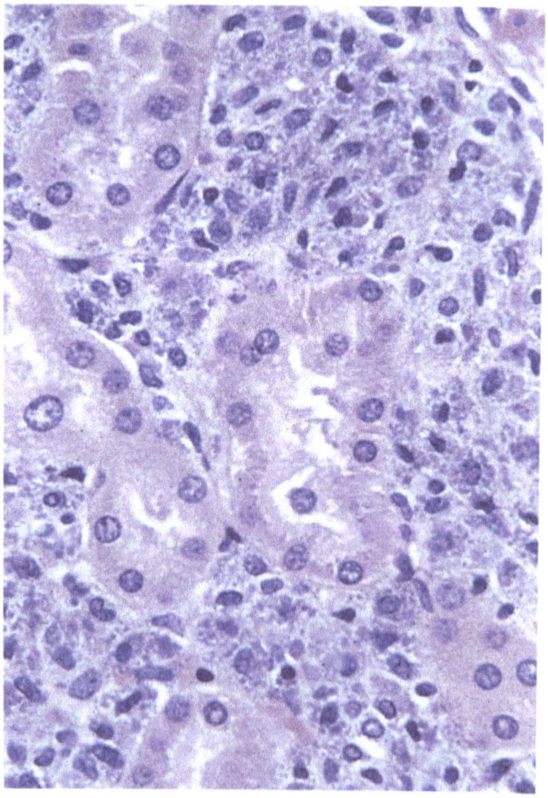

Synonym(s). Malignant mast cell tumor; malignant mastocytoma; mast cell tumor, malignant; mastocytoma, malignant

Histogenesis

Develops from peripheral mast cells or mast cells and their precursors present in the hematopoietic and/or connective tissue.

Diagnostic Features

- Comprised of nodules which may be widely distributed in lymphatic and other organs
- No evidence of invasion of the peripheral blood, although mast cell leukemia has been reported
- Composed mostly of well-differentiated mast cells showing the characteristic metachromatic granules with Giemsa, toluidine blue or other metachromatic stains
- Nuclei are uniform, large and round
- Cytoplasm is abundant, granular and slightly basophilic
- Metachromatic granules are often not visible with H&E
- Locally infiltrative growth is present

Fig. 17 *(above).* Malignant mast cell tumor of the skin/subcutis. Toluidine blue

Fig. 18 *(below).* Malignant mast cell tumor of the kidney. H&E

Differential Diagnosis

Sarcoma, Histiocytic

Nuclei are less regular and
Cytoplasm is eosinophilic and
No evidence of metachromatic granules in the cytoplasm is demonstrated by metachromatic stains (Giemsa and toluidine blue)

References

(See 2, 3, 15, 16, 18, 29)

Sarcoma, Histiocytic *(M)*

Histogenesis

Cells of the mononuclear phagocyte system.

Diagnostic Features

- Most commonly present in livers of male mice and in livers and uterus of female mice
- Cell population is uniform
- Areas of necrosis surrounded by palisaded cells may be present and are characteristic
- Multinucleated giant cells are often present
- Phagocytosis may be present
- May resemble malignant schwannoma in the uterus
- Cells are rounded or oval
- Cytoplasm is eosinophilic and present in abundance
- Nuclei are pleomorphic, and elongated or folded
- Mitotic figures may be numerous
- Metastasis and spread on serosal surfaces and in the vascular spaces is common
- Metastasis is commonly found in the vessels of the lungs

Differential Diagnosis

Lymphoma, Malignant (Pleomorphic)
No evidence of giant cells and
Eosinophilic cytoplasm is not abundant

Schwannoma, Malignant (Uterus)
No evidence of giant cells and
Eosinophilic cytoplasm is not abundant and
S-100 immunoreactive

Tumor, Mast Cell, Malignant
Nuclei are uniform and
Cytoplasm is granular and slightly basophilic and
Metachromatic granules in the cytoplasm are demonstrated by metachromatic stains (Giemsa and toluidine blue)

Comment

Lysozyme and MAC-2 are reliable and specific markers for histiocytic sarcoma of mice. Histiocytic sarcomas may spread in a lymphomatous manner but more often are hematogenous.

References

(See 1, 12, 13, 14, 16, 26, 27, 34, 38)

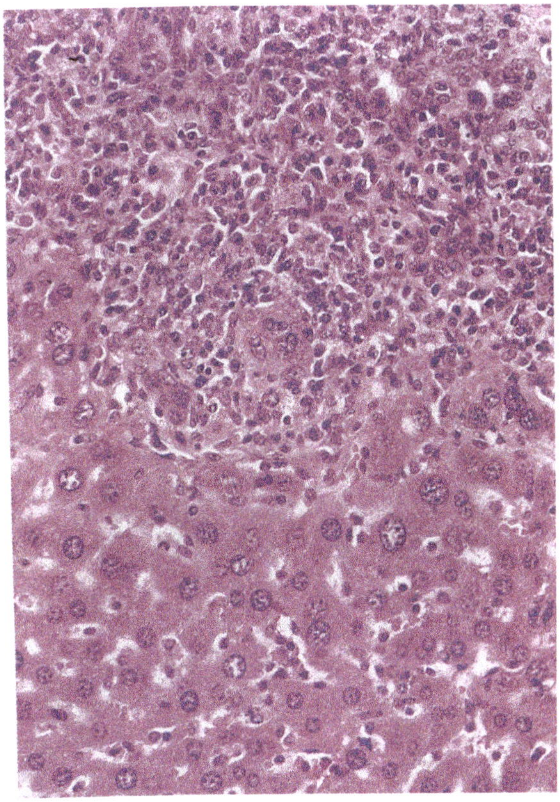

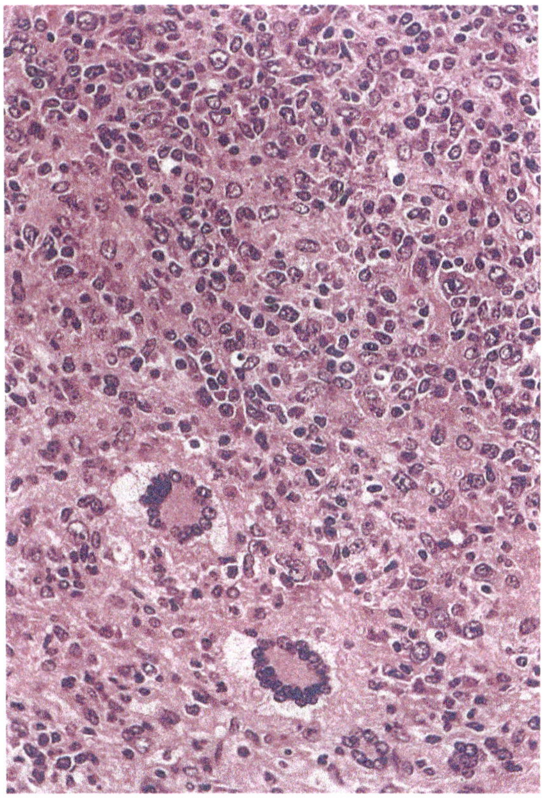

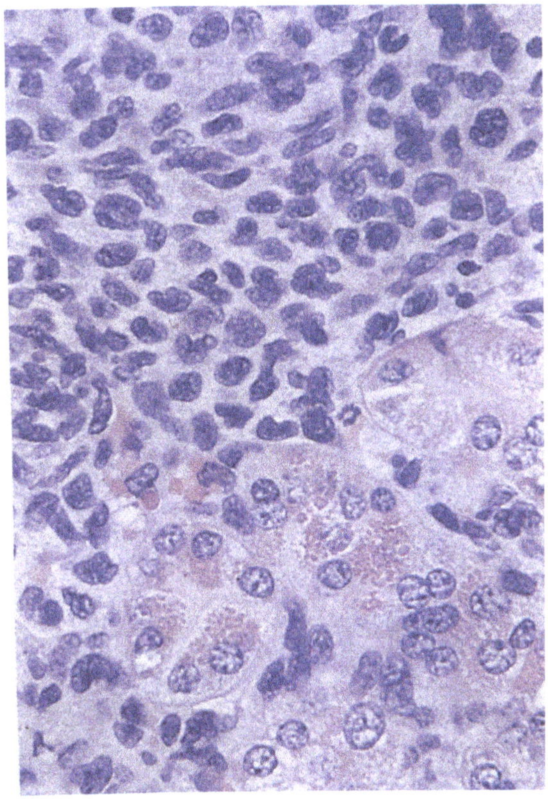

Fig. 19 *(upper left)*. Histiocytic sarcoma, liver. H&E

Fig. 20 *(upper right)*. Histiocytic sarcoma, skin/subcutis. H&E

Fig. 21 *(below)*. Histiocytic sarcoma of the salivary gland. H&E

Hyperplasia, Lymphoid *(H)*

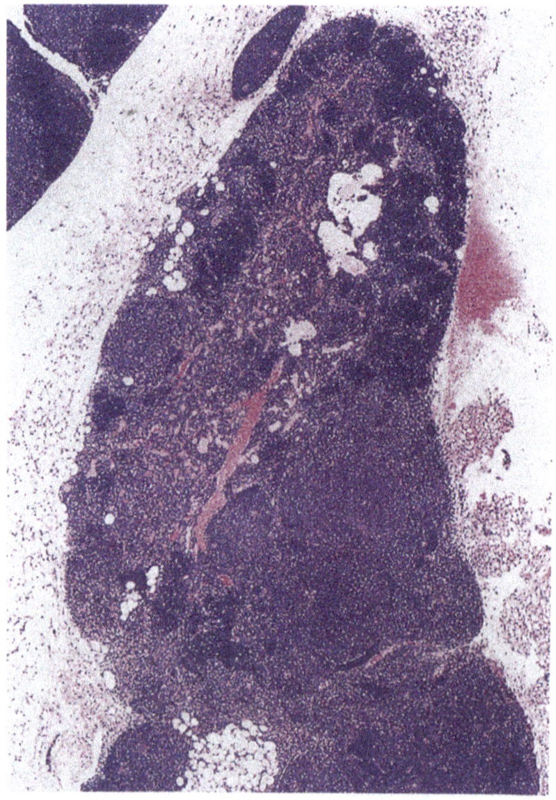

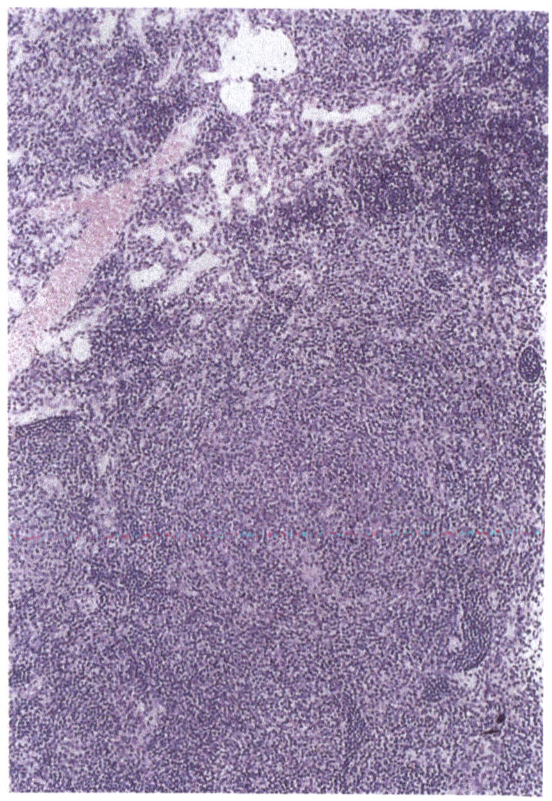

Histogenesis

Lymphoid component of the thymus.

Diagnostic Features

- May be present unilaterally
- Infiltration of the medulla with small mostly uniform lymphocytes or organization of lymphocytes in follicle-like nodules resembling germinal centers
- The cortex generally shows patchy atrophic changes
- Epithelial and lymphoid hyperplasia may be present together

Differential Diagnosis

Lymphoma, Malignant

Focal lesion arising in the cortex of one lobe, usually of the lymphoblastic type or Diffuse infiltration is present of both lobes and mediastinal lymph nodes and often invasion of mediastinal fat or Generalization throughout the hematopoietic system and other organs of the body is present

Comment

Generally present in older mice (>3–6 months) of some strains more frequently in female animals.

References

(See 4, 13, 14, 16)

Fig. 22 *(above)*. Lymphoid hyperplasia of the thymus. H&E

Fig. 23 *(below)*. Lymphoid hyperplasia of the thymus. H&E

Hyperplasia, Epithelial *(H)*

Synonym(s). Hyperplasia, epithelial tubules and cords

Histogenesis

Epithelial component of the thymus, most likely derived from branchial remnants.

Diagnostic Features

- Epithelial cells are arranged in tubules and cords
- Tubules may be partially cystic and filled with markedly eosinophilic colloid material
- Tubules and cords may be distributed throughout the thymus between and within the lobular structure, and may have focal areas with marked proliferation and cyst formation
- Lining of the cystic structures contains some secretory goblet cells
- Cells are cuboidal or columnar and may occasionally be ciliated
- Epithelial and lymphoid hyperplasia may be present together

Differential Diagnosis

Thymoma, Benign
Solid growth pattern of tubules and epithelial cords located centrally within the lobules of the thymus and
No evidence of areas of cyst formation filled with eosinophilic colloid

Thymoma, Malignant
Marked invasion of adjacent tissue is present.

Comment

Epithelial hyperplasias may be present as an age-associated lesion in a relatively high incidence in some strains. They have been reported to be more frequent and have a higher secretory activity in females than in males.

References

(See 4, 13, 14, 16)

Thymoma, Benign *(B)*

Epithelial

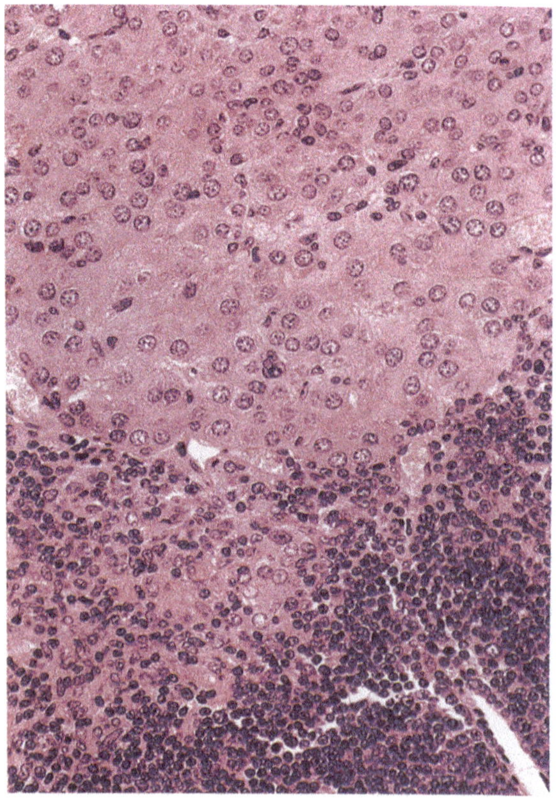

Synonym(s). Benign thymoma

Histogenesis

Epithelial component of the thymus.

Diagnostic Features

- Solitary distribution
- May be well-encapsulated.
- Primary involvement of thymic epithelial cells often with admixture of lymphocytes
- Solid growth of tubules and epithelial cords located centrally within the lobules of the thymus
- Various levels of differentiation may be present ranging from a predominantly regular thymic structure with medullary differentiation to a mixture of epithelial cells and lymphocytes without medullary differentiation
- Relative proportions of lymphocytes and epithelial cells may vary from case to case and within lesions
- Slight local invasion beyond the confines of the thymus may be present

Epithelial

- More than 80% of the lesion are thymic epithelial cells

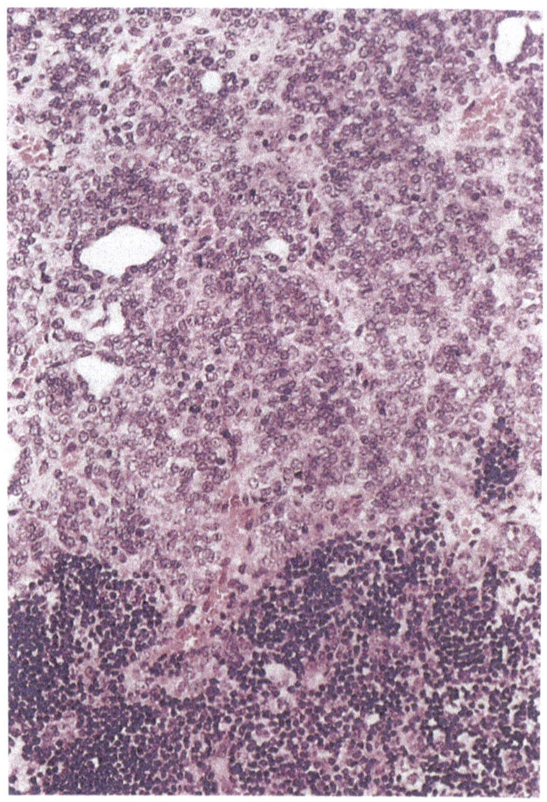

Fig. 24 *(above).* Benign thymoma of the thymus. H&E

Fig. 25 *(below).* Benign thymoma, thymus. H&E

Differential Diagnosis

Hyperplasia, Epithelial Tubules and Cords
Growths is present between and within the lobular structure of the thymus and
Areas of cyst formation filled with eosinophilic colloid are present

Thymoma, Malignant
Marked invasion of adjacent tissues is present

Lymphoma, Malignant
No evidence of an epithelial component

Hyperplasia, Lymphoid
No evidence of a epithelial component

Comment

Thymomas are relatively rare tumors in mice. The term "thymoma" is only utilized if the neoplasm has an epithelial component, often seen along with lymphoid components.

References

(See 4, 13, 16)

Thymoma, Malignant *(M)*

Epithelial

Synonym(s). Malignant thymoma

Histogenesis

Epithelial component of the thymus.

Diagnostic Features

- Primary involvement of thymic epithelial cells
- Various levels of differentiation may be present ranging from a mixture of epithelial cells and lymphocytes without medullary differentiation to lesions exclusively composed of epithelial cells
- Proportion of epithelial cells is high and often squamous differentiation is present
- Marked local invasion beyond the confines of the thymus is present

Epithelial

- More than 80% of the lesion are thymic epithelial cells

Differential Diagnosis

Hyperplasia, Epithelial Tubules or Cords/Thymoma, Benign

No evidence of marked invasion of adjacent tissues

Carcinoma, Squamous Cell

In the absence of a lymphocytic component, the diagnosis is made on the anatomic localization, thymomas must originate in the thoracic cavity

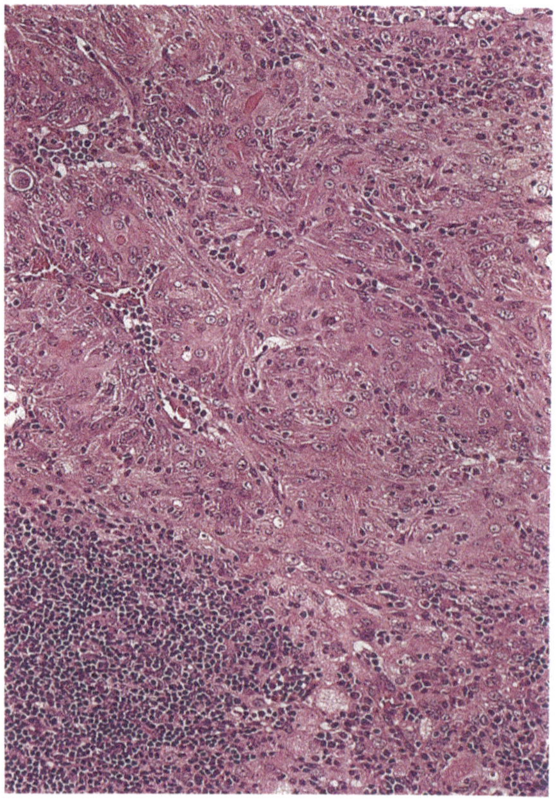

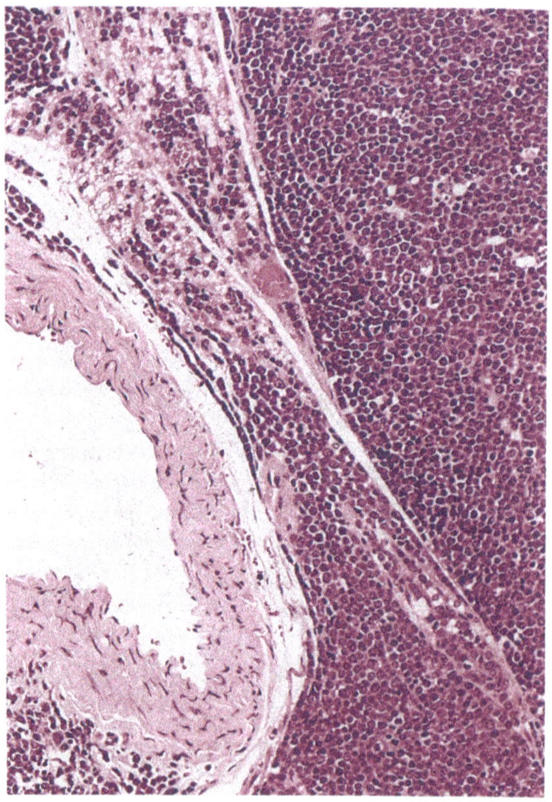

Fig. 26 *(above).* Malignant thymoma, thymus. B6C3F1 mouse, female, age 95 weeks. H&E (courtesy of Dr. Shinji Yamamoto)

Fig. 27 *(below).* Malignant thymoma, thymus. H&E

Lymphoma, Malignant
No evidence of an epithelial component

Comment

Thymomas are relatively rare tumors in mice. The term "thymoma" is only utilized if the neoplasm has an epithelial component.

References

(See 4, 13, 16)

Hyperplasia, Lymphoid *(H)*

Histogenesis

White pulp of spleen, which is composed of the periarteriolar lymphoid sheath (PALS) surrounded by the mantle zone (B-cells), and marginal zone each of which may undergo hyperplasia.

Diagnostic Features

- Lesion is confined to the organ of origin involving multiple follicles
- Represents a response to a stimulus such as inflammation, tumors, etc.

Differential Diagnosis

Lymphoma, Malignant

No evidence of a stimulus, such as inflammation, tumors, etc and
Originating in a single follicle or
Multiple organs are involved (i.e., other organs of the hematopoietic system and liver generally show an infiltration of malignant cells)

Comment

Differentiation of lymphoid hyperplasia of the spleen from early lymphoma, pleomorphic is often difficult.

References

(See 13, 14, 16, , 33, 35)

Extramedullary Hematopoiesis *(H)*

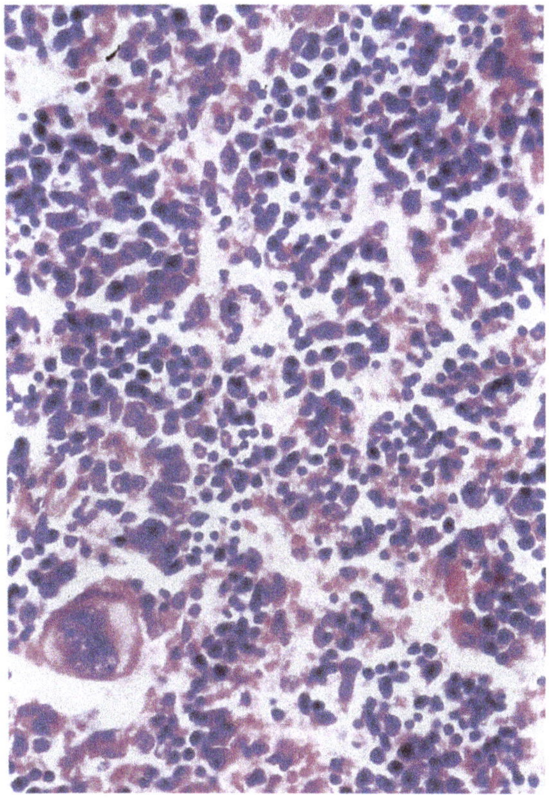

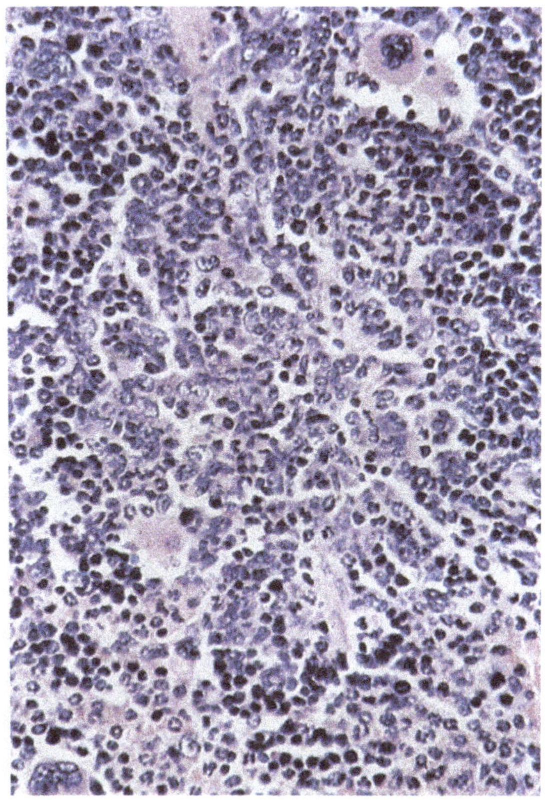

Histogenesis

Red pulp of spleen.

Diagnostic Features

- Normal finding in spleens of mice, variable in degree
- Increase in response to stimuli (reactive hematopoiesis) such as inflammation, anemia, immune stimulation, cytokine effects, thrombocytopenia or to certain neoplasms
- Increase involves the following cellular components
  - Red and white cells (extramedullary hematopoiesis) – liver and renal tumors
  - Red cells (erythropoiesis) – anemia
  - White cells (granulopoiesis) – inflammation
  - Plasma cells (plasmacytosis) – immune response or chronic inflammation
  - Megakaryocytes (megakaryocytosis) – inflammation
- In each case, proliferation results in differentiation to mature forms within the red pulp
- In such reactive phases, the spleen may increase in size up to 0.5 g
- Grossly, there is no white pulp mottling and the spleen is symmetrically enlarged, dark red and firm

Fig. 28 *(above)*. Extramedullary hematopoiesis (erythropoiesis) of the red pulp of the spleen. H&E

Fig. 29 *(below)*. Extramedullary hematopoiesis (granulopoiesis) of the red pulp of the spleen. H&E

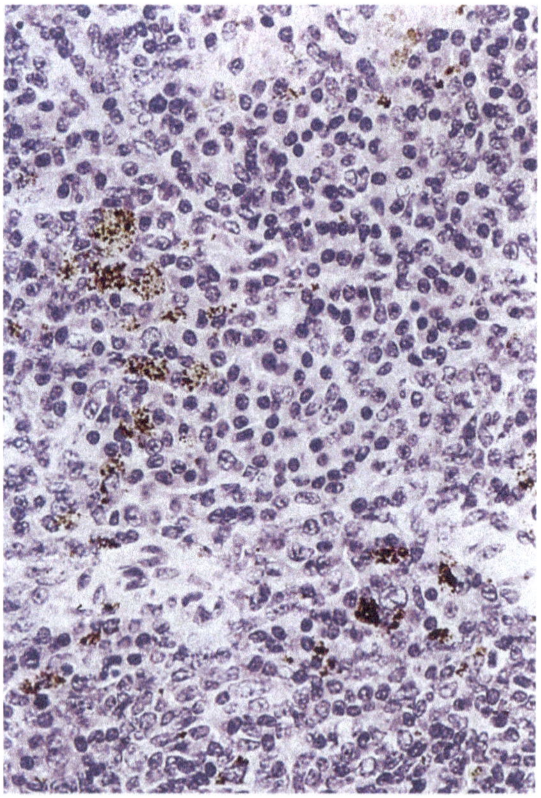

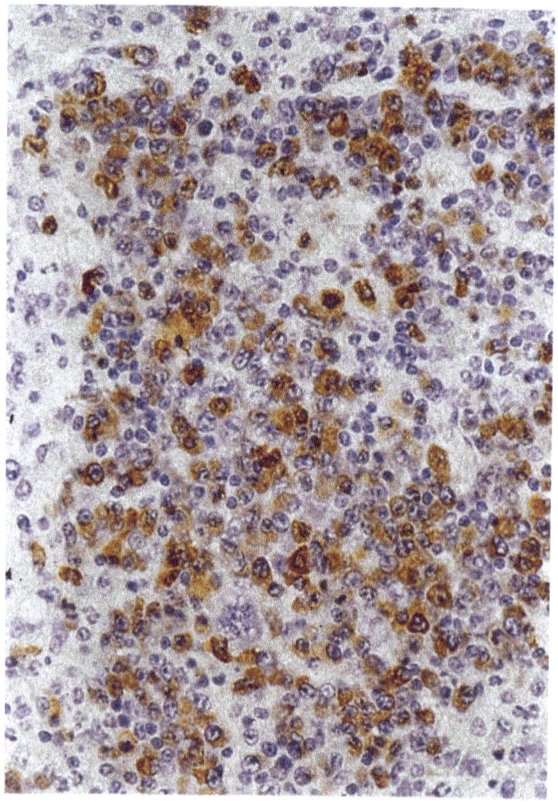

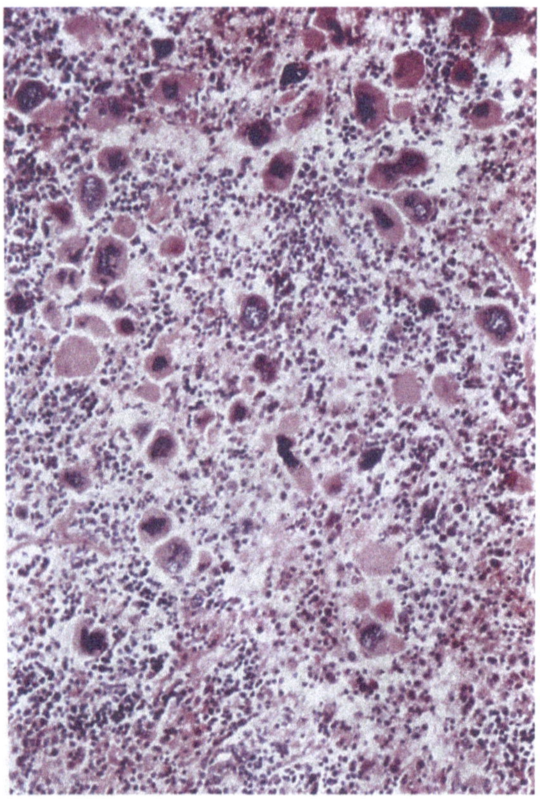

Fig. 30 *(upper left)*. Plasmacytosis of the spleen. H&E

Fig. 31 *(upper right)*. Plasmacytosis of the spleen. IgG

Fig. 32 *(below)*. Megakaryocytosis of the red pulp of the spleen. H&E

Differential Diagnosis

Leukemia, Granulocytic or Leukemia, Erythroid
Leukemias involve only a single lineage of cells with a high percentage of blast cells

Comment

A degree of extramedullary hematopoiesis is a normal finding in mice. Increased extramedullary hematopoiesis is a reaction to a stimulus. However, the hyperplasia orients itself along normal anatomic boundaries, is evenly distributed throughout the red pulp and has in most cases a mixed character.

References

(See 13, 14, 16, 33, 35)

Hyperplasia, Lymphoid *(H)*

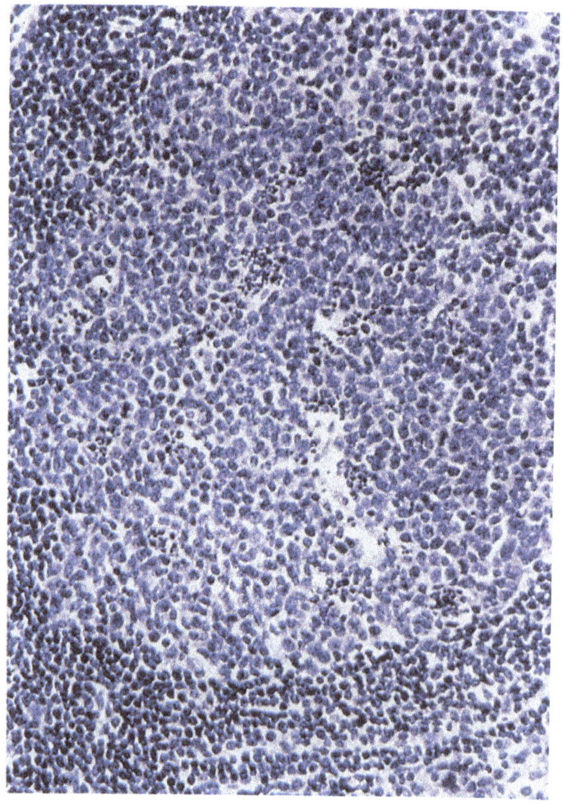

Fig. 33. Lymphoid hyperplasia of a lymph node. H&E

Histogenesis

Lymphoid cells.

Diagnostic Features

- Lesion may be present in the cortex or medulla
- Architecture is maintained
- During normal response to antigens, hyperplastic germinal centers comprised of immature B-cells develop in the cortex
- Hyperplasia of the paracortex may include lymphocytes and interdigitating reticular cells
- Immunoblasts and macrophages filled with nuclear debris may be present
- Small lymphocytes are present mixed with plasma cells and immunoblasts
- Paracortical infiltration may be present in early metastatic thymic T-cell lymphomas and other lymphomas or leukemias
- Hyperplasia of B-cells (follicular hyperplasia) in the B-cell area of the white pulp may be present in acute and chronic immune responses to antigens or administration of certain chemicals
- Germinal centers are enlarged
- Mitotic figures are present

Differential Diagnosis

Lymphoma, Malignant
Architecture is not maintained and
No evidence of a stimulus and
Cell population is homogeneous and
Multiple organs are involved and
The capsule is often penetrated and strands of cells are present in the surrounding fat

Comment

Lymphoid follicles or germinal centers may be evident to varying degrees, in part dependent on the plane of section of the node, the age of the animal and its health status.

References

(See 13, 14, 16, 32, 35)

Extramedullary Hematopoiesis *(H)*

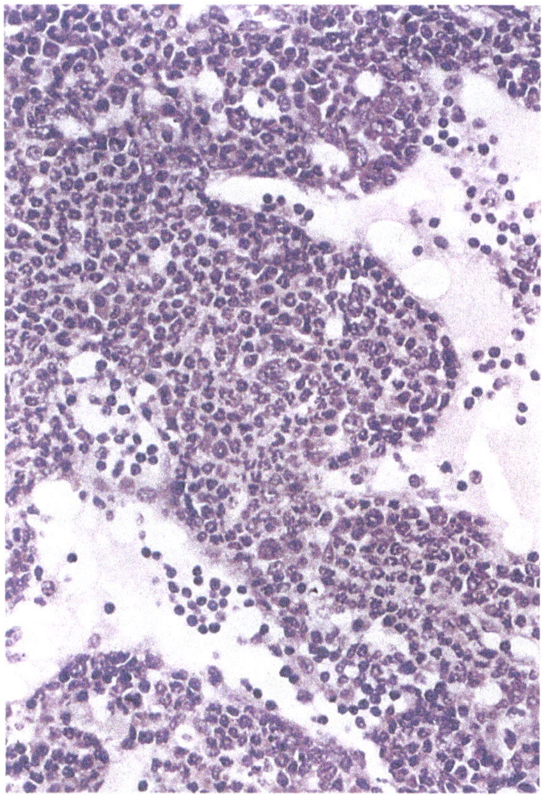

Fig. 34. Extramedullary hematopoiesis of a lymph node. H&E

Histogenesis

Cells of erythrocytic and granulocytic series.

Diagnostic Features

- Confined to a single node
- Architecture is maintained
- Cell population is mixed
- Primarily present in the medullary cords
- Presence of erythrocytic, granulocytic and megakaryocytic cells
- Cells usually represent various stages of development
- Extramedullary hematopoiesis is usually in response to inflammation, tumors, etc., and cell population is mixed in contrast to more homogeneous leukemia population

Differential Diagnosis

Leukemia, Granulocytic

Cells are immature and
No evidence of a stimulus and
Multiple organs are involved and
Capsule may be penetrated with strands of cells in the surrounding fat and
Regular architecture of the lymph node is distorted

Comment

Extramedullary hematopoiesis is often most prominent in the mesenteric lymph node.

References

(See 13, 14, 16, 32, 35)

Plasmacytosis *(H)*

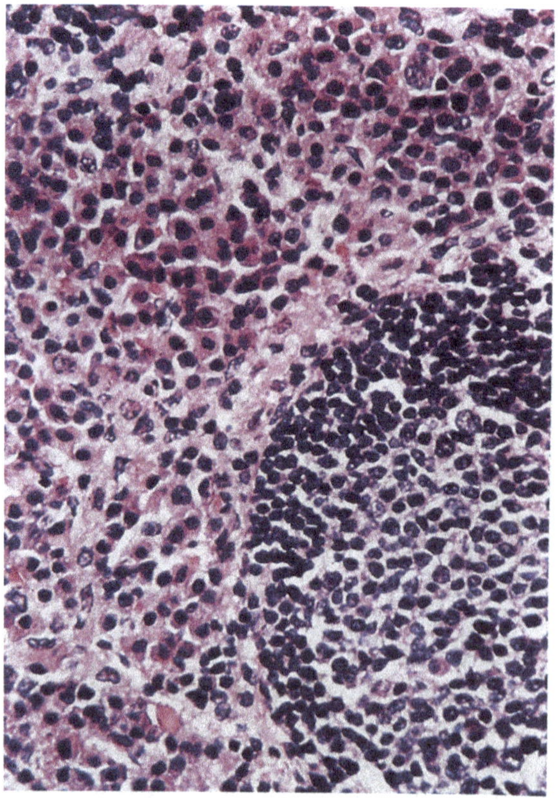

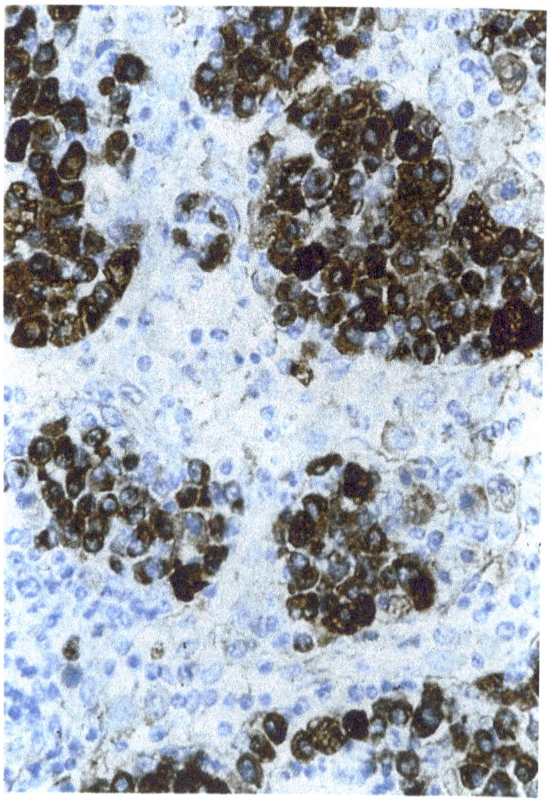

Histogenesis

Plasma cells.

Diagnostic Features

- Common in mandibular lymph nodes and to a lesser extent in mesenteric lymph nodes
- Often present in lymph nodes draining tissue with lesions caused by infectious agents and tumors
- Restricted to one organ
- Plasma cells are located predominantly in the medullary cords
- Some cells may contain Russell's bodies
- Cells are mature
- Cytoplasm is purple to intensely eosinophilic and a perinuclear halo (Golgi apparatus) is present
- Nucleus is eccentrically located and has a cartwheel nuclear chromatin pattern
- No evidence of mitotic figures
- Plasmacytosis is usually in response to inflammation
- Affected nodes are firm and may be enlarged up to five times normal size
- Spilling of cells in the adjacent tissues may be present

Fig. 35 *(above)*. Plasmacytosis of a lymph node. H&E

Fig. 36 *(below)*. Plasmacytosis of a lymph node. IgG

Differential Diagnosis

Lymphoma, Malignant (Plasmacytic)
Cells are immature and
Mitotic figures may be present and
No evidence of a stimulus and
Multiple organs are involved and
Capsule may be penetrated with strands of cells in the surrounding fat

References

(See 13, 14, 16, 32, 35)

Mastocytosis *(H)*

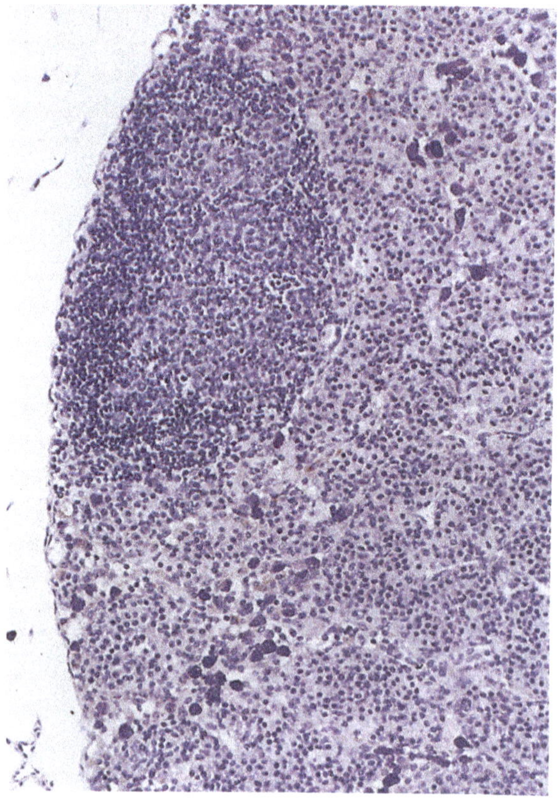

Fig. 37. Mastocytosis of a lymph node. Plasmacytosis is also evident. H&E

Histogenesis

Mast cells.

Diagnostic Features

- Multiple organs are involved
- Number of mast cells in a specific organ are increased and may vary from strain to strain
- Cells are present as individual non-cohesive cells
- Cells are mature
- Cytoplasmic granules are prominent
- No evidence of mitotic figures
- Etiology and pathogenesis are often unknown

Differential Diagnosis

Tumor, Mast Cell, Malignant or Tumor, Mast Cell, Benign
Cells are immature and
Cytoplasmic granules are not prominent and metachromatic stains may be necessary to visualize granules
Cells are clumped together and
Capsule may be penetrated with strands of cells in the surrounding fat

References

(See 13, 14, 16, 32, 35)

Sinus Histiocytosis *(H)*

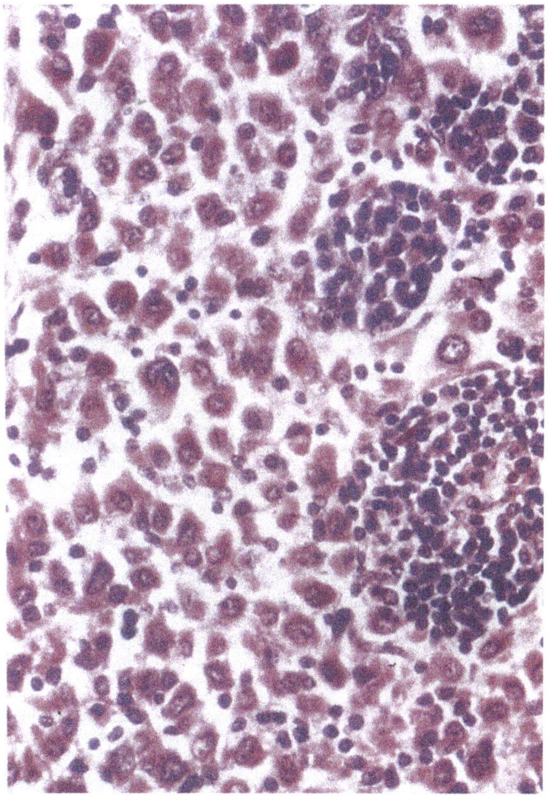

Fig. 38. Sinus histiocytosis of the lymph node. H&E

Histogenesis

Histiocytes.

Diagnostic Features

- Frequently present in lymph nodes draining inflammatory lesions
- Histiocytes are present in the sinuses
- Cytoplasm has a distinct eosinophilic appearance and may contain hemosiderin and other phagocytozed materials
- Appear as regular histiocytes
- No evidence of cellular cohesion
- No evidence of mitotic figures
- In histiocytosis, the histiocytes often contain pigment and show erythrophagocytosis
- Usually a response to inflammation, foreign debris, etc.

Differential Diagnosis

Sarcoma, Histiocytic

Cells are large and
Mitotic figures may be present and
Cells are often cohesive and
No evidence of a stimulus and
Multiple organs are involved and
No evidence of intracellular pigment or erythrophagocytosis

References

(See 13, 14, 16, 32, 35)

Hyperplasia *(H)*

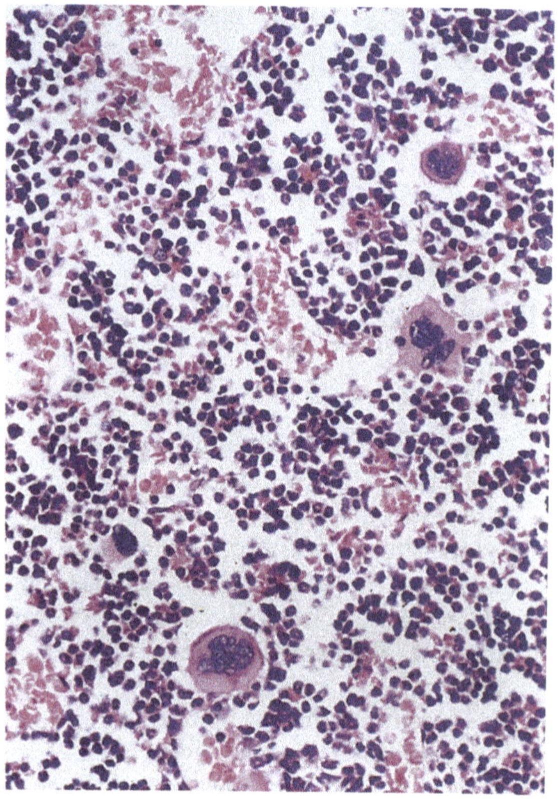

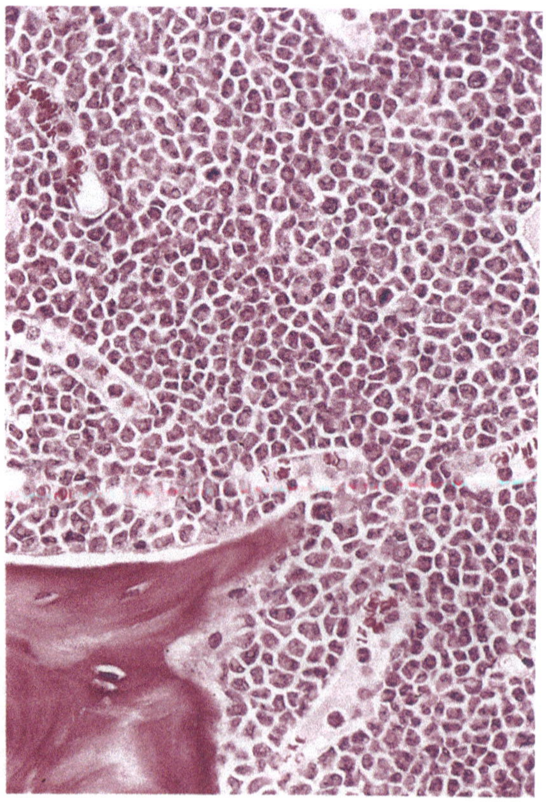

Histogenesis

May involve cells of the erythrocytic series, granulocytic (myeloid) series or megakaryocytes.

Diagnostic Features

- Granulocytic component predominates, but may involve an increase of both erythrocytic and granulocytic cells, as well as megakaryocytes, as a result of a shift in hematopoietic activity of a particular cell type
- Change is usually in response to a stimulus and similar changes are generally also present in the spleen

References

(See 13, 14, 16)

Fig. 39 *(above)*. Normal bone marrow. H&E

Fig. 40 *(below)*. Granulocytic hyperplasia of the bone marrow. H&E

Myelodysplastic Hematopoietic Disorder *(H)*

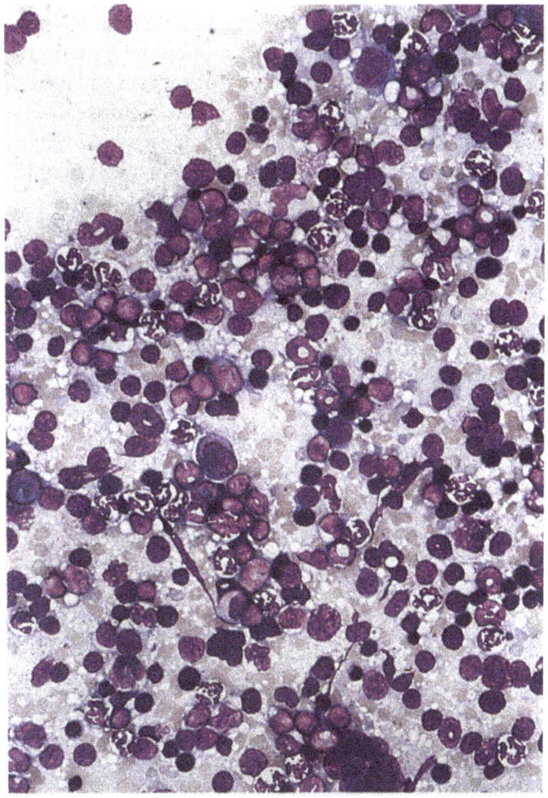

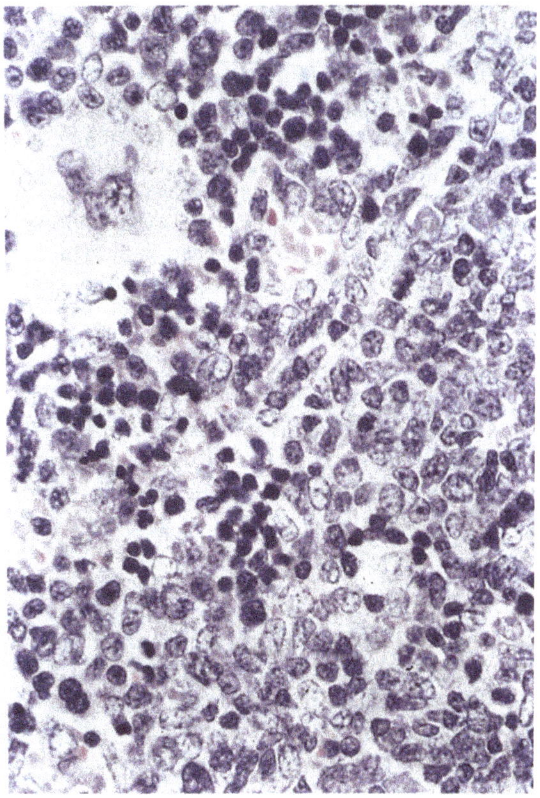

Histogenesis

Hematopoietic cells.

Diagnostic Features

- Increased number of granulocytes with hypersegmentation as well as irregular-shaped blast cells.
- Dyserythropoiesis with intranuclear cytoplasmic invagination and giant mitochondria containing irregular deposits of ferritines
- Disproportional development of nuclei with megakaryocytic cytoplasm or irregular large nuclei in megakaryoblasts
- Extensive apoptosis in both spleen and bone marrow

Comment

Some of the myelodysplastic syndromes are know to be a receptor disease. Transgenic mice carrying human GM-receptors are similar to human phenotypes. AZT-treated CBA mice showed myelodysplastic hematopoiesis. This lesion has not been described as a spontaneous lesion in the mouse; however, it is included in this classification as it probably represents analogous to humans a preneoplastic stage.
Synonyms are myelodysplasia, refractory anemia, dyserythropoiesis.

References

(See 20, 23)

Fig. 41 *(above)*. Splenic smear illustrating an increased amount of myeloid elements with hypersegmentation as well as atypical large mononuclear cells.

Fig. 42 *(below)*. Spleen showing myelodysplastic hematopoietic disorder. H&E

References

1. Della Porta G, Chieco-Bianchi L, Penelli N (1979) Tumors of the hematopoietic system. In: Turusov VS (ed) Pathology of tumors in laboratory animals, vol II. Tumors of the mouse. IARC Scientific Publications No. 23, Lyon, pp 527–576
2. Deringer MK, Dunn TB (1947) Mast cell neoplasia in mice. J Natl Cancer Inst 7: 289–298
3. Dunn TB (1969) Mast cell neoplasia in mice. Natl Cancer Inst Monogr 32: 285–287
4. Faccini JM, Abbott DP, Paulus GJJ (1990) Mouse histopathology. A glossary for use in toxicity and carcinogenicity studies. Elsevier, Amsterdam, pp 18–47
5. Fredrickson TN (1990a) Erythroleukemia, mouse. In: Jones TC, Ward JM, Mohr U, Hunt RD (eds) Monographs on pathology of laboratory animals. Hemopoietic system. Springer, Berlin Heidelberg New York Tokyo, pp 205–211
6. Fredrickson TN (1990b) Megakaryocytic leukemia, mouse. In: Jones TC, Ward JM, Mohr U, Hunt RD (eds) Monographs on pathology of laboratory animals. Hemopoietic system. Springer, Berlin Heidelberg New York Tokyo, pp 51–54
7. Fredrickson TN, Langdon WY, Hoffman PM, Hartley JW, Morse HC III (1984a) Histologic and cell surface antigen studies of hematopoietic tumors induced by Cas-Br-M murine leukemia virus. J Natl Cancer Inst 72: 447–454
8. Fredrickson TN, Morse HC III, Rowe WP (1984b) Spontaneous tumors of NFS mice congenic for ecotropic murine leukemia virus induction loci. J Natl Cancer Inst 73: 521–524
9. Fredrickson TN, Morse HC III, Yetter RA, Rowe WP, Hartley JW, Pattengale PK (1985) Multiparameter analyses of spontaneous nonthymic lymphomas occurring in NFS/N mice congenic for ecotropic murine leukemia viruses. Am J Pathol 121: 349–360
10. Fredrickson TN, Tang Y, Chattopadhyay SK, Morse HC III, Hartley JW (1993) Retrovirus-induced lymphoproliferation as a model for developing diagnostic criteria for malignant lymphoma in mice. Toxicol Pathol 21: 219–228
11. Fredrickson TN, Lennert K, Chattopadhyay SK, Morse HC 3rd, Hartley JW (1999) Splenic marginal zone lymphomas of mice. Am J Pathol 154: 805–812
12. Frith CH (1990) Histiocytic sarcoma, mouse. In: Jones TC, Ward JM, Mohr U, Hunt RD (eds) Monographs on pathology of laboratory animals. Hemopoietic system. Springer, Berlin Heidelberg New York Tokyo, pp 58–65
13. Frith CH, Ward JM (1988) Color atlas of neoplastic and non-neoplastic lesions in aging mice. Elsevier, Amsterdam, pp 77–86
14. Frith CH, Wiley LD (1981) Morphologic classification and correlation of incidence of hyperplastic and neoplastic hematopoietic lesions in mice with age. J Gerontol 36: 534–545
15. Frith CH, Sprowls RW, Breeden CR (1976) Mast cell neoplasia in mice. Lab Anim Sci 26: 478–481
16. Frith CH, Pattengale PK, Ward JM (1985) A color atlas of hematopoietic pathology of mice. Toxicology Pathology Associates, Little Rock Arkansas, p 12
17. Frith CH, McConnell RF, Johnson AN (1990) Erythroleukemia in a mouse. Lab Anim Sci 40: 418–419
18. Frith CH, Ward JM, Chandra M (1993) The morphology, immunohistochemistry, and incidence of hematopoietic neoplasms in mice and rats. Toxicol Pathol 21: 206–218
19. Frith CH, Ward JM, Frederickson T, Harleman JH (1996) Neoplastic lesions of the hematopoietic system. In: Mohr U, Dungworth DL, Capen CC, Carlton WW, Sundberg JP, Ward JM (eds) Pathobiology of the aging mouse, vol 1. ILSI Press, Washington, DC, pp 219–235
20. Furuta Y, Aizawa S, Suda Y, Ikawa Y, Nishikawa S, Hayashi S, Hirabayashi Y, Inoue T (1993) MDS-like experimental myelodysplasia: multilineage abnormal hematopoiesis in transgenic mice harboring the SV40 large T antigen under an immunoglobulin enhancer. Exp Hematol 21: 806–815
21. Harleman JH, Schuurman HJ, Kuper CF (1994) Carcinogenesis of the hematopoietic system. In: Waalkes MP, Ward JM (eds) Carcinogenesis. Raven Press, New York, pp 403–428
22. Hirabayashi Y, Inoue T, Yoshida K, Sasaki H, Kubo S, Kanisawa M, Seki M (1991) Murine acute leukemia cell line with megakaryocytic differentiation (MK-8057) induced by whole-body irradiation in C3H/He mice: cytological properties and kinetics of its leukemic stem cells. Int J Cell Cloning 9: 24–42
23. Inoue T, Hirabayashi Y, Mitsui H, Furata Y, Suda Y, Aizawa S, Ikawa Y (1994) Experimental model for MDS-like myelodysplasia in transgenic mice harboring the SV40 large-T antigen under an immunoglobulin enhancer. Leukemia 8: 202–205
24. Pattengale PK (1990) Classification of mouse lymphoid cell neoplasms. In: Jones TC, Ward JM, Mohr U, Hunt RD (eds) Monographs on pathology of laboratory animals. Hemopoietic system. Springer, Berlin Heidelberg New York Tokyo, pp 157–143

25. Pattengale PK (1994) Tumours of the lympho-haematopoietic system. Pathology of tumours in laboratory animals, vol 2. Tumours of the mouse, 2nd edn. IARC Scientific Publications No. 111, Lyon, pp 651–670
26. Pattengale PK, Frith CH (1983) Immunomorphologic classification of spontaneous lymphoid cell neoplasms occurring in female BALB/c mice. J Natl Cancer Inst 70: 169–179
27. Pattengale PK, Frith CH (1986) Contributions of recent research to the classification of spontaneous lymphoid cell neoplasms in mice. Crit Rev Toxicol 16: 185–212
28. Pattengale PK, Taylor CR (1983) Experimental models of lymphoproliferative disease. The mouse as a model for human non-Hodgkin's lymphomas and related leukemias. Am J Pathol 113: 237–265
29. Rehm S, Ward JM, Devor DE, Kovatch RM (1990) Mast cell neoplasm, mouse. In: Jones TC, Ward JM, Mohr U, Hunt RD (eds) Monographs on pathology of laboratory animals. Hematopoietic system. Springer, Berlin Heidelberg New York Tokyo, pp 201–204
30. Scolnick EM (1982) Hyperplastic and neoplastic erythroproliferative diseases induced by oncogenic murine retroviruses. Biochim Biophys Acta 651: 273–283
31. Seki M, Inoue T (1990) Granulocytic leukemia, mouse. In: Jones TC, Ward JM, Mohr U, Hunt RD (eds) Monographs on pathology of laboratory animals. Hemopoietic system. Springer, Berlin Heidelberg New York Tokyo, pp 46–50
32. Ward JM (1990a) Classification of reactive lesions of lymph nodes. In: Jones TC, Ward JM, Mohr U, Hunt RD (eds) Monographs on pathology of laboratory animals. Hemopoietic system. Springer, Berlin Heidelberg New York Tokyo, pp 155–161
33. Ward JM (1990b) Classification of reactive lesions, spleen. In: Jones TC, Ward JM, Mohr U, Hunt RD (eds) Monographs on pathology of laboratory animals. Hemopoietic system. Springer, Berlin Heidelberg New York Tokyo, pp 220–226
34. Ward JM, Sheldon W (1993) Expression of mononuclear phagocyte antigens in histiocytic sarcoma of mice. Vet Pathol 30: 560–565
35. Ward JM, Uno H, Frith CH (1993) Immunohistochemistry and morphology of reactive lesions in lymph nodes and spleen from rats and mice. Toxicol Pathol 21: 199–205
36. Ward JM, Mann PC, Morishima H, Frith CH (1999a) Thymus, spleen, and lymph nodes. In: Maronpot RR, Boorman GA, Gaul BW (eds) Pathology of the mouse. Reference and atlas. Cache River Press, Vienna, pp 333–360
37. Ward JM, Tadesse-Heath L, Perkins SN, Chattopadhyay SK, Hursting SD, Morse HC 3rd (1999b) Splenic marginal zone B-cell and thymic T-cell lymphomas in p53-deficient mice. Lab Invest 79: 3–14
38. Wogan GN (1983) Tumors of the mouse hematopoietic system: their diagnosis and interpretation in safety evaluation tests. Crit Rev Toxicol 13: 161–181
39. Yoshida K, Nemoto K, Nishimura M, Hayata I, Inoue T, Seki M (1986) Nature of leukemic stem cells in murine myelogenous leukemia. Int J Cell Cloning 4: 91–102
40. Yumoto T, Yoshida Y, Yoshida H, Ando K, Matsui K (1980) Prelymphomatous and lymphomatous changes in splenomegaly of New Zealand black mice. Acta Pathol Jpn 30: 171–186

Subject Index

www.ingramcontent.com/pod-product-compliance
Lightning Source LLC
Chambersburg PA
CBHW061151240926
51846CB00006B/986
* 9 7 8 3 6 4 2 0 8 4 2 2 5 *